Methods in Enzymology

Volume 326
APPLICATIONS OF CHIMERIC GENES
AND HYBRID PROTEINS
Part A
Gene Expression and Protein Purification

METHODS IN ENZYMOLOGY

EDITORS-IN-CHIEF

John N. Abelson   Melvin I. Simon

DIVISION OF BIOLOGY
CALIFORNIA INSTITUTE OF TECHNOLOGY
PASADENA, CALIFORNIA

FOUNDING EDITORS

Sidney P. Colowick and Nathan O. Kaplan

Methods in Enzymology

Volume 326

Applications of Chimeric Genes and Hybrid Proteins

Part A
Gene Expression and Protein Purification

EDITED BY

Jeremy Thorner

UNIVERSITY OF CALIFORNIA
BERKELEY, CALIFORNIA

Scott D. Emr

HOWARD HUGHES MEDICAL INSTITUTE
UNIVERSITY OF CALIFORNIA SAN DIEGO
LA JOLLA, CALIFORNIA

John N. Abelson

CALIFORNIA INSTITUTE OF TECHNOLOGY
PASADENA, CALIFORNIA

ACADEMIC PRESS
San Diego   London   Boston   New York   Sydney   Tokyo   Toronto

This book is printed on acid-free paper.

Academic Press
A Harcourt Science and Technology Company
525 B Street, Suite 1900, San Diego, California 92101-4495, USA

http://www.academicpress.com

Academic Press Limited
32 Jamestown Road, London NW1 7BY, UK

International Standard Book Number: 0-12-182227-3

PRINTED IN THE UNITED STATES OF AMERICA
00 01 02 03 04 05 06 MM 9 8 7 6 5 4 3 2 1

Table of Contents

Section I. Historical Overview

Section II. Gene Fusions as Reporters of Gene Expression in Prokaryotic Cells

Section III. Gene Fusions as Reporters of Gene Expression in Eukaryotic Cells

Section IV. Functional Tags for Biochemical Purification

Section V. Hybrid Proteins for Detection and Production of Antigens and Antibodies

Contributors to Volume 326

Article numbers are in parentheses following the names of contributors.
Affiliations listed are current.

JON BECKWITH (1), *Department of Microbiology and Molecular Genetics, Harvard Medical School, Boston, Massachusetts 02115*

JOSHUA A. BORNHORST (16), *Department of Chemistry and Biochemistry, University of Colorado, Boulder, Colorado 80309-0215*

LISA BREISTER (22), *Stratagene Cloning Systems, La Jolla, California 92037*

IRENA BRONSTEIN (13), *Tropix, Inc., PE Biosystems, Bedford, Massachusetts 01730*

CLAYTON BULLOCK (14), *Department of Pharmacology, College of Medicine, University of California, Irvine, California 92697*

ANDREW CAMILLI (5), *Department of Molecular Biology and Microbiology, Tufts University School of Medicine, Boston, Massachusetts 02111*

CHARLES R. CANTOR (19), *Center for Advanced Biotechnology and Departments of Biomedical Engineering and Pharmacology and Experimental Therapeutics, Boston University, Boston, Massachusetts 02215 and Sequenom, Inc., San Diego, California 92121*

JOHN M. CHIRGWIN (20), *Research Service, Audie L. Murphy Memorial Veterans Administration Medical Center and Departments of Medicine and Biochemistry, University of Texas Health Science Center at San Antonio, Texas 78229-3900*

SHAORONG CHONG (24), *New England Biolabs, Inc., Beverly, Massachusetts 01915*

R. JOHN COLLIER (33), *Department of Microbiology and Molecular Genetics, Harvard Medical School, Boston, Massachusetts 02115*

LISA A. COLLINS-RACIE (21), *Genetics Institute, Cambridge, Massachusetts 02140*

JOHN E. CRONAN, JR. (27), *Departments of Microbiology and Biochemistry, University of Illinois, Urbana, Illinois 61801*

MILLARD G. CULL (26), *Avidity, L.L.C., Eleanor Roosevelt Institute, Denver, Colorado 80206*

BRYAN R. CULLEN (11), *Howard Hughes Medical Institute and Department of Genetics, Duke University Medical Center, Durham, North Carolina 27710*

BRIAN D'EON (13), *Tropix, Inc., PE Biosystems, Bedford, Massachusetts 01730*

SALVATORE DEMARTIS (29), *Institute of Pharmaceutical Sciences, Department of Applied BioSciences, Swiss Federal Institute of Technology Zurich, CH-8057 Zurich, Switzerland*

ELIZABETH A. DIBLASIO-SMITH (21), *Genetics Institute, Cambridge, Massachusetts 02140*

ROY H. DOI (25), *Section of Molecular and Cellular Biology, University of California, Davis, California 95616*

CHARLES F. EARHART (30), *Section of Molecular Genetics and Microbiology, The University of Texas at Austin, Austin, Texas 78712-1095*

DOLPH ELLEFSON (31), *Department of Molecular Microbiology and Immunology, Oregon Health Sciences University, Portland, Oregon 97201*

JOSEPH J. FALKE (16), *Department of Chemistry and Biochemistry, University of Colorado, Boulder, Colorado 80309-0215*

CATHERINE FAYOLLE (32), *Unité de Biologie des Régulations Immunitaires, CNRS URA 2185, Institut Pasteur, Paris, Cedex 15, France*

CORNELIA GORMAN (14), *DNA Bridges, Inc., San Francisco, California 94117*

PIERRE GUERMONPREZ (32), *Unité de Biologie des Régulations Immunitaires, CNRS URA 2185, Institut Pasteur, Paris, Cedex 15, France*

NICHOLAS J. HAND (2), *Department of Molecular Biology, Princeton University, Princeton, New Jersey 08544*

FRED HEFFRON (6, 31), *Department of Molecular Microbiology and Immunology, Oregon Health Sciences University, Portland, Oregon 97201*

DANNY Q. HOANG (22), *Stratagene Cloning Systems, La Jolla, California 92037*

PHILIPP HOLLIGER (28), *MRC Laboratory of Molecular Biology, Cambridge CB2 2QH United Kingdom*

JOE HORECKA (7), *Department of Molecular Biology, NIBH, Tsukuba, Ibaraki 305-8566 Japan*

ADRIAN HUBER (29), *Institute of Pharmaceutical Sciences, Department of Applied Bio-Sciences, Swiss Federal Institute of Technology Zurich, CH-8057 Zurich, Switzerland*

SATOSHI INOUYE (12), *Yokohama Research Center, Chisso Corporation, Yokohama 236-8605 Japan*

RAY JUDWARE (13), *Tropix, Inc., PE Biosystems, Bedford, Massachusetts 01730*

GOUZEL KARIMOVA (32), *Unité de Biochimie Cellulaire, CNRS URA 2185, Institut Pasteur, Paris, Cedex 15, France*

CHRISTIAAN KARREMAN (9), *Institute of Oncological Chemistry, Heinrich Heine University, 40225 Duesseldorf, Germany*

DANIEL LADANT (32), *Unité de Biochimie Cellulaire, CNRS URA 2185, Institut Pasteur, Paris, Cedex 15, France*

EDWARD R. LAVALLIE (21), *Genetics Institute, Cambridge, Massachusetts 02140*

CLAUDE LECLERC (32), *Unité de Biologie des Régulations Immunitaires, CNRS URA 2185, Institut Pasteur, Paris, Cedex 15, France*

BETTY LIU (13), *Tropix, Inc., PE Biosystems, Bedford, Massachusetts 01730*

ZHIJIAN LU (21), *Genetics Institute, Cambridge, Massachusetts 02140*

COLIN MANOIL (3), *Department of Genetics, University of Washington, Seattle, Washington 98195*

CHRIS MARTIN (13), *Millennium Predictive Medicine, Cambridge, Massachusetts 02139*

DINA MARTIN (13), *Tropix, Inc., PE Biosystems, Bedford, Massachusetts 01730*

ROBERT A. MASTICO (34), *Astbury Centre for Structural Molecular Biology, University of Leeds, Leeds LS2 9JT, United Kingdom*

MARK MCCORMICK (23), *Novagen, Inc., Madison, Wisconsin 53711*

JOHN M. MCCOY (21), *Biogen, Inc., Cambridge, Massachusetts 02142*

ROBERT C. MIERENDORF (23), *Novagen, Inc., Madison, Wisconsin 53711*

DARIO NERI (29), *Institute of Pharmaceutical Sciences, Department of Applied Bio-Sciences, Swiss Federal Institute of Technology Zurich, CH-8057 Zurich, Switzerland*

FREDRIK NILSSON (29), *Institute of Pharmaceutical Sciences, Department of Applied BioSciences, Swiss Federal Institute of Technology Zurich, CH-8057 Zurich, Switzerland*

CORINNE E. M. OLESEN (13), *Tropix, Inc., PE Biosystems, Bedford, Massachusetts 01730*

JAE-SEON PARK (25), *Sampyo Foods Co., Ltd., Seoul 132-040, Korea*

DAVID PARKER (31), *Department of Molecular Microbiology and Immunology, Oregon Health Sciences University, Portland, Oregon 97201*

HENRY PAULUS (24), *Boston Biomedical Research Institute, Watertown, Massachusetts 02472-2829*

RONALD T. RAINES (23), *Departments of Biochemistry and Chemistry, University of Wisconsin-Madison, Madison, Wisconsin 53706*

LALITA RAMAKRISHNAN (4), *Department of Microbiology and Immunology, Stanford University School of Medicine, Stanford, California 94305-5124*

KELYNNE E. REED (27), *Department of Biology, Austin College, Sherman, Texas 75090*

DEEPALI SACHDEV (20), *University of Minnesota Cancer Center, Minneapolis, Minnesota 55455*

SOFIE REDA SALAMA (8), *Microbia, Inc., Cambridge, Massachusetts 02139*

TAKESHI SANO (19), *Center for Molecular Imaging Diagnosis and Therapy and Basic Science Laboratory, Department of Radiology, Beth Israel Deaconess Medical Center, Harvard Medical School, Boston, Massachusetts 02215*

PETER J. SCHATZ (26), *Affymax Research Institute, Palo Alto, California 94304*

THOMAS G. M. SCHMIDT (18), *Institut für Bioanalytik GmbH, D-37079 Göttingen, Germany*

HAE-SUN SHIN (25), *Sampyo Foods Co., Ltd., Seoul 132-040, Korea*

THOMAS J. SILHAVY (2), *Department of Molecular Biology, Princeton University, Princeton, New Jersey 08544*

ARNE SKERRA (18), *Lehrstuhl für Biologische Chemie, Technische Universität München, D-85350 Freising-Weihenstephan, Germany*

JAMES M. SLAUCH (5), *Department of Microbiology, University of Illinois, Urbana, Illinois 61801*

STEPHEN SMALL (10), *Department of Biology, New York University, New York, New York 10003*

DONALD B. SMITH (17), *Garden Cottage, Clerkington, Haddington, East Lothian, Scotland, United Kingdom*

GEORGE F. SPRAGUE, JR. (7), *Institute of Molecular Biology, University of Oregon, Eugene, Oregon 97403*

MICHAEL N. STARNBACH (33), *Department of Microbiology and Molecular Genetics, Harvard Medical School, Boston, Massachusetts 02115*

PETER G. STOCKLEY (34), *Astbury Centre for Structural Molecular Biology, University of Leeds, Leeds LS2 9JT, United Kingdom*

IAN TOMLINSON (28), *MRC Laboratory of Molecular Biology, Cambridge CB2 2QH United Kingdom*

AGNES ULLMANN (32), *Unité de Biochimie Cellulaire, CNRS URA 2185, Institut Pasteur, Paris, Cedex 15, France*

PETER VAILLANCOURT (22), *Stratagene Cloning Systems, La Jolla, California 92037*

RAPHAEL H. VALDIVIA (4), *Department of Molecular and Cell Biology, University of California, Berkeley, California 94702*

ADRIANUS W. M. VAN DER VELDEN (31), *Department of Molecular Microbiology and Immunology, Oregon Health Sciences University, Portland, Oregon 97201*

THOMAS R. VAN OOSBREE (23), *Novagen, Inc., Madison, Wisconsin 53711*

FRANCESCA VITI (29), *Institute of Pharmaceutical Sciences, Department of Applied Biosciences, Swiss Federal Institute of Technology Zurich, CH-8057 Zurich, Switzerland*

JOHN C. VOYTA (13), *Tropix, Inc., PE Biosystems, Bedford, Massachusetts 01730*

MICAH J. WORLEY (6), *Department of Molecular Microbiology and Immunology, Oregon Health Sciences University, Portland, Oregon 97201*

MING-QUN XU (24), *New England Biolabs, Inc., Beverly, Massachusetts 01915*

YU-XIN YAN (13), *Tropix, Inc., PE Biosystems, Bedford, Massachusetts 01730*

CHRISTOPHER C. ZAROZINSKI (33), *Department of Microbiology and Molecular Genetics, Harvard Medical School, Boston, Massachusetts 02115*

CHAO-FENG ZHENG (22), *Stratagene Cloning Systems, La Jolla, California 92037*

GREGOR ZLOKARNIK (15), *Aurora Biosciences Corporation, San Diego, California 92121*

Preface

The modern biologist takes almost for granted the rich repertoire of tools currently available for manipulating virtually any gene or protein of interest. Paramount among these operations is the construction of fusions. The tactic of generating gene fusions to facilitate analysis of gene expression has its origins in the work of Jacob and Monod more than 35 years ago. The fact that gene fusions can create functional chimeric proteins was demonstrated shortly thereafter. Since that time, the number of tricks for splicing or inserting into a gene product various markers, tags, antigenic epitopes, structural probes, and other elements has increased explosively. Hence, when we undertook assembling a volume on the applications of chimeric genes and hybrid proteins in modern biological research, we considered the job a daunting task.

To assist us with producing a coherent work, we first enlisted the aid of an Advisory Committee, consisting of Joe Falke, Stan Fields, Brian Seed, Tom Silhavy, and Roger Tsien. We benefited enormously from their ideas, suggestions, and breadth of knowledge. We are grateful to them all for their willingness to participate at the planning stage and for contributing excellent and highly pertinent articles.

A large measure of the success of this project is due to the enthusiastic responses we received from nearly all of the prospective authors we approached. Many contributors made additional suggestions, and quite a number contributed more than one article. Hence, it became clear early on that given the huge number of applications of gene fusion and hybrid protein technology—for studies of the regulation of gene expression, for lineage tracing, for protein purification and detection, for analysis of protein localization and dynamic movement, and a plethora of other uses—it would not be possible for us to cover this subject comprehensively in a single volume, but in the resulting three volumes, 326, 327, and 328.

Volume 326 is devoted to methods useful for monitoring gene expression, for facilitating protein purification, and for generating novel antigens and antibodies. Also in this volume is an introductory article describing the genesis of the concept of gene fusions and the early foundations of this whole approach. We would like to express our special appreciation to Jon Beckwith for preparing this historical overview. Jon's description is particularly illuminating because he was among the first to exploit gene and protein fusions. Moreover, over the years, he and his colleagues have continued to develop the methodology that has propelled the use of fusion-based techniques from bacteria to eukaryotic organisms. Volume 327 is focused on procedures for tagging proteins for immunodetection, for using chimeric proteins for cytological purposes, especially the analysis of membrane proteins and intracellular protein trafficking, and for monitoring and manipulating various aspects of cell signaling and cell physiology. Included in this volume is a rather extensive section on the green fluorescent protein (GFP) that deals with applications not covered in Volume 302. Volume 328 describes protocols for using hybrid genes and proteins to identify and analyze protein–protein and protein–nucleic interactions, for mapping molecular recognition domains, for directed molecular evolution, and for functional genomics.

We want to take this opportunity to thank again all the authors who generously contributed and whose conscientious efforts to maintain the high standards of the *Methods in Enzymology* series will make these volumes of practical use to a broad spectrum of investigators for many years to come. We have to admit, however, that, despite our best efforts, we could not include each and every method that involves the use of a gene fusion or a hybrid protein. In part, our task was a bit like trying to bottle smoke because brilliant new methods that exploit the fundamental strategy of using a chimeric gene or protein are being devised and published daily. We hope, however, that we have been able to capture many of the most salient and generally applicable procedures. Nonetheless, we take full responsibility for any oversights or omissions, and apologize to any researcher whose method was overlooked.

Finally, we would especially like to acknowledge the expert assistance of Joyce Kato at Caltech, whose administrative skills were essential in organizing these books.

JEREMY THORNER
SCOTT D. EMR
JOHN N. ABELSON

METHODS IN ENZYMOLOGY

VOLUME LV. Biomembranes (Part F: Bioenergetics)
Edited by SIDNEY FLEISCHER AND LESTER PACKER

VOLUME LVI. Biomembranes (Part G: Bioenergetics)
Edited by SIDNEY FLEISCHER AND LESTER PACKER

VOLUME LVII. Bioluminescence and Chemiluminescence
Edited by MARLENE A. DELUCA

VOLUME LVIII. Cell Culture
Edited by WILLIAM B. JAKOBY AND IRA PASTAN

VOLUME LIX. Nucleic Acids and Protein Synthesis (Part G)
Edited by KIVIE MOLDAVE AND LAWRENCE GROSSMAN

VOLUME LX. Nucleic Acids and Protein Synthesis (Part H)
Edited by KIVIE MOLDAVE AND LAWRENCE GROSSMAN

VOLUME 61. Enzyme Structure (Part H)
Edited by C. H. W. HIRS AND SERGE N. TIMASHEFF

VOLUME 62. Vitamins and Coenzymes (Part D)
Edited by DONALD B. MCCORMICK AND LEMUEL D. WRIGHT

VOLUME 63. Enzyme Kinetics and Mechanism (Part A: Initial Rate and Inhibitor Methods)
Edited by DANIEL L. PURICH

VOLUME 64. Enzyme Kinetics and Mechanism (Part B: Isotopic Probes and Complex Enzyme Systems)
Edited by DANIEL L. PURICH

VOLUME 65. Nucleic Acids (Part I)
Edited by LAWRENCE GROSSMAN AND KIVIE MOLDAVE

VOLUME 66. Vitamins and Coenzymes (Part E)
Edited by DONALD B. MCCORMICK AND LEMUEL D. WRIGHT

VOLUME 67. Vitamins and Coenzymes (Part F)
Edited by DONALD B. MCCORMICK AND LEMUEL D. WRIGHT

VOLUME 68. Recombinant DNA
Edited by RAY WU

VOLUME 69. Photosynthesis and Nitrogen Fixation (Part C)
Edited by ANTHONY SAN PIETRO

VOLUME 70. Immunochemical Techniques (Part A)
Edited by HELEN VAN VUNAKIS AND JOHN J. LANGONE

VOLUME 71. Lipids (Part C)
Edited by JOHN M. LOWENSTEIN

VOLUME 72. Lipids (Part D)
Edited by JOHN M. LOWENSTEIN

VOLUME 91. Enzyme Structure (Part I)
Edited by C. H. W. HIRS AND SERGE N. TIMASHEFF

VOLUME 92. Immunochemical Techniques (Part E: Monoclonal Antibodies and General Immunoassay Methods)
Edited by JOHN J. LANGONE AND HELEN VAN VUNAKIS

VOLUME 93. Immunochemical Techniques (Part F: Conventional Antibodies, Fc Receptors, and Cytotoxicity)
Edited by JOHN J. LANGONE AND HELEN VAN VUNAKIS

VOLUME 94. Polyamines
Edited by HERBERT TABOR AND CELIA WHITE TABOR

VOLUME 95. Cumulative Subject Index Volumes 61–74, 76–80
Edited by EDWARD A. DENNIS AND MARTHA G. DENNIS

VOLUME 96. Biomembranes [Part J: Membrane Biogenesis: Assembly and Targeting (General Methods; Eukaryotes)]
Edited by SIDNEY FLEISCHER AND BECCA FLEISCHER

VOLUME 97. Biomembranes [Part K: Membrane Biogenesis: Assembly and Targeting (Prokaryotes, Mitochondria, and Chloroplasts)]
Edited by SIDNEY FLEISCHER AND BECCA FLEISCHER

VOLUME 98. Biomembranes (Part L: Membrane Biogenesis: Processing and Recycling)
Edited by SIDNEY FLEISCHER AND BECCA FLEISCHER

VOLUME 99. Hormone Action (Part F: Protein Kinases)
Edited by JACKIE D. CORBIN AND JOEL G. HARDMAN

VOLUME 100. Recombinant DNA (Part B)
Edited by RAY WU, LAWRENCE GROSSMAN, AND KIVIE MOLDAVE

VOLUME 101. Recombinant DNA (Part C)
Edited by RAY WU, LAWRENCE GROSSMAN, AND KIVIE MOLDAVE

VOLUME 102. Hormone Action (Part G: Calmodulin and Calcium-Binding Proteins)
Edited by ANTHONY R. MEANS AND BERT W. O'MALLEY

VOLUME 103. Hormone Action (Part H: Neuroendocrine Peptides)
Edited by P. MICHAEL CONN

VOLUME 104. Enzyme Purification and Related Techniques (Part C)
Edited by WILLIAM B. JAKOBY

VOLUME 105. Oxygen Radicals in Biological Systems
Edited by LESTER PACKER

VOLUME 106. Posttranslational Modifications (Part A)
Edited by FINN WOLD AND KIVIE MOLDAVE

VOLUME 107. Posttranslational Modifications (Part B)
Edited by FINN WOLD AND KIVIE MOLDAVE

VOLUME 193. Mass Spectrometry
Edited by JAMES A. MCCLOSKEY

VOLUME 194. Guide to Yeast Genetics and Molecular Biology
Edited by CHRISTINE GUTHRIE AND GERALD R. FINK

VOLUME 195. Adenylyl Cyclase, G Proteins, and Guanylyl Cyclase
Edited by ROGER A. JOHNSON AND JACKIE D. CORBIN

VOLUME 196. Molecular Motors and the Cytoskeleton
Edited by RICHARD B. VALLEE

VOLUME 197. Phospholipases
Edited by EDWARD A. DENNIS

VOLUME 198. Peptide Growth Factors (Part C)
Edited by DAVID BARNES, J. P. MATHER, AND GORDON H. SATO

VOLUME 199. Cumulative Subject Index Volumes 168–174, 176–194

VOLUME 200. Protein Phosphorylation (Part A: Protein Kinases: Assays, Purification, Antibodies, Functional Analysis, Cloning, and Expression)
Edited by TONY HUNTER AND BARTHOLOMEW M. SEFTON

VOLUME 201. Protein Phosphorylation (Part B: Analysis of Protein Phosphorylation, Protein Kinase Inhibitors, and Protein Phosphatases)
Edited by TONY HUNTER AND BARTHOLOMEW M. SEFTON

VOLUME 202. Molecular Design and Modeling: Concepts and Applications (Part A: Proteins, Peptides, and Enzymes)
Edited by JOHN J. LANGONE

VOLUME 203. Molecular Design and Modeling: Concepts and Applications (Part B: Antibodies and Antigens, Nucleic Acids, Polysaccharides, and Drugs)
Edited by JOHN J. LANGONE

VOLUME 204. Bacterial Genetic Systems
Edited by JEFFREY H. MILLER

VOLUME 205. Metallobiochemistry (Part B: Metallothionein and Related Molecules)
Edited by JAMES F. RIORDAN AND BERT L. VALLEE

VOLUME 206. Cytochrome P450
Edited by MICHAEL R. WATERMAN AND ERIC F. JOHNSON

VOLUME 207. Ion Channels
Edited by BERNARDO RUDY AND LINDA E. IVERSON

VOLUME 208. Protein–DNA Interactions
Edited by ROBERT T. SAUER

VOLUME 209. Phospholipid Biosynthesis
Edited by EDWARD A. DENNIS AND DENNIS E. VANCE

VOLUME 210. Numerical Computer Methods
Edited by LUDWIG BRAND AND MICHAEL L. JOHNSON

Section I

Historical Overview

[1] The All Purpose Gene Fusion

By Jon Beckwith

The biological revolution of recent years has derived its greatest impetus from the development and utilization of a handful of techniques and approaches for manipulating DNA. These methods include, most prominently, DNA cloning, DNA sequencing, the polymerase chain reaction, and gene fusion. Given the advent of the first three technical developments only during the past 25 years, one might have thought that the use of gene fusions also appeared during this period. In fact, gene fusion as a method for studying biological problems can be traced back to the earliest days of molecular biology.

Many of the principles of the gene fusion approach appear in work on one of the classical genetic systems of molecular biology, the *rII* genes of the *Escherichia coli* bacteriophage T4. In the late 1950s and early 1960s, Seymour Benzer and colleagues charactered two adjacent but independently transcribed genes, *rIIA* and *rIIB,* which constituted the *rII* region. In 1962, Champe and Benzer described an *rII* mutation in which a deletion (*r1589*) had removed all transcription and translation punctuation signals between the two genes and, thus, fused them into a single transcriptional and translational unit.[1] The deletion covered the sequences coding for the carboxy terminus of the *rIIA* protein and for approximately 10% of the amino terminus of the *rIIB* protein.

Despite the absence of a substantial portion of the B protein, the gene fusion still exhibited B activity. This property of the *r1589* deletion was to provide a very important tool for understanding fundamental aspects of the genetic code. These insights were made possible by the understanding that missense mutations in the fusion that altered the A portion of the hybrid rIIA-B protein would be unlikely to affect B function, whereas mutations that caused termination of translation in the A portion would simultaneously result in loss of B function. Benzer and Champe[2] found a class of suppressible *rIIA* mutations that did have the effect of eliminating *rIIB* activity when introduced into the *r1589* deletion. These findings were essential to the classification of these mutations (*amber*) as mutations that cause protein chain termination. This was the first description of such mutations and the recognition that special signals were involved in the

---

[1] S. P. Champe and S. Benzer, *J. Mol. Biol.* **4,** 288 (1962).
[2] S. Benzer and S. P. Champe, *Proc. Natl. Acad. Sci. U.S.A.* **48,** 1114 (1962).

chain termination process. At the same time, Crick and co-workers[3] were characterizing a class of mutations that they suspected to be frameshifts. A key step in their analysis was the demonstration that these mutations, when introduced into the *rIIA* region of the *r1589* fusion, also eliminated *rIIB* activity. These experiments were important to the use of frameshift mutations to establish the triplet nature of the genetic code.

Several key concepts underlying the gene fusion approach can be found in these studies. First, the idea that it is possible to remove a significant portion of a terminus of a protein (amino terminus in this case) and still retain sufficient protein function has proved to be the case with a large number of proteins. Second, the possibility of fusing two different proteins together and retaining one or both activities was not self-evident. It seemed quite reasonable to imagine that the generation of a single polypeptide chain from two chains would result in mutual interference with proper folding and functioning of each protein. Third, and most importantly, the notion of using downstream protein activity to report on what was happening upstream—the reporter gene concept—was key to these studies. This, of course, is the key feature of the gene fusion approach.

This history has been described as though it was known at the time that the *rII* genes coded for protein. Extraordinarily enough, it was not shown until many years later that this was the case. Nevertheless, the genetic evidence was considered compelling enough at the time that the conclusions of these studies gained widespread acceptance among molecular biologists.

The next steps in the development of gene fusion approaches came from studies on the *lac* operon of *E. coli*. The first fusions of *lac* were obtained unwittingly as revertants of strong polar mutations in the *lacZ* gene.[4] Selection for restoration of the activity of the downstream *lacY* gene yielded many deletions that removed the polar mutation site, the promoter of *lac,* and fused the *lacY* gene to an upstream promoter of an unknown neighboring gene. In 1965, Jacob and co-workers[5] exploited this approach to select for fusions in which the *lacY* gene was put under the control of an operon involved in purine biosynthesis. This was the first report of a gene fusion in which the *regulation* of a reporter gene was determined by the gene to which it was fused; the Lac permease was regulated by the concentration of purines in the growth media.

Subsequently, Muller-Hill and Kania[6] showed that the properties of β-galactosidase allowed an even broader use of the gene fusion approach

[3] F. H. C. Crick, L. Barnett, S. Brenner, and R. J. Watts-Tobin, *Nature* **192,** 1227 (1961).
[4] J. R. Beckwith, *J. Mol. Biol.* **8,** 427 (1964).
[5] F. Jacob, A. Ullmann, and J. Monod, *J. Mol. Biol.* **42,** 511 (1965).
[6] B. Muller-Hill and J. Kania, *Nature* **249,** 561 (1974).

in this system. Using a very early chain-terminating mutation, they found that they could restore β-galactosidase activity by deleting the polar mutation site and fusing the remaining portion of the polypeptide to the upstream *lacI* gene product, the Lac repressor. It was even possible to obtain hybrid proteins with both repressor and β-galactosidase activity.

Generalizing the Approach

In all the cases described to this point, genetic fusions were obtained between two genes that were normally located close to each other on the bacterial chromosome or on an F' factor. This feature of early gene fusion studies presented quite strict limitations on the systems that could be analyzed by this approach. However, beginning first with some old-fashioned approaches to transposing the *lac* region to different positions on the chromosome,[7] we began to see that the gene fusion approach might be applied more widely. A graduate student in the author's laboratory, Malcolm Casadaban, then developed improvements on transposition techniques that enhanced the ability to fuse *lac* more generally to bacterial genes.[8] Malcolm continued these improvements in Stanley Cohen's laboratory at Stanford University and ultimately in his own laboratory at the University of Chicago.[9,10]

All the approaches described so far involved generation of fusions *in vivo*. The arrival of recombinant DNA techniques for cloning and fusing genes in the mid-1970s provided a tremendous boost to the use of gene fusions. It became possible to fuse genes from or between any organism pretty much at will.

Gene Fusions for All Seasons

For many years, the gene fusion tool was considered to be one useful mainly for studying gene expression and regulation by reporter gene expression. However, as the ease of generating such fusions grew, other uses became evident. In 1980, we reported the first case where fusing a reporter protein to another protein of interest allowed purification of the latter protein.[11] In this case, β-galactosidase was fused to a portion of the cytoplasmic membrane protein, MalF. The unusually large size of

[7] J. R. Beckwith, E. R. Signer, and W. Epstein, *Cold Spring Harbor Symp. Quant. Biol.* **31,** 393 (1966).
[8] M. Casadaban, *J. Mol. Biol.* **104,** 541 (1976).
[9] M. J. Casadaban and S. N. Cohen, *Proc. Natl. Acad. Sci. U.S.A.* **76,** 4530 (1979).
[10] M. J. Casadaban and J. Chou, *Proc. Natl. Acad. Sci. U.S.A.* **81,** 535 (1984).
[11] H. A. Shuman, T. J. Silhavy, and J. R. Beckwith, *J. Biol. Chem.* **255,** 168 (1980).

β-galactosidase allowed ready purification of the hybrid protein, which was then used to elicit antibody to MalF epitopes, facilitating its purification. We also showed that gene fusions of β-galactosidase could be used to study the signals that determine subcellular protein localization. Fusion of β-galactosidase to the MalF protein resulted in membrane localization of the former protein,[11] and fusion of β-galactosidase to exported proteins permitted the genetic analysis of bacterial signal sequences.[12,13]

Another important step in the evolution of uses of gene fusions came with the concept of signal sequence traps. The first development of this concept came out of the recognition that the bacterial enzyme alkaline phosphatase is active when it is exported to the periplasm but inactive when it is retained in the cytoplasm.[14] Thus, alkaline phosphatase without its signal sequence provides an assay for export signals via gene fusion approaches, i.e., alkaline phosphatase will only be active if one attaches a region of DNA that encodes a signal sequence, thus reallowing its export. Hoffman and Wright[15] and Colin Manoil and the author[16] reported systems—one plasmid, one transposon—that allowed the detection of signal sequences in random libraries of DNA or in a bacterial chromosome. This approach has been extended with use of numerous other reporter genes, including, most prominently, β-lactamase.[17]

Extending beyond the differentiation of exported vs cytosolic proteins, gene fusion techniques can be evolved to determine subcellular localization of proteins more generally. Clearly, the use of GFP fusions enhances this ability.[18] In addition, reporter proteins that sense specific features of organelle environment may provide a tool for detecting location and genetically manipulating signals for the localization process. The report of a GFP that responds to the pH of its environment may be a harbinger of things to come.[19] One might imagine GFP derivatives that respond to all sorts of cellular conditions, e.g., the redox environment.

Finally, gene fusions can be used for the study of protein structure, protein–protein interactions, and protein folding. The yeast two-hybrid system described by Fields and Song[20] in 1989 has become a powerful tool for analyzing aspects of quaternary structure of proteins and for detecting

[12] S. D. Emr, M. Schwartz, and T. J. Silhavy, *Proc. Natl. Acad. Sci. U.S.A.* **75,** 5802 (1978).
[13] P. Bassford and J. Beckwith, *Nature* **277,** 538 (1979).
[14] S. Michaelis, H. Inouye, D. Oliver, and J. Beckwith, *J. Bacteriol.* **154,** 366 (1983).
[15] C. Hoffman and A. Wright, *Proc. Natl. Acad. Sci. U.S.A.* **82,** 5107 (1985).
[16] C. Manoil and J. Beckwith, *Proc. Natl. Acad. Sci. U.S.A.* **82,** 8129 (1985).
[17] Y. Zhang and J. K. Broome-Smith, *Mol. Microbiol.* **3,** 1361 (1989).
[18] D. S. Weiss, J. C. Chen, J. M. Ghigo, D. Boyd, and J. Beckwith, *J. Bacteriol.* **181,** 508 (1999).
[19] G. Miesenböck, D. A. DeAngelis, and J. E. Rothman, *Nature* **394,** 192 (1998).
[20] S. Fields and O. Song, *Nature* **340,** 245 (1989).

novel protein–protein interactions. Whereas the structure of soluble proteins is accomplished relatively easily by X-ray crystallography techniques, the structure of membrane proteins still largely resists such approaches. Gene fusion techniques have been able to contribute to understanding important features of membrane protein structure. The signal sequence trap techniques have proved invaluable in the determination of the topological structure of integral membrane proteins,[21] i.e., fusion of the reporter protein to intra- or extracytoplasmic domains of membrane proteins usually reports the location of that domain accurately. Similarly, more recent techniques for detecting interactions between transmembrane segments of such proteins should allow the elucidation of additional structural features.[22,23] Although not so widely employed, gene fusion approaches can aid in the study of protein folding. Luzzago and Cesareni[24] used a cute fusion approach to isolate mutants affecting the folding of ferritin. Other such ideas must be waiting in the wings.

The realm of gene fusions has continually expanded. While this volume describes a host of different issues that can be studied with this technique, it seems certain that the expansion will continue.

[21] C. Manoil and J. Beckwith, *Science* **233**, 1403 (1986).
[22] J. A. Leeds and J. Beckwith, *J. Mol. Biol.* **280**, 799 (1998).
[23] W. P. Russ and D. M. Engelman, *Proc. Natl. Acad. Sci. U.S.A.* **96**, 863 (1999).
[24] A. Luzzago and G. Cesareni, *EMBO J.* **8**, 569 (1989).

Section II

Gene Fusions as Reporters of Gene Expression in Prokaryotic Cells

[2] A Practical Guide to the Construction and Use of *lac* Fusions in *Escherichia coli*

By Nicholas J. Hand and Thomas J. Silhavy

Introduction

The Lac system is without equal as a reporter system for the study of transcriptional and translational regulation in bacteria. In addition, the properties of β-galactosidase have enabled a number of elegant schemes to be developed using it as a tag to purify proteins of interest or in the production of antibodies. However, significant innovations in the use of small polypeptide epitopes in recent years have decreased the desirability of β-galactosidase for many of the applications for which it was formerly the molecule of choice. In particular, many of the biochemical uses of β-galactosidase developed in the past, and reviewed by Silhavy and Beckwith,[1] are no longer the logical first choice when weighed against newer technologies. These considerations notwithstanding, however, β-galactosidase remains of unparalleled usefulness as a tool in the hands of the bacterial geneticist. By virtue of the broad range over which its activity can be assayed, coupled with the low cost and robustness of the assays, β-galactosidase remains unrivaled as a transcriptional reporter. Using appropriate media, mutations that increase or decrease the expression of an operon fusion of interest can be isolated easily. Conversely, screening pools of random LacZ chromosomal insertions can identify targets of a regulator (either transcriptional or translational). Finally, β-galactosidase remains useful in the study of translational regulation, although certain caveats must be considered in studying LacZ protein fusions.

Rather than revisit techniques that are no longer of significant interest, we will discuss a more limited selection of those applications for which β-galactosidase is generally most useful. We will also present a limited set of up-to-date protocols for making specific or random *lac* fusions (both transcriptional and translational). The protocols presented here are those currently in use in our laboratory and are adapted from a number of sources.[1–5] Useful additional resources for basic issues not covered in this

[1] T. J. Silhavy and J. R. Beckwith, *Microbiol Rev.* **49,** 398 (1985).
[2] E. Bremer, T. J. Silhavy, and G. M. Weinstock, *J. Bacteriol.* **162,** 1092 (1985).
[3] R. W. Simons, F. Houman, and N. Kleckner, *Gene* **53,** 85 (1987).

article may be found elsewhere.[4,6-8] In addition, we will attempt to preserve some of the "LacZ lore" that is disappearing with the increased use of other protein tags and reporter systems. In particular, we will discuss the use of indicator and selector media.

Production of LacZ Fusions

Three separate partially overlapping nomenclatures exist to describe lac fusions. Transcriptional fusions are also known as promoter or operon fusions, whereas translational fusions are alternatively referred to as gene or protein fusions. Broadly speaking, two classes of LacZ fusions cover most uses: fusions created to specific *cis*-acting regulatory regions (including inframe protein fusions) and fusions created by random chromosomal insertion of the *lac* operon. We will approach the production of fusions of both types separately. For all intents and purposes, the craft of engineering the former, specific LacZ fusions by genetic means (using phage Mu, for example) has been replaced by more conventional molecular cloning techniques. In the latter case, where components of a regulon are being sought, specialized transposon and phage vectors have simplified the procedure greatly. The first part of this section presents a protocol for creating a fusion of a specific DNA fragment to the *lac* operon and the subsequent isolation of the fusion in single copy on the bacterial chromosome for analysis.

Transcriptional and Translational Fusions to Specific Genes

A vast array of plasmid vectors and corresponding λ phage exist for the purpose of creating operon and protein fusions. In the case where a specific gene is to be studied, essentially the same techniques apply to all vectors, with the only difference being the choice of the vector and phage. A later section cites specific merits and demerits of various plasmids and

[4] T. J. Silhavy, M. L. Berman, and L. W. Enquist, "Experiments with Gene Fusions." Cold Spring Harbor Laboratory, Cold Spring Harbor, NY, 1984.
[5] G. M. Weinstock, M. L. Berman, and T. J. Silhavy, *Gene Amplif. Anal.* **3,** 27 (1983).
[6] J. H. Miller, "Experiments in Molecular Genetics." Cold Spring Harbor Laboratory, Cold Spring Harbor, NY, 1972.
[7] J. H. Miller, "A Short Course in Bacterial Genetics: A Laboratory Manual and Handbook for *Escherichia coli* and Related Bacteria." Cold Spring Harbor Laboratory Press, Cold Spring Harbor, NY, 1992.
[8] J. Sambrook, E. F. Fritsch, and T. Maniatis, "Molecular Cloning: A Laboratory Manual." Cold Spring Harbor Laboratory, Cold Spring Harbor, NY, 1989.

TABLE I
MULTICOPY VECTORS FOR THE CONSTRUCTION OF TRANSCRIPTIONAL
AND TRANSLATIONAL FUSIONS

Vector[a]	Size (kb)	Marker(s)[b]	Fusion[c]	Fusion type	Notes
pRS308	8.0	Amp^R	*lac`Z_{SC}YA*	Either	pRS308 is used to recover existing fusions by *in vivo* recombination (see text for details)
pRS415, pRS528	10.8	Amp^R	*lacZYA*	Transcriptional	pRS415 is a derivative of pNK678, in which four tandem copies of the *rrnB* transcriptional terminator have been cloned between the *bla* gene and the *lac* operon, eliminating background expression from the plasmid
pRS551, pRS550	12.5	Amp^R, Kan^R	*lacZYA*	Transcriptional	Kanamycin-resistant derivatives of pRS415 and pRS528, respectively. Resulting single copy fusions (lysogens) are marked with Kan^R
pRS414, pRS591	10.7	Amp^R	*lac`ZYA*	Translational	pRS414 is derived from pRS415 by a 120-bp restriction fragment deletion, which removes the ribosome-binding site
pRS552, pRS577	12.4	Amp^R, Kan^R	*lac`ZYA*	Translational	Kanamycin-resistant derivatives of pRS414 and pRS591, respectively. Resulting single copy fusions (lysogens) are marked with Kan^R

[a] R. W. Simons, F. Houman, and N. Kleckner, *Gene* **53,** 85 (1987).
[b] Amp^R, ampicillin resistance; Kan^R, kanamycin resistance.
[c] Fusions designated *lacZYA* contain functional *lacZ, lacY,* and *lacA* genes and include the sequences necessary for translational initiation. Fusions designated *lac`ZYA* are deleted for the translation initiation sequences. The *lac`Z_{SC}YA* fragment on pRS308 is deleted for the *lacZ* sequence upstream of the *Sac*I site and therefore carries a 3′ fragment comprising roughly one-third of the *lacZ* gene, as well as functional *lacY* and *lacA* genes.

phages and differences in the analysis of transcriptional and translational fusions.

The number of vectors available for the creation of Lac fusions is positively bewildering, and a more comprehensive listing can be found elsewhere.[9] For practical purposes, only a few plasmids and phage strains are necessary. We have found that the excellent set of fusion vectors created by Simons *et al.*[3] meet most of our needs (see Table I and Fig. 1). For this reason, we will use the specific example of an operon fusion created on pRS415, recombined onto λRS45 (see Table II and Fig. 1), and integrated onto the chromosome.

[9] J. M. Slauch and T. J. Silhavy, *Methods Enzymol* **204,** 213 (1991).

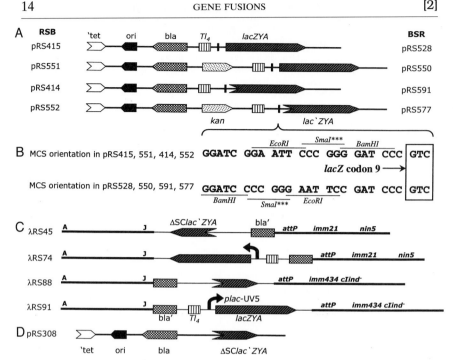

FIG. 1. Adapted from Simons et al.[3] (A) Schematic diagram of plasmid vectors for constructing lac fusions. All of the plasmids are based on a pBR322 backbone and carry a fragment of the 3' end of the tetA gene. In addition, all of the plasmids carry four copies of the rrnB transcriptional terminator (Tl₄) upstream (with respect to the lac operon) of the multiple cloning site. Complete details of the construction and inferred DNA sequence can be found in Simons et al.[3] pRS415 has the multiple cloning site (MCS) in the order EcoRI, SmaI, BamHI (RSB), whereas pRS528 is the same plasmid, with the MCS reversed (BSR) as shown in (B). Similarly, pRS551 and pRS550 are the same plasmid with the MCS in opposite orientations, and so on. Plasmids pRS415 and pRS551 (and the corresponding plasmids pRS528 and pRS550) are designed for making transcriptional fusions, and thus carry the sequences necessary for translational initiation. Plasmids pRS414 and pRS552 (as well as pRS591 and pRS577) are derivatives of the transcriptional fusion vectors designed for making translational fusions. These plasmids carry a 120-bp deletion that removes the ribosome-binding site, thus expression of the lac genes is dependent on an in vitro fusion in-frame to an open reading frame with a promoter and translational initiation sequences. (B) Sequence of the multiple cloning sites of the plasmids shown. The spacing shows the lacZ reading frame of the translational fusion vectors. *** Note that in the case of the Kan^R plasmids the SmaI site in the MCS is not unique, as the Tn903-derived sequence carrying the kanamycin resistance gene introduces a second SmaI site. (C) Schematic diagram of λ vectors for making single-copy derivatives of cloned lac fusions. λRS45 and λRS88 carry a region of homology from pRS308 (see D), with a truncated 5' fragment of the β-lactamase (bla) gene and a truncated 3' fragment of the lac operon. Fusions recombined onto this vector are sensitive to ampicillin. In the case of fusions with low levels of expression, lac activity may be difficult to distinguish from background on indicator agar (see text for details). For this reason, corresponding λ vectors with high lac activity have been constructed. λRS74 and λRS91 carry a complete lac operon

Primer Design Considerations and Polymerase Chain Reaction

The availability of the complete DNA sequence of *Escherichia coli*[10] and the advent of the polymerase chain reaction (PCR)[11] has made the cloning of any region of the genome a relatively trivial affair. PCR cloning strategies have the advantage that they do not depend on available restriction sites in the sequence to be cloned, as convenient sites can be added as 5′ "tails" on the oligodeoxynucleotide primers used for the amplification of the region to be studied.

A detailed discussion of PCR cloning is beyond the scope of this article and, furthermore, is not particularly useful as the strategy to be employed depending on the individual circumstances. We do feel, however, that it is worthwhile to present a few general considerations of primer design and fragment size in cloning into *lac* fusion vectors. Because the orientation of the DNA fragment is critical, it is worthwhile to employ a directional or forced cloning strategy if possible. Given the paucity of restriction sites in Lac fusion vector multiple cloning sites, it is seldom the case that a fragment of suitable size (usually less than 1 kb in length for promoter fusions) can be cloned in the right orientation using multiple cloning site restriction enzymes. Thus, we routinely add restriction site sequences to our amplification primers. If this is to be done, additional deoxynucleotides should be added 5′ of the new site, as many restriction enzymes cleave sites close to the ends of linear DNA molecules with substantially reduced efficiency.[12]

---

[10] F. R. Blattner, G. Plunkett III, C. A. Bloch, N. T. Perna, V. Burland, M. Riley, J. Collado-Vides, J. D. Glasner, C. K. Rode, G. F. Mayhew, J. Gregor, N. W. Davis, H. A. Kirkpatrick, M. A. Goeden, D. J. Rose, B. Mau, and Y. Shao, *Science* **277,** 1453 (1997).

[11] R. K. Saiki, S. Scharf, F. Faloona, K. B. Mullis, G. T. Horn, H. A. Erlich, and N. Arnheim, *Science* **230,** 1350 (1985).

[12] R. F. Moreira and C. J. Noren, *Biotechniques* **19,** 56, 58 (1995).

---

under the control of the strong, inducible *placUV5* promoter. Recombination of fusions with low activity onto these λ vectors can therefore be detected easily against a background of dark blue plaques. The fact that the inserts in λRS45 and λRS74 are in the opposite orientation with respect to those in λRS88 and λRS91 is not relevant. However, it is important to note that λRS88 and λRS91 are *cIind* and therefore form "locked-in" lysogens. Such lysogens cannot be induced to produce mature λ phage particles, so other strategies, such as generalized transduction using P1, must be employed to transfer these fusions to different strains (see text). (D) Schematic diagram of pRS308. This plasmid is designed for the recovery of fusions from lysogens (for details, see text). This plasmid can be used to recover not only fusions created using pRS and λRS vectors, but also fusions made with compatible vectors (those that have divergent *bla* and *lac* genes), e.g., pMLB1034[4] and λRZ5[14] (which has the occasionally undesirable property of yielding ampicillin-resistant lysogens).

TABLE II
VECTORS FOR TRANSFERRING FUSIONS TO THE CHROMOSOME IN SINGLE COPY

Vector[a]	Size (kb)	Marker[b]	Lysogen[c]	Inducible[d]	Notes
λRS45	43.3	*imm21*	*bla'*	Yes	Contains a 5.0-kb fragment from pRS308 (see Table I), including the 5' end of the *bla* gene, and a 3' fragment of the *lac* operon (partial sequence of *lacZ*). Fusions transferred by *in vivo* recombination are sensitive to ampicillin and can be detected by an increase in β-galactosidase activity
λRS74	46.2	*imm21*	*bla'*	Yes	Contains a 7.9-kb fragment including the entire *lac* operon under the control of the p*lac*UV-5 promoter. λRS74 lysogens have high β-galactosidase activity. Cloned fusions with low activity can thus be detected by recombining away the p*lac*UV-5 promoter
λRS88	47.6	*imm434*	*bla'*	No	λRS88 carries the *cI ind* mutation. Produces "locked-in" lysogens that cannot recovered by induction of λ, and must therefore be moved by P1 transduction if necessary. Contains the same fragment of pRS308 as λRS45, but in the opposite orientation
λRS91	50.5	*imm434*	*bla'*	No	Like λRS88, λRS91 carries the *cI ind* mutation and results in noninducible lysogens. Contains the same insert as λRS74, but in the opposite orientation. Like λRS74, λRS91 is used to detect transfer of fusions with low activity to the chromosome
λRZ5	~46	*imm*λ	Amp[R]	Yes	Unlike λRS vectors, this vector contains an insert that includes a 3' `*bla* gene fragment (rather than a truncated 5' piece) in divergent orientation with a portion of the *lac* operon, including the 3' end of *lacZ* and all of *lacY*. Fusions transferred to λRZ5 remain Amp resistant

[a] λRS vectors are fully described in R. W. Simons, F. Houman, and N. Kleckner, *Gene* **53,** 85 (1987). λRZ5 is described in K. S. Ostrow, T. J. Silhavy, and S. Garrett, *J. Bacteriol.* **168,** 1165 (1986).
[b] *imm,* phage immunity.
[c] Lysogens referred to as *bla'* carry a truncated 5' fragment of the *bla* open reading frame and are sensitive to ampicillin. Amp[R] lysogens have an intact *bla* gene and are ampicillin resistant.
[d] Inducibility refers to the ability to recover mature λ phage particles from the integrated lysogen by UV induction.

For amplification purposes, we routinely use primers of 30–33 nucleotides in length, with 6 nucleotides at the 5' end to promote efficient cleavage, 6 nucleotides comprising the restriction site to be added, and 18–21 nucleotides homologous to the region to which the primer is to anneal. The 3' dinucleotide should be WS-3' (i.e., W = {A/T}, S = {G/C}-3') to allow efficient amplification with a minimum of background (SS-3' yields more background bands, whereas WW-3′ and SW-3′ yield less product). We estimate the melting temperature (T_m) of the primers, based on the composition of the 3′ 18 nucleotides,[8] and we attempt insofar as is reasonably convenient to match the T_m of the two primers as closely as possible. A thorough discussion of primer design considerations can be found elsewhere.[13]

Because different thermostable DNA polymerases have different fidelities, it is necessary to fully confirm the DNA sequence of the cloned amplicon when a low-fidelity enzyme is used. High-fidelity enzymes avoid this complication, but often present additional difficulties, requiring more troubleshooting and higher DNA purity for successful amplification. Our preference is to use the more robust low-fidelity enzymes and confirm the sequence of the products obtained. This allows us to amplify DNA from single bacterial colonies without purification simply by picking a well-isolated colony with a sterile disposable plastic (P-200 type) pipette tip, adding it to the PCR reaction mix (minus polymerase), and boiling for 5 min to lyse the cells. The polymerase is then added and the PCR is started.

After the reaction is complete, the products are separated by electrophoresis and the bands of interest are cut from the gel and purified. The specifics of this part of the procedure are not relevant in this instance. Suffice it to say that the bands are cut, purified, and cloned into the appropriate plasmid (in this instance pRS415) using standard methods. We find that in the case of forced cloning it is unnecessary to dephosphorylate the linearized plasmid.

Choice of λ Phage

The next step of the procedure is to isolate single-copy lysogens carrying the fusion of interest. Whereas analysis of plasmid-borne fusions may be sufficient for some purposes, we believe it introduces a number of unnecessary problems, such as excessively high β-galactosidase activity, variable plasmid copy number, and gene dosage effects. More worrisome, on high-copy vectors the very regulatory factors in which one is interested may be titrated by the number of *cis*-acting sites, giving the impression that the promoter is unregulated. Furthermore, a final practical consideration is that maintaining a plasmid in a strain constrains the use of other plasmids in that strain (because the plasmids must come from different incompatibility groups and use different antibiotic resistance markers for their maintenance).

Different schools of thought exist on the choice of phage and plasmids for the integration of fusions onto the chromosome. Some researchers

[13] J. M. Robertson and J. Walsh-Weller, *Methods Mol. Biol.* **98,** 121 (1998).

prefer to use combinations that produce lysogens in which the fusion is linked to a drug resistance marker. This facilitates the movement of the fusions from one strain to another by generalized transduction. Paradoxically, marked fusions can complicate strain construction because of the loss of the ability to use the marker on the λ phage to move other mutations or maintain plasmids. For illustrative purposes, we will consider a case in which the lysogen is not marked.

Note: In all procedures involving pipetting of phage-containing solutions, aerosol-resistant pipette tips should be used to prevent contamination of pipettors. Because λ is sensitive to trace amounts of detergent, all glassware used should be rinsed thoroughly with deionized water to ensure that it is soap-free before use. Furthermore, λ is sensitive to light and should be stored in the dark. In all procedures involving plates, the plates in question are standard, disposable plastic (100 × 15 mm) petri plates, typically filled with 25 ml of solid media.

Preparation of Single Plaques

1. Pellet the cells from a fresh 5-ml overnight culture of MC4100 (or similar Δ*lac* strain) and resuspend in 2.5 ml of 10 mM MgSO$_4$.

2. In a small culture tube, mix 50 μl of cells with 10 μl of an appropriate dilution of the λRS45 phage (usually, dilutions between 10^{-3} and 10^{-5} of a high-titer phage stock yield well-isolated plaques).

Note: λ should be diluted in TMG buffer (per liter: 1.21 g Tris base, 1.20 g MgSO$_4 \cdot$ 7H$_2$O, 0.10 g gelatin, adjust to pH 7.4 with HCl and sterilize by autoclaving).

3. Incubate the tube at room temperature for 10 min.

4. Add 2.5 ml of molten (45°) LB top agar, and pour onto prewarmed LB plates. Immediately spread the top agar evenly by tilting the plates gently from side to side.

Optional step: Prior to the addition of the top agar, the cells and phage may be mixed with prewarmed (37°) LB broth. This yields bigger plaques, but the surface of the plates can be uneven, and the resulting plates are much sloppier.

5. Allow the plates to cool at room temperature (until the top agar has solidified) and then incubate the plates at 37° overnight, agar side down.

Preparation of λ Phage Lysates Carrying Lac Fusions

In Vivo Plasmid Recombination

1. Set up a 5 ml culture of the strain containing the plasmid-encoded fusion, and rotate or shake overnight at 30° or 37°.
2. Spin down the overnight culture carrying the plasmid (in this case pRS415 with a cloned insert) and resuspend in 2.5 ml of 10 mM MgSO$_4$.
3. Add 50 μl of the cell suspension to each of five sterile test tubes.
4. Using a sterile Pasteur pipette, pick individual well-isolated plaques from the phage plate as agar plugs and transfer one, two, three, and four plugs, respectively, to each of the first four tubes.
5. Mix the phage and cells by vortexing the tubes briefly, making sure that the plugs are immersed in the cell suspension. Incubate at room temperature for 5 min.
6. Add 2 ml of LB broth containing 10 mM MgSO$_4$ to each of the five tubes.
7. Shake or rotate the tubes at 37° for 4–6 hr or until lysis occurs. Lysis in phage-containing tubes should be assessed by comparison to the tube containing cells only.
8. As lysis occurs, add 100 μl of chloroform to each tube.

Note: After 6 hr, add chloroform, irrespective of whether there is obvious lysis (occasionally, high-titer lysates can be obtained without apparent lysis).

9. Vortex the tubes briefly and then centrifuge at 4500g for 10 min.
10. Transfer the supernatant to sterile screw-cap tubes (taking care to avoid transferring any of the chloroform) and store in the dark at 4°.

Lysates prepared in this way contain a mixed population of phage, some of which are recombinant Lac$^+$ phage (i.e., they have picked up the plasmid encoded *lac* fusion) and some of which are nonrecombinant Lac$^-$ parental phage. The next step in the procedure is to isolate recombinant Lac$^+$ plaques and use them to prepare high-titer phage lysates.

Selection of Lac$^+$ Plaques

1. Prepare 10^{-2}, 10^{-4}, 10^{-6}, and 10^{-8} dilutions in TMG buffer of the lysates produced on the plasmid-containing strains.
2. Pellet the cells from a fresh 5-ml overnight culture of MC4100 and resuspend in 2.5 ml of 10 mM MgSO$_4$.
3. In a small culture tube, mix 50 μl of cells with 10 μl of each dilution of the λ lysates.
4. Incubate the tubes at room temperature for 10 min.

5. To each tube of cells and phage add 2.5 ml of prewarmed (37°) LB broth containing 40 μl of freshly added X-Gal (20 mg ml^{-1}) stock solution per ml of LB broth.

6. Add 2.5 ml of molten (45°) LB top agar and plate immediately on prewarmed LB plates.

7. Allow the plates to cool at room temperature (until the top agar has solidified) and then incubate the plates at 37° overnight, agar side down.

Preparation of High-Titer Lac$^+$ λ Lysates

1. Examine the plates from the previous day. Plates from the 10^{-2} and 10^{-4} dilutions should exhibit confluent or near confluent lysis. Plates from the 10^{-6} dilution plate should have many individual plaques, whereas plates from the 10^{-8} dilution plate generally have less than 10 plaques.

2. Pellet the cells from a fresh 5-ml overnight culture of MC4100 and resuspend in 2.5 ml of 10 mM MgSO$_4$. Set up five test tubes containing 50 μl of the cell suspension per tube.

3. Pick four individual, well-isolated, blue, turbid plaques per fusion construct.

4. Place one plaque per tube in each of the first four tubes. Mix the phage and cells by vortexing the tubes briefly, making sure that the plugs are immersed in the cell suspension. Incubate at room temperature for 5 min.

5. Add 2 ml of LB broth containing 10 mM MgSO$_4$ to each of the five tubes.

6. Shake or rotate the tubes at 37° for 4–6 hr or until lysis occurs. Lysis in phage-containing tubes should be assessed by comparison to the tube containing cells only.

7. As lysis occurs, add 100 μl of chloroform to each tube.

Note: Single plugs from recombinant phage plaques rarely clear well. After 6 hr, add chloroform, irrespective of whether there is obvious lysis. Lysates thus obtained are almost invariably of a sufficiently high titer to be useful.

8. Vortex the tubes briefly and then centrifuge at 4500g for 10 min.

9. Transfer the supernatant to sterile screw-cap tubes (taking care to avoid transferring any of the chloroform) and store in the dark at 4°.

Production of Single-Copy Lac Fusion λ Lysogens

Single-copy Lac fusion-containing lysogens may be conveniently placed at two different sites in the bacterial chromosome. The most common choice of sites is at the wild-type λ attachment site (*att*) located at 17' on the *E. coli* chromosome. However, it may also be desirable to place fusions at the chromosomal locus of the gene under study. For the latter case, using a relatively short piece of DNA to construct the fusion *in vitro*, genes can be studied in the context of a much larger operon, or upstream regulatory region. In order to target fusions to a site other than the *att* site, lysogens need to be constructed by infecting an *att* deletion strain. In the absence of a chromosomal *att* site, the recombinant phage can only integrate onto the chromosome by homologous recombination at the locus encoding the sequence used to construct the fusion. For this reason, the strain must be *recA*+.

In order to minimize the formation of multiple lysogens (tandem direct repeats of the λ prophage), λ infections are performed at a low multiplicity of infection (MOI). Potential lysogens should be chosen from the plate in which the fewest number of phage was used in the infection. Further steps to avoid multiple lysogens are described.

Targeting Lac Fusions to the Chromosome

1. Prepare 10^{-2}, 10^{-4}, 10^{-6}, and 10^{-8} dilutions in TMG buffer of the lysates described earlier.

2. Pellet the cells from a fresh 5-ml overnight culture of MC4100 [or MC4100 Δ(*gal att bio*) strain, if the fusion is to be targeted to the native chromosomal site] and resuspend the pellet in 2.5 ml of 10 mM MgSO$_4$.

3. In a small culture tube, mix 50 μl of cell suspension with 10 μl of each dilution of the λ lysates.

4. Incubate the tube at room temperature for 5 min.

5. Add 1 ml of LB broth containing 10 mM MgSO$_4$.

6. Incubate the tubes at 37° for 1 hr.

7. Pellet the cells at 4500g and decant the supernatant.

8. Wash the pellet with 1 ml of 10 mM MgSO$_4$, vortex briefly, and pellet the cells again.

9. Decant the supernatant, vortex the tubes to resuspend the cells in the residual liquid, and plate the entire contents of each tube on a minimal lactose plate [in the case of Δ(*gal att bio*) strains, the plates must be supplemented with biotin to a final concentration of 1 μg ml^{-1}].

10. Incubate the plates at 37° overnight.

11. After incubation overnight, lysogens will form colonies on the minimal lactose plates (nonlysogens cannot grow as MC4100 carries a chromosomal deletion of the *lac* operon).

12. Purify several well-isolated colonies as potential lysogens on minimal lactose plates.

Note: Using minimal lactose plates as a selective medium will not work in the case of fusions with very low levels of expression. The simplest way of making lysogens of weakly expressed genes is to use a combination of plasmids and phage that link the fusion to a drug resistance marker. Potential lysogens selected as drug-resistant colonies on appropriate media can be tested by cross-streaking. Alternatively, unmarked lysogens of weakly expressed fusions can be made by plating on soft agar in the presence of appropriate selector phage.

Testing Lysogens by Cross-Streaking

1. On the back of an LB plate, draw two thin parallel lines in permanent ink using a ruler.
2. Hold the plate at an angle with the lines vertical.
3. Pipette 50 μl of the undiluted parent phage, which was used to make the recombinant phage lysogen [e.g., in this example λRS45 at 10^9 plaque forming units (pfu)/ml] onto the top of one line and allow the drop of lysate to run down along the line.
4. Pipette 50 μl of undiluted λvir onto the top of the second line and allow the drop of lysate to run down along the line.
5. Allow the plate to dry agar side down.
6. Cross-streak purified single colonies from putative lysogens in smooth single streaks, perpendicular to the lines drawn on the plate. Always cross-streak first across the parent phage line and then across the λvir line.
7. As controls, cross-streak a known lysogen, a nonlysogen, and a λ-resistant strain (e.g., a *lamB*$^-$ strain) on each plate.
8. Incubate the plates at 37° overnight.
9. Examine the plates following overnight incubation to determine which of the test strains are lysogens. Nonlysogens are sensitive to the parent λ strain; thus they will be confluent leading up to the first line of phage, with sparse single colonies after the streak crosses the line. The streak from a lysogenic strain will cross the first, parental λ strain line (because the lysogen is immune to the parent phage), but will show sensitivity to λvir. λ-resistant strains will grow across both lines.

Multiple Lysogens

Despite the precaution of infecting at a low MOI, multiple lysogens may still occur. This presents the uninitiated investigator with a strain with perplexing unstable Lac phenotypes. The solution to the problem of multiple lysogens is trivial. Fusions at the *att* site can be moved into different strain backgrounds by generalized transduction mediated by bacteriophage P1. We use a strain with a marked *att* deletion to facilitate this process [NJH140: *MC4100 nadA ::Tn10 Δ(gal att bio)*]. The deletion is first moved into the desired recipient background (selecting for the Tn10-encoded Tet^R marker and screening for cotransduction of the Gal$^-$ phenotype on galactose tetracycline MacConkey agar). Because the *nadA ::Tn10* and *bio* mutations flank the *attL* site and because both confer a Min$^-$ phenotype (inability to grow on minimal medium), subsequent transduction of a prophage integrated at *attL* can be selected by requiring growth on minimal agar. Furthermore, because of the distance between *nadA* and *bio* markers, only generalized transducing P1 particles carrying single lysogens are capable of rescuing both mutations and thereby conferring the selected Min$^+$ phenotype.

Special Considerations

Genes with low levels of expression may not yield easily identifiable blue plaques. For this reason, corresponding λ vectors have been constructed that express β-galactosidase from the strong, inducible *pLac*UV5 promoter (Table II). Using these vectors, it is possible to create fusions to genes that are expressed weakly under the plating conditions by simply selecting light blue plaques against a background of dark blue plaques. Expression of these fusions can then be examined under other conditions. Because the expression level of a gene of interest is not necessarily known in advance, our default choice is to use white plaque-producing phage vectors. These satisfy most fusion construction criteria. In rare cases where blue plaques are not obvious, the same plasmid can be recombined onto the corresponding *pLac*UV5-Lac$^+$-containing λ strain and the plaque assay repeated.

Recovering Single-Copy *lac* Fusions

In the past we have often used the phage λRZ5[14] to convert multicopy *lac* fusions to single-copy λ lysogens. This vector is equivalent to the λRS vectors in many respects, but it encodes an intact β-lactamase open reading frame, conferring resistance to ampicillin. Because many plasmids require ampicillin for their maintenance, we have found it useful in some cases to convert these fusions to Amp^s derivatives. Because the structure of λRZ5 and those of the λRS vectors are compatible, this can be accomplished by transforming pRS308 into a *recA$^+$* strain carrying the lysogen of interest. Plasmid DNA is prepared from the resulting strain, transformed into

[14] K. S. Ostrow, T. J. Silhavy, and S. Garrett, *J. Bacteriol.* **168,** 1165 (1986).

MC4100, and the transformants are plated on minimal lactose agar. Rare (approximately 10^{-4} per plasmid) recombination events will lead to the transfer of the fusion in the original strain to the plasmid. Strains carrying the fusion on the pRS308 vector can then simply be infected with the appropriate λRS phage as already described, and the fusion can be thus converted to a single-copy Amp^s lysogen.

Assaying β-Galactosidase Activity

The most common assay of β-galactosidase activity takes advantage of the chromogenic substrate o-nitrophenyl-β-D-galactoside (ONPG). Aqueous solutions of ONPG are colorless, but hydrolysis by β-galactosidase liberates the yellow compound o-nitrophenol. In the assay, the production of o-nitrophenol is monitored over time by reading the absorbance of the samples at 420 nm. Because ONPG is a very sensitive substrate, as little as one molecule of β-galactosidase per cell can be detected reproducibly.

In our laboratory, we follow a modified version of the procedure of Miller,[6] which uses chloroform and sodium dodecyl sulfate to permeabilize the cells, and therefore does not require mechanical disruption of the cells. In addition, we have adapted Miller's protocol for use in 96-well microtiter plates.[9,15] In practice, most investigators define the activity of the β-galactosidase in terms of the same arbitrary units described by Miller, and hence these have come to be known as "Miller units." These units were arbitrarily contrived such that a wild-type Lac$^+$ strain has approximately 1000 Miller units of β-galactosidase activity. It has been estimated[5] that 3 Miller units correspond to approximately one specific activity unit (in other words, 3 Miller units of β-galactosidase will hydrolyze 1nmol of ONPG per minute per milligram of total protein at 28° and pH 7.0).

Random Pools of lac Fusions

A number of strategies exist in E. coli for the construction of random pools of chromosomal elements. These strategies can be generally summarized as follows: a transposable genetic element capable of forming fusions is introduced into the strain of interest on a conditional vector, and transposition events are selected for under conditions favoring elimination of the donor vector. Examples of these include the TNTRPLAC hopper,[16] and miniTnLK10,[17] carried on conditionally defective λ phage vectors, and

[15] R. Menzel, Anal. Biochem. **181,** 40 (1989).
[16] J. C. Way, M. A. Davis, D. Morisato, D. E. Roberts, and N. Kleckner, Gene **32,** 369 (1984).
[17] O. Huisman and N. Kleckner, Genetics **116,** 185 (1987).

TABLE III
λ*plac*Mu VECTORS FOR MAKING RANDOM FUSIONS

Vector	Marker[a]	Fusion[b]	Fusion type	MuA⁻ derivative[c]
λ*plac*Mu50[d]	*imm*λ	*lacZYA'*	Transcriptional	λ*plac*Mu52[e]
λ*plac*Mu51[d]	*imm21*	*lacZYA'*	Transcriptional	λ*plac*Mu54[e]
λ*plac*Mu53[d]	*imm*λ, Kan^R	*lacZYA'*	Transcriptional	λ*plac*Mu55[e]
λ*plac*Mu1[f]	*imm*λ	*lac`ZYA'*	Translational	λ*plac*Mu5[e]
λ*plac*Mu3[f]	*imm21*	*lac`ZYA'*	Translational	λ*plac*Mu13[e]
λ*plac*Mu9[f]	*imm*λ, Kan^R	*lac`ZYA'*	Translational	λ*plac*Mu15[e]

[a] *imm*, phage immunity; Kan^R, kanamycin resistance.
[b] Fusions designated *lacZYA'* include the sequences necessary for translational initiation and contain functional copies of both *lacZ* and *lacY* genes, but are truncated in *lacA* (and are *lacA⁻*). Fusions designated *lac'ZYA'* are deleted for the translation initiation sequences.
[c] Derivatives of the λ*plac*Mu vectors that carry a nonsense mutation (*A*am1093) in the MuA transposase are completely transposition defective. Transposase function can be supplied *in trans* using λpMu507.
[d] From E. Bremer, T. J. Silhavy, and G. M. Weinstock, *J. Bacteriol.* **162**, 1092 (1985).
[e] From E. Bremer, T. J. Silhavy, and G. M. Weinstock, *Gene* **71**, 177 (1988).
[f] From E. Bremer, T. J. Silhavy, J. M. Weisemann, and G. M. Weinstock, *J. Bacteriol.* **158**, 1084 (1984).

the λ*plac* Mu series of hybrid phage vectors.[2,18,19] The first example takes advantage of λ vectors carrying mutations conditionally defective for multiple replicative functions. These vectors can be propagated in a bacterial host strain containing appropriate suppressor mutations. On infection into a nonsuppressing bacterial strain, these "suicide" vectors are unable to lysogenize, replicate, or kill the host; thus, in the absence of selection they would simply be lost. By selecting transmission of genes carried on a transposon within the vector, one demands that a transposition event occur. In addition, some of these Tn10 derivatives (e.g., miniTnLK10) have the transposase function supplied *in trans* on a plasmid, so that once inserted, after curing the plasmid, they remain extremely stable. These transposition events are relatively nonspecific and this approach can be used to isolate transposon insertions not only on the bacterial chromosome, but also on episomal DNA.

The λ*plac*Mu series of hybrid phage vectors (see Table III), in contrast, retain the ability to plaque on the recipient strain. These vectors also utilize transposition, but in this case the transposition event leads to the insertion of a λ prophage flanked by sequences from bacteriophage Mu that have been manipulated to carry a portion of the *lac* operon. Thus the prophage is capable of forming *lac* operon fusions if inserted in a transcribed region

[18] E. Bremer, T. J. Silhavy, J. M. Weisemann, and G. M. Weinstock, *J. Bacteriol.* **158**, 1084 (1984).
[19] E. Bremer, T. J. Silhavy, and G. M. Weinstock, *Gene* **71**, 177 (1988).

in the appropriate orientation (see Fig. 2). In this way, stable *lac* fusions can be made, taking advantage of the benefits offered by Mu transposition (high rate of transposition and lack of specificity) but without the problems of temperature sensitivity and instability associated with previous fusion systems based on defective Mu phage [Mu dI1(Ap, Lac), for example[20]].

Isolating Random lac Operon Chromosomal Insertions Using λplacMu53

1. Remove 1 ml of a fresh overnight 5-ml LB culture of MC4100 to a new culture tube.
2. Add 10^8 pfu of λ*plac*Mu53.
3. Incubate at room temperature for 30 min to allow infection.
4. Add 5 ml of LB broth and pellet the cells at 4500*g*.
5. Decant the supernatant to remove unabsorbed phage.
6. Repeat steps 4 and 5 twice more, for a total of three washes.
7. Prepare serial dilutions from 10^{-1} to 10^{-4} and plate 100 μl of each dilution per plate onto LB plates containing kanamycin and X-Gal.
8. Incubate at 30° or 37° overnight.

Colonies produced in this way carry a λ prophage flanked by a Mu sequence, which has inserted in the chromosome randomly (i.e., not all of the insertions will result in a Lac$^+$ phenotype). In theory, each independent colony represents a separate transposition event. However, because the λ*plac*Mu53 phage, like most *lac* fusion vectors, carries, for historical reasons, a *trpA–lacZ* protein fusion (W209),[21] if a *recA*$^+$ host such as MC4100 is used, the pool may be biased in favor of integration of the prophage at *trpA* by homologous recombination. In practice, we have not found this to be a significant problem, and in any case it can be avoided by simply using a *recA* host strain to isolate the pool.

An alternative to selecting lysogens based on kanamycin resistance is to select for the ability to grow on minimal lactose. Using this strategy demands that the prophage be inserted in the appropriate orientation in a transcribed region of the genome. Furthermore, because integration of the prophage at *trpA* by homologous recombination confers a *trpA*$^-$ phenotype, it removes the aforementioned concern. Selection on minimal lactose, however, biases the pool against genes whose expression is too low to permit growth on minimal lactose, either because the intrinsic level of expression of the gene is low or because it is actively repressed under the conditions used to isolate the pool.

[20] M. J. Casadaban and S. N. Cohen, *Proc. Natl. Acad. Sci. U.S.A* **76**, 4530 (1979).
[21] M. J. Casadaban, *J. Mol. Biol.* **104**, 541 (1976).

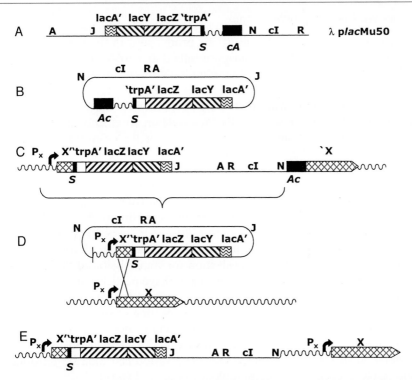

Fig. 2. Adapted from Bremer *et al.*[2] (A) Schematic diagram of λp*lac*Mu50. Details of the construction of this complex vector can be found in Bremer *et al.*[2] The wavy line indicates the *uvrD* gene that was used in the construction of the vector. (B) On infection, the λ sequences circularize. At this point it is most helpful to regard the vector as a transposon inserted in the *uvrD* sequence. (C) Transposition out of the *uvrD* sequence and onto the bacterial chromosome places the *lac* operon under the control of the promoter of gene X. Note that this transposition also disrupts the function of gene X. Because both Mu ends are intact, the vector is still capable of transposition (albeit at a low frequency). For this reason, it is usually desirable to isolate interesting fusions and transfer them to a clean genetic background. This is accomplished by using UV irradiation to stimulate excision of the λ prophage by illegitimate recombination. (D) Illegitimate recombination releases the λ prophage and some of the DNA flanking it. If transmission of the fusion is selected (by selecting for lac activity, for example), the sequence upstream of the *lac* operon can mediate transfer of the fusion to the chromosome by homologous recombination. Because transmission of the sequence to the left is under selection, it is less likely that additional DNA to the right of the prophage will also be packaged into the λ phage particle. (E) The fusion thus formed is now stable, as it lacks one of the Mu ends necessary for transposition.

Practical Use of λplacMu Pools

Because the λplacMu53 insertions produced with the method just described retain the ability to transpose (albeit at a 10^5-fold lower frequency
than Mud insertions[18]) and because multiple insertions may be present in
the same strain, it is advisable to take additional steps in studying interesting
insertions. All inserts with interesting properties should be transferred to
a new genetic background by P1 transduction to ensure that they only
contain single insertions, and then retested for the desired properties. A
further step should be taken to stabilize fusions to be retained for long-term
study. This step takes advantage of the ability of λplacMu53 to form plaques.

Induction of the prophage by ultraviolet (UV) light stimulates illegitimate recombination and promotes the formation of specialized transducing
phage particles that carry λ sequences and a portion of flanking bacterial
chromosome fused to the lac operon. lac genes carried by the prophage
are situated on the opposite side of the integrated prophage to the Mu
transposase (see Fig. 2). Thus, selection for transmission of the lac fusion
by UV-induced, specialized-transducing λ phage particles promotes loss of
one side of the Mu sequences, with concomitant loss of the ability of the
fusion to transpose. In Δlac strains, lysogenization of specialized-transducing λ phage particles of this type can only occur by homologous recombination mediated by the exogenous chromosomal DNA (ΔplacMu phages carry
a deletion of the λ attachment site). The fusion can then be introduced
into a clean background by homologous recombination.

UV Induction of λ Lysogens

1. Inoculate a single colony of the lysogen into 5 ml of LB broth.
2. Grow with aeration (shaking or rocking) to a cell density of approximately 2 to 4 × 10^8 cells/ml (generally between 3 and 4 hr at 37° or 6–8
hr at 30°).
3. Prepare a foil-covered culture tube with 4.5 ml of LB broth.
4. Pellet the cells by centrifugation and resuspend in 2.5 ml of 10
mM MgSO$_4$.
5. Transfer the cell suspension to an empty sterile petri dish and expose
the suspension to a UV irradiation dose of 350 erg mm^{-2} with the cover off.
6. Immediately transfer 0.5 ml of the cell suspension to the foil-covered
culture tube and shake or rotate at 37° until lysis occurs.
7. Add 100 μl of chloroform and vortex well.
8. Centrifuge at 4500g for 10 min to pellet the chloroform and cell debris.
9. Transfer the supernatant to sterile screw-cap tubes, taking care not
to transfer any of the chloroform. Store λ lysates at 4° in the dark.

Lysates produced in this way contain a heterogeneous population of phage particles. The desired specialized transducing phage particles can be selected by following the procedure described previously for targeting *lac* fusions to the chromosome. The recipient strain should be *rec⁺*. Lysogens should be tested by cross-streaking as described, with λ*cI* and λ*vir* as test phage. Most lysogens selected on minimal lactose will have lost the Mu sequences from the opposite side of the original integrated prophage. Mu requires two intact ends and the transposase encoded by the Mu A gene for transposition. Because sequences deleted by selecting for Lac⁺ lysogens include both the Mu A gene and one of the ends, the fusions thus formed are now unable to transpose. This can be tested by assaying the ability of a strain of interest to produce Lac⁺ colonies by infection of a *recA⁻* recipient (in the absence of recombination, formation of colonies would indicate that the transposase function has been retained).

A further consideration in studying *lac* fusions formed by λ*placMu* phage is that because of the size of the inserted DNA, they will lead to loss of expression of any transcriptionally linked downstream genes. For this reason it is important to also make single-copy *lac* operon fusions in the presence of the wild-type gene product (note that because the λ*placMu* vectors are Δ*att*, the single-copy lysogen must be made at the native chromosomal locus).

Uses of Random Pools of *lac* Operon Fusions

Pools of *lac* operon fusions are used in an attempt to identify genes whose transcription is differentially regulated under different growth conditions. The more strictly these growth conditions can be defined, the more likely it is that the use of a pool of random *lac* operon fusions will prove fruitful. For example, the expression of a great many genes is likely to be affected by comparing growth at 30° and 42°, whereas only a few genes will be affected by overexpression of a single gene.

Numerous strategies exist for identifying target genes using pools of random *lac* insertions. Insertions that have low levels of *lac* activity, and those with high activity, can be divided into separate pools, and the pools transformed with a plasmid carrying a gene of interest. Insertions that are differentially regulated in response to the presence of the plasmid can then be identified as "Lac-ups" or "Lac-downs," respectively.

A different approach has been used in our laboratory to identify genes upregulated in response to the overexpression of NlpE, a lipoprotein whose overproduction induces periplasmic stress.[22] In this case,[23] the pool of inserts

[22] W. B. Snyder, L. J. Davis, P. N. Danese, C. L. Cosma, and T. J. Silhavy, *J. Bacteriol.* **177**, 4216 (1995).
[23] P. N. Danese and T. J. Silhavy, *J. Bacteriol.* **180**, 831 (1998).

was produced in a strain with a plasmid carrying the *nlpE* gene under the control of the arabinose-inducible p*BAD* promoter. Insertions of λp*lac*Mu53 onto the chromosome were produced by selecting kanamycin resistance. The resulting colonies were restreaked on indicator media in the presence or absence of inducer (arabinose). Colonies with higher expression in the presence of arabinose were subjected to further analysis. Because this induced expression could be a response either to the presence of arabinose or to the overproduction of NlpE, these strains were retested with a plasmid that constitutively overexpresses NlpE. The fact that only a small number of the colonies initially selected for further study met this criterion underscores the importance of the use of appropriate controls in these kinds of studies.

Difficulties Associated with LacZ Protein Fusions

In the protocols outlined earlier, we considered transcriptional (operon) fusions. By simply choosing different vectors, it is possible to create translational (protein) fusions using essentially the same *in vitro,* transposon, or λp*lac*Mu strategies. However, the study of protein fusions is more complicated. Some LacZ protein fusions are toxic to the cell. For example, fusions to extracytoplasmic proteins jam the secretory apparatus on induction, leading to a generalized accumulation of extracytoplasmic precursor proteins, and ultimately to cell death. Other fusions are degraded or form inclusion bodies if their expression is induced. In many cases the Lac activity of fusions is lower than expected for reasons that are not immediately obvious. These complications have meant that LacZ protein fusions are used less commonly as a means of monitoring translational regulation, and small epitope tags have become increasingly common for this purpose. Despite these considerations, LacZ protein fusions remain very useful, particularly since, like *lac* operon fusions (and in contrast to epitope tags), they allow genetic strategies to isolate mutations that increase or decrease their expression.

Appropriate Use of Indicator Media

The usefulness of the *lac* operon as a reporter system is due in large part to the ability to sense and monitor changes in Lac activity over a variety of ranges on inexpensive media. This allows the detection of mutations that increase or decrease the activity of a particular fusion, even if the function of the gene to which the fusion is made does not have a phenotype that can be assayed readily (or indeed if the gene has no known function).

The most sensitive indicator of Lac activity is 5-bromo-4-chloro-3-

indolyl-β-D-galactoside, more commonly referred to as X-Gal or simply XG.[24] X-Gal is a colorless compound, which produces an insoluble blue dye when hydrolyzed. This allows the identification of β-galactosidase-expressing cells as blue colonies, whereas Lac$^+$ phage form, blue plaques. X-Gal works well in both rich and minimal media and can be used to detect levels of expression corresponding to as little as an average of one molecule of β-galactosidase per cell. Furthermore, because its uptake is not dependent on a functional lactose permease (*lacY*) gene product, X-Gal can be used as an indicator for fusions that do not carry *lacY*.

X-Gal is normally made as a 20-mg ml^{-1} stock solution in *N,N*-dimethylformamide and is typically spread using 100 μl of stock solution per plate. X-Gal is light sensitive; stock solutions should be stored at $-20°$ in the dark, and plates should not be prepared far in advance of their anticipated use.

A second type of indicator media, much less expensive than X-Gal media and invaluable for bacterial genetics, is lactose MacConkey agar. This rich medium, which was originally formulated to facilitate screening for Lac$^+$ gram-negative microorganisms, contains bile salts (to inhibit the growth of nonenteric bacteria) and neutral red as a pH indicator. Lac$^-$ bacteria, which can grow on lactose MacConkey agar, form white colonies. In contrast, cells that can ferment lactose form acid by-products, which turn the pH-sensitive dye red, and thus produce red colonies. Unlike X-Gal, phenol red is a diffusible dye, and therefore the colony phenotype should be read promptly after incubation, as differences will become less clear over time. In terms of sensitivity, cells producing 100 Miller units of β-galactosidase,[5] which are *lacY*$^+$, produce pale pink colonies, whereas those producing wild-type levels of LacZ (about 1000 Miller units) form dark red colonies, which are surrounded by a hazy precipitate of bile salts. Thus, lactose MacConkey agar is useful in examining strains with Lac activity between these values. Indeed, the trained observer can detect twofold changes in Lac activity on this media. Because rare red Lac$^+$ mutant colonies can be detected easily against a background of Lac$^-$ colonies (or Lac$^+$ colonies of lower activity), MacConkey medium is very useful in detecting mutations that increase the activity of reporter fusions.

The third type of media commonly used in bacterial genetics is lactose tetrazolium agar.[6] Like lactose MacConkey agar, this rich medium takes advantage of the acidic by-products formed by cells that can ferment lactose, although the way in which this works is a little more convoluted. All cells, which can grow on lactose tetrazolium agar, reduce the tetrazolium (2,3,5-triphenyl-2-*N*-tetrazolium chloride) to form a red, insoluble formizan dye.

[24] J. P. Horwitz, J. Chua, R. J. Curby, A. J. Tomson, M. A. DaRooge, B. E. Fisher, J. Mauricio, and I. Klundt, *J. Med. Chem.* **7**, 574 (1964).

The acid by-products secreted by Lac$^+$ colonies inhibit the reduction, preventing formation of the formizan. Therefore, Lac$^+$ colonies are white on tetrazolium, whereas Lac$^-$ colonies are red. Tetrazolium is a less sensitive indicator medium than MacConkey, requiring both a functional *lacY$^+$* gene and approximately 400 to 500 Miller units of β-galactosidase activity for the formation of Lac$^+$ (white) colonies.

Somewhat paradoxically, it is precisely this insensitivity that makes tetrazolium such a useful indicator. Tetrazolium agar can be used to screen for rare mutations that decrease the β-galactosidase activity of a fusion with high activity. Differences between strains with 300 Miller units of β-galactosidase activity and those with wild-type (1000 Miller units) activity are indistinguishable on minimal lactose or X-gal agar and are very difficult to detect on lactose MacConkey agar. However, these differences are clearly visible on lactose tetrazolium agar. Thus, rare (10^{-4}) "Lac-down" red colonies can be detected easily in a background of Lac$^+$ white colonies on this medium. Furthermore, because mutations with reasonably high Lac activity will register as Lac$^-$ on lactose tetrazolium agar, this facilitates screens for mutations, which lower but do not abolish expression of a particular fusion.

The lactose tetrazolium agar color phenotypes mentioned previously are pertinent to the study of well-isolated single colonies. A perplexing observation for the uninitiated is that in regions of heavy growth, the color phenotypes are reversed. Therefore, a Lac$^-$ strain will form a confluent white lawn when spread on lactose tetrazolium agar, but when streaked for single colonies, the same strain will exhibit a white color in the primary streak, but will produce isolated red colonies. In fact, this apparently confusing phenomenon allows lactose tetrazolium agar to be used as a selective medium. Because this rich medium contains a high concentration of lactose (1%), rare Lac$^+$ mutants appearing in a confluent lawn of Lac$^-$ cells will be able to continue to grow out of the lawn using the lactose as a carbon source after the surrounding Lac$^-$ cells have exhausted the other carbon sources in the media and have stopped growing. These rare mutants can be distinguished from the background, as they will appear red against the white lawn (note that, when restreaked on the same media, they will now form isolated white colonies, whereas cells from the lawn would form red single colonies). This same strategy can be used to isolate Lac$^+$ mutants on lactose MacConkey agar; however, because the phenol red dye in MacConkey agar is diffusible, the color distinctions are not as sharp as those produced by the insoluble formizan dye in lactose tetrazolium agar.

Lactose tetrazolium agar is an extremely useful tool in the repertoire of the bacterial geneticist. However, because of the difficulties described, the key to its effective use is to always streak known Lac$^+$ and Lac$^-$ strains for comparison with strains of interest.

Selecting Mutants in Fusion Strains

One of the most useful properties of Lac fusions is that they allow the formidable power of the genetic tools developed for the study of the *lac* operon itself to be brought to bear on the analysis of a gene or regulon, irrespective of whether the gene of interest itself has a known function or readily assayable phenotype. This is a fact often overlooked by investigators who view LacZ simply as a chromogenic reporter enzyme. It is the extensive study of the *lac* operon that makes Lac fusions so useful, even in comparison to trendier reporter systems such as the pervasive green fluorescent protein (GFP).

However, to take full advantage of the tools available to study Lac fusions, a functional *lacY*$^+$ gene is required. For this reason, we favor $\lambda placMu$-generated fusions over those generated by smaller transposon-derived vectors, which in many cases do not contain a full-length *lacY* gene.

The use of various galactosides enables one to select (rather than screen) for mutations satisfying virtually any Lac phenotype. Here we present a few illustrative examples that are particularly useful.

Selecting for Decreased Expression of Reporter Fusions

o-Nitrophenyl-β-D-thiogalactoside (TONPG) is a toxic galactoside. Cells that accumulate this metabolic poison cannot grow. TONPG is imported by the lactose permease encoded by the *lacY* gene, and its import requires relatively high levels of LacY. Thus, TONPG can be used to select for mutations that decrease expression of the *lacY* gene and, consequently, of any promoter to which it is fused.

TONPG-resistant colonies can arise in three ways: by mutations in *lacY* itself, by polar mutations in *lacZ* that decrease expression of *lacY,* or by mutations (either *cis* or *trans*) that decrease expression of the whole *lac* operon. The combined use of X-Gal and TONPG can distinguish all three of these classes. The first class (Z$^+$Y$^-$) will produce blue colonies on XG agar, whereas the second class will disrupt LacZ function (Z$^-$Y$^-$) and will thus yield white colonies. The third desired class of mutants will decrease, but not abolish, expression of both genes and will therefore form pale blue colonies on X-Gal indicator media.[25,26]

TONPG works best in media in which bacterial growth is slow, such as minimal media with succinate as a carbon source. The inhibitory concentration of TONPG necessary for the selection of mutants will depend on the level of expression of LacY and should be determined empirically for a

[25] M. L. Berman and J. Beckwith, *J. Mol. Biol.* **130,** 303 (1979).
[26] J. D. Hopkins, *J. Mol. Biol.* **87,** 715 (1974).

given fusion strain. Because a relatively high level of LacY expression is required, it may not be possible to apply TONPG selection to fusion strains that already have low activity.

Selecting for Increased Expression of Reporter Fusions

A different galactoside, phenylethyl-β-D-thiogalactoside (TPEG), is a competitive inhibitor of β-galactosidase,[27] and can be used to select for mutations that increase the expression of lac operon fusions with low levels of expression. Fusions with low β-galactosidase activity can grow on minimal lactose media. By adding TPEG to minimal media, the minimum inhibitory concentration sufficient to prevent growth of a particular fusion-containing strain can be determined empirically, thus allowing the selection of mutations that increase expression to levels sufficient to overcome this inhibition.[28]

Monitoring LacY Expression

Fusions that contain a functional $lacY^+$ gene have the additional advantage that the expression of LacY can be monitored crudely on indicator plates containing the α-galactoside, melibiose. Escherichia coli has specific transport proteins for the uptake of melibiose and other α-galactosides. In E. coli K-12 and its derivatives, however, the uptake of melibiose by these proteins is temperature sensitive. The lactose (β-galactoside) permease system can promiscuously permit the import of melibiose, thus at elevated temperatures ($\geq 40°$), $lacY^+$ strains are Mel$^+$.[29] Melibiose minimal, MacConkey, and tetrazolium agar can be used to quantify the expression of LacY, with the caveat that because the assay requires growth at high temperature, it may not be possible to use it in all cases. In the study of protein fusions, the ability to measure LacY expression has the advantage that the analysis of LacY activity is not subject to the previously described complications associated with monitoring the LacZ activity of hybrid proteins (because in the case of LacY, the activity of the wild-type protein is being measured). Moreover, because the transcription, but not the translation, of $lacY$ is coupled to that of the protein fusion, LacY can therefore effectively be considered a transcriptional reporter of the protein fusion.

Melibiose minimal media containing X-Gal can be used to screen for mutations that decrease the LacZ activity of a particular protein fusion, while allowing one to select for those mutations that do not abolish the

[27] F. Jacob and J. Monod, J. Mol. Biol. **3**, 318 (1961).
[28] M. N. Hall, J. Gabay, M. Debarbouille, and M. Schwartz, Nature **295**, 616 (1982).
[29] J. R. Beckwith, Biochim. Biophys. Acta **76**, 162 (1963).

expression of the downstream gene *lacY*. This permits an enrichment of interesting nonpolar, Lac-down mutations with specific effects on the fusion protein from the much larger class of nonsense mutations that simply abolish expression of both *lacZ* and *lacY*.[30]

Conclusion

Despite the usefulness of new reporter systems such as GFP, and the many epitope-tagging systems, the Lac system remains a powerful genetic tool both in studying gene regulation and in generating mutants in diverse complex biological systems. In comparison to these newer technologies, β-galactosidase, with its lengthy history (relative to the age of molecular genetics), seems almost old-fashioned. However, it is precisely this extensive history that makes the Lac system so useful. In the hands of an experienced investigator, exquisitely detailed questions can still be addressed with little more than agar plates and toothpicks. It seems likely, therefore, that Lac will remain an often-used weapon in the arsenal of the bacterial geneticist for the foreseeable future.

Acknowledgments

We thank N. Ruiz, A. Kaya, K. Gibson, T. Raivio, A. Greenberg, and J. T. Blankenship for critical reading and helpful discussion of this paper and S. DiRenzo for assistance in preparation of the manuscript. T.J.S. was supported by an NIGMS grant (GM34821).

[30] D. R. Kiino and T. J. Silhavy, *J. Bacteriol.* **158,** 878 (1984).

[3] Tagging Exported Proteins Using *Escherichia coli* Alkaline Phosphatase Gene Fusions

By Colin Manoil

Escherichia coli alkaline phosphatase (AP) requires export out of the reducing environment of the cytoplasm to form disulfide bonds and fold into an active form.[1] Hybrid proteins consisting of an alkaline phosphatase

[1] A. Derman and J. Beckwith, *J. Bacteriol.* **173,** 7719 (1991).

lacking its N-terminal cleavable signal sequence fused to a target protein will thus exhibit AP enzymatic activity only if the target protein provides an export signal that substitutes for the missing signal sequence.[2,3] Most cleavable signal sequences of gram-negative bacterial periplasmic and outer membrane proteins appear to satisfy this requirement, as do inner membrane-spanning sequences oriented with their N termini cytoplasmic and C termini periplasmic.[4,5]

Four well-established uses of AP gene fusions are (a) to monitor the expression and export from the cytoplasm of a given protein, (b) to analyze the membrane topology of bacterial integral inner membrane proteins, (c) to selectively mutate genes encoding exported proteins, and (d) to tag exported proteins with epitopes or protease cleavage sites.[4–7] A number of techniques for generating alkaline phosphatase gene fusions *in vitro* and *in vivo* have been described (e.g., see references 2, 3, 8–13). This article presents procedures for the generation and analysis of AP gene fusions using two recently constructed transposons. The AP gene fusions generated using these transposons can be readily converted into in-frame insertions encoding epitopes. The combined analysis of hybrid and epitope-tagged versions of a membrane or secreted protein allows a wide range of approaches to be used to characterize it. For example, the insertion tags have been used to identify permissive sites in proteins, to dissect structure–function relationships, to monitor membrane protein complex assembly, and to analyze the membrane topology of (unfused) membrane proteins (reviewed in Manoil and Traxler[6]).

New *phoA* Fusion Transposons

Restriction maps of two transposons that may be used to generate AP gene fusions and in-frame insertions are shown in Fig. 1. Both transposons are derived from the left IS*50* element of Tn*5*.[14] The first transposon

[2] C. Hoffman and A. Wright, *Proc. Natl. Acad. Sci. U.S.A.* **82**, 5107 (1985).
[3] C. Manoil and J. Beckwith, *Proc. Natl. Acad. Sci. U.S.A.* **82**, 8129 (1985).
[4] C. Manoil, J. J. Mekalanos, and J. Beckwith, *J. Bacteriol.* **172**, 515 (1990).
[5] B. Traxler, D. Boyd, and J. Beckwith, *J. Membr. Biol.* **132**, 1 (1993).
[6] C. Manoil and B. Traxler, *Methods* **20**, 55 (2000).
[7] M. R. Kaufman and R. K. Taylor, *Meth. Enzymol.* **235**, 426 (1994).
[8] M. Ehrmann, P. Bplek, M. Mondigler, D. Boyd, and R. Lange, *Proc. Natl. Acad. Sci. U.S.A.* **94**, 13111 (1997).
[9] M. Wilmes-Riesenberg and B. Wanner, *J. Bacteriol.* **174**, 4558 (1992).
[10] V. de Lorenzo, M. Herrero, U. Jakubzik, and K. Timmis, *J. Bacteriol.* **172**, 6568 (1990).
[11] J. Sugiyama, S. Mahmoodian, and G. Jacobson, *Proc. Natl. Acad. Sci. U.S.A.* **88**, 9603 (1991).
[12] A. Das and Y. Xie, *Mol. Microbiol.* **27**, 405 (1998).
[13] M. McClain and N. Engleberg, *Gene* **17**, 147 (1996).
[14] W. S. Reznikoff, *Annu. Rev. Microbiol.* **47**, 945 (1993).

IS*phoA*/in

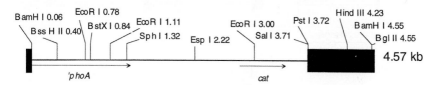

IS*phoA*/hah

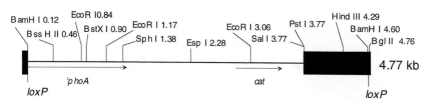

FIG. 1. Transposons for generating alkaline phosphatase gene fusions and in frame epitope insertions. Restriction maps of IS*phoA*/in and IS*phoA*/hah are shown. *phoA*, alkaline phosphatase gene; *cat*, chloramphenicol acetyltransferase gene; *loxP*, site-specific recombination sequence.

(IS*phoA*/in) may be used to generate *phoA* fusions to cloned genes, which may then be converted *in vitro* into 31 codon insertion mutations (Fig. 2).[15] The second transposon (IS*phoA*/hah) may be used to generate fusions to chromosomal genes (and cloned genes), which may be converted *in vivo* into 63 codon insertions (Fig. 3).[16] A transposon analogous to IS*phoA*/in, which generates β-galactosidase gene fusions, has also been constructed.[15]

This article presents procedures for creating and analyzing alkaline phosphatase gene fusions and in-frame insertions using the new transposons. The article is intended to serve as a companion to several earlier articles describing applications and procedures involving AP gene fusions[4,5,7] and a more recent review describing uses of in-frame insertions.[6] In the procedures that follow, the general references for microbiological and molecular biological methods are Miller[17] and Sambrook *et al.*[18]

[15] C. Manoil and J. Bailey, *J. Mol. Biol.* **267,** 250 (1997).
[16] J. Bailey and C. Manoil, unpublished results.
[17] J. H. Miller, "A Short Course in Bacterial Genetics." Cold Spring Harbor Laboratory Press, Cold Spring Harbor, NY, 1992.
[18] J. Sambrook, E. F. Fritsch, and T. Maniatis, "Molecular Cloning: A Laboratory Manual," 2nd Ed. Cold Spring Harbor Laboratory, Cold Spring Harbor, NY, 1989.

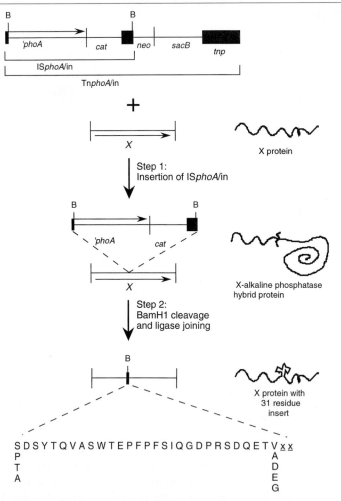

Fig. 2. Generation of *phoA* fusions and 31 codon insertions using IS*phoA*/in. In the first step, IS*phoA*/in is inserted into a plasmid-borne gene to generate a *phoA* translation gene fusion. In a second step, sequences between *Bam*HI restriction sites in the transposons are removed to yield the 31 codon insertion encoding the sequence shown. B, *Bam*HI; *phoA*, alkaline phosphatase gene; *cat*, chloramphenicol acetyltransferase gene; *neo*, neomycin phosphotransferase gene; *sacB*, levansucrase gene; *tnp*, Tn5 transposase gene; x, indeterminant residue encoded by target gene sequences duplicated during transposition

Generation of IS*phoA*/in Insertions in Cloned Genes

The use of IS*phoA*/in to generate alkaline phosphatase gene fusions and in-frame insertions is diagrammed in Fig. 2. The transposon can be used to create *phoA* fusions in the chromosome, although its primary applications have involved fusions and 31 codon insertions in cloned genes.[6]

Strains and Phage

    CC118: Δ(*ara-leu*)*7697 araD139* Δ(*lac*)*X74 galE galK thi rpsE phoA20 rpoB argE*(*am*)
    CC245: *supE supF hsdR galK trpR metB lacY tonA dam::kan*
    λTn*phoA*/in: *b221* Pam2 *c*I857 *rex*::Tn*phoA*/in

Target Gene

Ideally, the target gene to be mutated should be carried in a plasmid devoid of *Bam*HI sites, which is present as a monomer in a *recA*⁻ strain such as CC118. The *recA*⁻ mutation limits multimerization of the target plasmid, and multimers can complicate the identification of desired transposon insertion derivatives.

Growth of λTnphoA/in

The delivery vector for IS*phoA*/in is a phage derivative ("λ Tn*phoA*/ in") unable to replicate or lysogenize nonsuppressing strains of *E. coli.* Phage λTn*phoA*/in is grown as a plate stock using strain CC245 as host.

The *dam* mutation in CC245 blocks adenine methylation of IS*phoA*/in and promotes its transposition. After a phage stock has been prepared, it is important to determine its titer on both amber-suppressing (e.g., CC245) and -nonsuppressing (e.g., CC118) strains. This step provides a measure of the number of $P_{am} \rightarrow P^+$ revertants in the phage stock, as such revertants form plaques on nonsuppressing strains. Revertant phage can limit recovery of cells carrying transposon insertions by killing them, and it is prudent to make a fresh phage stock (from a single plaque) if the proportion of revertants in a stock is greater than about 10^{-5} of the total.

Generation of ISphoA/in Insertions.

Note: This protocol assumes that the plasmid to be mutagenized encodes ampicillin (Amp) resistance as a selectable marker.

    a. Mix 0.2 ml cells (carrying the target plasmid) grown to stationary phase in LB–Amp (100 μg/ml)–10 m*M* MgCl₂ with λTn*phoA*/in at a multiplicity of approximately 0.1–0.3 phage/cell. Incubate 10 min in a 37° water bath, and add 0.8 ml LB. Grow at 37° with aeration for 1–2 hr.

b. Plate 0.4 ml of the undiluted culture onto L agar–Amp (100 μg/ ml)–Cm (100 μg/ml) to select growth of cells carrying transposon insertions. Incubate at 37° for 1 day or at 30° for 2 days. [*Note:* The higher concentration of chloramphenicol (Cm) used in this step relative to later steps in the procedure reduces background cell growth.]

c. Scrape up colonies (typically 1000–2000 per plate) with a clean toothpick and isolate plasmid DNA using the alkaline lysis procedure (phenol–chloroform extraction is not necessary). Resuspend plasmid DNA in 10 μl TE (10 mM Tris–HCl, pH 7.0, 1 mM EDTA).

d. Transform CC118 cells (0.2 ml) with 5 μl plasmid DNA (~250 ng DNA) using a CaCl$_2$-based method. Add 0.8 ml LB and incubate with aeration for 1–2 hr at 37°. Plate 0.5 ml undiluted transformation mixture on L agar lacking added NaCl and containing 5% (w:v) sucrose, Amp (100 μg/ml), Cm (40 μg/ml) and 5-bromo-4-chloro-3-indoyl toluidine salt (XP) (40 μg/ml). Incubate at 37° overnight. [This step selects for transformants that have obtained plasmids carrying IS*phoA*/in insertions. Because *sacB* leads to a sucrose-sensitive phenotype,[19] the sucrose in the selective medium reduces the recovery of cells with Tn*phoA*/in insertions (see Fig. 2). In some cases, sucrose killing of *sacB*[+] cells may be greater at 30° than at 37°.[19] In addition, killing by sucrose of strains carrying *sacB*[+] in relatively low-copy number plasmids (e.g., lower in copy number than pBR322) may be inefficient.]

e. Screen blue (PhoA[+]) colonies for kanamycin sensitivity by patching on L agar–Km (30 μg/ml) and L agar–Cm (40 μg/ml). Incubate at 37° overnight. [This step checks that the transformant plasmids in the cells examined do not carry Tn*phoA*/in insertions, which confer kanamycin (Km) resistance.]

f. Score patches. Purify kanamycin-sensitive transformants by single colony purification on L agar–Cm (40 μg/ml) at 37°.

g. Start 5 ml LB–Cm (40 μg/ml) cultures from single colonies and incubate overnight with aeration at 37°.

h. Isolate plasmid DNA using alkaline lysis or an alternative method. Restriction map fusion inserts to determine whether they fall in the target gene. Sequence insertions of interest using primers hybridizing to the '*phoA* end of the transposon.[15]

---

[19] I. Blomfield, V. Vaughn, R. Rest, and B. Eisenstein, *Mol. Micro.* **5**, 1447 (1991).

Excision of ISphoA/in Sequences to Generate 31 Codon Insertion Mutations

IS*phoA*/in insertions may be converted into 31 codon insertions by cleavage with *Bam*HI, followed by phage T4 DNA ligase treatment (Fig. 2). This step eliminates most of the IS*phoA*/in DNA, leaving a sequence corresponding to the 31 codons.

Generation of ISphoA/hah Insertions in Chromosomal Genes

The use of IS*phoA*/hah to generate alkaline phosphatase gene fusions and in-frame insertions is represented in Fig. 3. The IS*phoA*/hah procedure differs from that of IS*phoA*/in in that transposon insertions are converted into in-frame insertions *in vivo* by site-specific recombination rather than *in vitro,* thus allowing the insertions to be generated directly in the chromosome. The IS*phoA*/hah–derived in-frame insertion is 63 codons long and encodes an influenza hemagglutinin epitope and a hexahistidine sequence ("hah"). Delivery vectors for both IS*phoA*/hah ("pHAH") and the *cre* recombinase gene ("pCRE") carry broad host range conjugal transfer origins, but are unable to replicate in recipient cells unless an auxiliary replication protein is present.[20] Because Tn5 exhibits a broad host range for transposition, it should be possible to use IS*phoA*/hah in a variety of bacteria other than *E. coli.* A detailed description of the construction and properties of IS*phoA*/hah will be presented elsewhere.

Strains and Plasmids

SM10 λ*pir: thi thr leu tonA lacY supE recA* :: RP4-2-Tc :: Mu km^R/λ*pir*[21]

CC1405: Δ(*lac*)U169 *araD139 dam* :: *kan* Δ(*phoA-proC*) *tsx-247* :: Tn 10 *rpsL thi*

pHAH: A derivative of pUT[20] carrying IS*phoA*/hah between *Xba*I and *Kpn*I sites in the polylinker

pCRE: A derivative of pUT (lacking the *tnp* gene)[20] carrying the *cre* gene from pRH133[22] between *Sal*I and *Eco*RI sites

[20] M. Herrero, V. de Lorenzo, and K. Timmis, *J. Bacteriol.* **172,** 6557 (1990).
[21] V. Miller and J. Mekalanos, *J. Bacteriol.* **170,** 2575 (1988).
[22] A. Wierzbicki, M. Kendall, K. Abremski, and R. Hoess, *J. Mol. Biol.* **195,** 785 (1987).

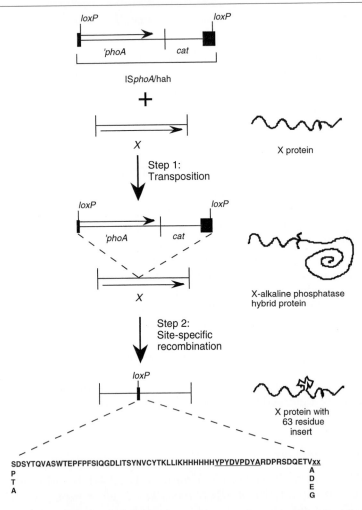

Fig. 3. Generation of *phoA* gene fusions and 63 codon insertions using IS*phoA*/hah. Step 1: Insertion of IS*phoA*/hah in the appropriate orientation and reading frame generates an alkaline phosphatase gene fusion. Step 2: Action of Cre recombinase at *loxP* sites eliminates most of the IS*phoA*/hah sequence, leaving a 63 codon insertion encoding the sequence shown. The influenza hemagglutinin epitope sequence is underlined. *x*, indeterminant residue encoded by target gene sequences duplicated during transposition.

Protocol for Generating Chromosomal Insertions of ISphoA/hah

a. Grow cultures of the recipient (CC1405) in LB and donor (SM10 λ*pir*/pHAH) in LB–Amp (100 μg/ml). Incubate overnight at 37° with aeration.

b. Dilute the donor strain 1 : 10 in fresh LB (no ampicillin) and grow for 45 min at 37° with aeration.

c. Mix 0.5 ml each of the donor and recipient strains and immediately filter the mixture (e.g., through a sterile Nalgene analytical test filter funnel, 0.45-μm nominal pore size). Wash with 1 ml 10 m*M* MgSO$_4$. Remove the filter and place it on a prewarmed LA plate using sterile forceps. Incubate 1–3 hr at 37°.

d. Remove the filter with sterile forceps and place it in a sterile test tube (e.g., 18 × 150 mm) containing 1 ml LB. Vortex thoroughly, checking that the cell mass has been transferred from the filter into the broth.

e. Plate undiluted, 10^{-1}, and 10^{-2} dilutions of the cells onto LA–Cm (50 μg/ml)–streptomycin (100 μg/ml)–XP (40 μg/ml). Incubate 1–2 days at 37°. (*Note:* this step selects for growth of recipient cells carrying IS*phoA*/hah insertions.)

f. Colonies that are significantly bluer than most (typically ~1% of the total) should be visible. The majority of such colonies should express active *phoA* fusions resulting from IS*phoA*/hah insertion.

Conversion of ISphoA/hah Insertions into 63 Codon Insertions

a. Grow cultures of recipient cells carrying IS*phoA*/hah insertions in LB and the donor strain (SM10 λ*pir*/pCRE) in LB–Amp (100 μg/ml). Incubate overnight with aeration at 37°.

b. Dilute the donor and recipient strains 1 : 10 in fresh LB (no ampicillin) and grow for 45 min at 37° with aeration.

c. Mix 20-μl drops of donor and recipient strains on the surface of a prewarmed LA plate. Incubate 2 hr at 37°.

d. Streak from the mating mixture spot onto LA–streptomycin (100 μg/ml)–XP (40 μg/ml). Incubate overnight at 37°.

e. Usually at least 10% of the resulting colonies are white or faint blue and are candidates for strains in which the IS*phoA*/hah insertions have been converted into 63 codon insertions. Several white colonies per cross should be streaked onto LA and LA–Cm (50 μg/ml) and incubated overnight at 37°. Strains carrying 63 codon insertions should fail to grow on LA–Cm.

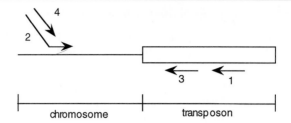

FIG. 4. PCR amplification of transposon insertion junction fragments. The end of the transposon and adjacent chromosomal sequences are amplified in a first step using primers 1 (which is transposon specific) and 2 (which is a semidegenerate mixture with a unique sequence at the 5' end). The product of the first step is amplified in a second step using a second transposon-specific primer (primer 3) and a primer (4) that recognizes the 5' end of primer 2.

Identification of ISphoA/hah Insertion Sites in Chromosomal Genes

DNA fragments, which include transposon–chromosome junctions, are amplified by polymerase chain reaction (PCR),[23,24] sequenced, and compared to the *E. coli* genome sequence to identify insertion sites. The amplification of junction fragments consists of two PCR steps (Fig. 4). In the first step, chromosomal DNA is amplified using two primers, one of which hybridizes to transposon sequences (primer 1) and a second one which is a degenerate mixture designed to hybridize at many sites in the genome (primer 2). Primer 2 also introduces a unique sequence at its 5' end. In the second step of the amplification, the products of the first step are amplified using a different primer hybridizing to the transposon (primer 3) and a primer hybridizing to the unique sequence introduced at the end of primer 2 (primer 4).

a. Primers

1	("Tn*phoA*-II")	GTGCAGTAATATCGCCCTGAGCA
2a	("CEKG 2A")[a]	GGCCACGCGTCGACTAGTACNNNNNNNNNNNNAGAG
2b	("CEKG 2B")[a]	GGCCACGCGTCGACTAGTACNNNNNNNNNNNNACGCC
2c	("CEKG 2C")[a]	GGCCACGCGTCGACTAGTACNNNNNNNNNNNNGATAT
3	("hah-1")	ATCCCCCTGGATGGAAAACGG
4	("CEKG 4")	GGCCACGCGTCGACTAGTAC

[a] Three different primers 2 are listed. Often one but not another will amplify a particular IS*phoA*/hah insertion junction.

[23] K. Chun, H. Edenberg, M. Kelley, and M. Goebl, *Yeast* **13**, 233 (1997).
[24] K. Gibson and T. J. Silhavy, *J. Bacteriol.* **181**, 563 (1999).

b. Set up the following reaction mixes for the first PCR step using primer 1 and one of the three primers 2.

DNA[a]	1 μl
10× PCR buffer (100 mM Tris–HCl, pH 8.3, 15 mM MgCl$_2$, 500 mM KCl)	2 μl
Deoxyribonucleotide mixture (2 mM each)	2 μl
Primer 1 (20 pmol/μl)	1 μl
Primer 2 (20 pmol/μl)	1 μl
H$_2$O (double distilled)	13 μl
Taq polymerase (Boehringer)	1 unit

       [a] Use 1 μl of a mixture made up of a fresh colony picked into 50 μl H$_2$O and heated at 95–100° for 5 min.

c. Carry out the first thermocycling step according to the following protocol. Keep reaction mixes on ice until the thermocycler reaches 95° and then begin.

$$94° \ 30 \ \text{sec} \qquad\qquad 94° \ 30 \ \text{sec}$$
$$95° \ 5 \ \text{min} \rightarrow 8× \ 30° \ 35 \ \text{sec} \rightarrow 30× \ 43° \ 35 \ \text{sec}$$
$$72° \ 45 \ \text{sec} \qquad\qquad 72° \ 45 \ \text{sec}$$
$$\downarrow$$
$$4°$$

d. Set up the following reaction mixes for the second PCR step.

DNA[a]	1 μl
10× PCR buffer (100 mM Tris–HCl, pH 8.3, 15 mM MgCl$_2$, 500 mM KCl)	2 μl
Deoxyribonucleotide mixture (2 mM each)	2 μl
Primer 3 (20 pmol/μl)	1 μl
Primer 4 (20 pmol/μl)	1 μl
H$_2$O (double distilled)	13 μl
Taq polymerase (Boehringer)	1 unit

       [a] Undiluted step 1 amplification mixture.

e. Carry out the second thermocycling step according to the following protocol. Again, keep reaction mixes on ice until beginning.

$$94° \ 30 \ \text{sec}$$
$$95° \ 2 \ \text{min} \rightarrow 30× \ 43° \ 35 \ \text{sec} \rightarrow 72° \ 5 \ \text{min}$$
$$72° \ 45 \ \text{sec} \qquad \downarrow$$
$$4°$$

f. Analyze the products of the second step by electrophoresis through 1.5% agarose. Amplified fragments are typically 150–1000 bp and can be checked by restriction digestion for known sites in the transposon or by additional PCRs to check that amplication of the fragment requires both primers 3 and 4.

g. Products of interest can be sequenced after unincorporated nucleotides and primers have been removed [e.g., using a product such as a QIAprep PCR purification column (Qiagen)]. Comparison with the *E. coli* genome sequence[25] can then be carried out to identify chromosomal insertion sites. *Note:* If more than one amplified fragment was present in the mixture, overlapping sequences with different 3' ends will be obtained. In such cases, target gene sequences near the transposon junction site can generally be used successfully for sequence searches.

Assay of Alkaline Phosphatase Activity

This assay is derived from those described earlier[26–28] and measures alkaline phosphatase activities in permeabilized *E. coli* cells. In some cases, it is important to measure the rate at which AP hybrid proteins are synthesized in conjunction with activity assays for normalization purposes.[28,29]

1. Grow duplicate cultures of cells to be assayed in LB supplemented with any antibiotics required to select for plasmid maintenance. Grow cultures overnight at room temperature with aeration.
2. Dilute cultures 1/100 into fresh media and incubate with aeration at 37° until cultures reach midexponential phase growth (usually 1.5–2 hr).
3. Centrifuge 1 ml of each culture in a microcentrifuge for 3–5 min (16,000g) at 4°. Wash cells once in cold 10 mM Tris–HCl, pH 8.0, 10 mM MgSO$_4$, and 1 mM iodoacetamide and resuspend the final pellet in 1 ml cold 1 M Tris–HCl, pH 8.0, and 1 mM iodoacetamide. (Iodoacetamide is included in buffers to prevent activation of cytoplasmic AP by oxidation on cell lysis.[27])
4. Dilute 0.1 ml cells into 0.4 ml 1 M Tris–HCl, pH 8.0, and measure OD$_{600}$.
5. Add 0.1 ml of washed cells (either undiluted or from the diluted cells in step 4, depending on the level of alkaline phosphatase activity

[25] http://www.ncbi.nlm.nih.gov/BLAST/unfinishedgenome.html.
[26] E. Brickman and J. Beckwith, *J. Mol. Biol.* **96**, 307 (1975).
[27] A. Derman and J. Beckwith, *J. Bacteriol.* **177**, 3764 (1995).
[28] C. Manoil, *Methods Cell Biol.* **34**, 61 (1991).
[29] J. San Millan, D. Boyd, R. Dalbey, W. Wickner, and J. Beckwith, *J. Bacteriol.* **171**, 5536 (1989).

expected) to 0.9 ml 1 M Tris–HCl, pH 8.0, 0.1 mM ZnCl$_2$, and 1 mM iodoacetamide in 13 × 100-mm glass tubes. Include a blank without cells. Add 50 μl 0.1% sodium dodecyl sulfate and 50 μl chloroform, vortex, and incubate at 37° for 5 min to permeabilize cells. Place tubes on ice for 5 min to cool.

6. Add 0.1 ml of 0.4% p-nitrophenyl phosphate (in 1 M Tris–HCl, pH 8.0) to each individual tube, agitate, and place in 37° water bath. Note time.

7. Incubate each until pale yellow and then add 120 μl 1 : 5 0.5 M EDTA, pH 8.0, 1 M KH$_2$PO$_4$, and place the tube in an ice-water bath to stop the reaction. Note time.

8. Measure OD$_{550}$ and OD$_{420}$ for each sample, using an assay mixture without cells as a blank.

9. Units activity $= \dfrac{[OD_{420} - (1.75 \times OD_{550})] \times 1000}{time~(min) \times OD_{600} \times volume~cells~(ml)}$

Acknowledgments

We thank J. Bailey, D. Dargenio, and L. Gallagher for numerous contributions to the procedures presented here and Tom Silhavy for communicating the method for identifying chromosomal insertion sites of transposons. Research was supported by grants from the National Science Foundation and the Cystic Fibrosis Foundation.

[4] Applications of Gene Fusions to Green Fluorescent Protein and Flow Cytometry to the Study of Bacterial Gene Expression in Host Cells

By RAPHAEL H. VALDIVIA and LALITA RAMAKRISHNAN

Introduction

The identification of bacterial genes that are expressed in response to environmental stimuli has enhanced our understanding of microbial biology. This is particularly true in the study of microbial pathogenesis, where investigators have focused their efforts in identifying and monitoring the change in expression of bacterial genes during infection. Many of these genes, including some encoding virulence factors, were identified by mimicking the presumed environment that pathogens encounter during infection (e.g., Fe deprivation, starvation, oxidative stress, acidity.)[1] However, because the host environment is complex, dynamic, and ill understood, only a small subset of genes has been identified by attempts to reproduce the environmental cues provided by the host. To address this problem, novel genetic strategies have been developed for the direct identification of bacterial genes expressed *in vivo*.[2] The advent of fluorescent protein markers has extended the resolution of these gene identification strategies to allow for the temporal and spatial characterization of *in vivo* expressed genes.[3] This undoubtedly will result in a greater understanding of the mechanisms by which a pathogen is able to colonize mucosal surfaces, penetrate into deeper tissues, and disseminate to new hosts.

This article describes new methodologies that use fluorescent proteins to dissect the molecular basis of microbial infections, including protocols for the use of green florescent protein (GFP) and flow cytometry to identify bacterial genes differentially expressed under varied environments (e.g., host cells and tissues). The latter part of this article explores applications of GFP to dissect genetic regulatory networks and methods to quantify bacterial gene expression in host cells. To illustrate these applications, we use examples from two intracellular bacterial pathogens: *Salmonella typhimurium* and *Mycobacterium marinum*.

Green Fluorescent Protein as a Selectable Marker to Identify Differentially Expressed Genes

Traditional "promoter trap" strategies have identified environmentally regulated genes by creating random gene fusions to reporter genes whose translated products have phenotypes that can be measured readily or provide a growth advantage. For example, the screening of random DNA fusions to *lacZ* (encoding β-galatosidase) identified bacterial genes whose expression changes in response to such varied stimuli as nutrient deprivation, heat stress, DNA damage, and varied carbon source.[1] Unfortunately, performing this type of genetic selections and screens has been limited to stimuli that can be mimicked on solid growth media, as clonal isolation requires the bacterial cell to form a colony in the continued presence of the inducing stimulus. Several groups have described the application of *Aquorea victorea* GFP as a reporter of gene expression in pathogenic

[1] B. B. Finlay, *in* "Molecular Biology of Bacterial Infections: Current Status and Future Perspectives" (C. E. Hormaeche and C. J. Smyth, eds.), p. 33. Cambridge Univ. Press, New York, 1992.
[2] J. M. Slauch, M. J. Mahan, and J. J. Mekalanos, *Methods Enzymol.* **235,** 481 (1994).
[3] R. H. Valdivia and S. Falkow, *Trends Microbiol.* **5,** 360 (1997).

bacteria.[4] GFP is a small fluorescent protein that does not require any exogenous cofactors or substrates for its fluorescence.[5] Therefore, GFP permits the monitoring of gene expression and protein localization in live samples and with single cell resolution. New GFP variants with enhanced solubility and altered emission and excitation spectra are now used routinely for most applications.[6] One such variant [GFPmut (A_{max} 488 nm) (E_{max} 510 nm)] has been described extensively and is the basis of most of the protocols described here.[7]

Bacterial cells expressing GFPmut can be identified easily by examining bacterial colonies under blue light or by looking at bacterial suspensions under a fluorescence microscope with settings and filter sets similar to those used for routine imaging of fluorescein-stained samples. Any electronic imaging systems capable of detecting fluorescence can be used to screen for GFP-expressing cells. For instance, fluorescent-activated cell sorters (FACS), can rapidly scan cells in suspension, record the fluorescent properties of each cell, and separate them accordingly.[8] Therefore, FACS allows GFP expression to be a "selectable" phenotype. Because a FACS is able to distinguish between relative levels of fluorescence intensity, it is possible to separate cells displaying differences in GFP expression as low as two- to threefold. Based on this basic premise, a variety of gene selection strategies can be devised to identify promoter elements differentially expressed under virtually any kind of environmental stimuli. This basic principle led to the design of a selection system termed differential fluorescence induction (DFI)[9] in which bacteria are isolated on the basis of a stimulus-dependent expression of GFP (Fig. 1). DFI provides several advantages over more conventional selection strategies: (1) selections can be performed under conditions where bacterial colonies do not form, (2) selections can be performed as a result of transient responses to complex stimuli, as GFP is very stable and accumulates in cells, and (3) the selection system is not biased by the absolute levels of gene expression so that cells can be separated on the basis of incremental changes in gene expression.

[4] R. H. Valdivia, B. P. Cormack, and S. Falkow, in "GFP: Green Fluorescent Protein Strategies and Applications" (M. Chalfie and S. Kain, eds.). Wiley-Liss, New York, 1998.
[5] M. Chalfie, Y. Tu, G. Euskirchen, W. W. Ward, and D. C. Prasher, Science 263, 802 (1994).
[6] A. B. Cubitt et al., TIBS 20, 448 (1995).
[7] B. P. Cormack, R. H. Valdivia, and S. Falkow, Gene 173, 33 (1996).
[8] D. R. Parks, L. A. Herzenberg, and L. A. Herzenberg, in "Fundamental Immunology" (W. E. Paul, ed.), p. 781. Raven Press, New York, 1989.
[9] R. H. Valdivia and S. Falkow, Mol. Microbiol. 22, 367 (1996).

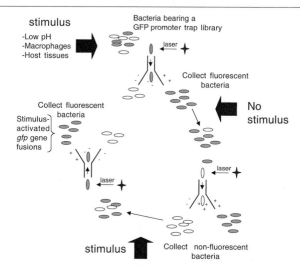

Fig. 1. Fluorescence-based selection strategy for the identification of differentially expressed genes. Bacteria bearing random *gfp* gene fusions are subjected to an inducing stimulus (e.g., low pH, host cells) and are analyzed by fluorescence-activated cell sorting. The sample stream crosses a laser in the FACS (emission 488 nm), which excites any GFP molecules in the bacteria. The bacterial fluorescence is registered by photomultiplier detectors in the cytometer, and cells with the desired levels of GFP (gray) are separated by a droplet-charging mechanism and deflection plates. Fluorescent bacteria are amplified in growth media in the absence of the inducing stimulus and are subjected to a second round of fluorescence-activated cell sorting. At this stage, bacteria that do not express the GFP tag (white) are collected. These nonfluorescent bacteria are amplified and subjected to a last round of cell sorting in the presence of inducing conditions. Alternating cell sorting for fluorescent and nonfluorescent bacteria in the presence and absence of inducing conditions rapidly enriches for bacteria bearing stimulus-activated *gfp* gene fusions.

GFP Expression Vectors

A variety of bacterial vectors and transposable elements designed to construct transcriptional and translational *gfp* fusions has been described. Ideally, one would study gene expression with single copy gene fusions in the same chromosomal context as the native gene being studied. Unfortunately, GFP, unlike other reporters of gene expression, such as *cat* or *lacZ*, is not an enzyme and therefore it does not provide the signal amplification that arises from multiple substrate cleavage. It is often necessary to artificially increase the copy number of *gfp* with plasmid-based vectors in order to obtain a measurable fluorescence signal. Table I shows a compilation of some of the vectors currently available for expressing *gfp* in a variety of bacterial species.

General Methods for Bacterial Detection by Flow Cytometry

Bacteria are at the lower end of the detection capabilities of most flow cytometers. Therefore, increasing the voltage of forward (FCS) and side scatter (SSC) photomultiplier (PMT) detectors and collecting the information with logarithmic amplifiers is required to detect most bacterial-size particles. The increased sensitivity of the scatter detectors also leads to an increase in electronic noise. Therefore, the FSC and SSC properties used to differentiate between bacteria and electronic noise have to be established with the aid of a third parameter. In some instances, autofluorescent pigments can be used to identify bacterial populations. Alternatively, bacterial cells can be stained with a variety of lypophilic dyes or antibodies conjugated to fluorescent molecules or transformed with GFP expression plasmids.[10] The subset of particles in the typical dot plot that exhibits a shift in fluorescence intensity defines the bacterial population. Adjustments can be made in the threshold of the PMT detectors so as to exclude particles that are not bacteria (background noise). After the light-scattering parameters are set for a particular organism, the analysis is simple, rapid (2–3 sec), and very reproducible.

Bacterial Isolation by Fluorescence-Activated Cell Sorting

FACS have the ability to collect cells based on their light-scattering and fluorescence properties. Sorting small cells, such as bacteria, is technically more challenging than sorting mammalian cells. Unlike larger cells, bacteria give poor FSC, making it necessary to "trigger" the sorting electronics of the cytometer with the SSC signal of the bacterium. If the objective of the experiment is to isolate bacteria that express an endogenous fluorescent marker, the fluorescence signal of the bacterium increases the triggering accuracy of the flow-sorting system. However, when no endogenous fluorescence marker is available, sorting may not be possible, especially when bacteria in populations are not homogeneous in their light-scattering properties. In these cases, an exogenously provided fluorescence parameter can be used as the trigger signal to increase the sorting efficiency.

Fluorescent Antibody Labeling for Flow Cytometry

Strains and Reagents. A bench top flow cytometer (e.g., FACScan or FACScalibur); FACS analysis tubes: 12 × 75-mm round-bottom tubes. Phosphate-buffered saline (PBS); antibody (Ab)-staining buffer [5% bovine serum albumin (BSA) in PBS]; midlog cultures of bacteria; Ab

---

[10] R. H. Valdivia and S. Falkow, *Curr. Opin. Microbiol.* **1**, 359 (1998).

TABLE I
GFP-Based Promoter Trap Vectors

Vectors	Markers[a]	gfp allele[b]	Comments	References[c]
Plasmids				
Transcriptional fusions				
pGFPUV2	ori$_{colE1}$ AP Gm	gfpUV		5
pASV3	ori$_{colE1}$ Ap	WT		6
pASV4	ori$_{colE1}$ Ap gfpcat	WT	Dual Gfp-Cat reporter system	6
pFPV25	ori$_{colE1}$ Ap mob	gfpmut3	Conjugative plasmid	7
Broad host range				
pANT-3	Ori$_{RSF1010}$ mob Km	gfpmut3	Conjugative plasmid	8
Integrative				
pIVET-GFP	ori$_{R6K}$ mob Ap asd-gfp	gfpmut3	Dual Asd-Gfp reporter system	9
pVIK165	ori$_{R6K}$ Km	gfpS65T	Complements asd auxotrophy	10
pFW-gfp	Sp	WT	Integrative vector for streptococci	11
Shuttle vectors				
pAD123	ori$_{colE1}$ Cm, Ap ori$_{TA1060}$	gfpmut3	Bacillus plasmid	12
pFPV27	ori$_{colE1}$ ori$_M$ Km	gfpmut3	Mycobacterial plasmid	13
MYGFP1	ori$_{colE1}$ori$_M$ Km	WT	Mycobacterial plasmids	14
MYGFP2				14
Translational fusions				
pSG1151	ori$_{colE1}$ Ap Cm	gfpmut1		15
pSG1156	ori$_{colE1}$ Ap Cm	gfpUV		
Transposons and gfp cassettes				
Transcriptional fusions				
miniTn5gfpgusA	Km	gfpUV	Dual Gfp-GusA reporter system	16
mini-Tn10-kan-gfp	Km			17

52

mini-Tn3-km-*gfp*	Km	*gfp*mut2		18
pTn3*gfp*	Ap	*gfp*mut2		19
pUTmini-Tn5*gfp*	Tc	*gfp*mut2		19
Tn5GFP1	Km	WT		20
mTn5*gfp*	Km	*gfp*mut3		21
pGreen		*gfp*mut1	Hypertransposing *tnpR* allele	22
pGreenTIR		*gfp*mut1	*gfp* cassette flanked by polylinker; pGreen+T7 phage translational enhancer	22
Translational fusions				
AT2GFP	*dhfr* (TmR)	*gfp*S65T	*In vitro* Ty transposon	23
GS	*supF*	*gfp*S65T	Trihybrid protein fusions	23
TyK'GFP+,	Km	WT	*In vitro* Ty transposon	24
pAG408 mTn5*gfp*	Km Gm	*gfp*UV		6

[a] ori, origin of replication; mob, mobilization region; Ap, ampicillin resistance; Km, kanamycin resistance; Gm, gentamycin resistance; Tm, trimethoprim resistance; Sp, spectinomycin resistance; Cm, chloramphenicol resistance; Tc, tetracycline resistance.

[b] GFP WT (1), *gfp* S65T (2), *gfp*UV (3), *gfp*mut1-3(4).

[c] References: (For references (1)–(4), see footnote b.) (1) M. Chalfie, Y. Tu, G. Euskirchen, W. W. Ward, and D. C. Prasher, *Science* **263**, 802 (1994); (2) A. B. Cubitt *et al.*, *TIBS* **20**, 448 (1995); (3) A. Crameri, E. A. Whitehorn, E. Tate, and W. P. Stemmer, *Nature Biotechnol.* **14**, 315 (1996); (4) B. P. Cormack, R. H. Valdivia, and S. Falkow, *Gene* **173**, 33 (1996); (5) S. Dhandayuthapani, W. G. Rasmussen, and J. B. Baseman, *Gene* **215**, 213 (1998); (6) A. Suarez *et al.*, *Gene* **196**, 69 (1997); (7) R. H. Valdivia and S. Falkow, *Mol. Microbiol.* **22**, 367 (1996); (8) A. K. Lee and S. Falkow, *Infect. Immun.* **66**, 3964 (1998); (9) M. Handfield *et al.*, *Biotechniques* **24**, 261 (1998); (10) V. S. Kalogeraki and S. C. Winans, *Gene* **188**, 69 (1997); (11) A. Podbielski, B. Spellerberg, M. Woischnik, B. Pohl, and R. Lutticken, *Gene* **177**, 137 (1996); (12) A. K. Dunn and J. Handelsman, *Gene* **226**, 297 (1999); (13) L. P. Barker, D. M. Brooks, and P. L. Small, *Mol. Microbiol.* **29**, 1167 (1998); (14) S. Dhandayuthapani *et al.*, *Mol. Microbiol.* **17**, 901 (1995); (15) P. J. Lewis and A. L. Marston, *Gene* **227**, 101 (1999); (16) C. Xi, M. Lambrecht, J. Vanderleyden, and J. Michiels, *J. Microbiol. Methods* **35**, 85 (1999); (17) S. Stretton, S. Techkarnjanaruk, A. M. McLennan, and A. E. Goodman, *Appl. Environ. Microbiol.* **64**, 2554 (1998); (18) C. Josenhans, S. Friedrich, and S. Suerbaum, *FEMS Microbiol. Lett.* **161**, 263 (1998); (19) A. G. Matthysse, S. Stretton, C. Dandie, N. C. McClure, and A. E. Goodman, *FEMS Microbiol. Lett.* **145**, 87 (1996); (20) R. S. Burlage, Z. K. Yang, and T. Mehlhorn, *Gene* **173**, 53 (1996); (21) D. M. Cirillo, R. H. Valdivia, D. M. Monack, and S. Falkow, *Mol. Microbiol.* **30**, 175 (1998); (22) W. G. Miller, and S. E. Lindow, *Gene* **191**, 149 (1997); (23) G. V. Merkulov and J. D. Boeke, *Gene* **222**, 213 (1998); and (24) L. A. Garraway *et al.*, *Gene* **198**, 27 (1997).

specific to the bacteria being analyzed. (*Note:* for many enteric bacteria, commercially available antisera directed to lipopolysaccharide can be used. Similarly, polyclonal serum directed against *Mycobacterium bovis* BCG can be used to stain most mycobacteria. In all instances, the amount of antibody used must be titrated to minimize nonspecific binding.) Second-step monoclonal Ab conjugated to a fluorophore (for available reagents, see www.jacksonimmuno.com; use fluorophores with emission spectra distinct from GFP). The mAb chosen should be specific to the species in which the antibacterial serum was raised.

For each strain to be sorted, resuspend $\sim10^7$ bacteria in 0.5 ml Ab-staining buffer and split into two microcentrifuge tubes. To one tube add polyclonal antisera specific for the bacterial species being tested and incubate for >1 hr with rotation at room temperature. Pellet the bacteria and wash with 1 ml of Ab-staining buffer. Resuspend the opsonized bacteria and the untreated bacterial samples (negative control) in 100 μl of Ab-staining buffer containing a secondary mAb conjugated to phycoerythrin (PE). Incubate at room temperature for 1 hr. Wash twice with PBS and resuspend in 0.5 ml of PBS. For bacteria that aggregate (e.g., mycobacteria), pass the bacterial suspension through a 27-gauge needle and syringe 10 times to break clumps. Transfer to FACS analysis tubes and analyze immediately by flow cytometry. [*Note:* PE and GFP show a slight overlap in their fluorescence emission after excitation at 488 nm. If *gfp* expression is also being assessed, use the second-stage Ab-PE control to electronically subtract overlap in fluorescence emission between PE and GFP. "Triggering" on the PE signal, use the FSC and SSC parameters to set a gate around the smallest (i.e., low FSC and SSC) PE-positive particles.]

Before initiating cell-sorting experiments, run a series of controls to determine the sorting accuracy of the FACS system being used. Mix midlog cultures of GFP-labeled and unlabeled bacteria at a ratio of 1 : 1, 1 : 10, and 10 : 1. Collect approximately 2×10^6 fluorescent events from a 1 : 10 mix of GFP-labeled/unlabeled bacteria into a FACS analysis tube half-filled with PBS and reanalyze the sorted population. If the sorting conditions are working correctly, 90–95% of the events collected should lie within the fluorescence gates used for sorting. Alternatively, if insufficient number of bacteria are recovered for immediate reanalysis (<10^6), bacteria can be grown for several generations and analyzed at a later time.

Safety Considerations

Appropriate precautions should be taken when handling pathogenic organisms, particularly when cell sorting airborne pathogens. Consult your institution's biosafety board as to the appropriate pathogen handling and containment protocols. [*Note: S. typhimurium* is a food-borne pathogen and *M. marinum* is a skin pathogen. Gloves should be worn at all times.] Treatment of the flow system with a 0.5% SDS solution followed by sterilization with 70% ethanol or 10% bleach is usually sufficient to remove any adherent organisms and disinfect the machine.

Troubleshooting Cell-Sorting Experiments

1. The detection of bacterial FSC signals can be enhanced in some cytometers by removing the neutral density filters that are usually present in front of the light-scatter photodiode detectors.
2. The light-scattering properties of bacteria can change depending on growth conditions and exposure to different environmental stresses. Therefore, the control strains used to set the detection parameters should be grown under the same condition as the strains to be tested.
3. Cross-contamination occurs frequently and can be particularly problematic when sorting rare bacterial subpopulations. Sterilize between samples.
4. Sorting efficiencies may be increased by diluting the bacterial cultures ($\sim$10^7 organism/ml) and decreasing the flow rate of the sample stream. After all parameters are optimized, a FACS will separate bacteria on the basis of their fluorescence at a rate of $\sim$10^3–10^4 per second.
5. Large discrepancies between the electronic count of sorted events and the number of viable bacteria obtained suggest (1) loss of viability during collection or (2) inclusion of electronic noise in the sorting gate. This latter possibility is more common when sorting nonfluorescent bacteria. Consider using the "Fluorescent Antibody Labeling for Flow Cytometry" protocol.
6. Cytometers. Analysis of bacterial GFP expression can be achieved with most bench top flow cytometers equipped with argon ion lasers tuned to 488 nm. Fluorescence can be detected with band-pass filter sets centered at 510–515 nm (e.g., 515/40 or 530/30).[11] Typically, the laser power does not need to exceed 100 mW (higher laser power contributes to higher noise levels in the forward scatter detectors). FACS are more specialized machines that are often available as shared equipment in immunology core facilities and are often operated by a full-time trained technician, who may not be accustomed to performing bacterial cell sorts. It is recommended that the researcher discuss the experimental setup and objectives with the FACS operator prior to experimentation.

[11] L. Lybarger and R. Chervenak, *Methods Enzymol.* 189 (1999).

Construction of Promoter Libraries

The following protocol provides a simple method to construct promoter libraries from *S. typhimurium* (St) and *M. marinum* (Mm).[12,13] Total DNA is isolated from the test organism, and small restriction enzyme fragments generated by partial digestion with *Sau*3AI are inserted at a unique *Bam*HI site adjacent to a promoterless *gfp* gene. The following section describes the construction of St and Mm promoter libraries in the promoter trap vectors pFPV25 and pFPV27, respectively. These libraries will be used to illustrate different variations of flow-sorting methods that can be used to identify St and Mm genes expressed differentially during infection.

Protocol

Strains and Reagents. All recombinant DNA manipulations and plasmid amplification can performed in standard *recA* laboratory strains of *Escherichia coli* (Ec). St and Ec are grown on Luria-Bertani (LB) broth or LB agar. Plasmid and genomic DNA can be isolated by a variety of methods (http://research.nwfsc.noaa.gov/protocols.html). DNA restriction and modifying enzymes can be obtained from several commercial sources (e.g., NEB, Stratagene, Promega). Use as recommended by the manufacturer. For pFPV25 and pFPV27 maintenance, supplement the growth media with 100 μg/ml ampicillin (Ap) and 50 μg/ml kanamycin, respectively. Mm is grown at 33° in 7H9 media (Difco, Detroit, MI) supplemented with 0.5% glycerol, 10% oleic acid dextrose complex (OADC) [OADC consists of 5% BSA, 1.0% oleic acid in 0.2 N NaOH 5%, glucose 2%, NaCl 0.85%], 0.01% cycloheximide, and 0.25% Tween 80. Electroporation protocols are available at www.bio-rad.com.

Preparation of DNA Inserts. Pipette 10–15 μg of total bacterial DNA in 95 μl of 1× *Sau*3A restriction buffer into five reaction tubes and place on ice. To each tube add 5 μl of twofold serial dilutions of *Sau*3AI restriction enzyme (2 to 0.125 U) in 1× *Sau*3AI restriction buffer. Transfer the restriction digests to a heating block set at 37° and incubate for 20 min. Stop the reactions by adding 20 μl of loading buffer (20% glycerol, 50 m*M* EDTA, 10 m*M* Tris–HCl, pH 8.0) and place on ice. Load the entire DNA digests on a 1% agarose gel. With a clean razor blade, slice gel fragments containing *Sau*3AI fragments ranging between 0.5 and 1.5 kb. The DNA fragments can be recovered from the agarose gel by a variety of methods, including commercially available DNA extraction kits (e.g., Geneclean, QuiaQuick).

---

[12] R. H. Valdivia and S. Falkow, *Science* **277,** 2007 (1997).
[13] L. Ramakrishnan, N. A. Federspiel, and S. Falkow, *Science* **288,** 1436 (2000).

Pool the excised *Sau*3AI fragments and adjust the DNA concentration to 0.1 mg/ml in TE buffer (10 mM Tris–HCl, pH 7.5, 1 mM EDTA).

Vector Preparation. Digest 1 μg of pFPV25 or pFPV27 with 2 U of *Bam*HI at 37° for 1 hr (100 μl total volume). Add 2 U of calf intestine alkaline phosphatase (CIAP) to the digestion mixture and incubate at 37° for an additional hour. Heat inactivate the CIAP at 65° for 15 min and extract the digestion mixture twice with TE-saturated phenol/choloroform (1:1, v/v) to remove any residual *Bam*HI and CIAP activity. Ethanol precipitate the DNA and resuspend in 20 μl of TE.

Ligation Reactions. Aliquot 2 μl (~0.1 μg) of CIAP-treated *Bam*HI vector into four small tubes and add 42 μl of 1× T4 DNA ligase buffer. To each tube add twofold serial dilutions of size-fractionated, *Sau*3AI-digested bacterial DNA fragments in a 5 μl total volume. Set the highest molar ratio of insert/vector to 1. Add 1 U of T4 DNA ligase to each of the four tubes and incubate the ligation reactions at 16° for 4 hr. Ethanol precipitate the DNA and resuspend in 5 μl TE buffer. Electroporate Ec with 1 μl from each ligation and plate on LB–agar supplemented with the appropriate antibiotic.

Determine the lowest ratio of insert/vector that provides greater than 2-3 K transformants. (*Note:* this step decreases the probability of multiple *Sau*3AI fragments inserting into the promoter trap vectors.) Transform Ec with the rest of the ligation reactions and maintain as separate pools of 2-3 K independent colonies. For each pool, scrape all colonies from the transformation plates and inoculate 500 ml of LB broth and grow at 37° for 3–5 hr. Harvest cells by centrifugation and isolate plasmid DNA. Promoter libraries can be maintained as plasmid DNA and used as needed by electroporation of St or Mm. The number of St or Mm transformants used as representative for each promoter library pool should be ~10- to 20-fold excess of the number of members in the original plasmid pool. Collect the transformed cell and resuspend in storage media (growth media + 15% glycerol). Aliquot 50 μl of the resuspended cells and store at −20°.

The complexity of the libraries made can be determined at different stages of construction by a variety of methods. GFP synthesis can be assessed by observing bacterial colonies under a blue light source. A simple method to achieve is to use the mercury arc lamp from standard fluorescence microscopes as a light source: Remove the lens objectives and use standard fluorescein dichroic filter sets (e.g., excitation 450–490 nm, emission 520 nm). Place the agar plate on the microscope stage. Brightly fluorescent colonies can be detected at a frequency of ~0.1%. If the bacterial colonies are observed under low magnification, an even greater diversity of fluorescence intensities can be observed. To determine the heterogeneity of DNA inserts in pFPV25 or pFPV27, recover plasmid DNA from 10 to 15 random bacterial colonies and assess the diversity of plasmid sizes by restriction digestion or polymerase chain reaction (PCR). For diagnostic purposes and ease of subcloning, these promoter trap vectors have several unique restriction sites flanking the *Bam*HI site where *Sau*3AI sites were inserted.

Identification of Bacterial Genes Expressed under Diverse Environmental Stimuli

The following protocol describes the isolation of St-bearing acid-inducible *gfp* gene fusions from a promoter library. The enrichment process consists of alternating cycles of cell sorting of fluorescent bacteria in the presence of an inducing stimulus and of nonfluorescent organisms in the absence of stimulation (Fig. 1).

Protocol

Strains and Reagents. St strain SL1344 (+promoter library in pFPV25); St (pFPV25); St (pFPV25.1). [*Note:* These latter bacterial strains are the negative (empty vector) and positive controls (constitutive *rpsM::gfp*)[9] required to set the correct sorting parameters.] LB broth acidified to pH 4.5 with HCl and LB broth at neutral pH. Supplement growth media with 100 μg/ml Ap. FACS analysis tubes: 12 × 75-mm round-bottom clear tubes.

1. Thaw an aliquot of St containing the promoter trap library and inoculate 5 ml of LB broth per pool. In addition, inoculate two tubes with 5 ml of LB broth with St (pFPV25) and St (pFPV25.1). Dilute cells to an $OD_{600} \sim 0.2$ and incubate at 30° for 2–3 hr without shaking. Remove 100 μl of the bacterial culture and inculate 5 ml of LB broth, pH 4.5. Incubate at 30° for 2 hr without shaking. Place cells on ice. Dilute the bacterial cultures (density $\sim 10^7$ bacteria/ml in LB broth, pH 4.5) and transfer 1.5 ml from each culture (controls strain and promoter libraries) to FACS analysis tubes.

2. Scan the control bacterial cultures and determine the proper light-scattering parameters for St as described in "Bacterial Detection by Flow Cytometry." Scan the promoter library, collect fluorescence intensity information for approximately 5×10^4 bacteria, and display this information as a contour plot. [*Note:* the cytometer's data analysis software allows the collected information to be displayed as histograms, density, contour, or dot plots. Contour lines displayed as log intervals can be used to highlight the presence of small (<1%) subpopulations of bacteria.] Compare this plot with a similar contour plot from St (pFPV25) and determine the minimal fluorescence intensity that distinguishes between GFP-expressing St and nonfluorescent

St. Depending on the complexity of the library, the proportion of bacteria in the promoter library that display fluorescence above background can range from 5 to 10% of the total library population. Sort approximately 30-60 K fluorescent bacteria (10- to 20-fold coverage of input promoter library). For collection, place a microcentrifuge tube in a FACS collection tube. This will allow the plating of bacteria without having to first centrifuge the collected material. The collection tube should be filled at least to two-thirds its capacity with nutrient broth to prevent the loss of sorted cells due to the buildup of static electricity during sorting.

3. Plate 100-μl aliquots of the collected samples on LB–agar plates. Incubate overnight at 37°. Pool bacteria and resuspend in storage media. Bacteria can be used either fresh for the next round of sorting or stored at $-20°$ in 50-μl aliquots.

4. Thaw new cells from the input promoter (IP) library and cells from the first acid passage cell sort (AP-1). Inoculate two 5-ml tubes of LB broth. Dilute cells to an $OD_{600} \sim 0.2$ in LB broth and incubate at 30° for 2–3 hr without shaking. Prepare control strains as described previously.

5. Remove a 100-μl aliquot of cells from each culture tube and inoculate 5 ml of LB broth (pH 7.0). Incubate at 30° for 2 hr on a heating block. Transfer 1.5 ml from each culture to FACS analysis tubes, scan the promoter libraries, and collect fluorescence intensity information for approximately 20 K events. Compare histogram plots from St (pFPV25), the IP library, and the AP-1 library. There should be a marked enrichment for fluorescent bacteria in the AP-1 library as compared to the IP library (see Fig. 2).

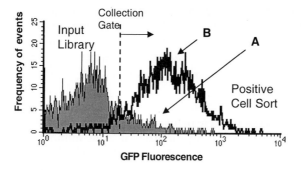

FIG. 2. Positive selection by fluorescence-activated cell sorting. Bacteria bearing random *gfp* gene fusions are subjected to a stimulus, and bacteria displaying fluorescence greater than that of non-*gfp*-expressing bacteria are sorted (A). The histogram plot illustrates the typical enrichment for fluorescent bacteria (B) seen after passage through FACS.

6. Most of these bacteria contain constitutive *gfp* promoter fusions. The subset of bacteria containing *gfp* fusions that are active only at pH 4.5 is isolated by sorting nonfluorescent cells at neutral pH. Set the cell sorter such that the least fluorescent 10% of the bacterial population is sorted. Collect at least 30 K bacteria and store as described earlier. Label cells as acid passage 2 (AP-2).

7. Thaw bacteria from AP-2 and treat cells as described previously. Expose bacteria to LB broth at pH 4.5 and collect 20 K bacteria from the most fluorescent 1–10% of the total population.

8. Plate serial dilutions of the collected cells such that single colonies are obtained on LB–agar plates. Test individual colonies for *gfp* expression in response to acidic media as follows: Inoculate 2 ml of LB broth with each candidate clone and grow cells at 30° for 3–4 hr. Remove 100 μl of cells and inoculate two tubes, one containing 2 ml of LB broth at neutral pH and the other with LB broth at pH 4.5. Place at 30° for 2 hr. Differences in bacterial fluorescence can be quantitated with a bench top flow cytometer (see Fig. 3). Alternatively, gross differences in fluorescence can be observed by placing a drop from each culture on a microscope slide and observing the bacteria under a fluorescence microscope.

9. Isolate plasmid DNA from bacterial clones that exhibit acid-dependent expression of *gfp*. Use primers flanking the *Bam*HI site where *Sau*3AI fragments were introduced to obtain DNA sequence information about the captured promoter.

Monitoring FACS Selections

The efficiency of the FACS selections can be assessed by monitoring the loss of heterogeneity of *Sau*3AI inserts present in the libraries after each selection step. This can be achieved by performing PCR reactions on samples from sorted bacterial pools with oligonucleotide primers flanking the *Bam*HI site. Each independent *Sau*3AI insert will give a characteristic PCR product. The IP library, for example, will yield a smear of PCR products, ranging from 0.5 to 1.5 kb. As the complexity of the pools decreases with each selection round, the number of potential PCR products will decrease to a few discrete bands. The number of discrete bands and their relative abundance is roughly proportional to the number and frequency of individual promoters that satisfied the selection criteria imposed by the cell sorter. The number of independent stimulus responsive clones isolated from each pool will be dependent on the library size and on the relative frequency of promoters that are activated by the particular stimulus being studied. For the isolation of acid-inducible promoters we have found

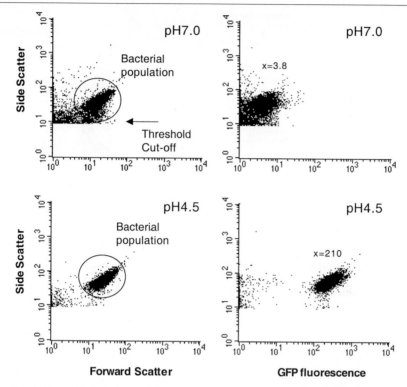

Fig. 3. Bacterial detection and fluorescence measurements by flow cytometry. Dot plots of side scatter (SSC) versus forward scatter and fluorescence intensity of an St bearing an acid-inducible *gfp* gene fusion (*aas::gfp*), which was isolated by the procedure described in the text. Note that the SSC detectors have been adjusted to exclude the majority of electronic noise and other nonbacterial particles ("Threshold cut-off"). The CellQuest (Beckton-Dickinson) program was used to determine the mean (×) fluorescence intensity of the bacterial population.

that the yield of acid-inducible clones ranged from 0–3 per pool of 3 K-independent *Sau*3AI inserts.[9]

Because the fluorescence selection gates are placed arbitrarily (in this case, the top 10% of the population under acidic conditions and lowest 10% under neutral pH), promoters that satisfy these strict conditions will be rapidly enriched over other acid-inducible promoters that are less induced by low pH. Therefore, promoters of different strengths can be isolated by varying the stringency of the collection gates (Fig. 4).

Potential Artifacts. The expression of *gfp* (or any other reporter gene) is influenced by the overall metabolic health of the cell. If a control strain bearing a constitutive *gfp* gene fusion displays major changes in fluorescence

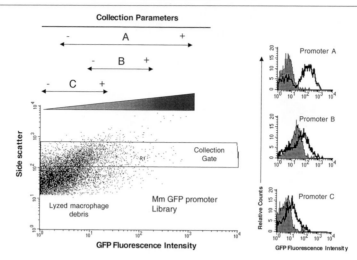

Fig. 4. Adjustments in fluorescence collection gates permit the identification of promoters of various strength. Mm bearing a GFP promoter library was recovered after Triton X-100 lysis of infected J774 cells. The "collection gate" contains Mm than can be separated from macrophage debris by their SSC signal. Promoters of different strengths can be identified by sorting Mm of different fluorescence intensities after passage through macrophages. The collection gates (A–C) show the parameters used for "positive" and "negative" sorts as described in the text. The intracellular-dependent induction of representative *gfp* gene fusions isolated from collection gates A–C was determined by the "Detergent-Release Assay" and displayed as histogram plots.

intensity in response to an applied stimulus, it is likely that the conditions chosen cause global changes in gene expression as a result of a loss in viability. This can lead to artifactual results where a condition may be interpreted as either "repressing" or "inducing" the expression of a promoter.

Identification of Bacterial Genes Differentially Expressed during Host Infections

The same principle of cell-sorting fluorescent organisms between inducing and noninducing conditions can be applied to far more complex stimuli, such as those encountered by bacterial pathogens within host cells and tissues. To demonstrate that the methodology shown is widely applicable, we describe protocols for the identification of *in vivo*-activated promoters by two different intracellular pathogens: *S. typhimurium* and *M. marinum.*

In this experimental setup, cultured macrophages are infected with either St or Mm bearing a library of random *gfp* gene fusions. Bacteria that express the fluorescence marker are separated by cell sorting either fluorescent, intact macrophages, or bacteria released from lysed macrophages. These bacteria are recovered and grown under *"ex vivo"* conditions (tissue culture media alone). Bacteria that do not fluoresce under these conditions are isolated by FACS and are used for a second round of macrophage infections. As in the selection for acid-inducible promoters, repeated cell sorting cycles rapidly enrich for bacteria bearing *gfp* fusions to promoters that are induced in the intracellular environment of host cells.[12]

Strains and Reagents. St or Mm (+promoter library in pFPV25 or pFPV27, respectively); St (pFPV25); St(pFPV25.1). Mm strain M containing pFPV27 alone and M containing pBEN (pFPV27 derivative with a constitutive *hsp60::gfp*). (*Note:* these bacterial cultures are the negative and positive controls required to set the correct sorting parameters.) Murine macrophage cell lines (J774.1 or RAW 264.7). Maintain cell lines in Dulbecco's modified Eagle's medium (DMEM) (GIBCO-BRL) supplemented with 10% fetal calf serum (FCS) and 1 mM glutamine in a 37°, 5% CO_2 humidified incubator. For Mm infections, maintain cultured macrophages at 33° (optimal growth temperature for Mm[14]). "High salt LB"—LB broth supplemented with 0.3 M NaCl.

Protocol 1: Cell Sorting Infected Macrophages

1. Thaw an aliquot of St bearing the *gfp* promoter library and inoculate 5 ml of LB broth. Incubate at 30° for 5 hr without aeration (standing culture). As controls, also inoculate St (pFPV25.1) and St (pFPV25). For each library pool and for each control, subculture the bacterial cultures into 5 ml of high-salt LB. Grow standing overnight at 37°. (*Note:* The combination of high salt and microaerophilic growth conditions enhances the invasiveness of St.)

2. Measure the OD_{600} of the bacterial culture (1 $OD_{600} \sim 1.5 \times 10^9$ bacteria). Add 5×10^6 bacteria to tissue culture dishes seeded with 5×10^5 RAW 264.7 cells to give a multiplicity of infection (MOI) of 5:1. Centrifuge the infected wells for 5 min ($\times 1500g$) to maximize bacterial contact with host cell and incubate for 45 min at 37° in a 5% CO_2 incubator. Wash the infected RAW 264.7 cells three times with prewarmed DMEM to remove all nonadherent bacteria. Incubate the infected cells for 4 hr.

3. Wash the infected macrophages four times with warm DMEM. Aspirate all excess media, add 5 ml of cold DMEM, and place the infected RAW 264.7 cells on ice. Gently scrape the RAW 264.7 cells from

---

[14] L. Ramakrishnan and S. Falkow, *Infect. Immun.* **62,** 3222 (1994).

each tissue culture well with a disposable cell scraper and resuspend the cells in DMEM. Transfer the infected cells to FACS analysis tubes and place on ice.

4. Scan cells by FACS. RAW 264.7 cells infected with SL1344 (pFPV25.1) will appear as brightly fluorescent cells. The parameters for the sorting macrophages infected with fluorescent bacteria are very similar to those used for routine cell sorting of mammalian cells. Consult the FACS operator to optimize the sorting parameters.

5. From the cell sample infected with the promoter trap library, collect all RAW 264.7 cells that exhibit fluorescence above that of cells infected with SL1344 (pFPV25) alone. This number will vary, depending on the efficiency of invasion and the complexity of the library. A helpful guideline is to sort the top 1% of the fluorescent RAW 264.7 cells (see Fig. 5). The fluorescent RAW 264.7 can be sorted directly on LB broth + Ap. Plate the collected cells on LB–agar and incubate overnight at 37°. Pool the bacterial colonies and store at −20°.

Protocol 2: Cell Sorting Bacteria from Infected Macrophage Lysates

Macrophages infected with fluorescent bacteria are lysed to release intracellular bacteria prior to FACS. This is particularly helpful for flow sorting bacteria that clump in culture or enter macrophages as aggregates, as passage through mammalian cells helps disperse clumps into single cells.[15] In addition, bacteria released from lysed cells can be sorted according to their fluorescence intensity. This permits the isolation of promoters of varied strengths in response to the macrophage intracellular environment (Fig. 4). We illustrate this method with an example to obtain *gfp*-expressing Mm from infected macrophages.

6. Inoculate 10 ml of 7H9-OADC with Mm bearing the *gfp* promoter library and the control strains Mm (pFPV27) and Mm (pBEN). Add approximately 5×10^6 bacteria (1 $OD_{600} \sim 1.5 \times 10^8$ bacteria) to tissue culture dishes seeded with 10^6 J774 macrophages. Incubate at 33° in a 5% CO_2 incubator for 20–24 hr. Wash the infected cells three times with prewarmed DMEM to remove all nonadherent bacteria. Aspirate excess media and add 0.5 ml of 0.1% Triton-X-100 in PBS to the flask. Incubate for 10 min on ice and scrape off the cell debris. Disrupt cells vigorously, add 2 ml of DMEM medium, and transfer the lysates to FACS analysis tubes (also see "Detergent-Release Assay").

[15] L. P. Barker, D. M. Brooks, and P. L. Small, *Mol. Microbiol.* **29,** 1167 (1998).

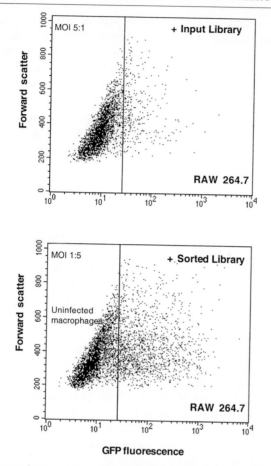

FIG. 5. Enrichment for macrophage-inducible *gfp* gene fusions. (Top) A dot of RAW264.7 cells infected at a high MOI (5 : 1) with St bearing a GFP promoter library (Input Library). Bacteria were collected and processed by FACS as described in the text. Bacteria that displayed no fluorescence in the absence of host cells were used to infect RAW26.4 cells at a low MOI (1 : 5). [*Note:* the forward scatter (FSC) signal was collected with linear amplifiers, as RAW264.7 cells are substantially larger than bacteria.]

7. Scan the positive and negative controls by FACS. Analyze the lysed J774 cells infected with Mm (pBEN) and adjust the FSC and SSC detectors to eliminate most of the nonfluorescent particles. Mycobacterial aggregates will display higher scatter and fluorescence. Some of the macrophage debris will scatter light in the range of single mycobacteria.

8. Scan the promoter library sample and set the sorting gates (e.g., the most fluorescent 10% of the population. This is an artificial percentage as the cell debris has similar light-scattering properties as single mycobacteria. Bacteria of varied fluorescence intensities can be recovered to obtain a wider spectrum of host-induced promoters). Collect enough bacteria in 7H9-OADC media to provide a 10-fold coverage of the IP library. This represents the pool of bacteria that contain transcriptionally active *gfp* gene fusion in the host cell environment. Plate the collected Mm and store as described previously.

At this stage, the two protocols converge. Bacteria that were isolated based on their fluorescence during their residence in host cells are now exposed to *ex vivo* conditions, so that constitutive promoters are separated from those that are only expressed within host cells.

1. Inoculate growth media with the macrophage-passaged promoter trap libraries as described previously. Measure the OD_{600} of the bacterial cultures and use approximately 10^6 bacteria to inoculate 5 ml of DMEM supplemented with 10% FCS and 1 mM glutamine. Incubate at 37° in 5% CO_2 for 3–4 hr. Dilute bacteria 1/10 in cold DMEM and place on ice. (*Note:* It is important that the bacterial libraries are exposed to the same conditions as during the macrophage infections, otherwise bacterial promoters that are responsive to factors in the DMEM, FCS, or glutamine may be selected preferentially. For Mm, expose bacteria to DMEM supplemented with 10% OADC complex and Tween 80 for 20–24 hr. This allows optimal growth of mycobacteria while retaining the DMEM and serum.)

2. Scan the bacterial cells by FACS. Bacteria in the population will vary in fluorescence intensities. Sort bacteria with the lowest fluorescence (lower 10% of the total population) as described previously. Plate the sorted bacteria on agar plates. Pool the bacterial colonies and store at −20°.

3. Thaw the pools of sorted bacteria and repeat steps 1–5 for St and steps 6–8 for Mm. For St infections of RAW264.7 cells, decrease the MOI to 1:5. The high MOI (5:1) used in the initial selection step increases the probability that all members of the promoter library will have infected a RAW 264.7 cell. Potential false positives are eliminated during this last selection step, where RAW 264.7 cells are infected at a low MOI (1:5).

Bacteria sorted by either of the protocols just described will show a marked increase in fluoresce as compared to bacteria from the IP library when grown in macrophages (Fig. 5). Test independent clones for intracellular-dependent *gfp* expression.

Measuring Bacterial Gene Expression within Host Cells:
Detergent-Release Assay

1. Infect macrophages with the control and test strains at an MOI ~5:1 (steps 1 and 2 of protocol 1 and step 1 of protocol 2).

2. For St, wash the infected cells twice with PBS to remove the majority of nonadherent bacteria. Place the tissue culture wells on ice and remove 1 ml of the supernatant. Transfer the supernatant to FACS analysis tubes and place on ice. (*Note:* This fraction contains mostly bacteria that have replicated extracellularly.) Aspirate the remaining supernatant from the infected RAW 264.7 cells. Wash four times with PBS. Add 0.1 ml 1% Triton-X (v/v) in PBS and incubate for 10 min. Lyse the cells thoroughly by repeated pipeting and scraping. Add 0.9 ml cold DMEM and transfer the lysate to FACS analysis tubes. For Mm, the lysis can be done at room temperature, as the bacteria are slow growing and will not replicate appreciably in the course of the lysis.

3. Scan the cells in a bench top flow cytometer. The scan will show cellular debris, bacteria, and intact macrophages. To determine what constitutes the bacterial population, use the positive and negative controls as described. (*Note:* bacteria grown extracellularly and intracellularly may scatter light differently. Therefore, each gate has to be set independently. Figure 6A shows an example of the typical scatter parameters for bacteria released from infected macrophages.)

4. Collect fluorescence information from 5 K bacteria. To determine levels of induction in response to macrophages, compare the mean fluorescence intensity from detergent-released bacteria with the mean fluorescence of bacteria present in the supernatant (Fig. 6B).

Identification of Bacterial Genes Expressed within Host Tissues

The cell sorting of fluorescent bacteria from crude macrophage lysates can be extended to identify bacterial genes expressed in infected animal tissues. The following protocol describes the basis for the identification of Mm genes expressed selectively during chronic Mm infections. This protocol can be adapted to identify *in vivo*-expressed genes in a variety of pathogenic bacteria.

Strains and Reagents. Leopard frogs (*Rana pipiens*). Mm bearing a *gfp* promoter trap library, Mm (pBEN), and Mm alone. Tissue lysis media: 7H9-OADC broth supplemented with 1% Triton X-100. (*Note:* all protocols involving experimental animals must be in accordance with local and federal guidelines. Consult with your institution as to the appropriate animal maintenance and euthanasia methods to be used.)

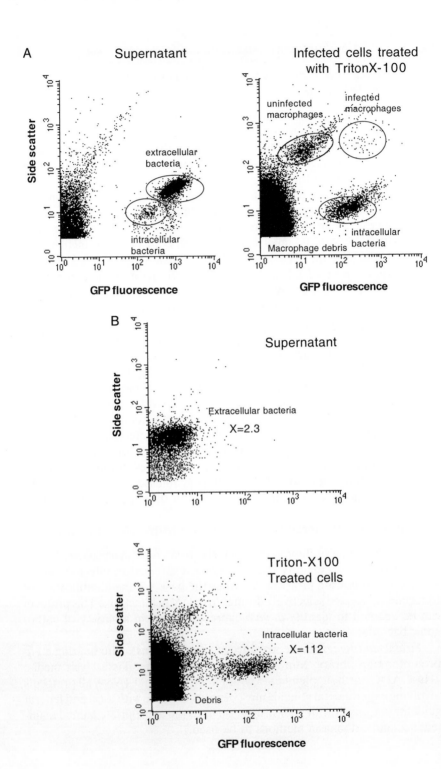

A   Supernatant   Infected cells treated with TritonX-100

Side scatter

extracellular bacteria

intracellular bacteria

GFP fluorescence

uninfected macrophages

infected macrophages

intracellular bacteria

Macrophage debris

GFP fluorescence

B

Side scatter

Supernatant

Extracellular bacteria

X=2.3

GFP fluorescence

Side scatter

Triton-X100 Treated cells

Intracellular bacteria

X=112

Debris

GFP fluorescence

1. Infect one frog per library pool and control strain by intraperitoneal injection with $\sim 10^7$ Mm. After 6 weeks, sacrifice the animals and remove their livers. Homogenize 250–500 mg tissue in a sealed plastic bag containing 5 ml of tissue lysis media by rolling a glass pipette over the bag. Remove particulate tissue debris by passage through a sterile disposable 30-μm filter.

2. Adjust the sorting parameters by FACS analysis of homogenized liver tissues of frogs infected with a 1:1 mixture of Mm and Mm (pBEN). The light-scatter profile of fluorescent bacteria present in liver homogenates may differ from the light-scatter profile of 7H9-grown bacteria. Collect fluorescent events from all scatter gates. To assess the efficiency of cell sorting, spread the collected samples on 7H9-OADC agar plates with or without Kan (the *gfp* expression plasmid pBEN has a Kan^R marker). Establish and record the sorting parameters that will yield greater than 90% purity before proceeding with FACS collections of the promoter libraries.

3. The library is sorted as in "Bacterial Isolation by Fluorescence-Activated Cell Sorting." In chronic granulomatous infections, the number of bacteria per gram tissue may be very low ($\sim 10^5$).[16] Therefore, it can take as long as 3 to 4 hr to collect 500–1000 bacteria. Collect the bacteria in 200 μl and plate the entire output at various dilutions without centrifugation.[13]

4. Pool the bacterial colonies, resuspend in storage media, and store frozen in small aliquots. These cells can be processed for further sorting experiments as described earlier. The choice of *ex vivo* conditions is arbitrary, and different conditions (minimal media, DMEM + FCS, or J774 cells) may yield different genes.

---

[16] L. Ramakrishnan, R. H. Valdivia, J. H. McKerrow, and S. Falkow, *Infect. Immun.* **65**, 767 (1997).

FIG. 6. Detergent-release assay for monitoring bacterial gene expression in infected RAW264.7 cells. (A) RAW 264.7 cells were infected with St tagged with a *rpsM::gfp* (pFPV25.1) fusion. Dot plots of supernatants and detergent-lyzed fractions show the heterogeneity of cells present in the crude fractions. Note that intracellular and extracellular bacteria exhibit different light-scattering properties. Also note the small decrease in the expression of the S13 ribosomal protein (*rpsM* gene product) in macrophages, probably as a result of the adverse growth conditions within phagocytic cells. (B) Flow cytometric quantitation of RAW264.7-dependent expression of a macrophage-inducible *gfp* gene fusion (*ssaH::gfp*) that was isolated as described in the text.

Applications of *gfp* Gene Fusions to Monitor
Host–Pathogen Interactions

Bacteria bearing *gfp* fusions to genes or proteins of interest provide useful tools to study diverse aspects of the interactions between the bacterium and its host. For example, *gfp* gene fusion technology has been applied to (i) analyze the subcellular distribution and fate of virulence factors in the host,[17] (ii) determine the tropism of bacterial pathogens and symbionts with host tissues,[18,19] and (iii) monitor bacterial gene expression *in situ*.[20] The following section provides examples and protocols of how to image and quantitate bacterial gene expression at the single cell level either by fluorescence microscopy or by flow cytometry.

Imaging of Bacterial Gene Expression within Live Host Cells

The imaging of bacteria expressing *gfp* in association with host cells and tissues can be achieved by standard epifluorescence microscopy, as well as more advanced systems, such as laser-scanning confocal microscopy or fluorescence deconvolution systems, when greater image resolution is required. For most microscopy applications, samples can be fixed with a variety of cross-linking agents [e.g., 2% (w/v) glutaraldehyde in PBS]. Other fixatives, such as methanol or acetone, which dissolve membranes, lead to a loss of fluorescence signal, especially if the GFP tag being imaged is in a soluble form. Imaging of live interaction between pathogenic organisms and mammalian cells requires more complex setups in order to maintain the cultured mammalian cells at 37° in buffered tissue culture media. For descriptions of such setups, please refer to protocols in Chalfie and Kain.[21] In our experience, prolonged exposure to the intense light sources required to detect low levels of fluorescence is detrimental to the cells and may alter the normal outcome of the interaction between the pathogen and the host. Therefore, we recommend time-lapse acquisition of fluorescence and transmitted light images. Figure 7 provides an example of such time-lapse acquisition of differential interference contrast and fluorescence images of *S. typhimurium* interacting with a host macrophage.

---

[17] C. A. Jacobi, A. Roggenkamp, A. Rakin, R. Zumbhil, L. Leitritz, and J. Heesemann, *Mol. Microbiol.* **30,** 865 (1998).

[18] H. Zhao, R. B. Thompson, V. Lockatell, D. E. Johnson, and H. L. Mobley, *Infect. Immun.* **66,** 330 (1998).

[19] D. J. Gage, T. Bobo, and S. R. Long, *J. Bacteriol.* **178,** 7159 (1996).

[20] S. Moller, C. Sternberg, J. B. Andersen, B. B. Christensen, J. L. Ramos, M. Giskov, and S. Molin, *Appl. Environ. Microbiol.* **64,** 721 (1998).

[21] M. Chalfie and S. Kain, "GFP: Green Fluorescent Protein Strategies and Applications." Wiley-Liss, New York, 1998.

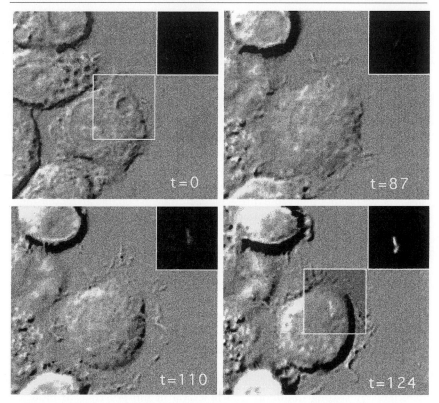

Fig. 7. Time-lapse differential interference contrast and fluorescence microscopy of RAW264.7 infected with St bearing a macrophage-inducible *gfp* fusion (*ssaH::gfp*). Two bacteria (white box) can be seen in a vacuole at *t* = 0 min. As infection progresses, the phagosome shrinks and intracellular St activate the expression of *ssaH::gfp* (fluorescence panel).

Flow Cytometry-Based Quantification of Host-Induced Bacterial Gene Expression

The measurement of *gfp* expression by flow cytometry can be used to compare the fluorescence of bacteria grown in the absence of host cells to the fluorescence of bacteria that have been released from infected cells.[9] Because this approach normalizes *gfp* expression to a per cell basis, the assay is highly reproducible and, unlike enzymatic assays, is impervious to variations in the number of bacteria added. For protocols, refer to "Measuring Bacterial Gene Expression within Host Cells: Detergent-Release Assay."

In Vivo Detection of Bacterial Association with Host Cells

Because *gfp* expression does not affect interactions between bacterial pathogens and their hosts adversely,[22] GFP-tagged bacteria can be used to identify cells in the host that may preferentially interact with the pathogen during an infection. Host cell tropism can be determined quantitatively by two-color FACS analysis[11] of infected tissues. For example, crude tissue homogenates can be incubated with Ab–fluorophore conjugates that are specific to antigens present in host cells to determine whether GFP-tagged bacteria associate with immune cells such as T cells, B cells, macrophages, or neutrophils. Given the availability of well-characterized immunological reagents for flow analysis of the immune system heterogeneity and responses, we predict that *gfp*-expressing bacteria will become an increasingly popular tool to unravel the immunology of host–pathogen interactions.

Genetics by Flow Cytometry: Dissecting Genetic Regulatory Circuits

The application of flow cytometry as a tool to identify differentially expressed genes can be extended to dissect regulatory networks. The principle for such identification is similar to the DFI selections described earlier. In short, bacteria bearing a *gfp* gene fusion of interest is mutagenized (chemical, UV, or transposon) and processed by FACS either in the presence or in the absence of the inducing stimulus. Mutations in negative regulators can be identified by sorting fluorescent bacteria in the absence of inducing conditions, whereas mutations in positive regulators can be identified by sorting nonfluorescent bacteria in the presence of inducing conditions. We have used such an approach to determine that the *ompB* locus of St is necessary for the expression of a subset of macrophage-inducible genes.[23]

Concluding Remarks

The methods described here provide the conceptual and technical basis that should allow for FACS-based selections to be tailored to most bacterial species that are amenable to genetic manipulation. With the rapid pace at which new bacterial genomes are being sequenced, the application of either complete *gfp* fusion libraries or *gfp* fusions to specific virulence genes will allow the spatial and temporal measurements of bacterial gene expression with single cell resolution. New GFP variants with short half-lives and spectrally distinct excitation and emission spectra promise even more excit-

---

[22] R. H. Valdivia, A. E. Hromockyj, D. Monack, L. Ramakrishnan, and S. Falkow, *Gene* **173**, 47 (1996).

[23] R. H. Valdivia, M. Rathman, and S. Falkow, unpublished results.

ing prospects in the study of bacterial gene regulation *in vivo*, for they will permit the simultaneous monitoring of multiple bacterial (and host) genes during infection.

One disadvantage of the FACS-based selection system described here is the need for specialized central FACS facilities and access to trained personnel. Traditionally, FACS have not been used for sorting bacteria, and the small size of bacteria places them at the lower limit of detection of most available machines. The sorting of nonfluorescent bacteria can be particularly challenging. As an alternative, individual bacterial clones can be analyzed for gene expression by fluorimetry, microscopy, or with a bench top cytometer. Although labor intensive, these modifications can still be amenable to high-throughput analysis and eliminate the steps of flow sorting nonfluorescent bacteria.

Different permutations of FACS enrichment cycles can be used to match particular experimental requirements and objectives. For example, if the investigator's objective is to identify genes repressed in response to a stimulus, the DFI enrichment cycle can be modified by altering either the cell-sorting order or the fluorescent status of the cells being collected after exposure to the stimulus. Initial trials may be necessary to determine what the best modifications, if any, of the methods described here may be required for each investigator's particular needs.

Acknowledgments

We thank T. Knaak for flow sorting expertise, M. Amieva for invaluable help with microscopy and Stanley Falkow for encouragement and support in the development of the work presented here.

[5] IVET and RIVET: Use of Gene Fusions to Identify Bacterial Virulence Factors Specifically Induced in Host Tissues

By JAMES M. SLAUCH and ANDREW CAMILLI

Introduction

Transcriptional gene fusions facilitate the study of gene regulation greatly. This article describes the adaptation of gene fusion technology to the study of bacterial pathogenesis. Our starting premise is that many

TABLE I
IVET Systems and Selections

Organism	Selection/screen	Reference
Salmonella typhimurium	purA-lacZY (pIVET1)	2, 3
Vibrio cholerae	γδ resolvase (pIVET5)	4
Listeria monocytogenes	lacZ	5
Salmonella typhimurium	cat-lacZY (pIVET8)	6
Pseudomonas aeruginosa	purEK	7
Yersinia entercolitica	cat	8
Salmonella typhimurium	GFP	9
Staphylococcus aureus	γδ resolvase	10
Escherichia coli	cat	11
Streptococcus gordonii	amylase-cat	12
Candida albicans	Flp recombinase	13

virulence factors are coordinately regulated by *in vitro* environmental signals that presumably reflect cues encountered in host tissue.[1] In order to study regulation in infection models, we have developed *in vivo* expression technology (IVET) to directly select those genes that are induced *in vivo.*[2] A subset of those genes that are induced in host tissues include virulence genes that are specifically required for the infection process.

IVET or variations on a theme have been used to identify *in vivo*-induced genes in a variety of both prokaryotic and eukaryotic pathogens (Table I[2–13]). These IVET systems fall into three categories: (1) selection systems based on metabolic (e.g., *purA*) or antibiotic (e.g., *cat*) reporters, (2) recombination-based systems, and (3) GFP-based systems. This article

---

[1] M. J. Mahan, J. M. Slauch, and J. J. Mekalanos, in *"Escherichia coli* and *Salmonella typhimurium:* Cellular and Molecular Biology"* (F. C. Neidhardt, ed.), Chap. 154. American Society for Microbiology, Washington, DC, 1996.

[2] M. J. Mahan, J. M. Slauch, and J. J. Mekalanos, *Science* **259,** 686 (1993).

[3] J. M. Slauch, M. J. Mahan, and J. J. Mekalanos, *Methods Enzymol.* **235,** 481 (1994).

[4] A. Camilli, D. T. Beattie, and J. J. Mekalanos, *Proc. Natl. Acad. Sci. U.S.A.* **91,** 2634 (1994).

[5] A. D. Klarsfeld, P. L. Goossens, and P. Cossart, *Mol. Microbiol.* **13,** 585 (1994).

[6] M. J. Mahan, J. W. Tobias, J. M. Slauch, P. C. Hanna, R. J. Collier, and J. J. Mekalanos, *Proc. Natl. Acad. Sci. U.S.A.* **92,** 669 (1995).

[7] J. Wang, A. Mushegian, S. Lory, and S. Jin, *Proc. Natl. Acad. Sci. U.S.A.* **93,** 10434 (1996).

[8] G. M. Young and V. L. Miller, *Mol. Microbiol.* **25,** 319 (1997).

[9] R. H. Valdivia and S. Falkow, *Science* **277,** 2007 (1997).

[10] A. M. Lowe, D. T. Beattie, and R. L. Deresiewicz, *Mol. Microbiol.* **27,** 967 (1998).

[11] M. A. Khan and R. E. Isaacson, *J. Bacteriol.* **180,** 4746 (1998).

[12] A. O. Kili, M. C. Herzberg, M. W. Meyer, X. Zhao, and L. Tao, *Plasmid* **42,** 67 (1999).

[13] P. Staib, M. Kretschmar, T. Nichterlein, G. Kohler, S. Michel, H. Hof, J. Hacker, and J. Morschhauser, *Mol. Microbiol.* **32,** 533 (1999).

provides a methodological overview for the paradigms of metabolic and recombination-based systems. Although we use our studies in *Salmonella typhimurium* and *Vibrio cholerae* as examples, we discuss the relevant points that should allow adaptation of these systems to other pathogens.

pIVET1: Rationale and Design

The original IVET selection system is based on the fact that a mutation in a biosynthetic gene can dramatically attenuate the growth and persistence of a pathogen in host tissues. Growth of the auxotrophic mutant in the host can then be complemented by operon fusions to the same biosynthetic gene, thus selecting for those promoters that are expressed in host tissues. Several biosynthetic genes can, theoretically, be used in this type of selection system. We have initially concentrated on a *purA* system in *S. typhimurium*.[2,3] The *purA* gene encodes an enzyme, adenylosuccinate synthetase, required for synthesis of adenosine 5'-monophosphate (AMP). *Salmonella typhimurium* strains mutant in the *purA* gene are extremely attenuated in their ability to cause mouse typhoid or persist in animal tissues.[14] Thus, one can select, in the animal, genes that are transcriptionally active *in vivo* by fusing their promoters to the *purA* gene. Only those strains that contain fusions to promoters that are transcriptionally active enough to overcome the parental PurA deficiency in the animal will be able to survive and replicate in the host. Fusion strains that answer this selection are then screened for those that are transcriptionally inactive on normal laboratory medium. This subset of strains contains fusions to genes that are "on" in the mouse and "off" outside the mouse. In other words, these genes are specifically induced in the host. We call these genes *ivi,* for *in vivo* induced.[2]

In constructing our fusion system, we met several criteria, which have proved to be very important for the success of the selection. First, we wanted to construct the fusions in single copy in the chromosome. This avoids any complications that can arise from the use of plasmids. Second, we wanted to construct the fusions without disruption of any chromosomal genes. If the gene of interest encodes a product required for the infection process, then a fusion that disrupts the gene would not be recovered in the selection. Third, we wanted a convenient way to monitor the transcriptional activity of any given fusion both *in vitro* and *in vivo*.

We met these criteria by constructing a synthetic operon composed of a promoterless *purA* gene and a promoterless *lac* operon in the suicide

[14] W. C. McFarland and B. A. Stocker, *J. Microbial. Pathol.* **3,** 129 (1987).

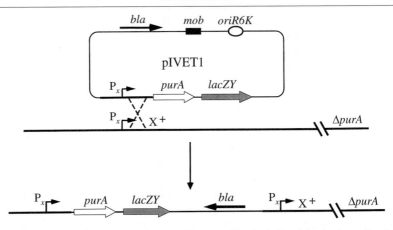

FIG. 1. Positive selection for genes that are specifically induced in the host. Random, partial Sau3AI-restricted S. typhimurium chromosomal DNA fragments were cloned into the BglII site of pIVET1. Replication of the plasmid is dependent on the Pi protein, which must be supplied *in trans*. The pool of *purA-lac* transcriptional fusion plasmids was mobilized into a *purA* deletion strain of S. typhimurium that lacks the Pi replication protein. Selection for ampicillin resistance in this strain demands integration of the plasmid into the bacterial chromosome, generating a duplication of S. typhimurium material in which one promoter drives the *purA* gene and the other promoter drives the expression of the putative virulence gene, X.

plasmid pGP704,[15] creating pIVET1[3] (Fig. 1). The promoterless *purA* gene was obtained using the polymerase chain reaction (PCR) from the chromosome of *Escherichia coli* K-12. The promoterless *lac* operon fragment contains the W205 *trp-lac* fusion that effectively removes the transcription start site of the *lac* operon, resulting in a $lacZ^+$, $lacY^+$ transcriptional fusion.[16] The BglII site in pIVET1 provides a convenient place to insert random fragments of Sau3AI-digested chromosomal DNA. pIVET1 also contains the *cis*-acting site (*mob*) that allows broad host range conjugal transfer by plasmid RP4. Cloning of chromosomal fragments 5′ to the *purA* gene results in transcriptional fusions in which S. typhimurium promoters drive the expression of a wild-type copy of *purA* and *lacZY*. Replication of pGP704 derivatives requires the replication protein, Pi, the product of the *pir* gene, supplied *in trans*. The construction of the fusions is done in E. coli strains that make Pi, such as DH5αypir or SM10ypir, which also contains the functions required for broad host range conjugal transfer of pGP704 derivatives.[17] The *pir* gene can also be cloned into other species.

[15] V. L. Miller and J. J. Mekalanos, *J. Bacteriol.* **170,** 2575 (1988).
[16] D. H. Mitchell, W. S. Reznikoff, and J. R. Beckwith, *J. Mol. Biol.* **93,** 331 (1975).
[17] R. Simon, U. Priefer, and A. Puhler, *Biotechnology* **1,** 784 (1983).

Construction of *purA-lac* Fusions

Random *Sau*3AI fragments of chromosomal DNA are cloned into the *Bgl*II site, 5' to the *purA* gene in pIVET1. In our experiments, we isolate DNA from *S. typhimurium* strain JS120 (Δ*purA3141::kan*). It is important that the *purA* allele in the strain used as a source of DNA be the same allele as in the strain used in the selection. If chromosomal DNA is cloned from wild-type cells, for example, then the wild-type *purA* gene will be cloned at some frequency and will give false-positive results. We size fractionate the *Sau*3AI-digested DNA and isolate approximately 5-kb fragments for introduction into pIVET1. This insert size allows efficient homologous recombination and also increases the probability of cloning the 5' end of the transcript (see later). We ensure that the resulting plasmid pool size is >42,000 clones. This represents, with 99% probability, a fusion to every 1 kb of chromosome in either orientation. In other words, given that the average gene is approximately 1 kb, we have created a fusion to every gene in the cell. The best efficiency of cloning is obtained in *E. coli* strains constructed for this purpose. However, we have to subsequently introduce the clones into *S. typhimurium*, which has a significant restriction barrier. We have determined empirically that moving the libraries by conjugal transfer from *E. coli* to *S. typhimurium* is more efficient than, for example, passaging the isolated plasmid library through a restriction minus, Pi⁺ *S. typhimurium* strain.

The pool of fusions is introduced into a Δ*purA* strain of *S. typhimurium* that lacks the *pir* gene. Selection for resistance to ampicillin demands the integration of the plasmids into the chromosome by homologous recombination with the cloned *Salmonella* DNA. This results in single copy diploid fusions in which one promoter drives the expression of the *purA-lac* fusion and the other promoter drives the expression of the wild-type gene (Fig. 1).

There are several important points to be made about this integration event. First, the cloned chromosomal sequences provide the only site of homology for integration into the chromosome. The *purA* gene was obtained by PCR from the *E. coli* chromosome, and *E. coli* and *S. typhimurium* chromosomes are sufficiently divergent to prevent recombination.[18] Also, *S. typhimurium* does not contain a *lac* operon. Second, only those clones that contain the 5' end of the operon of interest (in other words, the promoter) will generate both a functional fusion and a duplication that maintains synthesis of the wild-type gene (the event drawn in Fig. 1). Other types of clones will often result in constructs that will not answer the

---

[18] J. M. Smith, *in* "*Escherichia coli* and *Salmonella typhimurium*: Cellular and Molecular Biology" (F. C. Neidhardt, ed.), Chap. 146. American Society for Microbiology, Washington, DC, 1996.

selection. For example, those constructs that contain an internal fragment of an operon will generate a fusion under the appropriate regulation, but will disrupt the expression of the wild-type gene. If the product of the wild-type gene is required for the infection process, then this construct will be selected against in the animal. In other cases, there will not be a properly placed promoter to drive the expression of *purA*. For example, the fusion can be in the wrong orientation with respect to the gene. This fusion will never be expressed and will, therefore, be selected against in the animal.

The Red Shift

The transcriptional activity of any given fusion can be determined by assaying the ability of the fusion strain to utilize lactose as a carbon source. We routinely use Lactose MacConkey agar. In fusions constructed with pIVET1, the Lac activity measured on a MacConkey plate correlates very well with the PurA activity. In other words, fusion strains that are Lac^+ (red) are Pur^+, fusions that are $Lac^{+/-}$ (pink) are semiauxotrophic, and fusions that are Lac^- (white) are Pur^-.

When the initial preselected pool of *purA-lac* fusion strains is plated on Lactose MacConkey agar, the majority of fusions (50–60%) should be Lac^-. Most of these Lac^- colonies represent cases in which the fusion is in the wrong orientation with respect to the promoter. PurA activity is required for all aspects of *S. typhimurium* infection of a mouse. Therefore, we can carry out selections for genes that are expressed in virtually any tissue. For example, the pool of fusion strains can be intraperitoneally (ip) injected (10^6 organisms total) into a BALB/c mouse. Note that the inoculum size is sufficient to represent all of the fusions in the pool. Three to 5 days after infection, we remove the spleen, one of the major sites of systemic infection for *S. typhimurium*. Bacteria recovered from the splenic extract are then grown overnight, and the process is repeated. Bacteria recovered from the spleen of the second mouse are plated on Lactose MacConkey agar. In the mouse-selected pool, >85–90% of the colonies should be Lac^+ and approximately 5% are Lac^-. This dramatic shift toward Lac^+ cells (termed "the red shift") indicates that there was the expected selection for Pur^+ in the mouse. The majority of these strains contain fusions to genes that are constitutively expressed at high levels. However, because we were interested in genes that are specifically induced *in vivo,* we choose the Lac^- strains for further analysis. Fusions in these cells were transcriptionally active enough in the mouse to overcome the parental PurA deficiency. However, *in vitro,* they are transcriptionally inactive. Thus, the promoters driving these fusions are induced in mouse tissues.

Observation of this so-called red shift is the only indication that the selection took place. This is important for any IVET selection and exemplifies the importance of the *lac* operon in the system. The various parameters that affect any given selection, e.g., the length of incubation time in the host or the number of times the pool is taken through the selection, should be determined empirically with the success of the selection monitored by the red shift.

It should be noted that the level of Lac activity in fusions constructed with vectors other that pIVET1 may not coincide with the activity of the gene that is the basis of the selection; i.e., the profile of Lac$^+$ to Lac$^-$ in the preselected pool will vary depending on the IVET vector. In this regard, the red shift observed after selection with other vectors may be more or less than that observed with pIVET1. However, genes that presumably show the greatest *in vivo* induction are still those that have the least transcriptional activity *in vitro*. In those cases where the host-selected pool still has a significant number of white colonies on MacConkey agar or in organisms that will not grow on MacConkey agar, it may be necessary to screen for Lac activity using a different assay medium, such as one that contains the chromogenic substrate X-Gal. In this case, one observes a "blue shift" after passage through the host, where the fusions that are the lightest blue among the blue-shifted survivors are the fusions of interest. The *lac* system has many advantages, including a variety of substrates for use in assays and selections.[19] The main advantage is the ability to select Lac$^+$ on plates to genetically analyze the regulation of *in vivo*-induced genes. However, this may not be feasible in all organisms. Indeed, in some organisms it may be necessary to use a reporter system other than *lac*. However, the importance of a convenient method to monitor transcriptional activity cannot be overemphasized.

Cloning the Fusions

In order to identify *in vivo*-induced genes, we clone the fusions and determine the DNA sequence of the region of chromosome adjacent to the fusion joint. There are several ways to recover the integrated plasmids. In *S. typhimurium,* we clone the fusions, in a single step, using the generalized transducing bacteriophage, P22.[20] Briefly, a bacteriophage P22 lysate is made on the fusion strain of interest and used to transduce a recipient strain that contains the replication protein, Pi, which is required for autonomous replication of pIVET1. After introduction of the chromosomal fragment containing the integrated fusion construct, the plasmid circularizes

[19] J. M. Slauch and T. J. Silhavy, *Methods Enzymol.* **204,** 213 (1991).
[20] M. J. Mahan, J. M. Slauch, and J. J. Mekalanos, *J. Bacteriol.* **175,** 7086 (1993).

by homologous recombination at the region of duplication. This results in a plasmid clone of the fusion. Several other methods can be used to recover plasmids. Because the integrated plasmids are always undergoing homologous recombination, some fraction is present as free plasmid. These extrachromosomal plasmids can be recovered using standard plasmid isolation techniques (see later). These free plasmids can also be recovered by triparental mating into *E. coli*.[21] We have also recovered plasmid by isolating chromosomal DNA, shearing the DNA by simple vortexing, and transforming Rec[+] Pi[+] restriction *E. coli*. This is essentially equivalent to a transduction. The recircularization of the plasmid from the chromosomal fragment is efficient enough that plasmids are recovered at relatively high frequency.

Sibling plasmids recovered from any given selection are identified most easily by restriction analysis. For unique plasmids, primers homologous to the 5' end of the *purA* sequence are used to sequence 400–600 bp of chromosomal DNA adjacent to the fusion joint. These sequences are then used to search the DNA database to identify the gene.

Results of IVET Selections

We have characterized approximately 75 *S. typhimurium in vivo*-induced fusions from a variety of selections. Here we simply categorize these fusions as examples of IVET outcomes. Approximately 45% of the fusions are to genes that have known functions. These include genes whose products are directly involved in intermediary metabolism as well as genes involved in the synthesis of the outer surface of the bacteria, protein synthesis, and other metabolic functions. This is an important class of genes in that their identification provides information about the internal host environment in which *S. typhimurium* must adapt and grow. We must ultimately understand the metabolism of the pathogen in order to gain a complete view of pathogenesis. Thirty percent of the fusions represent genes that have unknown functions. However, these genes have orthologs in *E. coli* and are not likely to be involved directly in virulence, although this may be too simple of an assumption.

The IVET selection has also identified a number of genes whose products are known or suspected to be involved in *S. typhimurium* pathogenesis. At least 10% of the recovered fusions are in genes that have been implicated directly in virulence. Indeed, IVET selections performed in our laboratory or by Heithoff and colleagues[22] have identified members of all of the

---

[21] P. B. Rainey, D. M. Heithoff, and M. J. Mahan, *Mol. Gen. Genet.* **256**, 84 (1997).
[22] D. M. Heithoff, C. P. Conner, P. C. Hanna, S. M. Julio, U. Hentschel, and M. J. Mahan, *Proc. Natl. Acad. Sci. U.S.A.* **94**, 934 (1997).

major virulence regulons known in *S. typhimurium* except the SPI1 invasion system. We believe this is due to the transient expression of this system in the small intestine. Finally, 24% of the fusions represent genes that have unknown function and do not have orthologs in *E. coli*. We have termed these genes *Salmonella* specific. These genes are potentially important for virulence, but this has not been tested explicitly.

We anticipated that IVET would identify two broad classes of genes. The first class includes those whose products are required for adaptation to the host environment, whereas the second class are those whose products are specifically involved in the interaction with the host and host immune system (true virulence factors). Our results to date are consistent with this premise. However, it is often difficult to distinguish between metabolism and virulence. One cannot simply conclude that a previously unidentified gene must be a true virulence factor (see later).

In Vivo Assays

Integration of the IVET plasmid results in a tandem duplication in the chromosome (Fig. 1). This introduces a caveat to the IVET selection that must be addressed experimentally. Subsequent homologous recombination between sister chromosomes, which arise from DNA replication, can result in amplification of the fusion construct (Fig. 2). Note that this event can repeat itself, resulting in increased copy number. This event results in potential false positives that can be selected in the animal as follows. Amplification will increase the copy number of fusion. Eventually, the copy number is high enough to result in a PurA⁺ phenotype, allowing survival in the

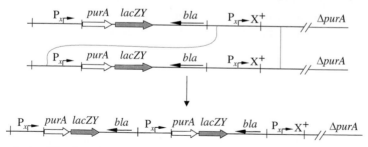

FIG. 2. Amplification of pIVET1 fusions by homologous recombination. Vertical lines demarcate the region of homology that constitutes the tandem duplication. Dotted lines indicate homologous recombination. Only one product of recombination events is shown. The reciprocal event generates a chromosome that has lost the fusion construct. Note that the product shown now has a larger region of homology and further recombination events can occur to amplify the fusion construct.

animal. Note that this is a simple multiplication of the basal level of activity from the fusion and does not require a change in expression from the promoter driving the fusion. However, this event is somewhat unstable and recombination can resolve the event back to one copy of the plasmid. Thus, a given fusion can be PurA$^+$ in the animal by amplification and resolve to give Lac$^-$ on a MacConkey plate. This type of fusion would answer our selection and screen and would represent a false positive. We estimate that the upper limit on amplification is approximately 10 copies of the fusion; fusions that have very low basal levels of activity cannot amplify enough to confer a Pur$^+$ phenotype. Thus, survival by amplification is limited to the class of fusions with a certain level of expression.

Given this caveat, one must prove that a selected fusion is truly induced in the animal. One method is to directly measure the transcriptional activity of the isolated fusions *in vivo*. This assay is performed after moving the fusion construct to a Pur$^+$ background. This essentially removes the selection for amplified fusions in the animal, and hence the β-galactosidase produced from the fusion is a true indication of transcriptional activity. Comparison of *in vivo* β-galactosidase activity with the activity of the same strain grown *in vitro* proves that the fusions are induced in the animal. This technique has been described in detail.[23] Briefly, BALB/c mice were injected ip with approximately 10^5 cells of an individual *in vivo*-induced fusion strain. Six days after infection, the mice were sacrificed and their spleens were removed and homogenized. The bacterial cells were largely separated from the splenic material by differential lysis of animal cells in deionized H$_2$O. The β-galactosidase activity in each sample was determined by a kinetic assay, using the fluorescent substrate fluorescein di-β-D-galacto-pyranoside (Molecular Probes. Inc.) and a spectrofluorometer. The activity is reported per colony-forming unit (CFU) in the bacterial suspension. The units of β-galactosidase were obtained by comparing the activity to a standard curve determined with purchased β-galactosidase (Sigma). As a control for this experiment, we use a random Lac$^+$ fusion strain from the preselected pool. The fusion in this strain is not induced significantly in animal tissues compared to growth in laboratory medium; the fusion is highly expressed in both conditions. This technique is limited to tissues where at least 10^4-10^5 bacteria can be recovered.

Proof of *in vivo* induction is accomplished most easily using a competition assay (Fig. 3). A bacterial strain containing an IVET fusion to an *in vivo*-induced gene of interest is competed against a strain containing a fusion to a promoter that is sufficiently active to allow for survival of the strain in any host tissue. These two strains have very different phenotypes

---

[23] J. M. Slauch, M. J. Mahan, and J. J. Mekalanos, *Biotechniques* **16**, 641 (1994).

Inoculate 1:1 mix of potential *in vivo* induced fusion
and constitutively expressed fusion

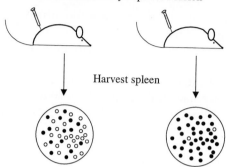

Harvest spleen

Fusion is *in vivo* induced   Fusion is NOT *in vivo* induced
Competes well   or survives via amplification
                                     Does not compete

Fɪɢ. 3. Screen to prove *in vivo* induction using IVET fusions. The potential *ivi* fusion strain is mixed 1:1 with a constitutive fusion strain that will act as a benchmark. An example of a screen is shown to prove induction during systemic stages of infection. If the *ivi* fusion is induced, it will compete with the constitutive strain. If it is not induced or can only survive via amplification, few Lac⁻ colonies will be recovered. The *in vivo*-induced fusion strain is phenotypically Lac⁻ (O) on Lactose MacConkey Agar. The constitutive fusion strain is Lac⁺ (●).

on Lactose MacConkey agar. Fusion to the potential *in vivo*-induced gene is phenotypically Lac⁻, whereas the constitutive fusion strain is Lac⁺. Mice are inoculated either orally or intraperitoneally with an equal mixture of the two strains. After 3–5 days, mice are sacrificed and the appropriate tissues are removed, homogenized, and plated on Lactose MacConkey agar. If the gene of interest is induced, then the fusion strain will survive and compete with the constitutive fusion strain. Thus, an approximately equal number of Lac⁻ and Lac⁺ colonies will be recovered. In contrast, if the gene of interest is not sufficiently induced in the animal, it will not survive well and the vast majority of recovered bacteria will be Lac⁺. If the potential *in vivo*-induced fusion is surviving via amplification, the recovered colonies will also be mostly Lac⁺. Although the amplified fusions are relatively unstable (the hallmark of the event), the vast majority (>99%) of the cells containing an amplification give phenotypically Lac⁺ colonies when they are initially recovered from the animal. In striking contrast, a truly *in vivo*-induced fusion survives well in the animal, but is phenotypically Lac⁻ when recovered from the animal. Given that many fusion strains can survive by amplification, simply showing that significant CFU are recovered from an animal after infection with the fusion strain is not sufficient proof of induction in the host.

We have taken this competition assay one step further to identify genes that are induced in specific tissues in the animal. For example, we have isolated a number of *in vivo*-induced fusions from the small intestine after oral inoculation. Thus, these are fusions to genes that are induced in the small intestine. We reasoned that a subset of these genes would include those whose products are specifically required for some aspect of the early infection process. This subset of genes would be expressed during the early stages but not during the systemic stages of the disease. Accordingly, we screened for fusions that were induced only in the small intestine and not in the spleen by performing competition assays. Mice were inoculated either orally or ip with an equal mixture of the two strains. After 3–5 days, the mice were sacrificed and the small intestine was removed from orally infected animals, while the spleen was removed from ip-infected animals. In the case of a gene that is specifically induced in the small intestine, fusion would be induced in the oral infection, the strain would be able to compete with the constitutive fusion strain, and both fusion strains would be recovered from the small intestine. In contrast, this fusion would not be induced during the systemic stages of the disease, would not survive, and would not be significantly recovered from the spleen after ip infection. A strain containing a fusion to the gene that we now call *gipA* (growth in Peyer's patches) clearly answered this selection (Table II).

The *sitABCD* operon encodes a putative iron transport system in

TABLE II

Competition between *In vivo*-Induced IVET Fusions and a Constitutive
IVET Fusion

In vivo-induced gene	Route of inoculation[a]	Organ	Median competitive index[b]
gipA	Oral	Small intesting	3.0[c]
		Spleen/liver	0.08[c]
sitABCD	Oral	Small intestine	0.27[c]
		Spleen/liver	1.3

[a] Groups of greater than five mice each were inoculated with a 1 : 1 mixture of the *in vivo*-induced fusion strain and the constitutive fusion strain. The dose was approximately 10^7 oral and 10^2 ip. Approximately 5 days postinoculation, the appropriate organs were removed, homogenized, and plated on Lactose MacConkey agar to distinguish the *in vivo*-induced fusion strain (Lac⁻) and the constitutive fusion strain (Lac⁺).

[b] Competitive index = output (*in vivo*-induced fusion strain/constitutive fusion strain)/input (*in vivo*-induced fusion strain/constitutive fusion strain).

[c] Results are significantly different from inoculum. $P < 0.05$ by students' t test.

SPI1.[24,25] An IVET fusion to this operon has an expression profile opposite to that of *gipA*. Because iron is apparently available in the small intestine, the Fur-regulated *sitABCD* fusion is only induced after invasion of the intestinal epithelium.[24] Thus, the fusion is outcompeted after oral infection, but competes well with the constitutive fusion in an ip infection (Table II). The virulence defects conferred by mutations in either *sitABCD* or *gipA* are perfectly consistent with these expression profiles. Strains containing mutations in *gipA* are specifically defective in growth or survival in Peyer's patches (data not shown). The mutants can initially colonize the small intestine and invade. They are also phenotypically wild type with respect to their ability to grow in systemic tissues. Strains containing mutations in *sitABCD* have the opposite phenotype in that they are primarily defective in systemic stages of growth.[24]

Subsequent Characterization of *ivi* Operons

An important question is whether these *in vivo*-induced genes have any role in pathogenesis. As described, fusions were constructed in diploid, to maintain the function of the wild-type gene. To test their role in pathogenesis, one must disrupt the gene and test the mutant in virulence assays. Mutations in the majority of *ivi* genes that we have tested confer a virulence defect, although sometimes this defect is subtle. We believe that this is one of the strengths of IVET; the system identifies virulence factors that would not be uncovered using techniques that might require a more dramatic virulence defect, such as signature tagged mutagenesis.[26]

Once potential virulence genes have been identified, it is important to determine the factors that regulate their expression. The system is designed to facilitate this analysis and one can use standard genetic techniques described elsewhere.[19] Fusions that are of the most interest are phenotypically Lac⁻ on laboratory medium. Therefore, to isolate mutations that result in constitutive expression of these operons, one can simply select Lac⁺.[19] This type of analysis can yield both *cis-* and *trans*-acting regulatory elements. This procedure is slightly complicated by the fact that the fusions are diploid. Therefore, selection for increased expression of the *lac* operon can result in amplification of the fusion construct that starts with recombination between the duplicated chromosomal fragment. These amplification events can be prevented *in vitro* by introducing a *recA* mutation. However, subsequent genetic analysis has to take the duplication into consideration.

[24] A. Janakiraman and J. M. Slauch, *Mol. Microbiol.* **35,** 1146 (2000).
[25] D. Zhou, W. D. Hardt, and J. E. Galan, *Infect. Immun.* **67,** 1974 (1999).
[26] M. Hensel, J. E. Shea, C. Gleeson, M. D. Jones, E. Dalton, and D. W. Holden. *Science* **269,** 400 (1995).

Variations on a Theme

The pIVET1 system is based on the *purA* gene. Although *purA* auxotrophy is known to attenuate a number of pathogenic organisms,[14,27–30] this may not always be the case. In addition, a *purA* mutation may be difficult to obtain in some organisms. Any number of biosynthetic genes can, in theory, be used in IVET selection. We have also developed plasmids based on drug resistance. For example, we have constructed a synthetic operon in which chromosomal promoters drive the expression of the CAT gene, encoding chloramphenicol acetyltransferase and a promoterless *lac* operon (pIVET8).[6] Mice are treated with chloramphenicol (Cm) and challenged with the pool of fusion strains. Successful selection is monitored, as always, by plating the selected pool of fusion strains on Lactose MacConkey agar and noting the red shift. In our experiments with a BALB/c mouse animal model, we have noted that there can be tremendous variability with this type of selection. Therefore, we have treated mice with a wide range of Cm and empirically determined the selection efficiency by monitoring the degree of red shift. This chloramphenicol selection should be of general use, particularly in systems where it is difficult to obtain auxotrophic mutations. Both the *purA* system[24] and the CAT selection[4] can be used in tissue culture systems.

Recombinase-Based *in Vivo* Expression Technology

An alternative IVET strategy to the selection-based methods described earlier is the recombinase-based *in vivo* expression technology (RIVET), which functions as a screen for *ivi* genes. The basic principle of RIVET relies on the use of a transcriptional reporter encoding a site-specific DNA recombinase, which, when expressed, catalyzes an irreversible recombination event whose heritable product can subsequently be screened or selected for *ex post facto*. The absence of the requirement for selective forces during infection facilitates use of RIVET to identify different classes of *ivi* genes. For instance, RIVET can identify *ivi* genes that are expressed transiently or constitutively in host tissues, and at low or high transcription levels. Moreover, because the screen is based on an inducible and irreversible recombination event (as described later), the methodology can be applied to temporal and spatial determinations of gene expression in host tissues.

[27] V. S. Baselski, S. Upchurch, and C. D. Parker, *Infect. Immun.* **22,** 181 (1978).
[28] G. Ivanovics, E. Marjai, and A. Dobozy, *J. Bacteriol.* **85,** 147 (1968).
[29] H. B. Levine and R. L. Maurer, *J. Immunol.* **81,** 147 (1958).
[30] S. C. Straley and R. L. Harmon, *Infect. Immun.* **45,** 649 (1984).

Construction of Recombinase Fusions

To identify *ivi* genes of the gram-negative intestinal pathogen *Vibrio cholerae*, we constructed random transcriptional fusions to a recombinase that can excise a tetracycline resistance (Tc^R) gene from the genome when expressed. The recombinase fusion vector pIVET5 is virtually identical to pIVET1 (see earlier), except that *purA* has been replaced with a promoterless copy of *tnpR*, which encodes the resolvase enzyme of the bacterial transposon Tnγδ.[31] Like the *purA* gene in pIVET1, the *tnpR* gene in pIVET5 forms an operon with the downstream *E. coli*-derived *lacZY* genes. Size-fractionated 1- to 3.5-kb fragments from *Sau*3AI partially digested *V. cholerae* genomic DNA, isolated from the El Tor biotype strain C6709-1, are ligated into the *Bgl*II site in pIVET5. The recombinant pIVET5 plasmid pool is first transformed into the conjugal donor strain *E. coli* SM10λpir and is then integrated by insertion–duplication into the genome of the Sm^R, Tc^R, and Lac$^-$ *V. cholerae* strain, AC-V66, by bacterial mating as described previously. AC-V66 was constructed by allelic replacement of the endogenous *lacZ* gene in C6709-1 with an artificial substrate cassette for resolvase, called *res1-tet-res1*.[4] When expressed, resolvase specifically binds to the directly repeated *res1* sequences and mediates recombination and excision of *tet-res1*, leaving behind one complete *res1* sequence in the genome (Fig. 4). This reaction is analogous to resolution of the cointegrate intermediate formed by Tnδγ during its replicative mode of transposition.[31] Because the excised *tet-res1* element lacks an origin of replication, it is diluted out in subsequent generations, giving rise to Tc^s daughter cells.

Identification of *ivi* Genes

The transcriptional activity of any given fusion can be assayed qualitatively either by colony color on β-galactosidase indicator plates or by growth phenotype in media containing Tc. The initial pool of *tnpR-lacZY* fusion strains is plated on LB agar plus Sm, Ap, and X-Gal; we have shown that ~5% of the resulting colonies are dark blue, 35% are blue, 59% are light blue, and 1% are white in *V. cholerae*. The light blue strains express background levels of β-galactosidase, whereas the white strains correspond to mutational loss of *lacZ* expression. We found that the light blue and Tc^R and, conversely, the blue to dark blue and Tc^s phenotypes correlate with each other, although not absolutely. Thus, to identify *V. cholerae ivi* genes, we use both phenotypes to collect strains harboring transcriptionally silent

[31] R. R. Reed, *Cell* **25,** 713 (1981).

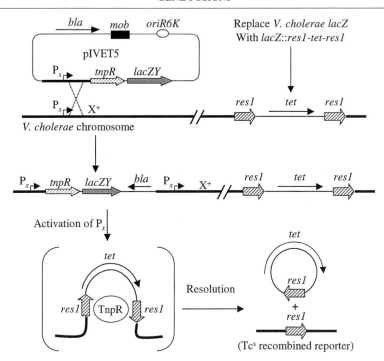

Fig. 4. Construction of gene fusion strains. The pIVET5 plasmid is a derivative of the broad host range mobilizable suicide vector, pGP704 [V. L. Miller and J. J. Mekalanos, *J. Bacteriol.* **170,** 2575 (1988)], whose replication is dependent on the Pi protein supplied *in trans.* Random, partial *Sau*3AI-restricted *V. cholerae* chromosomal DNA fragments (shown by the thick line) were cloned immediately 5' of the promoterless *tnpR,* and the recombinant plasmids were conjugated into a *lacZ::res1-tet-res1* strain of *V. cholerae* that lacks the Pi replication protein [AC-V66 (A. Camilli and J. J. Mekalanos, *Mol. Microbiol.* **18,** 671 (1995)]. Selection for ampicillin resistance demands integration of the plasmid into the bacterial genome, generating a duplication of *V. cholerae* sequence in which the native promoter drives expression of *tnpR* and the cloned promoter drives expression of the putative *ivi* gene, X. Expression of *tnpR* results in resolution of the cointegrate, i.e., site-specific recombination at the directly repeated *res1* sequences resulting in excision of *res1-tet.* Bacterial cells that resolve are Tc^s and are detected by replica plating.

gene fusions (Fig. 5). Light blue colonies are collected in pools of 200 using sterile toothpicks to transfer colony material onto LB agar plus Tc. Bacteria are grown for 2 hr at 37° and are then harvested by washing the plate with LB broth plus glycerol for archival storage at −80°. A small aliquot of each pool is inoculated into LB broth and grown overnight at 30°. Next, ~10^5 CFU are inoculated intragastrically into 5-day-old CD-1 mice as described.[4] After 24 hr, *V. cholerae* are recovered from the mice by plating homogenates of the small intestine on LB agar plus Sm. After growth at 37° overnight,

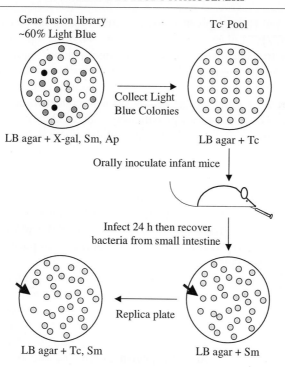

Fig. 5. Screening for genes that are specifically induced in the host. *V. cholerae* chromosomal fusion strains were plated on LB agar supplemented with 5-bromo-4-chloro-3-indolyl-β-D-galactoside (X-Gal), and colonies were screened for those that had low expression of the fusion [light blue]. The light blue colonies were patched onto medium supplemented with tetracycline and grown for 2 hr to enrich the unresolved fusion strains. Each pool was inoculated intragastrically into 5-day-old CD-1 mice (10^5 cells total). After 24 hr, bacterial cells were recovered from the small bowel and plated on medium supplemented with streptomycin. Strains that had resolved during the infection were screened for by replica plating to medium supplemented with tetracycline.

colonies are then replica plated onto LB agar plus Tc in order to identify Tc^s strains containing putative *ivi::tnpR* gene fusions (Fig. 5). We found that an average of 0.5% (range 0–3% per pool) of *V. cholerae* input strains resolved during the course of infection.

Cloning and Retesting the Fusions

In order to both eliminate siblings and confirm that each fusion is to a *bona fide ivi* gene, the integrative plasmid is recovered for insert size analysis and for reintegration back into the unresolved AC-V66 background. Despite the inability to replicate extrachromosomally in *V. cholerae*, each integrated pIVET5 plasmid is readily recoverable (as a plasmid) by virtue of its ability to excise from the chromosome at low frequencies (reverse of the integration reaction). Plasmid DNA is purified from 2 ml of culture using Qiagen Tip20 columns (Qiagen Inc., Valencia, CA), with the modification of adding 1 μg of carrier tRNA prior to concentrating the DNA by isopropanol precipitation. Approximately half of the recovered plasmid DNA is electroporated into DH5α*λpir* with selection for Ap^R colonies. Plasmid DNA is recovered, digested with *Bam*HI, and separated on a large 0.7% agarose gel to determine the *V. cholerae* DNA insert size. Identically sized inserts recovered from fusion strains from the same pool are considered siblings. Nonsibling plasmids are transformed into SM10*λpir* as an intermediate step and are then integrated back into strain AC-V66 by bacterial mating. We have found that recovery of the pIVET5 derivates from *V. cholerae* directly into SM10*λpir* often results in plasmids harboring large spontaneous deletions.

Each reconstructed fusion strain is retested for its Lac phenotype during *in vitro* growth as well as its ability to lose Tc^R via resolution both *in vitro* and *in vivo*. To measure resolution of individual strains, each strain is grown overnight in LB broth plus Sm, Ap, and Tc and is inoculated into LB broth or infant mice. After 24 hr of *in vitro* or *in vivo* growth, bacteria are recovered on LB agar plus Sm, grown overnight, and replica plated to determine the percentage of colony-forming units that are Tc^s (% Tc^s CFU). We found that ~20% of reconstructed strains were *bona fide ivi* fusion strains, i.e., they formed light blue colonies on agar medium with X-Gal and they resolved to Tc^s at a higher rate during infection than during growth *in vitro*. Resolution results obtained for five strains are shown in Table III.[32] Because the extent of resolution for any particular strain varies slightly between experiments, multiple experiments need to be done to derive statistically robust values.

Characterization of *ivi* Genes

To identify *ivi* genes, a primer homologous to the 5' end of *tnpR* is used to sequence ~400 bp of chromosomal DNA adjacent to the fusion joint. These sequences are then used to search available databases for homologous sequences. In the vast majority of *ivi* fusion strains, we found that *tnpR* is fused within an open reading frame (ORF) that is in the same transcriptional orientation as *tnpR*. However, in a few cases, the fusion is within an ORF that is in the opposite transcriptional orientation. Presum-

---

[32] A. Camilli and J. J. Mekalanos, *Mol. Microbiol.* **18,** 671 (1995).

TABLE III
RECOMBINATION LEVELS *in vitro* AND *in vivo*

	% Tc^s CFU^a after growth		
	---	---	---
Strain	*In vitro*	*In vivo*	N $(P<)^b$
AC-V231	0	87	4 (0.07)
AC-V232	0	69	4 (0.07)
AC-V228	0	40	5 (0.04)
AC-V235	6	36	5 (0.04)
AC-V237	2	14	8 (0.05)

^{a} Mean from four to eight independent experiments.
^{b} N equals the number of independent experiments done with each experiment consisting of two animals, and P equals the probability that each strain recombined to a greater extent (% Tc^s CFU) in infant mice than when grown *in vitro* using the Wilcoxon paired sample test.

ably, these latter *ivi* genes represent antisense transcripts. We found both known and unknown genes in *V. cholerae,* and almost half of the unknown genes encode putative polypeptides with no homologies or homology to proteins of unknown function.

It is likely that many *ivi* genes perform useful functions for *V. cholerae* during survival and multiplication in the host, and it is further likely that a subset of these may play essential roles in these processes in one or more animal models of disease. To test for an essential role of *V. cholerae ivi* genes during infection, the partial nucleotide sequence of each gene is used to design a forward PCR primer to be used in conjunction with the *tnpR* sequencing primer to amplify a fragment internal to the coding sequence of each gene. Each internal fragment is ligated into the mobilizable suicide plasmid pGP704[15] and is integrated into the corresponding *ivi* gene on the *V. cholerae* genome to generate an insertional mutation. These insertion mutants can then be tested for virulence in the infant mouse model of cholera by competition assays using the parental strain. For example, an insertion mutation in the AC-V238 *ivi* gene attenuates virulence in this animal model. Lack of attenuation of virulence in these experiments should not be taken as proof that the *ivi* gene plays no role during infection. For example, it is likely that redundant functions exist in bacteria for some *ivi* genes. Moreover, it is likely that some *ivi* genes play subtle roles during infection and/or play roles that may not be detectable in commonly used animal models.

Temporal and Spatial Patterns of *ivi* Gene Expression

Once *ivi* genes have been identified, RIVET can be used as a tool to determine where and when they are induced transcriptionally during the infectious process. For example, we have used a *V. cholerae* gene-*tnpR* fusion strain to show specific gene expression in one host compartment but not in another,[4] demonstrating the utility of this method for spatial studies of pathogen gene expression in a host animal. This was accomplished simply by infecting both host compartments with the reporter strain, recovering bacteria from each compartment after 24 hr, and determining the percentage Tc^s CFU in each population of recovered cells. To determine the temporal pattern of *ivi* induction during infection, a group of animals are infected with the reporter strain, bacteria are collected from individual animals at various times postinoculation, and the percentage Tc^s CFU is determined by replica plating (see example later).

Many *ivi* genes may have moderate-to-high basal levels of transcription during *in vitro* growth but are transcribed at still higher levels during infection. Most such genes in *V. cholerae* were excluded from detection in our initial screen because of the exceptional sensitivity of RIVET. Moreover, analysis of the temporal and spatial patterns of expression of such genes using RIVET is similarly prohibited. For example, constructing *tnpR* fusions to *ctxA*, which encodes the catalytic subunit of cholera toxin, or to *tcpA*, which encodes the pilin subunit of the toxin coregulated pilus (TCP), results in immediate resolution due to the moderate basal levels of transcriptional activity of these genes during growth of *V. cholerae in vitro*. To circumvent this problem, we have reduced the sensitivity of RIVET by incorporating three new alleles of *tnpR* that contain downmutations in the Shine–Dalgarno (SD) sequence and that result in reduced levels of translation for any given level of transcription.[33] This "tunable" RIVET can be used to identify and subsequently monitor expression of virtually any gene that is induced transcriptionally during infection.

For example, *tcpA* transcriptional fusions to all three mutant *tnpR* alleles were constructed in *V. cholerae* AC-V584 by insertion–duplication so as to regenerate a wild-type copy of *tcpA* in addition to the fusion. Strain AC-V584 is the El Tor biotype strain E7946 modified to contain the *lacZ::res1-tet-res1* cassette in its genome. One of the three alleles was found not to mediate resolution during the construction of, nor subsequent *in vitro* growth of, the resulting fusion strain (AC-V585). To determine whether AC-V585 could report expression of *tcpA* during infection, 10^5 CFU were intragastrically inoculated into an infant mouse and the extent of resolution of bacteria recovered from the small bowel after 20 hr was

[33] S. H. Lee, D. L. Hava, M. K. Waldor, and A. Camilli, *Cell* **99**, 625 (1999).

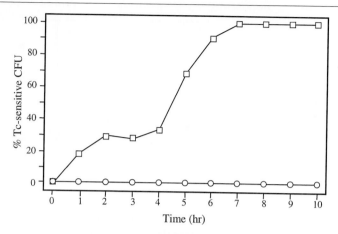

FIG. 6. The temporal pattern of transcriptional induction of the *V. cholerae tcpA* gene during infection of the suckling mouse small bowel. A *V. cholerae tcpA::tnpR* fusion strain containing the *res1-tet-res1* substrate cassette for resolvase was inoculated intragastrically into 5-day-old infant mice. At the times indicated on the *x* axis, bacteria were recovered from a small bowel and plated onto medium supplemented with streptomycin. Colonies were replica plated onto medium supplemented with tetracycline to determine the percentage of tetracycline-sensitive colony-forming units (% Tc^s CFU). An increase in the percentage Tc^s CFU is due to resolution and correlates with transcriptional induction of the gene fusion. Temporal patterns of resolution mediated by the *tcpA::tnpR* fusion are shown for a wild-type strain background (■) and for a *toxT* null background.

assayed. The recovered population of bacteria was 99% resolved, indicating that this particular *tnpR* allele is a good reporter for *tcpA*. This experiment also directly demonstrates that *tcpA* is an *ivi* gene.

To further dissect transcriptional induction of *tcpA* during infection, the temporal pattern of resolution of AC-V585 was determined over a 10-hr period. A group of eight infant mice were infected with 10^5 CFU each, and *V. cholerae* were recovered from one small bowel every hour and assayed for resolution. A biphasic pattern of resolution was observed, which consists of early and late induction periods, with an inflection point at approximately 3 hr postinoculation (Fig. 6). This temporal pattern of expression is very reproducible in independent experiments and is virtually identical to the pattern observed for another El Tor biotype strain containing the same *tnpR* fusion to *tcpA*.[33] In contrast, this pattern differs considerably from that obtained for another *V. cholerae ivi* gene, *vieB*. In the case of *vieB,* resolution is not observed until 3 hr postinoculation, and it then increases steadily, reaching a maximum level by 10 hr (Fig. 7).[34]

The temporal patterns of resolution observed for *tcpA* and *vieB* result

[34] S. H. Lee, M. J. Angelichio, J. J. Mekalanos, and A. Camilli, *J. Bacteriol.* **180,** 2298 (1998).

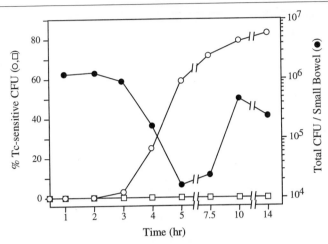

FIG. 7. The temporal pattern of transcriptional induction of the *V. cholerae vieB* gene and total viable bacterial counts during infection of the suckling mouse small bowel. Resolution was assayed as described in Fig. 6. The temporal pattern of resolution mediated by the *vieB::tnpR* fusion is shown for a wild-type strain background (O) and for a *tcpA* null background (□). Total *V. cholerae* colony-forming units present in a small bowel at each time indicated on the *x* axis were determined for the wild-type strain background by plating on medium supplemented with streptomycin (●).

not only from the kinetics of transcriptional induction of each gene fusion, but also from the population dynamics of *V. cholerae* in the small bowel. The latter need not be considered when monitoring pathogen gene induction in host compartments where the influx and/or efflux of the test population of bacteria is minimal. However, in the infant mouse model of cholera, there is a large efflux of bacteria out of the small bowel and into the large bowel at ~3–5 hr postinoculation (Fig. 7).[35] This change in resident population in the small bowel must be considered when interpreting data in Figs. 6 and 7. In particular, it was found that in the case of *vieB*, two populations of bacteria are present: a minority adherent population that remains in the small bowel and in which *vieB* is induced and a majority lumenal population that eventually passes out of the small bowel and in which *vieB* is uninduced. Thus, as the unresolved lumenal population exits the small bowel, a rapid rise in percentage Tc[s] CFU in the small bowel results (Fig. 7).[35] These results led us to hypothesize that *vieB* transcriptional induction is dependent on prior colonization of the mucosal layer. To test this hypothesis, the temporal pattern of induction of *vieB::tnpR* was determined in a *tcpA* null strain background, which is unable to colonize the small bowel, as TCP is

[35] M. J. Angelichio, J. Spector, M. K. Waldor, and A. Camilli, *Infect. Immun.* **67**, 3733 (1999).

absolutely essential for colonization. No induction of the fusion was observed (Fig. 7), supporting our hypothesis that colonization is required, presumably in order to receive the proper environmental signals, to induce *vieB*.

A second experiment that exemplifies the usefulness of RIVET to study the regulation of virulence gene expression is the determination of the temporal pattern of *tcpA* induction in a *toxT* null strain background. ToxT is a transcriptional regulator that plays an essential role in coordinately regulating virulence gene expression *in vitro,* including that for *tcpA*.[36] Because a *toxT* null strain is attenuated severely for virulence in the infant mouse model of cholera, it is likely that similar regulation occurs *in vivo.* In support of this, we found no *tcpA* transcriptional induction in the *toxT* background (Fig. 6).

Variations on a Theme

The pIVET5 system is based on the *tnpR* gene, encoding resolvase, and several alternative alleles of *tnpR* that exhibit reduced translation efficiencies. The second component is a substrate cassette for resolvase, *res1-tet-res1*. Resolvase requires only a supercoiled DNA cointegrate substrate and Mg^{2+} for activity and therefore may be applied in any genetically manipulable bacterium (or other microorganism) in which *tnpR* can be expressed stably. The particular mutant RBS alleles of *tnpR* being used in *V. cholerae* may not be useful, or as useful, for reducing the sensitivity of RIVET in other bacteria, and new alleles may need to be isolated. The resolvase substrate cassette is easily modifiable for use in other microorganisms simply by exchanging the *tet* gene for another selectable or screenable marker, which functions in the microorganism of study. Experience suggests that the substrate construct must be in single copy in the chromosome under conditions where it cannot excise by homologous recombination. In addition, Tn$\gamma\delta$, or very closely related transposons, must not be present in the bacterium of interest. Other modifications are also possible. For example, we have inserted the counterselectable marker sacB downstream of *tet* to allow for direct selection of resolved bacteria after recovery from infected animal tissues. The *sacB* gene encodes a sucrose-hydrolyzing enzyme from *Bacillus subtilis,* which, when expressed in a variety of bacterial species, including *V. cholerae,* produces toxic by-product levels during growth on media containing sucrose. The presence of a counterselectable marker within the excisable cassette simplifies the recovery of resolved

---

[36] V. J. DiRita, *Mol. Microbiol.* **6,** 451 (1992).

strains from animal tissues when performing large-scale screening for *ivi* genes.

Summary

IVET was designed to identify those bacterial genes that are induced when a pathogen infects its host. A subset of these induced genes encode virulence factors, products specifically required for the infection process. The paradigm IVET system is based on complementation of an attenuating auxotrophic mutation by gene fusion and is designed to be of use in a wide variety of pathogenic organisms. In *S. typhimurium,* we have used this system successfully to identify a number of genes that are induced in a BALB/c mouse and that, when mutated, confer a virulence defect.

The RIVET system is based on recombinase gene fusions, which, on induction during infection, mediate a site-specific recombination, the product of which can be screened for after recovery of bacteria from host tissues. In *V. cholerae,* we have used this system successfully to identify genes that are induced transcriptionally during infection of the gastrointestinal tract of infant mice. RIVET is also uniquely designed for postidentification analysis of *in vivo*-induced genes: (1) it has been used to analyze the temporal and spatial patterns of virulence gene induction during infection and (2) it has been used to dissect the regulatory requirements of *in vivo* induction with respect to both bacterial regulatory factors and host-inducing environments.

The IVET system has several applications in the area of vaccine and antimicrobial drug development. This technique was designed for the identification of virulence factors and thus may lead to the discovery of new antigens useful as vaccine components. The IVET system facilitates the isolation of mutations in genes involved in virulence and, therefore, should aid in the construction of live-attenuated vaccines. In addition, the identification of promoters that are expressed optimally in animal tissues provides a means of establishing *in vivo*-regulated expression of heterologous antigens in live vaccines, an area that has been problematic previously. Finally, we expect that our methodology will uncover many biosynthetic, catabolic, and regulatory genes that are required for growth of microbes in animal tissues. The elucidation of these gene products should provide new targets for antimicrobial drug development.

Acknowledgments

IVET was developed in collaboration with Drs. Michael J. Mahan and John J. Mekalanos. This work was supported by National Institutes of Health Research Grants AI37530 (J.M.S.) and AI40262 (A.C.).

[6] Identification of Exported Bacterial Proteins via Gene Fusions to *Yersinia pseudotuberculosis* Invasin

By MICAH J. WORLEY and FRED HEFFRON

Introduction

Virtually all virulence factors localize to the cell envelope or are secreted to the extracellular environment. This group includes proteins that mediate critical interactions among eukaryotic and pathogen cell surfaces, toxins, proteases, components of type I and type III exporters, and almost all components of all fimbrial operons.[1]

Most exported bacterial proteins travel through the general secretory pathway (GSP). These proteins interact with components of the GSP via a signal sequence at their amino terminus. Signal sequences share little primary sequence homology, but conservation of their overall biochemical characteristics permits functional complementation. GSP-exported proteins that contain a signal sequence followed by a consensus cleavage site are processed proteolytically at the inner membrane, and mature proteins are released into the periplasmic space. These proteins can remain in the periplasm, insert into the outer membrane, or be secreted across it. However, proteins that contain noncleaveable signal sequences permanently associate with the inner membrane.[2,3]

The primary methods for the identification of exported proteins, fusions to alkaline phosphatase (*pho*A),[4–8] and β-lactamase (*bla*)[9–11] do not distinguish between cleaveable and noncleaveable signal sequences. Thus, noncleaveable signal sequences and generally hydrophobic stretches of amino acids can be identified with these methods at a high frequency.[10,12–15] We

[1] B. B. Finlay and S. Falkow, *Microbiol. Mol. Biol. Rev.* **61,** 136 (1997).

[2] A. P. Pugsley, *Microbiol. Rev.* **57,** 50 (1993).

[3] P. N. Danese and T. J. Silhavy, *Annu. Rev. Genet.* **32,** 59 (1998).

[4] C. S. Hoffman and A. Wright, *Proc. Natl. Acad. Sci. U.S.A* **82,** 5107 (1985).

[5] C. Manoil and J. Beckwith, *Proc. Natl. Acad. Sci. U.S.A* **82,** 8129 (1985).

[6] R. K. Taylor, C. Manoil, and J. J. Mekalanos, *J. Bacteriol.* **171,** 1870 (1989).

[7] D. R. Blanco, M. Giladi, C. I. Champion, D. A. Haake, G. K. Chikami, J. N. Miller, and M. A. Lovett, *Mol. Microbiol.* **5,** 2405 (1991).

[8] M. Giladi, C. I. Champion, D. A. Haake, D. R. Blanco, J. F. Miller, J. N. Miller, and M. A. Lovett, *J. Bacteriol.* **175,** 4129 (1993).

[9] J. K. Broome-Smith and B. G. Spratt, *Gene* **49,** 341 (1986).

[10] H. Smith, S. Bron, J. Van Ee, and G. Venema, *J. Bacteriol.* **169,** 3321 (1987).

[11] J. K. Broome-Smith, M. Tadayyon, and Y. Zhang, *Mol. Microbiol.* **4,** 1637 (1990).

[12] M. Friedlander and G. Blobel, *Nature* **318,** 338 (1985).

have described a new gene fusion system that exclusively selects cleaveable signal sequences, facilitating the identification of outer membrane proteins and putative virulence factors.[16]

This system relies on *Yersinia pseudotuberculosis* outer membrane protein invasin. When invasin inserts into the outer membrane, its carboxyl terminus becomes surface exposed and can mediate bacterial entry into eukaryotic cells. Expression of invasin alone will confer an invasive phenotype on *Escherichia coli*. However, an amino-terminal truncated invasin (ΔInv) that lacks its native signal sequence cannot reach the outer membrane and promote invasion.[17] We have found that heterologous translational fusions composed of cleaveable signal sequences and ΔInv will preferentially restore the invasive phenotype.[16] These clones can be selected easily with the antibiotic gentamicin because internalization by eukaryotic cells protects bacteria from this antibiotic (Fig. 1). This system has several features that may make it particularly useful in the study of genetically intractable organisms.

Materials and Methods

The invasin complementation procedure can be divided into three steps: (1) a recombinant library in *E. coli* must be generated, (2) invasive clones from the library must be selected with gentamicin protection assays, and (3) fusion junctions between Δ*inv* and the cloned DNA must be determined. See Fig. 1 for an overview of the general strategy. Access to a tissue culture facility and basic molecular biology reagents are required for this method.

Construction of a Recombinant Library in *E. coli*

Vector Characteristics

The plasmid vector used in this procedure, pICOM II, contains a fragment of invasin truncated at a *Cla*I site at bp 917 of the full-length sequence (GenBank accession number M17448). pICOM II is a derivative of pWKS30,[18] a low copy number-cloning plasmid that has the same poly-

---

[13] C. Manoil and J. Beckwith, *Science* **233,** 1403 (1986).
[14] J. Calamia, and C. Manoil, *J. Mol. Biol.* **224,** 539 (1992).
[15] E. Lee and C. Manoil, *J. Biol. Chem.* **269,** 28822 (1994).
[16] M. J. Worley, I. Stojiljkovic, and F. Heffron, *Mol. Microbiol.* **29,** 1471 (1998).
[17] R. R. Isberg, D. L. Voorhis, and S. Falkow, *Cell* **50,** 769 (1987).
[18] R. F. Wang and S. R. Kushner, *Gene* **100,** 195 (1991).

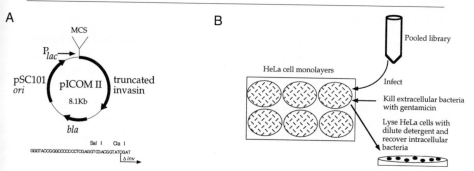

FIG. 1. (A) pICOM II (invasin complementation) design. A deletion of *Y. pseudotuberculosis inv* that lacks its amino terminus was cloned via *Cla*I and *Nru*I sites into the *Cla*I and *Eco*RV sites of pWKS30,[18] a low copy number plasmid with the pBluescript KS polylinker. This invasin derivative retains the sequences necessary to insert into the outer membrane and facilitate invasion. However, because it lacks a signal sequence, it cannot be exported past the inner membrane. Fusions can be expressed from a cloned gene's natural promoter or from the vector promoter. (B) Selection strategy. A chromosomal DNA library is generated in pICOM II, pooled, and used to infect HeLa cells. DNA fragments fused to invasin that allow it to be exported past the inner membrane and released into the periplasmic space allow the invasin derivative to insert into the outer membrane and facilitate invasion. These clones enter the HeLa cells and become protected from the antibiotic gentamicin.

linker as pBluescript KS. See Fig. 1 for additional information that may be helpful in constructing a library with this vector.

DNA Isolation

The appropriate procedure for the isolation of chromosomal DNA will of course depend on the organism being studied. The invasin procedure may be most useful in the study of genetically intractable organisms, many of which can be difficult to grow, let alone isolate DNA from. Thus for some organisms, this first step may be the most difficult. However, general methods can be found in Ausubel *et al.*[19] and Sambrook *et al.*[20] The plasmid vector can be purified by conventional plasmid DNA isolation procedures or with commercially available kits.

[19] F. M. Ausubel, R. Brent, R. E. Kingston, D. D. Moore, J. G. Seidman, J. A. Smith, and K. Struhl, "Current Protocols in Molecular Biology." Wiley, New York.
[20] J. Sambrook, E. F. Fritsch, and T. Maniatis, "Molecular Cloning: A Laboratory Manual." Cold Spring Harbor Laboratory Press, Cold Spring Harbor, NY, 1989.

DNA Fragmentation

Once isolated, the chromosomal DNA must be fragmented and ligated into the plasmid vector (pICOM II). For the most representative libraries, the chromosomal DNA should be fragmented with nebulization or sonication. After fragmentation with these methods, the ends of the fragments need to be repaired with the Klenow fragment of *E. coli* DNA polI. Then specific restriction sites need to be added and finally the fragments can be ligated to the dephosphorylated vector. A detailed protocol on nebulization of genomic DNA is available from the Institute for Genomic Research at http://www@tigr.org. For detailed protocols on sonication of genomic DNA, as well as general molecular biology techniques involved in library construction, consult Ausubel *et al.*[19] and Sambrook *et al.*[20]

The easiest method for generating a library, albeit not the most representative, is to partially digest the chromosomal DNA separately with the restriction enzymes *Taq*I and *Hpa*II, which are compatible with the *Cla*I site in the vector. The partially digested chromosomal DNA then needs to be size selected for fragments greater than 100 bp to ensure that positive clones will contain enough sequence to work and be informative after sequencing. Although there are many methods available for size fractionating DNA,[19,20] we have used QIAquick (Qiagen) polymerase chain reaction (PCR) purification columns for this purpose.

Transformation and Selection

We have used the *E. coli* cloning strain XL-2 blue as a recipient. However, any common *E. coli* cloning strain is probably suitable, as only a few clinical isolates of *E. coli* are capable of entering mammalian cells. Choice of an *E. coli* recipient should be guided by concern about the restriction of foreign DNA and stability of the DNA, as well as the number of transformants per microgram of DNA that can be obtained. Competent cells that are commercially available from Stratagene and Invitrogen may yield the largest libraries.

Transformants are recovered on Luria Bertani (LB) agar plates supplemented with 200 μg/ml ampicillin (amp) or carbenicillin. We estimate that genes encoding proteins with cleaveable signal sequences released from the inner membrane are about 1% of total, and translational fusions are obtained in one out of every six transformants. Therefore, the frequency at which invasive clones are expected is approximately 1 in 10^3. The size of the library can be estimated by plating serial dilutions. It is advisable to generate and recover transformants in multiple independent groups and to only analyze a few survivors from each group to minimize the chances of analyzing siblings.

Library Enrichments

After a library has been generated, it needs to be subjected to sequential invasion assays to select for clones that contain proteins that normally localize past the inner membrane.

Culture Methods

While the cell culture and invasion assay information provided herein should be sufficient, we have published more detailed protocols previously.[21] Bacteria expressing invasin will proficiently invade a variety of different mammalian cell types. We have used the common HeLa epithelial cell line (ATCC). HeLa cell culture is maintained in Dulbecco's modified Eagle's medium (DMEM; GIBCO-BRL), supplemented with 10% fetal calf serum (GIBCO-BRL) and 1 mM sodium pyruvate (GIBCO-BRL). HeLa cell culture is incubated at 37° with 5% CO_2 in a standard water-jacketed tissue culture incubator with 100% humidity. HeLa cells for library enrichments should be grown into nearly confluent monolayers in six-well tissue culture plates (Falcon).

Invasion Assays

The library should be removed from the selective plates with cotton swabs and resuspended in phosphate-buffered saline (PBS). Independent library groups may be enriched in parallel, but they should remain separate throughout the enrichments to minimize sibling isolation. Approximately 10^7 colony-forming units (cfu) should be used to infect HeLa cell monolayers. The concentration of a library group suspension can be determined with a spectrophotometer, assuming that an OD_{600} of 1.4 = 10^9 cfu/ml. HeLa cells should be washed once with 37° PBS and overlaid with 37° DMEM without serum prior to infection. Bacteria should be centrifuged onto the HeLa cell monolayers at 1000g for 10 min at ambient temperature.

The infection should be allowed to proceed for 2 hr at 37° with 5% CO_2. After this incubation, the medium is aspirated with a vacuum or a pipette, and the monolayers are washed three times with 37° PBS to remove extracellular bacteria. The monolayers are then incubated for another 2 hr with DMEM containing 75 μg/ml gentamicin (GIBCO-BRL), an antibiotic

---

[21] R. Tsolis and F. Heffron, *Methods Cell Biol.* **45,** 79 (1994).

that selectively kills extracellular bacteria. After 2 hr of additional incubation, the medium is aspirated, and the monolayers are washed three times with 37° PBS to remove the antibiotic. After this incubation, intracellular bacteria are released from the HeLa cells with 1% Triton X-100 (Sigma) in PBS, a detergent that selectively solubilizes eukaryotic membranes. The coculture should sit in Triton X-100 for 5 min. Then, the bottoms of the wells are scraped with a pipette tip and the suspension is pipetted up and down several times. This suspension should be vortexed briefly in a microtube and the released bacteria recovered on LB-amp-200 plates.

The next day, the survivors of one invasion assay are pooled and used to infect the HeLa cell culture again. The library should be enriched sequentially in this manner. We found no false positives after three rounds of enrichment; however, the diversity of the clones recovered may decrease with the number of enrichments. Plating of serial dilutions can be used to ascertain the number of bacteria recovered after each enrichment. The fraction of the inoculum recovered is expected to increase with each enrichment, indicating that invasive clones are being selected from the library (Fig. 2).

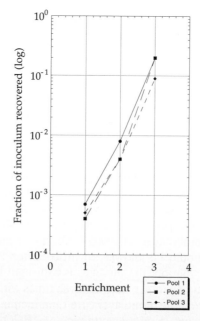

FIG. 2. Fraction of the inoculum recovered from each of three sequential invasion assays for three independent groups of a *S. typhimurium* library. An increase in the fraction of the inoculum recovered as seen is expected and indicates that the selection is working. This selection is so powerful that fewer rounds of selection are probably adequate.

When the library enrichments are complete, individual survivors should be streaked to isolation on LB plates containing ampicillin. Invasion assays should be performed on a few individual survivors from each library group to determine if the ΔInv export defect in these individual clones is being complemented. The procedure for the individual invasion assays is essentially the same as for the library enrichments described earlier, except that the inoculum should come from a liquid LB-amp overnight culture of an individual clone. All of the clones we have obtained after three rounds of enrichment were at least two log more invasive than the parent strain (Fig. 3).

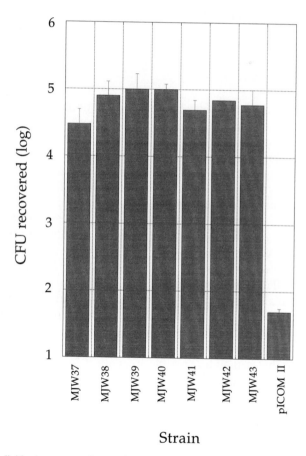

FIG. 3. Individual assays on clones selected arbitrarily from a *S. typhimurium* library after three sequential rounds of enrichment. All clones analyzed were at least two logs more invasive than the parent vector.

Sequence / Reading Frame Analysis

After generating a group of clones that are invasive in individual assays and that give unique restriction patterns, plasmid DNA needs to be isolated for sequence determination. The oligonucleotide 5'-AGGAGCCAGC-CAATCAAGAGAG-3' (recognizes bp 998 to 977 of *inv*) can be used to determine the sequences of inserts fused to Δ*inv*. The fusion junction can be determined by searching for the sequence 5'-CCGAATCG-3', which matches the end of Δ*inv*. If a gene fragment fused to Δ*inv* allows the CGA of the *Cla*I site to form a codon (XXX CGA), a hybrid protein will be produced. In addition to commercially available software, sequence analysis can be performed using a suite of programs available online at http://www.ncbi.nlm.nih.gov/.

Comments

The invasin system is applicable to the study of diverse organisms and may be especially useful in the study of genetically intractable bacteria. The invasin system will work well with most bacterial organisms because of the high level of conservation of components of the GSP and of protein motifs that promote export.[2] In addition, this system is not limited by transposon host range or the availability of a gene delivery system. Also, the vector promoter can facilitate the transcription of fusions that might not otherwise be expressed. Thus, the bacterial pathogens *Chlamydia trachomatis, Mycobacterium leprea,* and *Treponema pallidum* are all ideal candidates for study with this system. To date, the invasin system has not identified any false positives and thus provides strong functional evidence as to the subcellular location of a protein.

While originally intended for use with gram-negative bacteria, the invasin system appears to exclusively select secreted proteins when used to study gram-positive bacteria (M. Worley and F. Heffron, unpublished results). The competitive nature of the selection employed by this system is so powerful that it may be possible to adapt it for the identification of eukaryotic membrane proteins as well.

Acknowledgments

Work in Fred Heffron's laboratory is supported by National Institutes of Health Grant AI37201. M.J.W. is supported by a National Institutes of Health predoctoral fellowship.

Section III

Gene Fusions as Reporters of Gene Expression in Eukaryotic Cells

[7] Use of Imidazoleglycerolphosphate Dehydratase (His3) as a Biological Reporter in Yeast

By Joe Horecka *and* George F. Sprague, Jr.

Introduction

Hybrid genes that fuse transcriptional regulatory sequences upstream of DNA encoding a reporter protein are powerful tools for characterizing *cis*- and *trans*-acting factors that regulate gene expression. Some reporter proteins have a biological readout that can be scored as cell growth, which makes them especially useful tools for genetic studies. In the yeast *Saccharomyces cerevisiae,* imidazoleglycerolphosphate dehydratase (IGP dehydratase, the product of the *HIS3* gene) is such a reporter protein. IGP dehydratase functions in histidine biosynthesis[1] and, thus, expression of the reporter can determine whether a strain is a histidine prototroph or auxotroph. Moreover, because the enzyme is inhibited competitively by 3-amino-1,2,4-triazole (AT),[2,3] it is possible to obtain a semiquantitative measurement of the level of expression of the reporter simply by assessing growth on media containing different concentrations of AT. These two attributes have made IGP dehydratase an important biological reporter for yeast.

There are a number of applications for *HIS3* reporters in yeast. The most powerful applications enable the investigator to select mutants by demanding growth under conditions in which the reporter is not normally expressed. Indeed, *HIS3* reporters have been used with great success in mutant hunts aimed at identifying *trans*-acting factors that regulate transcription. For example, mutations that activated a *GAL1-HIS3* reporter in glucose-grown cells identified new genes involved in *GAL* gene repression.[4] Similarly, mutants selected as activators of a *FUS1-HIS3* reporter identified both new genes[5] and informative alleles of known genes[6,7] that function in

---

[1] G. R. Fink, *Science* **146,** 525 (1964).
[2] T. Klopotowski and A. Wiater, *Arch. Biochem. Biophys.* **112,** 562 (1965).
[3] K. Struhl and R. W. Davis, *Proc. Natl. Acad. Sci. U.S.A.* **74,** 5255 (1977).
[4] J. S. Flick and M. Johnston, *Mol. Cell. Biol.* **10,** 4757 (1990).
[5] B. J. Stevenson, B. Ferguson, C. De Virgilio, E. Bi, J. R. Pringle, G. Ammerer, and G. F. Sprague, Jr., *Genes Dev.* **9,** 2949 (1995).
[6] B. J. Stevenson, N. Rhodes, B. Errede, and G. F. Sprague, Jr., *Genes Dev.* **6,** 1293 (1992).
[7] B. Yashar, K. Irie, J. A. Printen, B. J. Stevenson, G. F. Sprague, Jr., K. Matsumoto, and B. Errede, *Mol. Cell. Biol.* **15,** 6545 (1995).

METHODS IN ENZYMOLOGY, VOL. 326

the mating pheromone response pathway. *HIS3* reporters can also be used as components in more complex genetic strategies. For example, two output branches of the pheromone response pathway are cell cycle arrest and *FUS1* transcription induction. Mutants specifically defective for cell cycle arrest were identified as strains that could grow in the presence of pheromone and simultaneously activate a *FUS1-HIS3* reporter, which ensured that the pathway up to the branch point remained intact[8] (see also reference 9). The use of *HIS3* reporters for growth selection has also been incorporated with great success into the yeast two-hybrid system and its variations for detecting protein–protein and protein–nucleic acid interactions. *HIS3* reporters can also be used in screens to identify mutants that decrease reporter expression and thereby render cells histidine auxotrophs (e.g., see refs. 10 and 11).

This article presents materials and methods for the construction and use of *HIS3* reporters in yeast. The work plan is straightforward: subclone transcriptional regulatory sequences from *your favorite gene* (*"YFG1"*) upstream of the *HIS3*-coding region in a reporter plasmid, stably integrate the *YFG1-HIS3* reporter at an innocuous locus in the yeast genome, and then assay reporter activity. The *YFG1-HIS3* reporter should be the strain's only source of *HIS3* expression, and we describe materials and methods to alter the native *HIS3* locus, if need be. We also point out potential pitfalls associated with *HIS3* reporters and how to avoid them or at least identify them when they appear.

Finally, our interests lie in the use of *HIS3* reporters to probe signal transduction pathways and *trans*-acting factors that regulate gene expression in yeast, and this bias likely shows through here. Nonetheless, much of the information presented in this article should be helpful to all applications of *HIS3* reporters. For use of *HIS3* reporters to detect one-, two-, and many-hybrid interactions in yeast, the reader should also consult chapters in this volume devoted specifically to those applications. Researchers who wish to use *YFG1-HIS3* reporters to study *cis*-acting mutations within *YFG1* regulatory sequences are referred to an example of this application published by Hagen *et al.*[11]

[8] J. Horecka and G. F. Sprague, Jr., *Genetics* **144,** 905 (1996).
[9] M. C. Edwards, N. Liegeois, J. Horecka, R. A. DePinho, G. F. Sprague, Jr., M. Tyers, and S. J. Elledge, *Genetics* **147,** 1063 (1997).
[10] L. Bruhn and G. F. Sprague, Jr., *Mol. Cell. Biol.* **14,** 2534 (1994).
[11] D. C. Hagen, L. Bruhn, C. A. Westby, and G. F. Sprague, Jr., *Mol. Cell. Biol.* **13,** 6866 (1993).

Materials

Plasmids

Plasmids are described that are useful for creating *YFG1-HIS3* reporters, detecting *HIS3* transcripts, and manipulating the *HIS3* locus. Plasmid maps are presented in Fig. 1.

pSL1470 carries a *FUS1-HIS3* reporter that can be targeted to integrate at the *HIS3* locus. The *FUS1* sequences can be excised with *Xba*I/*Sal*I digestion and replaced with upstream sequences from a gene of the researcher's choice. pSL1470 was created in numerous steps, but its overall structure is simple (unless otherwise noted, sequence coordinates listed in this article are relative to the "A" nucleotide of the corresponding gene's start codon). The *HIS3* 5′ flank corresponds to a 2.55-kb *Eco*RI/*Nsp*V genomic DNA segment upstream of the *HIS3* ORF (-2.6 kb to c. -50 bp). The *Nsp*V site was destroyed by ligation to an *Acc*I site present in the pUC19 polylinker and is followed by pUC19 sequences that introduce a unique *Xba*I site (5′-<u>TTCGA</u>CTCTAGA-3′; *Nsp*V remnant underlined,

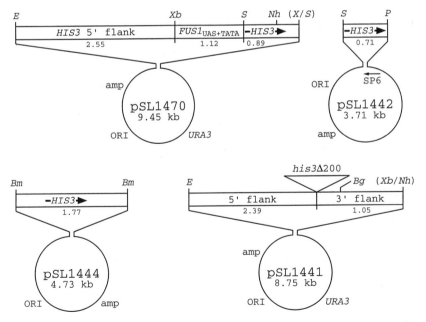

FIG. 1. Plasmids for making *YFG1-HIS3* reporters, detecting *HIS3* transcripts, and manipulating the *HIS3* locus. Details of plasmid structure and use are described in the text. Restriction fragment sizes (in kb) are listed below the inserts. *Bm, Bam*HI; *Bg, Bgl*II; *E, Eco*RI; *Nh, Nhe*I; *P, Pst*I; *S, Sal*I; *X, Xho*I; and *Xb, Xba*I.

*Xba*I site italicized). Downstream of the *HIS3* 5' flank is a 1.12-kb *FUS1* promoter segment (−1.12 kb to −3 bp) fused 22 bp upstream of the *HIS3*-coding region. The *FUS1-HIS3* junction is interrupted by a *Sal*I linker and is identical to that described by Stevenson *et al.*[61] The 0.89-kb *HIS3* coding region extends to the genomic *Xho*I site located 0.2 kb downstream of the *HIS3* stop codon. The entire insert is carried in the *Eco*RI and *Sal*I sites of YIp5[12]; *HIS3* downstream *Xho*I and YIp5 *Sal*I sites were destroyed by ligation to each other.

pSL1442 is a *HIS3* riboprobe construct for detecting *HIS3* transcripts. It corresponds to the 0.71-kb *Sal*I/*Pst*I segment present in pSL1470 subcloned into the same sites of pSP64 (Promega). The insert is almost pure *HIS3*-coding sequences, extending from 22 bp upstream of the start codon to 35 bp downstream of the stop codon.

pSL1444 carries the 1.77-kb *Bam*HI wild-type *HIS3* genomic DNA segment in the *Bam*HI site of pBS-KS(+) (Stratagene). The orientation of the insert is such that coding strands of *HIS3* and the vector's *lacZ* sequences are the same polarity.

pSL1441 is a YIp5-based plasmid that carries the *his3Δ200* allele. It was created by subcloning a 3.44-kb *Eco*RI/*Xba*I DNA segment from plasmid YRp14-Sc2605*his3-Δ200*,[13] corresponding to the *his3Δ200* allele and flanking genomic sequences, into the *Eco*RI and *Nhe*I sites of YIp5.

Yeast Media

Standard recipes are used for yeast media.[14] Plates containing 5-fluoroorotic acid (5-FOA) for *URA3* counterselection are prepared according to Boeke *et al.*[15] The preparation of synthetic complete medium lacking histidine and containing AT (SC-His + AT) is similar to the preparation of SC-His with the following modifications. When adding water during media preparation, omit a volume corresponding to the volume of 1 *M* AT stock solution (see recipe later) that will be added after autoclaving. Drop a magnetic stir bar into the flask and autoclave. Remove from the autoclave and stir at room temperature until cooled to ~65°. Add an appropriate volume of sterile 1 *M* AT stock solution down the side of the flask and then stir a few minutes. Pour plates as usual. Let plates cure by storing inverted, covered with a clean cloth

---

[12] K. Struhl, D. T. Stinchcomb, S. Scherer, and R. W. Davis, *Proc. Natl. Acad. Sci. U.S.A.* **76**, 1035 (1979).

[13] M. T. Fasullo and R. W. Davis, *Mol. Cell. Biol.* **8**, 4370 (1988).

[14] F. Sherman, *Methods Enzymol.* **194**, 3 (1991).

[15] J. D. Boeke, F. LaCroute, and G. R. Fink, *Mol. Gen. Genet.* **197**, 345 (1984).

or plastic sheet at room temperature for 2 to 4 days. Bag plates and seal well. Store at 4° for up to several months.

Make 100 ml of 1 *M* AT stock solution by dissolving 8.41 g AT powder (Sigma) in ~90 ml water. Adjust the volume to 100 ml, filter sterilize with a 0.45-μm pore-size filter, and then transfer to an autoclaved bottle. This stock solution can be stored at 4° for at least 6 months with no detectable loss of activity. Higher concentration AT stock solutions (up to 2.5 *M*) can be made if needed.

Methods

Making a YFG1-HIS3 Reporter Based on pSL1470

There are a number of ways to create chimeras that place *HIS3*-coding sequences under the control of heterologous regulatory sequences. Two methods are described for creating *HIS3* fusions based on pSL1470. Similar methods could be used to construct *HIS3* fusions based on other plasmids (Table I). Whatever strategy and vector chosen, however, the final chimera should take the form of UAS + TATA + *HIS3* coding sequences. Ideally, the chimera should be constructed so that it can be stably integrated at an innocuous location in the yeast genome.

The *HIS3*-coding DNA following the *FUS1-HIS3* junction in pSL1470 lacks *HIS3* UAS and TATA sequences; thus, these functions must be provided by the heterologous DNA used to replace the *FUS1* DNA. In addition to this requirement, the heterologous DNA cannot contain restriction sites incompatible with targeting the final construct for integration at *HIS3*. We cut at the unique *Nhe*I site located in the *HIS3* ORF (Fig. 1), but other suitable sites are predicted in this region, including *Bcl*I, *Bgl*II (two sites near each other), and *Kpn*I. The heterologous DNA must be free of at least one of these four sites.

The first method for creating a *YFG1-HIS3* fusion in pSL1470 involves fusing *YFG1* UAS + TATA DNA upstream of the *HIS3*-coding region. The 5' end point of the *YFG1* segment may correspond to a position upstream of an experimentally determined UAS element. If a particular regulatory sequence remains to be identified, one can choose a 5' end point that either corresponds to the junction with coding sequences from the upstream neighboring gene or, if such a segment would be unusually large, some reasonable length of *YFG1* 5' noncoding sequences. The 3' end point must encompass *YFG1* TATA and transcription start sequences. If these have not been determined, a safe approach might be to choose a 3' end point similar to that chosen for the *FUS1* UAS + TATA DNA present in pSL1470, which is −3 bp with respect to the *FUS1* start codon. Using

TABLE I
Plasmids for Constructing *YFG1-HIS3* Reporters

Plasmid	*HIS3* 5' end point[a]	*HIS3* TATA?	Cloning sites[b]	Transformation[c]	Integrated reporter[d]	Stability[e]
pSL1470[f]	−22	No	Xb, S	SC-Ura Two-step replacement	*his3::YFG1-HIS3*	Stable
pBM1436[4]	−5	No	Bm, ScI	SC-Ura Two-step replacement	*rad16::YFG1-HIS3*	Stable
pBM1499[4]	−118	Yes	Bm, E	SC-Ura Two-step replacement	*rad16::YFG1-HIS3*	Stable
pHIS3NB[g]	−98	Yes	N, Xb, Sp, Sm, E	YPD + G418 One-step replacement	*pdc6::APT1::YFG1-HIS3*	Stable
pHIS3NX[g]	−98	Yes	N, Sp, Bm, Sm, E	YPD + G418 One-step replacement	*pdc6::APT1::YFG1-HIS3*	Stable
pHISi-1[h]	−98	Yes	E, Sm, ScI, M, ScII, Xb	SC-His Plasmid integrant	*his3::pHISi::YFG1-HIS3*	Semi
pHISi[i]	−98	Yes	E, Sm, ScI, M, ScII, Xb	SC-Ura Plasmid integrant	*his3::pHISi::YFG1-HIS3* or *ura3::pHISi::YFG1-HIS3*	Semi

[a] A 5' end point of *HIS3* upstream sequences following the cloning sites; "A" of the *HIS3* start codon is +1.

[b] Bm, *Bam*HI; E, *Eco*RI; M, *Mlu*I; N, *Not*I; S, *Sal*I; ScI, *Sac*I; ScII, *Sac*II; Sm, *Sma*I; Sp, *Spe*I; Xb, *Xba*I.

[c] Selective media and integration method for yeast transformation.

[d] Genotype of the integrated reporter. pBM1436 and pBM1499 reporters disrupt *RAD16* and are often described in the literature as integrating "downstream of *LYS2*."

[e] Stable reporters lack tandem-repeat sequences following one- or two-step gene replacement. Semistable reporters retain tandem repeats of plasmid and genomic sequences; reporter pop out is possible.

[f] Described in this article.

[g] *YFG1-HIS3* fusions constructed in pHIS3NB and pHIS3NX are subsequently transferred to pINT1, a *pdc6::APT1* yeast integrating plasmid; *APT1* confers G418 resistance. The three plasmids are described in A. H. Meijer, P. B. Ouwerkerk, and J. H. Hoge, *Yeast* **14**, 1407 (1998).

[h] pHISi-1 (CLONTECH Laboratories, Inc.) integrates downstream of the *his3-Δ200* deletion end point. Transformation requires high enough *YFG1-HIS3* expression to confer histidine prototrophy.

[i] pHISi (CLONTECH Laboratories, Inc.) carries a *URA3* selectable marker. pHISi can integrate either downstream of the *his3-Δ200* deletion end point or at a nonfunctional, nondelete *ura3* allele.

standard techniques,[16] the *YFG1* UAS + TATA segment is subcloned into the *Xba*I and *Sal*I sites of pSL1470, replacing the *FUS1* sequences. The subcloning is straightforward if the *YFG1* segment is flanked by *Xba*I and *Sal*I sites, or any combination of sites with compatible sticky ends; a simple strategy is to engineer such sites by polymerase chain reaction (PCR) amplification or oligo-directed mutagenesis.

The second method is similar to the first, but it involves creating chimeras that contain *YFG1* UAS sequences fused to heterologous TATA sequences upstream of the *HIS3*-coding region. Two examples where this configuration might be preferable are (1) when studying a *YFG1* UAS element isolated from its native promoter or (2) when the native *YFG1* TATA and transcription start sites are either too weak or too strong to be compatible with the range of AT concentrations used to score *HIS3* expression. Non-*YFG1* TATA sequences could, in principle, be derived from any yeast gene. We have used the *CYC1* minimal promoter[17,18] with good results (see later). Another good source might be the natural *HIS3* TATA sequences. Whatever the TATA DNA used, it should not lead to significant transcription on its own and it should not include a start codon, which is already present downstream of the *Sal*I linker in pSL1470. With these considerations in mind, a *YFG1* UAS + heterologous TATA fusion is constructed by standard techniques and used to replace the *Xba*I/*Sal*I *FUS1* segment present in pSL1470. Two possible strategies are (1) a single-step strategy involving a three-part ligation of *YFG1* UAS + heterologous TATA + *Xba*I/*Sal*I cut pSL1470 DNAs and (2) a two-step strategy involving first the construction of a *YFG1* UAS + heterologous TATA fusion on a separate vector, followed by its transfer into the *Xba*I and *Sal*I sites of pSL1470.

Creating Host Yeast Strains

Depending on where the *YFG1-HIS3* reporter will be integrated, the host yeast strain should be either *HIS3* or harbor a *his3* mutation, preferably a deletion allele. In either case, the host strain must not contain additional mutations affecting histidine biosynthesis (e.g., *his4*), which would prevent biological scoring of the *YFG1-HIS3* reporter. Some reporters are integrated by two-step gene replacement,[19] which requires that the host strain be *ura3* to select for plasmid *URA3* sequences during transformation.

[16] F. M. Ausubel, R. Brent, R. E. Kingston, D. D. Moore, J. G. Seidman, J. A. Smith, and K. Struhl, "Current Protocols in Molecular Biology." Wiley Interscience, New York, 1987.
[17] L. Guarente, *Methods Enzymol.* **101**, 181 (1983).
[18] L. Guarente and M. Ptashne, *Proc. Natl. Acad. Sci. U.S.A.* **78**, 2199 (1981).
[19] R. Rothstein, *Methods Enzymol.* **194**, 281 (1991).

YFG1-HIS3 reporters based on pSL1470 are integrated by two-step gene replacement at the *HIS3* locus, essentially swapping the native *HIS3* regulatory region for that of *YFG1*. The host strain must be wild type at *HIS3* to introduce pSL1470-based reporters. Strains carrying any known *his3* allele, including *his3Δ200*, can be converted to *HIS3* by one-step gene replacement[19] using pSL1444. To convert *his3* strains to *HIS3*, cut pSL1444 with *Bam*HI to release the *HIS3* insert. Transform yeast[20] and select on SC-His medium. Choose several independent His[+] transformants and confirm the gene replacement by Southern blotting.[21] Point mutant alleles and wild-type *HIS3* should be identical in structure, but be on the lookout for unexpected bands in the His[+] strain that would indicate that the *his3* locus had not been repaired.

YFG1-HIS3 reporters based on other plasmids that do not integrate at the *HIS3* locus must be introduced into a *his3* host strain. A *his3* deletion allele is preferred because it precludes gene conversion of the *his3* locus by *HIS3* sequences present in the *YFG1-HIS3* reporter. A good choice is the *his3Δ200* allele, whose deletion (-205 bp to $+835$ bp[13]) removes all *HIS3*-coding sequences. pSL1441 can be used to convert *HIS3* or *his3* point mutant strains to *his3Δ200* by two-step gene replacement.[19] Digestion of pSL1441 with *Bgl*II creates a double-strand break that targets integration downstream of the *HIS3* locus. *his3Δ200* recombinants can be scored phenotypically if the starting strain was *HIS3*. Otherwise, the deletion can be scored by Southern blotting.

Integrating YFG1-HIS3 Reporters

We describe a method for integrating *YFG1-HIS3* reporters based on pSL1470. Procedures for integrating reporters based on other plasmids (Table I) can be obtained from the corresponding references.

YFG1-HIS3 reporters based on pSL1470 are integrated at *HIS3* by two-step gene replacement.[19] Cutting the plasmid within the *HIS3*-coding region creates a double-strand break that targets integration at the genomic *HIS3*-coding region. We cut with *Nhe*I (Fig. 1), but *Bcl*I, *Bgl*II, or *Kpn*I, all of which cut within the *HIS3*-coding sequences of pSL1470, should work if the *Nhe*I site occurs within the UAS + TATA *YFG1* DNA. The cut plasmid is transformed into a suitable *HIS3 ura3* host strain and transformants are selected on SC-Ura medium. Several independent, streak-purified Ura[+] isolates are then plated on 5-FOA medium to select for recombinants that have popped out *URA3*-containing vector sequences. 5-FOA-resistant

[20] D. Gietz, A. St. Jean, R. A. Woods, and R. H. Schiestl, *Nucleic Acids Res.* **20**, 1425 (1992).
[21] E. M. Southern, *J. Mol. Biol.* **98**, 503 (1975).

strains will fall into two classes: (1) those harboring the *YFG1-HIS3* reporter, resulting from recombination between repeated *HIS3* 5' flank DNA segments, and (2) those that are wild type at *HIS3*, resulting from recombination between repeated *HIS3*-coding DNA segments. Distinguish between these by Southern blotting. Depending on the expected regulation of the *YFG1-HIS3* reporter, phenotypic screening could be used to identify *YFG1-HIS3* recombinants. For example, we took advantage of the haploid-specific expression pattern of *FUS1*[22] to screen for *FUS1-HIS3* recombinants. In particular, 5-FOA-resistant colonies from pSL1470 transformants were mated to a *his3* haploid of the opposite mating type, and the resulting diploids were tested for *his3* complementation. **a**/α *his3*/*HIS3* diploids were His⁺, whereas **a**/α *his3*/*FUS1-HIS3* diploids, which were expected not to express *FUS1-HIS3*, scored as His⁻.

Confirming Regulation of YFG1-HIS3 Reporters

Once integrated, the *YFG1-HIS3* reporter should be tested to confirm that *HIS3* transcripts are regulated as expected. One method is to use Northern blot analysis[16] to measure *HIS3* transcripts directly, comparing them to native *YFG1* transcripts in the same experiment. Plasmid pSL1442 can be used to label either an RNA riboprobe or a DNA probe for detecting *HIS3* transcripts. To make a *HIS3* riboprobe, linearize pSL1442 with *Sal*I and use SP6 RNA polymerase. To make a *HIS3* DNA probe, cut pSL1442 with *Sal*I and *Pst*I and isolate the 0.71-kb *HIS3* fragment.

Depending on the regulation of *YFG1*, phenotypic tests could also be used to confirm *YFG1-HIS3* regulation. *FUS1-HIS3* is again a good example. Wild-type haploids carrying the *FUS1-HIS3* reporter are His⁺. Mutation of *ste4*, which abolishes native FUS1 expression, converts *FUS1-HIS3* strains to His⁻, indicating that the reporter behaves as expected. In this example, a condition predicted to decrease *FUS1-HIS3* expression was tested, but one could also test whether a condition predicted to activate the reporter would lead to increased AT resistance.

Plate Assays for HIS3 Reporter Expression

Yeast strains harboring a *YFG1-HIS3* reporter can be assayed biologically for *HIS3* expression by scoring growth on SC-His plates containing various concentrations of AT. We use AT concentrations ranging from 1

---

[22] G. McCaffrey, F. J. Clay, K. Kelsay, and G. F. Sprague, Jr., *Mol. Cell. Biol.* **7**, 2680 (1987).

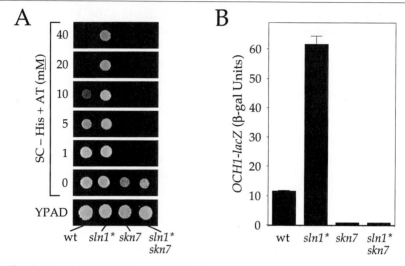

FIG. 2. Use of *OCH1-HIS3* and *OCH1-lacZ* reporters to characterize mutants that affect *OCH1* expression. An *OCH1-HIS3* reporter was constructed on pSL1470 as described in the text and then integrated at *HIS3* by two-step gene replacement in a *HIS3 ura3 mfa2::OCH1-lacZ* strain. (A) Spotting plate assay of *OCH1-HIS3* expression. (B) β-Galactosidase assay[8] of *OCH1-lacZ* expression. The *sln1** allele activates *OCH1* expression and was obtained as a spontaneous mutant that allowed growth of an *OCH1-HIS3* strain on SC-His + 20 m*M* AT plate medium. Deletion of *SKN7* inactivates *OCH1* expression and completely suppresses the *sln1** mutant.

to 40 m*M* (e.g., see Fig. 2A), and the use of up to 100 m*M* has been reported.[23] Thus, the potential range of AT concentrations that can be used extends over at least two orders of magnitude, but conceivably, AT concentrations below and above this range might be used in some applications.

The three basic plating assays are (in order of decreasing sensitivity): spotting, streaking, and replica plating. In general, the more cells applied to the plate, the less accurate the assay; when a large number of cells are applied, it can be difficult to distinguish between slow growth and no growth. For each of the assays described in this article, plates are incubated at an appropriate temperature (30° for wild-type strains) for 1 to 3 days before scoring.

Spotting is a sensitive plate assay for estimating *YFG1-HIS3* reporter expression and can be adapted for high throughput. Dilute fresh cells from

[23] D. A. Mangus, N. Amrani, and A. Jacobson, *Mol. Cell. Biol.* **18,** 7383 (1998).

a liquid culture or plate stock to a concentration of ~500 cells/5 μl in sterile water. Cell density can be determined using either a hemocytometer or a spectrophotometer (an OD_{600} of 1.0 is about 1.0 × 10^7 cells/ml). Use a micropipettor to apply 5 μl of cell suspension to plate media and allow the spot to dry before inverting the plate. For high throughput, cell suspensions can be made in wells of a microtiter plate and then replica transferred to plate media with a steel-pin replica plating block.

Streaking is a convenient alternative to spotting when testing a small number of samples. It is slightly less accurate than spotting because of the large number of cells in the initial application, but sparse regions of the streak can be scored accurately if one uses a consistent technique. Cells to be tested are taken from a fresh plate stock and streaked to plate media with up to six sectors per plate.

Replica plating by sterile velveteen transfer[14] is the least sensitive technique due to the large number of cells transferred. However, because of its convenience, it is the technique of choice for tetrad analysis and screening many individual colonies, such as when performing a cloning experiment.

Pitfalls

Imidazoleglycerolphosphate dehydratase, the enzyme encoded by *HIS3*, is the only AT-sensitive factor in yeast.[2,3] Because of this specificity, there are relatively few pitfalls associated with *HIS3* reporter expression and AT resistance. This section discusses pitfalls associated with two determinants of AT resistance in yeast: *HIS3* expression levels and intracellular AT concentrations. In all cases, assaying an independent reporter makes identifying rare problems straightforward. One can incorporate a *YFG1-lacZ* reporter or take advantage of physiological reporters when available. For example, when activated pheromone response pathway mutants were selected with a *FUS1-HIS3* reporter, they were subsequently cross-checked for increased mating pheromone production, which is also controlled by the pheromone response pathway.[6]

Increasing the copy number of a *YFG1-HIS3* reporter is a trivial way to increase *HIS3* expression and produce AT resistance. It is largely for this reason that we stress the importance of constructing a stably integrated *YFG1-HIS3* reporter. In some contexts, however, the copy number of even an integrated reporter might increase under AT selection. For example, when a diploid strain was used to select for potential dominant mutants that activate an *OCH1-HIS3* reporter, the majority of the AT-resistant strains failed to coactivate an *OCH1-lacZ* reporter.[24] In contrast, the major-

---

[24] J. Horecka, unpublished data.

ity of strains selected from haploids coactivates both. The basis for the discrepancy might lie in ploidy. Diploid cells can undergo mitotic nondisjunction to become aneuploid (e.g., trisomic) for the chromosome that harbors *YFG1-HIS3*. Another potential class of artifacts associated with increased *YFG1-HIS3* expression are *cis*-acting mutants arising in *YFG1* upstream sequences. Such mutants would be rare, but could arise, for example, in selections designed to identify *trans*-acting factors that activate *YFG1-HIS3* expression.

Strains can also become AT resistance by acquiring a mutation that increases the expression or activity of Atr1, a multidrug resistance transporter.[25] Cells that overexpress Atr1 accumulate less intracellular AT and are AT resistant.[26,27] Mutations that activate Atr1 could occur in *ATR1* itself or in genes that encode transcription factors that regulate *ATR1* expression.[27]

Finally, a pitfall unrelated to AT resistance is the potential for undesirable phenotypes to arise from disruption of a particular genomic locus with the *YFG1-HIS3* reporter. Target loci for the reporter plasmids listed in Table I are dispensable for most applications, but the researcher should be aware of this pitfall when designing reporter plasmids and troubleshooting a strain that is not behaving as expected.

Example of the Method

For reasons to be published elsewhere, we sought mutations that activate transcription of *OCH1*, a cell wall gene encoding α-1,6-mannosyltransferase.[28] To accomplish this, we created a yeast strain harboring an integrated *OCH1-HIS3* reporter and used it to select for mutants that activate the expression of the reporter. An *OCH1-HIS3* reporter plasmid based on pSL1470 was created in two steps. First, an *OCH1* UAS + *CYC1* TATA fusion was constructed by fusing *OCH1* 5′ noncoding sequences (-780 to -196 followed by a synthetic *Xho*I site) to the *CYC1* minimal promoter (from the genomic *Xho*I site at -250 to $+4$). Second, an *OCH1-CYC1* DNA segment (from OCH1 -620 to CYC1 -3) from the first step was amplified by PCR using oligonucleotides

[25] A. Goffeau, J. Park, I. T. Paulsen, J. L. Jonniaux, T. Dinh, P. Mordant, and M. H. Saier, Jr., *Yeast* **13**, 43 (1997).

[26] S. Kanazawa, M. Driscoll, and K. Struhl, *Mol. Cell. Biol.* **8**, 664 (1988).

[27] S. T. Coleman, E. Tseng, and W. S. Moye-Rowley, *J. Biol. Chem.* **272**, 23224 (1997).

[28] K. Nakayama, T. Nagasu, Y. Shimma, J. Kuromitsu, and Y. Jigami, *EMBO J.* **11**, 2511 (1992).

that introduced flanking *Xba*I and *Sal*I sites and then inserted into the same sites of pSL1470, creating plasmid pJH60. The *OCH1-HIS3* reporter was introduced into a *HIS3 ura3* yeast strain as described earlier, cutting pJH60 with *Nhe*I to target integration at *HIS3*. Two-step gene replacement yielded the integrated *his3::OCH1-HIS3* reporter. The host strain also carried an integrated *OCH1-lacZ* reporter, providing an independent measure of *OCH1* expression.

The wild-type strain expresses *OCH1-HIS3* to a level that allows growth on SC-His media containing up to 10 m*M* AT (Fig. 2A). To confirm that the reporter was regulated as expected, we tested the effects of a *skn7* mutation. *SKN7* encodes a transcription factor that shares homology with bacterial two-component response regulators, and deletion of *SKN7* leads to a defect in both native *OCH1* and *OCH1-lacZ* expression.[29] The *OCH1-HIS3 skn7* strain grew poorly on SC-His medium lacking AT and not at all on SC-His medium containing 1 m*M* AT (Fig. 2A), indicating proper regulation of the *OCH1-HIS3* reporter.

We selected for spontaneous mutants that activate *OCH1-HIS3* expression by streaking a wild-type strain to SC-His + 20 m*M* AT plate medium. The mutants were subsequently assayed for *OCH1-lacZ* expression to identify AT-resistant mutants that specifically activated *OCH1* expression; 16 of 19 strains coactivated both reporters.[24] One of the AT-resistant papillae that arose owed its phenotype to a mutation in the *SLN1* gene, which we designate as *sln1**[29]; *SLN1* encodes a histidine kinase homologous to bacterial two-component sensor kinases.[30,31] A plate assay showed that the *sln1** mutation activated *OCH1-HIS3* expression to a level that allowed growth on SC-His plates containing 40 m*M* AT, the highest concentration tested (Fig. 2A). By analogy to bacterial two-component systems, we reasoned that *sln1** mutation activated *OCH1* expression through the response regulator Skn7. This hypothesis was supported by results from a plating assay (Fig. 2A), where *sln1* skn7* and *skn7* strains scored the same. *OCH1-lacZ* reporter data were in good agreement with those from *OCH1-HIS3* plating tests and provided a more quantitative measure of *OCH1* expression in the four strains (Fig. 2B).

Acknowledgments

C. Boone helped construct plasmids pSL1470 and pSL1442. Work from the GFS laboratory was supported by a grant from the NIH.

[29] J. Horecka, G. F. Sprague, Jr., and Y. Jigami, unpublished data.
[30] T. Maeda, S. M. Wurgler-Murphy, and H. Saito, *Nature* **369**, 242 (1994).
[31] I. M. Ota and A. Varshavsky, *Science* **262**, 566 (1993).

[8] Use of Fusions to Human Thymidine Kinase as Reporters of Gene Expression and Protein Stability in *Saccharomyces cerevisiae*

By SOFIE R. SALAMA

Introduction

Protein stability plays an important role in the regulation of a number of proteins, and specific protein degradation plays a critical role in regulating a variety of cellular processes. For example, the importance of protein degradation in controlling cell cycle progression has been revealed (reviewed in King *et al.*[1]). These findings have led to a large research effort aimed at understanding the signals targeting key cell cycle regulators (such as the cyclins) for degradation, the mechanisms by which these signals are identified, and the mechanisms by which these proteins are rapidly and specifically degraded. Experiments in the budding yeast, *Saccharomyces cerevisiae,* have been particularly helpful in advancing our knowledge in this area. This is due, in part, to the powerful molecular genetic tools available in this organism. In many cases, such as B-type cyclins, both the degradation signal and the degradation pathway are remarkably conserved between yeast and vertebrates.[2] In this article, how chimeric proteins between human cytosolic thymidine kinase (hTK) and a protein of interest (X) can be used to identify the signals that destabilize protein X is described as is how the resulting *hTK-X* chimeras can be used in a genetic screen to identify mutants defective in the degradation of protein X. As an example, experiments using chimeras between a yeast G1 cyclin, Cln2, and hTK[3] are presented.

Construction of hTK Chimeras

A first step in understanding how a protein is degraded in a rapid and/or regulated manner is to identify the signal that targets the protein for degradation. In many cases, this signal will confer protein instability when fused to a heterologous protein.[3-7] hTK is an excellent candidate for such

[1] R. King, R. Deshaies, J. Peters, and M. Kirschner, *Science* **274**, 1652 (1996).
[2] M. Pagano, *FASEB J.* **11**, 1067 (1997).
[3] S. Salama, K. Hendricks, and J. Thorner, *Mol. Cell. Biol.* **14**, 7953 (1994).
[4] Y. Barral, S. Jentsch, and C. Mann, *Genes Dev.* **9**, 399 (1995).
[5] M. Glotzer, A. Murray, and M. Kirschner, *Nature* **349**, 132 (1991).

a study for a number of reasons. It is a small, cytosolic, monomeric enzyme of about 25 kDa whose activity can be measured by a simple *in vitro* assay.[8] This enzyme is particularly useful for studies in yeast because *S. cerevisiae* lacks any endogenous TK,[9,10] thereby permitting sensitive detection of a heterologously expressed TK. Furthermore, addition of sequences at its C terminus or removal of the C-terminal 40 amino acids does not perturb its enzyme activity.[11] Finally, yeast strains can be constructed that require TK activity for viability, thus allowing an *in vivo* test of the stability of hTK derivatives (see later).

Protocol: Generation of hTK-X Yeast Expression Constructs

1. Select a yeast expression vector. For initial studies of hTK chimeras in yeast, high levels of constitutive expression are desirable to facilitate detection of the protein chimeras. Toward this end, we used the multicopy (2 μm DNA-based) vector pAD4M,[12] which contains a polylinker region between the promoter and terminator of *S. cerevisiae ADH1* and *LEU2* for selection in yeast for the generation of hTK-Cln2 constructs (Fig. 2A). However, for the genetic selection described in this article, integration of the hTK chimera into the yeast genome is desirable. In addition, it may be necessary to modulate the expression of the hTK chimera at the RNA level to achieve the correct amount of expression for the genetic selection. Toward this end, a series of vectors described by Mumberg *et al.*[13] that allow for expression from a variety of constitutive promoters, which give different levels of RNA expression, could be very useful. In addition, these pRS-based vectors are compatible with the pRS400 series of integrating vectors.[14] Thus, constructs in these vectors can be subcloned easily into the appropriate integrating plasmid.

2. Generation of hTK-gene X fusion fragments. The polymerase chain reaction (PCR) can be used to generate precise in-frame fusions between

[6] P. Loetscher, G. Pratt, and M. Rechsteiner, *J. Biol. Chem.* **266,** 11213 (1991).
[7] J. Yaglom, M. H. Linskens, D. M. Rubin, B. Futcher, and D. Finley, *Mol. Cell. Biol.* **15,** 731 (1995).
[8] L.-S. Lee and Y.-C. Cheng, *J. Biol. Chem.* **251,** 2600 (1976).
[9] A. R. Grivell and J. F. Jackson, *J. Gen. Microbiol.* **54,** 307 (1968).
[10] L. Bisson and J. Thorner, *J. Bacteriol.* **132,** 44 (1977).
[11] M. G. Kauffman and T. J. Kelly, *Mol. Cell. Biol.* **11,** 2538 (1991).
[12] G. A. Martin, D. Viskochil, G. Bollag, P. C. McCabe, W. J. Crosier, H. Haubruck, L. Conroy, R. Clark, P. O'Connell, R. M. Cawthon, M. A. Innis, and F. McCormick. *Cell* **63,** 843 (1990).
[13] D. Mumberg, R. Muller, and M. Funk, *Gene* **156,** 119 (1995).
[14] R. S. Sikorski and P. Hieter, *Genetics* **122,** 19 (1989).

hTK and the protein sequence of interest. In a three-primer method described by Yon and Fried,[15] a PCR product containing the desired in-frame fusion is generated by using a "joiner" oligonucleotide, which contains 15 nucleotides corresponding to the five C-terminal amino acids of hTK followed by 15 nucleotides corresponding to the five N-terminal amino acids of the amino acid sequence of interest. In a reaction with a sense primer that corresponds to the N-terminal region of hTK and an antisense primer that corresponds to the C-terminal region of the amino acid sequence of interest from gene X, the "joiner" oligonucleotide is used at a concentration that is 1% that of the other two primers with plasmid templates containing hTK (pMGK24[11]) and Gene X (see Fig. 1).

3. Subclone the hTK-gene X PCR fragment into the yeast expression vector. It is critical to sequence the resulting fusion product before proceeding with further analysis.

Thymidine Kinase Assays to Measure the Relative Steady-State Levels of hTK Chimeras

An easy way to measure the steady-state level of the resulting hTK-X chimeras, and to screen for those causing protein destabilization, is to measure the thymidine kinase activity of extracts prepared from yeast strains that have been transformed with the desired hTK constructs (Figs. 2A and 2B). Confirmation that the hTK chimeras have enzymatic activity is also important for the genetic selection described later. TK activity is measured by following the ATP-dependent conversion of [*methyl*-^{3}H]thymidine to [*methyl*-^{3}H]dTMP in a filter-binding assay in which the charged [*methyl*-^{3}H]dTMP binds the filter, whereas the uncharged [*methyl*-^{3}H]thymidine does not.

Protocol: Preparation of Yeast Extracts for TK Assays

Solutions and Materials

TK buffer: 190 mM Tris–HCl, pH 7.5, 1.9 mM MgCl$_2$
Lysis buffer: TK buffer + 1 mM dithiothreitol (DTT) + 0.5 mM phenylmethylsulfonyl fluoride (PMSF)
Acid-washed glass beads (diameter: 0.45–0.5 mm)

Note: This lysis buffer was chosen for compatibility with the TK assay reaction buffer described later.

---

[15] J. Yon and M. Fried, *Nucleic Acids Res.* **17**, 4895 (1989).

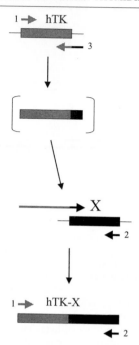

FIG. 1. Construction of precise hTK-X fusion constructs using PCR. To construct hTK-X fusions, a sense primer corresponding to the N terminus of *hTK* (primer 1), an antisense primer corresponding to the C terminus of the *gene X* sequence to be tested (primer 2), and an antisense primer corresponding to the C terminus of *hTK* and the N terminus of the *gene X* sequence to be tested (the "joiner oligo," primer 3) are used. Primer 3 is added at a concentration 1% that of the other two primers in the reaction.[15] Early in the reaction, primers 1 and 3 generate a product using an *hTK*-containing plasmid as template, which also contains a sequence corresponding to the N terminus of *gene X*. As the reaction proceeds, this PCR product can be used as a primer with primer 2 and a plasmid template containing *Gene X* to generate the *hTK-X* fragment, which can be amplified further using primers 1 and 2.

1. Grow 500-ml cultures of yeast strains harboring various hTK constructs as well as a yeast strain harboring the parent vector to an A_{600} of 1.0 in the appropriate selective media.
2. Harvest cells by brief centrifugation (5 min at 5000 rpm at 4° in a Sorvall GSA rotor). Discard the supernatant.
3. Wash the cells by resuspending in 25 ml ice-cold TK buffer and centrifuge again (5 min at 5000 rpm at 4° in a Sorvall SS34 rotor). Discard the supernatant.
4. Resuspend the cells with 1 ml ice-cold TK buffer. Transfer to 2-ml screw-capped microcentrifuge tubes. Pellet 15 sec in a microcentrifuge at maximum speed. Discard the supernatant.

A

Thymidine Kinase Construct   **Plasmid**

hTK   pSS3

hTKΔ40   pSS4

hTKΔ40-Cln2   pSS9

hTKΔ40-Cln2ΔPEST   pSS13

hTKΔ40-Cln2PEST   pKBH2

FIG. 2. Analysis of the steady-state levels of hTK-Cln2 chimeras by thymidine kinase assays and immunoblotting. (A) Plasmids expressing hTK and derivatives. To study the role of the C-terminal, PEST-containing region of Cln2 on protein stability, these constructs (left), diagrammed schematically (middle), were inserted between the promoter and the terminator of the *S. cerevisiae ADH1* gene in the expression vector pAD4M. (B) TK activity of strains expressing hTK and hTK-Cln2 chimeras. Cytosolic yeast extracts of strains expressing hTK (solid circles), hTKΔ40 (open circles), hTKΔ40-Cln2ΔPEST (squares), hTKΔ40-Cln2 (triangles), or the empty expression vector (diamonds) were assayed for TK activity by measuring the ATP-dependent conversion of [*methyl*-³H]thymidine to [*methyl*-³H]-dTMP. Each datum point represents the average of reactions performed in duplicate. (C) Immunoblots of hTK and hTK-Cln2 chimeras. An equivalent amount (10 μg) of the yeast extracts described in (B) expressing hTK (lanes1 and 4), hTKΔ40 (lane 2), hTKΔ40-Cln2PEST (lane 3), hTKΔ40-Cln2 (lane 5), or hTKΔ40-Cln2ΔPEST (lane 6) were subjected to SDS–PAGE (12.5% polyacrylamide), transferred electrophoretically to a nitrocellulose filter, and the filter replica was probed with anti-hTK antibodies (1 : 2000 dilution). Adapted from Salama *et al.*[3]

5. Resuspend the cells in 500 μl ice-cold lysis buffer as well as possible. Add glass beads equal to a volume of approximately 500 μl.

6. Lyse the cells by vortexing on the highest setting in the cold room for seven cycles in which the samples are vortexed for 30 sec and then placed on ice for 1 min.

7. Clarify the extract by centrifugation at 7000g for 6 min (6500 rpm at 4° in a Sorvall HB-4 rotor). *Take the supernatant* and transfer to a Beckman TL100.3 microcentrifuge tube.

8. Remove the membranes by centrifugation at 100,000g for 30 min (60,000 rpm at 4° in a Beckman TL100.3 rotor).

9. Aliquot the clarified supernatant and remove an aliquot to determine the protein concentration. Quick freeze the remainder in liquid N₂ and store at −80° until ready to use. The protein concentration of the extracts should range from 10 to 20 mg/ml and can be determined by a Bradford assay.[15a]

[15a] M. M. Bradford, *Anal. Biochem.* **72**, 248 (1976).

B

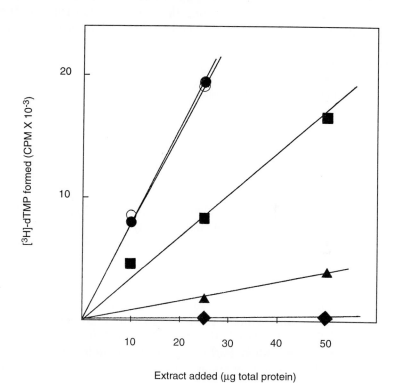

C

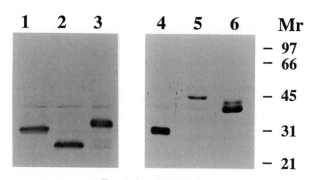

FIG. 2. (*continued*)

Protocol: TK Assays

This assay was modified from that of Lee and Cheng[8] for TK isolated from mammalian cells. For washing the filters, we constructed a porous plastic basket that fits inside a 500-ml beaker and washed the filters en masse. However, a sampling manifold, such as the Millipore 1225 sampling manifold, would work as well.

Solutions and Materials

10× TK assay buffer: 10× TK buffer + 1% bovine serum albumin (BSA) + 25% glycerol (store at −20°)

10× ATP-regenerating system: 19 mM ATP, 30 mM phosphocreatine, 67.5 μg/ml creatine kinase (store in small aliquots at −80°)

1 M DTT

1 M NaF

10× [*methyl*-³H]thymidine: 220 μM thymidine + 20 μl/0.1 ml [*methyl*-³H]thymidine (1 mCi/ml, 46 Ci/mmol)

Whatman DE81 2.3-cm-diameter DEAE-cellulose filter disks

95% ethanol

Liquid scintillation vials, fluid, and counter

1. Assemble reactions on ice in 100 μl total volume. The final reaction mixture should contain 1× TK assay buffer, 10 mM DTT, 10 mM NaF, 22 μM [*methyl*-³H]thymidine (910 mCi/mmol), and 1× ATP-regenerating system. Perform reactions in duplicate.
2. Start the reaction by adding the yeast extract to be tested (10–50 μg). Mix gently and incubate for 30 min at 37°.
3. Stop the reaction by spotting 50 μl onto a DE81 filter disk.
4. Place the disks in a basket suspended in a 500 ml beaker and wash (for 5 min each) with 10 ml/disk H₂O and then with 95% ethanol for three additional washes.
5. Air-dry the disks, place in plastic scintillation vials, add 5 ml scintillation fluid (Packard Scint-AXF), and count in a liquid scintillation counter.
6. Compare the relative activity of the hTK constructs by plotting counts per minute versus microgram yeast extract (see Fig. 2B).

Analysis of Steady-State Levels by Immunoblotting

Either a lower steady-state level of the fusion protein or a lower specific activity of the hTK-X fusion could result in a lower steady-state level of

TK activity. To distinguish between these two possibilities, an immunoblot[16] can be performed using the protein extracts prepared earlier. For studies with hTK-Cln2 chimeras, we used antibodies raised against hTK (Fig. 2C). Alternatively, a tag such as HA or myc could be incorporated into the *hTK-X* constructs to allow detection using commercially available antibodies. It is important that the epitopes being recognized by the chosen antibody are common to all of the chimeras to allow for comparison of relative protein levels. For analysis of hTK-Cln2 chimeras, we found that 10 μg yeast extract per sample was sufficient to produce a strong signal using our polyclonal anti-hTK serum and a chemiluminescence detection system (ECL; Amersham).

Analysis of Protein Stability by Pulse-Chase Analysis

Once a sequence is found that lowers the steady-state levels of the hTK-X chimera, it is important to check that the lower levels are due to increased degradation rather than decreased transcript levels or decreased translation efficiency. The most direct method to check whether the degradation rate of a protein is changed is to perform pulse-chase analysis.[17] This technique relies on immunoprecipitation of the desired constructs. Immunoprecipitations from yeast extracts are notoriously dirty, but performing the immunoprecipitation under denaturing conditions and washing extensively can reduce this problem (Fig. 3).

Protocol: Radiolabeled Pulse-Chase of Yeast Cells

Solutions

Lysis buffer: 1% SDS, 50 mM Tris, pH 7.4, 1 mM EDTA, 1 mM PMSF
Wash buffer: 600 μl 2× PBS, 60 μl 20% Triton X-100, 50 μl 10% fixed Staph A cells, 190 μl H$_2$O
Chase cocktail: 0.3% L-cysteine, 0.4% L-methionine, 100 mM (NH$_4$)$_2$SO$_4$

1. Grow cultures overnight in synthetic minimal media (SMM)[18] with needed supplements containing *only* 100 μM ammonium sulfate (use yeast nitrogen base *without* ammonium sulfate and *do not* use sulfate salts of amino acids) to an A_{600} of 0.2–0.4. Approximately 2 A_{600} units of cells are required for each time point.

[16] E. Harlow and D. Lane, "Antibodies: A Laboratory Manual." Cold Spring Harbor Laboratory, Cold Spring Harbor, NY, 1988.
[17] A. Varshavsky, *Methods Enzymol.* **327** [41] 2000.
[18] F. Sherman, *Methods Enzymol.* **194**, 3 (1991).

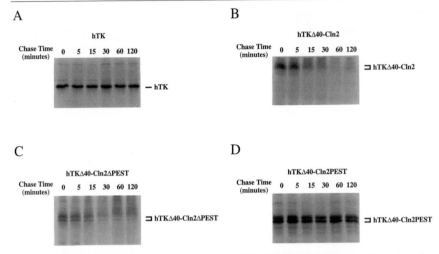

FIG. 3. Pulse-chase analysis of the turnover of hTK and hTK-Cln2 chimeras. A yeast strain expressing hTK (A), hTKΔ40-Cln2 (B), hTKΔ40-Cln2ΔPEST (C), or hTKΔ40-Cln2PEST (D) was radiolabeled with [35S]Met and [35S]Cys for 10 min. After the addition of excess unlabeled Cys, Met, and free SO_4^{2-}, cell samples were withdrawn at the indicated times and extracts were prepared. hTK-related antigens were immunoprecipitated from these extracts with anti-hTK antibodies and analyzed by SDS–PAGE and autoradiography. Adapted from Salama et al.[3]

2. Harvest cells in a tabletop centrifuge (2000 rpm for 5 min).

3. Resuspend cells to a concentration of 5 A_{600} units/ml in SMM lacking methionine and without ammonium sulfate (other needed supplements can be added).

4. After preincubation for 15 min at 30° with shaking, start the pulse-labeling by adding 30 $\mu Ci/A_{600}$ unit of a mixture of [35S]Met and [35S]Cys (Expre35SS Label; NEN) and incubate for an additional 10 min at 30° with shaking.

5. Start the chase by adding 1/100 volume of chase cocktail.

6. At 0, 5, 15, 30, 60, and 120 min, remove a 400-μl aliquot to an ice-cold microcentrifuge tube (screw-capped tubes help reduce radioactive contamination) containing an equal volume of 20 mM NaN₃. Invert to mix and incubate samples on ice until all time points have been taken.

7. Pellet cells in a microcentrifuge (15 sec at maximum speed). Remove the supernatant (very radioactive!). Wash with 1 ml 10 mM NaN₃. Pellet cells and remove supernatant.

8. Resuspend the cells in 300 μl lysis buffer. Add glass beads to the meniscus. Lyse the cells by vortexing at maximum strength for 90 sec.
9. Immediately after vortexing, incubate at 95° for 5 min.
10. After a brief spin to remove condensation, transfer the supernatant via a flat-ended Pipetman tip to a tube containing 900 μl wash buffer. Vortex briefly to mix.
11. Spin in a microcentrifuge for 10 min at maximum speed. Remove the supernatant to a fresh tube. Remove 5 μl for scintillation counting. Store the samples at $-20°$ until ready to immunoprecipitate.

Protocol: Immunoprecipitations under Denaturing Conditions

1. Aliquot 0.1 A_{600} unit or 1 $\times$ 10^7 cpm (from scintillation counting described earlier) per immunoprecipitation into a microcentrifuge tube.
2. Add 500 μl 2$\times$ IP buffer–SDS (300 mM NaCl, 2% Triton X-100, 30 mM Tris–HCl, pH 7.5, 2 mM NaN$_3$), SDS to a final concentration of 0.2% (the radiolabeled samples are 0.25% SDS), and H$_2$O to a final volume of 1 ml.
3. Add antibody (usually 0.5–5 μl) and 5 μl 10% protein A-Sepharose/ μl antibody and incubate for 2 hr at room temperature or 8+ hr at 4°.
4. Pellet the protein A-Sepharose beads by spinning for 15 sec in a microcentrifuge at maximum speed. Discard the supernatant.
5. Wash the beads by successive resuspension with 1 ml wash solution and recentrifugation as follows: twice in IP buffer (150 mM NaCl, 1% Triton X-100, 0.1% SDS, 15 mM Tris–HCl, pH 7.5, 2 mM NaN$_3$), twice in urea buffer (2 M urea, 200 mM NaCl, 2 mM NaN$_3$, 1% Triton X-100, 100 mM Tris–HCl, pH 7.5), twice in IP buffer containing 500 mM NaCl, and once in Tris buffer (50 mM NaCl, 2 mM NaN$_3$, 10 mM Tris–HCl, pH 7.5).
6. Resuspend the beads in 50 μl SDS–PAGE sample buffer, boil for 5 min, resolve the immunoprecipitates on an SDS–polyacrylamide gel, and visualize by autoradiography or phosphorimager analysis.
7. To calculate an approximate half-life of the hTK chimeras, the amount of radioactivity remaining at each time point after initiation of the chase is quantified with a phosphorimager and the logarithm of the values so obtained are plotted versus time (see Fig. 4). Although such decay curves are often not linear (especially at later time points), one can assume first-order kinetics between the zero time point and the first couple of time points. One can then determine an approximate half-life based on lines generated from these three data points using the formula $t_{1/2} = -[\ln 2/(2.3\ m)]$, where m is the slope of the line.

Decay of TK Constructs

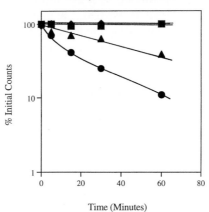

Time (Minutes)

FIG. 4. Comparison of the degradation rate of hTK and hTK-Cln2 chimeras. Decay curves for hTK (squares), hTKΔ40-Cln2PEST (diamonds), hTKΔ40-Cln2ΔPEST (triangles), and hTKΔ40-Cln2 (circles). Approximate half-lives based on these plots were calculated as described in the text: hTK and hTKΔ40-Cln2PEST >2 hr, hTKΔ40-Cln2ΔPEST ~32 min, and hTKΔ40-Cln2 ~12 min. Adapted from S. Salama et al.[3]

In Vivo Analysis of hTK Chimera Stability

To measure the stability of hTK chimeras in vivo, they can be tested for their ability to complement a temperature-sensitive mutation in CDC21/ TMP1, the gene encoding for thymidylate synthase.[10,19] Cdc21-1[ts] mutants cannot grow at the restrictive because they are unable to synthesize dTMP, an essential precursor for DNA synthesis. Although yeast lack their own thymidine kinase, introduction of hTK into a cdc21-1[ts] strain (YSS21) allows growth at nonpermissive temperature as long as thymidine (TdR) is provided in the growth media (Fig. 5).[3]

Protocol: Complementation of cdc21-1 by hTK Chimeras

1. Transform YSS21 (MATa cdc21-1 ade his3-Δ200 leu2-Δ1 lys2 pho80::HIS3 trp1-Δ63 ura3-52) with hTK chimeras to be tested. Select transformants on the appropriate selective media at 25°.
2. Streak the resulting transformants onto selective media supplemented with 100 μg/ml thymidine.
3. Incubate the plates for 2–3 days at the restrictive temperature (35°).

---

[19] L. H. Hartwell, J. Bacteriol. 115, 966 (1973).

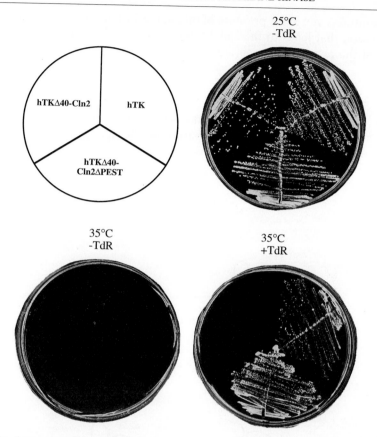

FIG. 5. Complementation of thymidylate deficiency *in vivo* by hTK and hTK-Cln2 chimeras. Individual colonies of yeast strain YSS21, which carries a temperature-sensitive mutation (*cdc21-1*) in the structural gene for thymidylate synthase and harboring a plasmid expressing hTK (right), hTKΔ40-Cln2 (left), or hTKΔ40-Cln2ΔPEST (bottom), were streaked on SCGlc-Leu medium either lacking (−TdR) or containing (+TdR) 100 μg/ml thymidine at either the permissive (25°) or the restrictive (35°) temperature, as indicated. The particular colony of the hTK-containing cells that was streaked on the thymidine-containing plate at 35° had apparently become petite. It is known that mitochondrial DNA replication is especially sensitive to dTMP deficiency.[23] Adapted from Salama *et al.*[3]

Selection of Mutants Defective in hTK-X Degradation

An hTK-X chimera that does not provide enough activity to allow growth of a *cdc21-1* strain, such as hTK-Cln2 (Fig. 5), can provide the basis for a positive genetic selection for mutations defective in hTK-X degradation. Whereas the starting strain cannot grow at the restrictive temperature, even in the presence of thymidine, isolates that have sustained a mutation that leads to the stabilization of the hTK-X chimera can now grow at the restrictive temperature (Fig. 6).

Protocol: Selection of Mutants That Stabilize hTK-X

1. Construct the appropriate yeast strain for selection. Whereas the strains constructed earlier (YSS21 harboring the appropriate hTK-X expression plasmid) could be used to isolate mutants, integration of the construct avoids false signals related to plasmid copy number that will inevitably arise. To integrate the hTK-X construct, the *promoter-hTK-X-terminator*

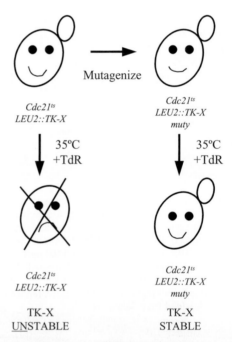

Fig. 6. Selection scheme for mutations stabilizing hTK-X chimeras. A yeast strain carrying a temperature-sensitive mutation (*cdc21-1*) in the structural gene for thymidylate synthase and an integrated copy of the *hTK-X* construct (integrated at LEU2 in this example) is mutagenized. The mutagenized cells are then plated on SCGlc-Leu medium containing 100 μg/ml thymidine (TdR) at the restrictive temperature (35°). Only those cells that sustain a mutation that stabilizes the hTK-X chimera should be able to grow at the nonpermissive temperature in a thymidine-dependent manner.

fragment can be cloned into a yeast integrating plasmid (YIp), such as the ones mentioned earlier. The resulting plasmid can then be integrated into the yeast genome by targeting to the genomic locus of the plasmid's auxotrophic marker.[20]

2. Generate mutants. A variety of procedures can be used to generate mutants. Classical mutagenesis techniques such as chemical mutagenesis by ethylmethane sulfonate (EMS) or UV mutagenesis[18] allow for coverage of the entire genome, but can present challenges in identifying the affected gene. Insertional mutants created by shuttle mutagenesis with bacterial transposons[21,22] have the advantage of enabling one to rapidly identify the affected locus in a mutant, but have the disadvantage of generating a more limited number of mutants.

3. Isolate mutants defective in hTK-X degradation. Plate the mutagenized strain onto media containing 100 μg/ml thymidine at a density of approximately 1×10^5 viable cells/10-cm plate. Incubate plates for 3–5 days at the *cdc21-1* restrictive temperature (35°) and select colonies that grow under these conditions. Retest these colonies for their ability to grow at 35° on thymidine-containing media.

4. Check the candidate mutants for their ability to grow at 35° in the absence of thymidine to eliminate *CDC21* revertants.

Isolates that show thymidine-dependent growth at 35° potentially have mutations that stabilize hTK-X and protein X. At this stage, the mutants can be tested for their ability to stabilize hTK-X using the assays described in this article. For a mutant to be of true interest, however, it should also stabilize protein X. Therefore, one should try to examine the effect of these mutations on protein X using biochemical or, if possible, genetic methods as soon as possible in the analysis of these mutants.

Acknowledgments

I thank Jeremy Thorner for his critical reading of this manuscript; Jeff Silva for helpful advice; and the other members of Microbia for their support. The work on hTK-Cln2 chimeras described here was performed in the laboratory of Jeremy Thorner at the University of California, Berkeley, and was supported by a predoctoral fellowship from the Howard Hughes Medical Institute, by NIH Research Grant GM21841, and by facilities provided by the Cancer Research Laboratory of the University of California, Berkeley.

[20] R. Rothstein, *Methods Enzymol.* **194,** 281 (1991).
[21] M. Hoekstra, H. Seifert, J. Nickoloff, and F. Heffron, *Methods Enzymol.* **194,** 329 (1991).
[22] N. Burns, B. Grimwade, P. Ross-Macdonald, E. Choi, K. Finberg, G. Roeder, and M. Snyder, *Genes Dev.* **8,** 1087 (1994).
[23] C. S. Newlon and W. L. Fangman, *Cell* **5,** 423 (1975).

[9] Use of Fusions to Thymidine Kinase

By CHRISTIAAN KARREMAN

Introduction

There are four commonly used natural genes that can be selected for and against in mammalian cells. All four genes code for proteins that are part of the nucleotide metabolic pathways and, by manipulating the growth medium, cells will either selectively survive their expression or die from it. These four genes are the thymidine kinase (*tk*) gene, the human gene for hypoxanthine phosphoribosyltransferase (HPRT),[1,2] the prokaryotic counterpart of HPRT, xanthine–guanine phosphoribosyltransferase, *gpt*,[3,4] and the cytosine deaminase (*codA*) gene of *Escherichia coli*.[5,6]

The use of the *tk* gene was established very early and circumstances were described to select cells either expressing the gene or not.[7–10] The physiological substrate of tk is thymidine, which is phosphorylated as part of the salvage pathway. The use of medium containing 5-bromodesoxyuridine, an analog of thymidine that tk also recognizes, allows the selection of *tk⁻* cells, as the incorporation of this nucleotide is lethal.

Cells expressing the *tk* gene can be selected in medium that makes the salvage pathway the only source of phosphorylated thymidine nucleotides. This is normally done by blocking *de novo* synthesis with the drug aminopterin and providing thymidine in the medium. Because aminopterin not only blocks the thymidine synthesis but also the purine pathway, hypoxanthine must also be added (HAT medium).

The use of *tk* as a selective marker in transfection experiments was also established very early using either the *tk* gene of herpes simplex virus (HSV-*tk*) or that of chicken.[11–13] A problem is that the use of *tk*, both in

[1] R. J. Albertini, J. P. O'Neill, J. A. Nicklas, N. H. Heintz, and P. C. Kelleher, *Nature* **316**, 369 (1985).
[2] S. C. Lester, S. K. LeVan, C. Steglich, and R. DeMars, *Somatic Cell. Genet.* **6**, 241 (1980).
[3] C. Besnard, E. Monthioux, and J. Jami, *Mol. Cell. Biol.* **7**, 4139 (1987).
[4] K. J. Spring, J. S. Mattick, and R. H. Don, *Biochim. Biophys. Acta* **1218**, 158 (1994).
[5] C. A. Mullen, M. Kilstrup, and R. M. Blaese, *Proc. Natl. Acad. Sci. U.S.A.* **89**, 33 (1992).
[6] K. Wei and B. E. Huber, *J. Biol. Chem.* **271**, 3812 (1996).
[7] S. Bacchetti and F. L. Graham, *Proc. Natl. Acad. Sci. U.S.A.* **74**, 1590 (1977).
[8] W. C. Topp, *Virology* **113**, 408 (1981).
[9] E. H. Szybalska and W. Szybalski, *Proc. Natl. Acad. Sci. U.S.A.* **48**, 2026 (1962).
[10] J. W. Littlefield, *Science* **145**, 709 (1964).
[11] M. Wigler, S. Silverstein, L. S. Lee, A. Pellicer, Y. C. Cheng, and R. Axel, *Cell* **11**, 223 (1977).

positive and in negative selection, is restricted to cells not expressing the endogenous enzyme. For practical purposes, this means that tk-negative substrains have to be generated before the *tk* gene can be used. This has to be done for every cell line of interest, and the resulting cell lines have to be checked thoroughly afterward. The accompanying increase of time and work has hampered the use of *tk* not only as a selectable marker, but also for other applications, such as the use of HSV-tk in enzymatic expression assays.[14]

The application of *tk* as a negative marker changed dramatically when substrate analogs were developed that HSV-*tk* recognizes, but for which the mammalian protein has no apparent affinity. A number of these analogs (prodrugs) have no apparent effect on cells, but their phosphorylated metabolites are highly toxic. Among the more commonly used are FIAU [1-(2-deoxy-2-fluoro-β-D-arabinofuranosyl)-5-iodouracil], acyclovir{9-[(2-hydroxyethoxy)methyl]guanine}, and ganciclovir {GANC-2-amino-9-[2-hydroxy-1-(hydroxymethyl)ethoxymethyl]-9H-purine-6(1H)-on}.

The negative selection with acyclovir and GANC is reported to be faster than that using FIAU,[15,16] and in this article, only results with GANC will be given.

The use of HSV-*tk* as a negative selectable marker was widely accepted in areas such as homologous recombination[16] and suicide strategies in tumor gene therapy.[17] In the latter application, one characteristic of tk is very important, the bystander effect: not only cells expressing the gene are killed, but also cells directly surrounding the expressing cell.[18] This process is dependent on the existence of gap junctions[19] and is thought to be the result of the transport of toxic metabolites.

Use of *tk* as a positive marker has never reached the popularity it has as a negative marker. This is a result of the fact that minus mutants of the recipient cells are still needed. In this area, *tk* cannot compete with such

---

[12] F. Colbere-Garapin, S. Chousterman, F. Horodniceanu, P. Kourilsky, and A. C. Garapin, *Proc. Natl. Acad. Sci. U.S.A.* **76**, 3755 (1979).

[13] M. Perucho, D. Hanahan, L. Lipsich, and M. Wigler, *Nature* **285**, 207 (1980).

[14] J. J. Jonsson and R. S. McIvor, *Anal. Biochem.* **199**, 232 (1991).

[15] E. Borrelli, R. Heyman, M. Hsi, and R. M. Evans, *Proc. Natl. Acad. Sci. U.S.A.* **85**, 7572 (1988).

[16] S. L. Mansour, K. R. Thomas, and M. R. Capecchi, *Nature* **336**, 348 (1988).

[17] Z. Ram, K. W. Culver, S. Walbridge, R. M. Blaese, and E. H. Oldfield, *Cancer Res.* **53**, 83 (1993).

[18] S. M. Freeman, K. A. Whartenby, J. L. Freeman, C. N. Abboud, and A. J. Marrogi, *Semin. Oncol.* **23**, 31 (1996).

[19] C. Denning and J. D. Pitts, *Hum. Gene Ther.* **8**, 1825 (1997).

simple selectable markers as the resistance genes against a number of antibiotics. The most commonly used are genes that confer resistances to neomycin[20] and hygromycin.[21]

A number of approaches clearly require functional cassettes that can be selected positively in one phase of the experiment and negatively in another.[22,23] The optimal solution would be the use of one single gene for such a task. An obvious next step was the combination of HSV-*tk* with other genes by gene fusion. These HSV-*tk* fusions can be selected negatively with GANC and positively with a simple antibiotic selection marker. The choice of these second genes was for the most commonly used resistance markers against neomycin or hygromycin.[24,25]

A problem with these new fusion genes is that many cell lines already have a resistance to neomycin or hygromycin. Furthermore, not all cells are readily selectable with these drugs (especially true with hygromycin). In these cases, another positive selectable gene should be fused to HSV-*tk*.

What Genes to Fuse with HSV-*tk*?

The gene to combine with HSV-tk is dependent on the cell line to be used later in the experiments. Some cell lines have a higher than average resistance to some drugs, which makes the selection of transfected cells rather difficult. The selection is between cells having a natural resistance and those having an increased resistance due to extra resistance genes. Because this can be a long and tedious process, the choice of the positive moiety of the fusion gene is one best made after some experience with the cell line.

In this regard, there seems to be no real problem with the HSV-*tk* gene itself. All cells will readily accept its products and the cytotoxicity of GANC is very low. So in case of an exceptionally high resistance to the metabolites of GANC, concentrations of GANC can always be increased to levels sufficient to kill all cells.

How to Fuse the Genes?

Rationale

There are two possible orientations, 5' and 3', to the other gene. Luckily, HSV-*tk* can be fused at both ends without losing its activity.[24,25] Full-length

[20] A. Jimenez and J. Davies, *Nature* **287**, 869 (1980).
[21] L. Gritz and J. Davies, *Gene* **25**, 179 (1983).
[22] S. Karreman, H. Hauser, and C. Karreman, *Nucleic Acids Res.* **24**, 1616 (1996).
[23] S. Karreman, Thesis TU-Braunschweig, FRG (1997).
[24] F. Schwartz, N. Maeda, O. Smithies, R. Hickey, W. Edelmann, A. Skoultchi, and R. Kucherlapati, *Proc. Natl. Acad. Sci. U.S.A.* **88**, 10416 (1991).
[25] S. D. Lupton, L. L. Brunton, V. A. Kalberg, and R. W. Overell, *Mol. Cell. Biol.* **11**, 3374 (1991).

HSV-*tk* can be used, complete from the first codon to the last. No amino acid linker (spacer) between HSV-tk and the second part of the protein seems to be necessary for the negative selection to function, at least in the cases tested so far. However, it is often very handy to have extra restriction sites situated directly on the border of the two moieties of the fusion protein. These can be introduced with no apparent negative effects either on the expression or on the functionality of HSV-*tk*.

If no similar information is available for the other gene, both the 5' and the 3' position should be tried for the best results. That there can be a large difference between fusion genes that differ only in their respective orientation is obvious in the example of the fusion of the puromycin resistance gene (PAC) with the cytosine deaminase gene (*CodA*) of *E. coli*.[26] This latter gene is an alternative for the HSV-*tk* gene in the construction of positive/negative selectable fusion genes (see later). The fusion of *CodA* with PAC at the 3' end, CODAPAC, is completely nonfunctional. However, the fusion in the other orientation, PACCODA, is fully functional. In this case no spacer was used and it is possible that the apparent mutual obstruction could have been elevated with the use of a flexible stretch of amino acids.

Method and Example

The most simple way to perform the actual fusion is by polymerase chain reaction (PCR).[27] As an example, fusion of the HSV-*tk* gene and the gene for puromycin resistance (PAC; at the 5' end!) was performed in a one-step PCR protocol[26] using the following three primers:

5'-GGGGCGGCCGCACC**ATGACCGAGTACAAGCCCAC**-3'
5'-**GACCCGCAAGCCCGGTGCC**ACTAGTATGGCTTCGTACCCCGGCCATC-3'
5'-GGGTCGACTCAGTTAGCCTCCCCCATCTC-3'

These primers were designed for high expression and easy cloning. Next to homology with the genes they contain the perfect Kozak sequence overlapping the ATG and a *Not*I site in the first oligonucleotide and a *Sal*I site in the third oligonucleotide (restriction sites are underlined in the respective oligonucleotides). On the border of PAC sequences (bold) and HSV-*tk* sequences (normal typeface), an *Spe*I site was incorporated. The following protocol was used.

Chemicals

10× PCR buffer (Boehringer Mannheim GmbH, FRG)
dNTPs, 10 m*M* each, pH 7.5

[26] C. Karreman, *Gene* **218,** 57 (1998).
[27] C. Karreman, *BioTechniques* **24,** 736 (1998).

Primer No. 1: 33-mer, stock 17 ng/μl
Primer No. 2: 47-mer, stock 2.4 ng/μl
Primer No. 3: 29-mer, stock 15 ng/μl
Plasmid PAC 1/1000 dilution of normal plasmid preparation ("mini")
Plasmid HSV-*tk* 1/1000 dilution of normal plasmid preparation ("mini")
Taq polymerase 3.5 U/μl (Sigma-Aldrich Chemie GmbH Deisenhofen, FRG)

Reaction Mix

6 μl of 10× buffer
1.2 μl of dNTPs
3 μl of primer No. 1 (5 pmol)
3 μl of primer No. 2 (5 pmol)
3 μl of primer No. 3 (5 pmol)
1 μl of plasmid PAC
1 μl of plasmid HSV-*tk*
0.5 μl of *taq* polymerase
41.3 μl H$_2$O; cover with 40 μl mineral oil

PCR Protocol

1 min 95°
1 min 59°
1 min/1000 bases 72° (1909 bp so in this case 2 min)
35 cycles

The resulting fragment is cloned into the pCRTOPO vector (Invitrogene, B.V. Groningen, the Netherlands).

The sequencing of the DNA is recommendable if *Taq* polymerase is used as its error rate is rather high. There are other polymerases that can be used that have a significant higher specificity. After correct clones are identified, the fusion product can be cloned into an expression vector and used. A good vector will express the fusion gene under the control of a strong promoter and will also code for an independent, second positive selection marker. The latter will enable the selection of transfectants even if the positive moiety of the fusion is not functional.

Positive Selection of Fusion Proteins

Rationale

Of course, the positive selection of fusion genes is comparable with the normal selection of drug resistances. However, there is one important difference. Normally one is just interested in getting resistant colonies; now the resulting colonies must have an extra property: they can be subsequently killed. At this stage it is important to realize that colonies transfected with integrating vectors are very inhomogeneous.[28] Some will survive drug levels only minimally higher than levels nontransfected cells can survive; other colonies will grow well in medium containing a 10-fold higher dose. The reason is clear. The integration site of the construct will dictate its expression level. The positive selection applied will not only kill those cells that are not transfected, but also those expressing the fusion protein at levels insufficient to mediate resistance. Depending on the level of antibiotic used, this can be a large or small percentage of the total transfectants.[29] At the molecular level, this means there is a large difference in the average number of protein molecules in cells selected under different protocols. Obviously, one can influence this number drastically by selecting with different concentrations of the drug in the medium.

The number of HSV-tk molecules a cells has will determine the sensitivity of the cell in the negative selection. As a result, the way the positive selection is performed is of great influence on the negative selection (see Fig. 1). As a rule the differences are even more pronounced if the promoter used is weaker, as the spread of expression levels is then in the lower regions.

Method/Protocol for Positive Selection

A good way to determine the behavior of the fusion gene is to perform a test transfection/selection.

1. Use a transfection protocol optimized for the cells under study.
2. After 2 days, use trypsin if working with adherent cells and seed the cells 1/10 in a six-well plate.
3. In these wells, add antibiotic to different final concentrations, starting with 75% of the normal concentration for the cells and going up in a series, e.g., 0.75, 1, 1.5, 2, 2.5, 5× normal doses.
4. Select for colonies with regular medium changes and color the colonies with crystal violet.

   a. Discard the medium.
   b. Add 3 ml of the following solution: 5.0 g crystal violet, 8.5 g NaCl, 143 ml formaldehyde (35–37%), and 500 ml ethanol. Fill with H_2O until 1000 ml.

---

[28] R. C. Hoeben, A. A. Migchielsen, R. C. van der Jagt, H. van Ormondt H, and A. J. van der Eb, *J. Virol.* **65,** 904 (1991).

[29] M. Duch, K. Paludan, J. Lovmand, M. S. Sorensen, P. Jorgensen, and F. S. Pedersen, *Hum. Gene Ther.* **6,** 289 (1995).

Efficient positive moiety in fusion   Inefficient positive moiety in fusion

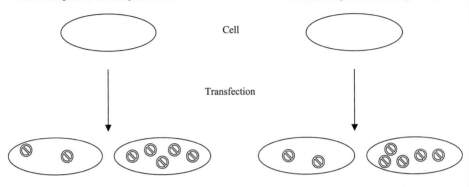

Cell

Transfection

Mixture of low and high producers   Mixture of low and high producers

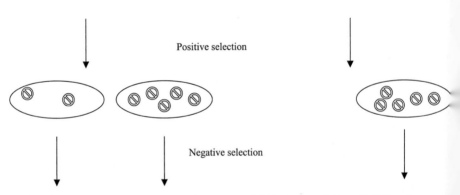

Positive selection

Negative selection

Mixture of cells wherein some survive the initial concentration of GANC and give rise to false positives   Cells in these population will all die rather uniformly at a given concentration of GANC

FIG. 1. Influence of the efficiency of the positive moiety on negative selection. (Left) A flow diagram is given in which a fusion gene with a very efficient positive selectable moiety is used for the transfection of mammalian cells. Almost all cells transfected will survive the first round off selection. The resulting mixture of cells will consist of high as well as low producing cells. The latter will be very resistant to GANC in the subsequent negative selection. (Right) The same kind of experiment with an inefficient positive marker. The initial mixture of high and low producing cells is changed enormously by the positive selection, resulting in fewer colonies, but these will be, on the average, high producers. These colonies can be killed easily with low levels of GANC.

    c. Incubate for 5 min at room temperature; pour the crystal violet solution back in the bottle; it can be used many times.

    d. Rinse with plenty of tap water.

5. Count the colonies.

This approach provides the certainty that even if the fusion protein does not have the activity of the unfused marker, one will obtain clones in the well with the 0.75× usual concentration.

If this fails, the positive moiety is probably not effective enough to work with. If a plasmid with a second positive marker was used it is still possible to evaluate the functioning of the negative moiety and to obtain some information on positioning effects. Maybe a (longer) spacer between the positive and the negative moiety should be used.

If all works well, it is possible to get some idea of how high one can go with the selection. The number of clones will go down rapidly as the selective pressure goes up; those clones able to withstand the highest concentrations are those killed most easily during the negative selection.

Choose a concentration of antibiotic as high as possible for further experiments. It should of course still give enough colonies for the proper application. Stick to this concentration after it is set, as otherwise the subsequent negative selection becomes unpredictable. The standardization of selection protocols can be very helpful in minimizing the later occurrence of false positives.

Negative Selection

Rationale

As mentioned before, the average concentration of the fusion protein in the cell can be influenced directly by the positive selection method. The actual concentrations in one particular cell and its descendants in a single colony are quite different than in cells from another colony. This is especially noticeable when clone mixtures are tested. To kill every last cell of a whole population may be a problem. This is especially the case when the positive selection was on a very effective marker, where even low levels of fusion protein are sufficient for cell survival.

Method for Negative Selection

The following method tests for the efficiency of GANC on colony pools.

1. If it is possible, select the cells positively on two different markers (see Method described under Positive selection).

2. After selection, mix at least 250 of the resulting colonies. If the vector has an extra marker, make another pool from colonies that were selected using this second marker. Seed both these pools in a six-well plate (two plates total).

3. Dilute the cells before seeding so that cells are spread apart to minimize the bystander effect, as otherwise the high producers in the cell pool will kill the low producers. This will look very good in the test, but will give false positives in later experiments (1/20 will do in most cases; very fast growing cells must be diluted even higher).

4. When using GANC, add this to a final concentration of, respectively, 0, 1, 2, 4, 8, and 16 μM. Add the original antibiotic (positive selection) as well in every well.

5. Wait for 1 week and color the resulting cells with crystal violet.

The plate with cells selected on the fused marker will give an idea of how well the fusion gene is expressed, and the second plate on how often the fused marker is downregulated or lost, as the cells can do this without losing resistance to the second marker. For most practical approaches it appears that the loss or switching off of the fusion marker is so rare that it is no real problem; the variance in expression between the various cell clones is a bigger problem.

Example for Coupling between Positive and Negative Selection

The best example demonstrating the influence of the positive marker in the final negative selection is the comparison of two very related fusions, HSV-*tk* with two genes for Blasticidin S deminase, *bsr* and BSD.[30]

The origins of these genes are prokaryotic in the case of *bsr* and eukaryotic in the case of BSD. Both catalyze the same reaction and both can be used in mammalian cells. BSD has a wider application as it confers resistance to more cell lines; in some, *bsr* is nonfunctional.[31] This has probably to do with the actual deaminase levels reached, as the codon usage of *bsr* is suboptimal for mammalian cells. In NIH 3T3 cells, both genes allow direct selection for colonies, albeit that the BSD gene is a very effective selection marker, at least as good as the *neo* gene, and that *bsr* is rather ineffective.

Fusions of these two genes with the HSV-*tk* gene were put under the control of two promoters (the HSV-*tk* promoter and the SV40 promoter) and tested positively and negatively. The positive selection was performed at a final concentration of 1.5 U/ml of Blasticidin S. The relative clone

---

[30] C. Karreman, *Nucleic Acids Res.* **26**, 2508 (1998).
[31] M. Kimura, A. Takatsuki, and I. Yamaguchi, *Biochim. Biophys. Acta* **1219**, 653 (1994).

numbers after selection are depicted in Fig. 2. In the case of BSD, the clone numbers of transfections with the HSV-*tk* fusion are comparable with transfections of the unfused resistance marker under control of the SV40 promoter. The *tk-bsr* fusions do not seem to be as efficient as the *bsr* gene itself. Behind the HSV-*tk* promoter, *tk-bsr* does not give any clones, but downstream of the SV40 promoter the fusion gives at least a sufficient number of clones to work with. In the latter case the *bsr* construct gives about 20% of the colonies obtained with the corresponding BSD construct. The verdict seems clear: the BSD gene is superior to the *bsr* gene.

The negative selection of cells harboring the two constructs with the fusion downstream of the SV40 promoter shows a marked difference be-

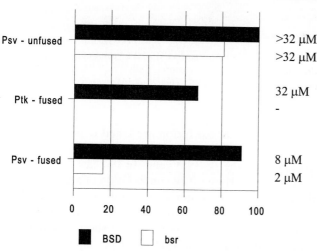

FIG. 2. Relative clone numbers after equimolar transfections of the various plasmids into NIH 3T3. The genes used are BSD and *bsr,* either unfused or fused with HSV-*tk*. The genes are under control of either the SV40 promoter (*Psv*) or the HSV-*tk* promoter (*Ptk*). (Left) BSD is more effective than *bsr* in NIH 3T3 and fused genes are not quite as effective as unfused versions. The combination of the *tk* promoter with the HSV-*tk-bsr* fusion does not give any colonies at all under standard selection. Only under the SV40 promoter are a few colonies obtained, about 20% when compared with HSV-*tk*-BSD. (Right) The concentration of GANC is given needed to kill cells harboring the constructs. Cells without HSV-*tk* can take concentrations of 32 μM GANC without any ill effect. The constructs that were so simple and efficient during the positive selection now have a major drawback. The fusions between HSV-*tk* and BSD are not easily selected against, especially when expressed under the control of a rather weak promoter. The use of an inefficient positive marker will not give these problems. The number of clones after selection will be lower than after transfection with a construct based on an efficient marker. However, these colonies are more uniformly sensitive GANC.

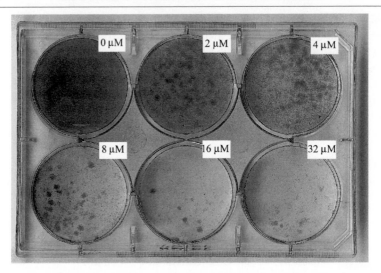

FIG. 3. Negative selection of NIH 3T3 cells transfected with a vector expressing an HSV-*tk*-BSD fusion gene under control of the HSV-*tk* promoter. Cells were selected with 1.5 U/ml Blasticidin S and various concentrations of GANC. The latter are given for each well.

tween them. Cells expressing the *tk-bsr* fusion are killed uniformly at very low levels of GANC (2 μM), whereas one needs 8 μM GANC for this to happen with cells that contain the *tk*-BSD gene. In the case of the HSV-*tk* promoter, this is even more extreme. The combination of a very efficient marker (BSD) with the HSV-*tk* promoter allows cells with the resulting low levels of protein to survive the selective pressure. These cells give problems when negative selection is applied. Figure 3 shows the negative selection of cells transfected with the *tk*-BSD construct under control of the HSV-*tk* promoter. As can be seen, almost all cells die at GANC concentrations of about 8 μM. However, some grow out and form colonies even at 16 and 32 μM. Now the verdict is not so clear; cells expressing *bsr* fusions may be harder to get but they are much easier to kill.

Conclusion of Negative Selection

For the reason outlined earlier, negative selection should always be tested with colony pools. A pool of at least 250 colonies must be used to have a reliable estimate of the variation between cells. Another good practice is to use strong promoters for the expression of fusion genes as these will partly overcome the problems that exist because of the "enzyme level" versus "efficiency per molecule" situation.

HSV-*tk* versus Other Negative Genes

Two other genes for negative selection are used with rising frequency in cell culture and suicide gene therapy. The first is the cytosine deaminase gene (*CodA*) from *E. coli* that confers lethality toward 5-fluorcytosine (5FlC). No analog of this gene is present in higher eukaryotes and the prokaryotic gene can be selected positively and negatively.[5,6] The positive selection of *CodA* is accomplished by blocking the *de novo* synthesis of the pyrimidines by PALA [*N*-(phosphonacetyl)-L-aspartate], thereby making the cell dependent on the conversion of cytosine to uracil. An additional problem is that the cell does not use this uracil to form uridine as the normal equilibrium favors uracil. This can be circumvented by adding extra iosine to the medium to shift the equilibrium toward uridine. The effects of all this on other processes in the cell are unknown. Coupled with the fact that PALA is not available commercially makes the positive selection rather unattractive, which is reason enough to go for fusion constructs analogous to *tk*.

Like the HSV-*tk* gene, *CodA* is fusionable at the 5' and 3' end without a spacer[26] and can be selected in all cell lines tested so far. This makes *CodA* an equal partner of HSV-*tk*. It also means that two completely independent positive/negative selectable markers can be used in the same cell, i.e., TKNEO and PACCODA. The only difference noticed so far is the time needed for selection. When HSV-*tk* is expressed at a reasonable level, GANC will kill most cell lines in 3 days; the *CodA*/5FlC selection takes about a week.

The bystander effect of cytosine deaminase is higher than that of tk[32] and therefore the popularity of this gene is growing rapidly.

The second gene used in (tumor) gene therapy is the PNP (purine nucleoside phosphorylase) (*DeoD*) gene of *E. coli*. The eukaryotic analog of this gene exists in mammalian cells, but the prokaryotic protein recognizes some substrates the endogenous protein does not. The lethal adenosine analog commonly used is 6-methylpurine-2'-deoxyriboside (MeP-dR).[33] The use of this gene is restricted to gene therapy as the products formed by PNP are readily diffusible and will reach lethal concentrations in the medium in cell culture.[34] For PNP, no gap junctions appear to be necessary for a bystander effect. The potency is so high that the medium of expressing cells can be filtered and used to kill another batch of cells.

[32] D. K. Hoganson, R. K. Batra, J. C. Olsen, and R. C. Boucher, *Cancer Res.* **56,** 1315 (1996).
[33] E. J. Sorscher, S. Peng, Z. Bebok, P. W. Allan, L. L. Bennett, Jr., and W. B. Parker, *Gene Ther.* **1,** 233 (1994).
[34] B. W. Hughes, S. A. King, P. W. Allan, W. B. Parker, and E. J. Sorscher, *J. Biol. Chem.* **273,** 2322 (1998).

This may be a desired effect in tumor gene therapy where only a small percentage of the tumor cells is actually transfected by the suicide vector. Even this low efficiency will give good end results as the primary producing cells not only kill themselves, but also a large number of cells surrounding them. However, for a number of applications in cell and tissue culture this toxicity makes PNP unsuitable.

At the moment nothing is known about the properties of PNP in regard to fusion with other genes.

Acknowledgments

This work was supported by the Deutsche Forschunganstalt für Luft- und Raumfahrt e.V. (DLR), Grant 01KV95380. The author expresses his gratitude to Drs. S. Karreman and U. Horsten for critically reading the manuscript.

[10] In Vivo Analysis of lacZ Fusion Genes in Transgenic Drosophila melanogaster

By STEPHEN SMALL

Introduction

In recent years, transgenic Drosophila lines containing lacZ fusion genes have been used extensively for studying developmental mechanisms. This approach has identified cis-acting elements (enhancers and basal promoter elements) that control when and where a particular gene will be expressed (e.g., references 1–3). lacZ reporter genes have also been used for the analysis of mechanisms that control translation and intracellular localization of RNA transcripts (e.g., reference 4). Finally, lacZ fusion genes have been used to study intracellular transport mechanisms. For example, a lacZ-kinesin fusion gene has been used to demonstrate that a reorientation of microtubule arrays is critical for establishing the anterior–posterior axis during oogenesis.[5]

[1] Y. Hiromi, A. Kuroiwa, and W. J. Gehring, Cell 43, 603 (1985).
[2] K. Harding, T. Hoey, R. Warrior, and M. Levine, EMBO J. 8, 1205 (1989).
[3] T. Goto, P. Macdonald, and T. Maniatis, Cell 57, 413 (1989).
[4] P. M. Macdonald and G. Struhl, Nature 336, 595 (1988).
[5] I. Clark, E. Giniger, H. Ruohola-Baker, L. Y. Jan, and Y. N. Jan, Curr. Boil. 4, 289 (1994).

lacZ reporter genes have been used most successfully for unraveling the *in vivo* transcription mechanisms that control embryonic pattern formation (e.g., references 6, 7). These studies combine genetic, biochemical, and transgenic experiments as follows: By crossing a line containing a *lacZ* transgene into *Drosophila* mutants, regulatory genes involved in activation and repression of the transgene can be identified. In many cases, these genes encode transcription factors, and DNA-binding studies are performed to determine whether they bind directly to sites within the enhancer. Individual binding sites are then subjected to *in vitro* mutagenesis, and *lacZ* reporter genes containing mutant forms of the enhancer are tested *in vivo*. If an identified binding site is important for the function of the enhancer, this will be reflected in changes in its expression pattern.

Analysis of *lacZ* transgenes requires an efficient method for introducing exogenous DNA into the genome. This was first accomplished by Rubin and Spradling,[8] who took advantage of the transposition properties of P-elements, which are natural transposons found in some strains of *Drosophila* (see reference 10 for review). Their work is the basis for the protocol for generating transgenic flies that is presented here. Also described is our current method for the detection of *lacZ* mRNA expression in transgenic embryos.

DNA Constructs Used for Genetic Transformation of *Drosophila*

The transformation strategy worked out by Rubin and Spradling[8] requires the simultaneous injection of two DNA constructs. The first carries the *lacZ* reporter gene of interest and a selectable marker gene, both of which are flanked by sequences from the 5' and 3' ends of P-elements. These sequences include the terminal repeats that are critical for transposition of the native elements. The second plasmid (the "helper" plasmid) contains a nearly intact P-element that encodes functional transposase activity, but cannot itself integrate due to truncations that remove one of the terminal repeats.[8] The helper plasmid currently used in our laboratory is pUChs$\pi^{\Delta 2-3}$ (also called pTURBO), which was generated by Don Rio (personal communication), and generates high levels of transposase activity.

To make transgenic lines that are not mosaic, it is necessary that transformation occurs in cells of the germ line. This is accomplished easily in *Drosophila* because P-element transposase is active only in the germ line due to a specific splicing event.[9] However, the DNA must be introduced into

[6] D. Stanojevic, S. Small, and M. Levine, *Science* **254,** 1385 (1991).
[7] S. Small, A. Blair, and M. Levine, *EMBO J.* **11,** 4047 (1992).
[8] G. M. Rubin and A. C. Spradling, *Science* **218,** 348 (1982).
[9] F. A. Laski, D. C. Rio, and G. M. Rubin, *Cell* **44,** 7 (1986).

very young embryos to take advantage of the fact that early embryogenesis occurs within a syncytium, in which all nuclei share a common cytoplasm. Nuclei that are progenitors of the germ line migrate toward the posterior of the embryo and are encapsulated by cytoplasmic membranes to become the pole cells. Efficient germ line transformation can only take place if the DNA is injected before the formation of these cells. To ensure this, embryos are injected as soon as possible after egg deposition, and the injection process is carried out at 18°, at which temperature the speed of development is slowed significantly.

Many available vectors facilitate the construction of *lacZ* reporter constructs for transgenic experiments (see reference 10 for review). All transformation vectors contain the critical P-element ends and easily selectable marker genes that aid in the identification of transgenic flies. Some commonly used markers include the eye color genes *white* (*w*) and *rosy* (*ry*), and genes that confer resistance to alcohol (*adh*) or to G418 (*neoR*). The vectors used most often in our laboratory are modifications of the CaSpeR vector,[11] which contains a mini-*white* (mini-*w*) gene as a selectable marker.[12] The *w* gene is normally involved in the production and distribution of red and brown pigment in the adult *Drosophila* eye.[13] Neither pigment is visible in strong *w* mutants, and the eyes appear white. Introducing constructs carrying the mini-*w* gene into *w* mutants causes a variable rescue of the white eye color that ranges from very light yellow to red. This variation seems to be caused by different positions of insertion in individual lines. In many cases, the eyes of homozygotes appear significantly darker than those of their heterozygous siblings, a difference that can be used to great effect in genetic experiments. However, because the eye colors of some transformant lines appear very light, there are probably cases where transgenic lines are missed because of undectable expression in the adult eye. To circumvent this problem, vectors have been generated in which multiple binding sites for glass, an eye-specific trancriptional activator, have been inserted upstream of the mini-*w* gene.[14] In general, this modification causes much stronger pigment production, thereby facilitating the identification of transformants. However, these darker eye colors are not as useful for differentiating between flies carrying one versus two copies of a given transgene.

[10] M. Ashburner, in "*Drosophila*, a Laboratory Handbook." Cold Spring Harbor Laboratory Press, Cold Spring Harbor, NY, 1989.
[11] V. Pirrotta, *Biotechnology* **10**, 437 (1988).
[12] V. Pirrotta, H. Steller, and M. P. Bozzetti, *EMBO J.* **4**, 3501 (1985).
[13] E. Hadorn and H. Mitchell, *Proc. Natl. Acad. Sci. U.S.A.* **37**, 650 (1951).
[14] M. Fujioka, Y. Emi-Sarker, G. L. Yusibova, T. Goto, and J. B. Jaynes, *Development* **126**, 2527 (1999).

Generating *Drosophila* Lines That Carry P-element Transgenes

DNA Preparation and Loading Injection Needles

Materials

Glass capillary needle puller (Camden Instruments No. 753 microelectrode puller)

Glass capillary tubes (Precision Instruments 1BBL w/FIL 1.0 mm 4 IN)

Procedure. We have tried several methods of DNA preparation for injection constructs and have determined that many different methods will work for this purpose. In our experience, however, the best method is to grow a large culture for each clone (~300 ml of 2× YT medium) and purify the DNA by two rounds of EtBr-CsCl density gradient centrifugation using standard methods.[15] After removal of the DNA band from the second spin, the sample is diluted 1:3 with TE (10 mM Tris, pH 8.0, 1 mM EDTA) before repeated extraction with sec-butanol to remove all traces of EtBr. This step causes a significant reduction in sample volume, which should be replaced by adding TE after the final extraction. The sample is mixed very well and is precipitated with 2.5 volumes of 100% EtOH. The DNA is then reprecipitated twice (using 0.3 M sodium acetate, pH 5.2) in a large volume (5 ml) before a final coprecipitation with the helper plasmid. For coprecipitation, we normally use 10 μg of the reporter and 2 μg of helper plasmid. After spinning down the DNA pellet, we wash with 70% EtOH and dissolve in 20 μl of injection buffer (0.1 mM KPO$_4$, pH 6.8, 5 mM KCl).

Needles for injection are prepared from capillary tubes using a commercial needle puller. Many researchers prewash the needles, but we find this is unnecessary given the following precautions. Pulled needles are stored in a dry chamber. For this purpose, we use a 150-mm petri dish with a line of Playdoh across the middle for suspending the needles so that their tips are not broken before loading. To prepare each needle for loading, we wipe clean the barrel near the blunt end with a Kim-Wipe, which is damp with 70% EtOH, and use a fine three-corner file to make a transverse scratch across the clean area. The needle is then broken along this scratch, creating a fresh blunt end onto which the DNA solution will be loaded. Needles are transferred to a petri dish–Playdoh chamber that contains several wet Kim-Wipes to ensure that the DNA solution does not evaporate after loading. The DNA sample itself is centrifuged for several minutes immediately before loading into injection needles. This step removes partic-

---

[15] J. Sambrook, E. Fritsch, and T. Maniatis, *in* "Molecular Cloning, A Laboratory Manual." Cold Spring Harbor Laboratory Press, Cold Spring Harbor, NY, 1989.

ulate matter that might clog the needle during the injection process. Each needle is loaded by touching a 0.5- to 0.7-μl droplet of the DNA solution to the blunt end. The DNA should transfer by capillary action to the pointy end in a few minutes. After loading, the wet chamber containing the loaded needles is sealed with Parafilm and stored at room temperature until needed for injection. We generally load two to four needles for each construct to be injected and never use loaded needles that have been stored overnight.

Embryo Collection and Dechorionation

Solutions and Materials

Drosophila fly stock for injection (yw[67])

Plastic culture bottles with holes punched to allow air flow

Apple juice agar plates (2.5% Bacto-agar, 1× clear apple juice from frozen concentrate, melted, and poured into 35-mm petri dish top plates)

Yeast paste [Baker's yeast (ICN No. 101400 dissolved in water to a consistency of peanut butter)]

Small basket made of stainless steel mesh (No. CX-200-C from Small Parts, Inc., Miami Lakes, FL)

Embryo wash (70 mM NaCl, 0.03% Triton X-100)

Procedure. As mentioned previously, transgenesis in the germ line requires that injection occurs before formation of the pole cells. These cells are formed at nuclear division cycle 10, which corresponds roughly to about 1 hr and 20 min after the embryos are laid.[16] Thus it is critical that the embryos be collected within a 30- to 45-min period after deposition by the females. We set up several cultures of *yw*[67] flies containing ~100 males and 200–300 females per bottle for this purpose. A bit of yeast paste is spread onto a fresh apple juice plate, and the plate is placed on top of the bottle and secured with a piece of tape. Cultures are kept at 25°, 60% humidity. Adult flies feed on the yeast paste, and females lay their fertilized embryos on the apple juice plates. After 30–45 min, plates containing freshly laid embryos are removed, and the embryos are processed for injection.

Embryos are dislodged from the apple juice plates using the embryo wash solution and a natural hair artist's brush and are transferred into a steel mesh basket. After rinsing with embryo wash, the baskets are submerged in 5.25% sodium hypochlorite (commercial bleach) for 2.5 min. This treatment

---

[16] J. Campos-Ortega and V. Hartenstein, "The Embryonic Development of *Drosophila melanogaster*." Springer Verlag, Berlin, 1985.

dissolves the chorion, but leaves the vitelline membrane intact. Embryos are then washed extensively by running distilled H_2O through the basket, and carried to an 18° room for injection.

Embryo Injection

Materials

Dissecting microscope (Nikon SMZ-2B)
Inverted microscope (Nikon TMS with a Plan 20/0.50 phase-contrast lens)
Micromanipulator (Narishige MN-151 joystick manipulator)
Glass coverslips (24 × 50 mm) coated along one edge with a thin layer of embryo glue, which is made by dissolving the glue from No. 4124 Tesa tape (Beiersdorf AG, Hamburg) in heptane overnight, centrifuging the solute to clarify the glue, and adjusting the concentration by adding heptane to an OD 420 of 0.1
Agar blocks: 2.5% Bacto-agar in H_2O, melted and poured into a 150-mm petri plate, cut into blocks ~10 × 40 mm just before use

Procedure. The dechorionated embryos are removed carefully from the steel mesh basket with a dry artist's brush and placed along one edge of the top of an agar block. Using a straight probe, individual embryos are moved to a position near the opposite edge of the agar block and arranged in a line along that edge. Each embryo is oriented perpendicular to the edge with its anterior end facing the edge and spaced so that ~2 embryos widths separate each one from its neighbor. A skilled worker can line up 40 embryos along the edge of the block in about 6 min. After the embryos are lined up, a coverslip coated along one side with glue is positioned above the embryos and lowered so that the glue touches them. The embryos stick readily to the glue, making it possible to transfer the whole line of embryos onto the coverslip. Individual embryos are now oriented so that their posterior ends face the edge of the coverslip.

Embryos must be dehydrated before injection or they will burst when penetrated with the injection needle. Dessication is carried out by placing the coverslip containing the embryos into a drying chamber (a small tightly sealed plastic container filled about halfway with Drierite). The optimal drying time must be determined empirically because it depends on several factors, including how wet the embryos are after washing and the humidity of the injection room. In general, embryos should be dried until they do not leak out cytoplasm when injected. However, they will also die if dried for too long. In our experience, reasonable drying times range from 12 to 20 min. After incubation in the drying chamber, the line of embryos is covered with a layer of No. 700 Halocarbon oil (Halocarbon Products,

River Edge, NJ), which prevents further dessication and allows for gas exchange during embryonic development.

We inject embryos using an inverted microscope with an attached micromanipulator that is used to position the tip of the needle in the microscope field. Pressure for delivery of the DNA is provided by a 50-ml syringe and 18-gauge needle, which is connected to the injection needle by ~2 ft. of plastic tubing (1 mm ID, 0.5 mm wall thickness).

Before injection, the pointy tip of the needle must be broken carefully to allow the correct flow of DNA solution. This is accomplished by positioning a microscope slide on the stage so that one of its edges is perpendicular to the mounted needle. While applying pressure to the 50-ml syringe pump, the edge of the slide is moved toward the needle tip until it touches it lightly, causing the tip to break. Breaking the tip correctly causes a visible outflow of the DNA solution when the needle is touching the edge of the slide, but no flow when the needle is not in contact. The flow will commence when the needle is submerged under oil. For injection, a coverslip containing the embryos is placed onto a 25 × 75-mm microscope slide so that the edge of the coverslip with the embryos extends beyond the edge of the slide. The slide is then placed onto the microscope stage so that the line of embryos is perpendicular to the needle. Embryos are injected individually by moving the stage back and forth, which impales the posterior end of each one onto the stationary needle. When the needle penetrates the embryo, DNA is delivered by applying pressure to the pump. Transfer of the DNA into the embryo can normally be visualized as a slight swirling of the cytoplasm. It is not possible to control the exact amount of DNA delivered to each embryo using the equipment described here. However, if one can see the cytoplasmic swirling and remove the needle from the embryo without losing any cytoplasm, transformation is very likely to take place.

The injected embryos (still under oil) are then transferred into humidified chambers (plastic containers with wet paper towels or sponges) and incubated at 18° until they hatch as first instar larvae. Special care must be taken to ensure that the oil covering the embryos does not run off the coverslips during the incubation period, as this will cause dessication and death of the embryos. This is accomplished by tilting the container so that the oil is held at the edge of the coverslips by surface tension. In these conditions, the embryos will hatch in approximately 48–60 hr. Individual first instar larvae are then picked off the coverslip using a straight probe and transferred to a vial containing normal fly food. Because some of the larvae are likely to be quite weak, a bit of water and yeast is added and mixed into the top layer of the food to enrich the media. The developing larvae are raised at 25°, 60% humidity until they hatch as adults (G_0). Separate crosses are then set up between individual G_0 adults and two to three males or females from the yw^{67} stock to establish individual lines. If a transformation event has taken place in the germ line of any of the G_0 adults, offspring from these crosses will contain individuals with colored eyes due to the function of the mini-*w* gene contained in the transformation vector. As mentioned earlier, detectable eye colors range from light yellow to a very bright red. In our experience, there is no good correlation between the intensity of eye color in the adults and the expression levels driven by the *lacZ* reporter genes during embryogenesis.

As a general rule, we inject at least 300 embryos for each construct. A skilled worker can do this in less than 3 hr. When an injection experiment is successful, approximately 30–40% of the embryos (100–120) will hatch to form larvae. Of these, 50–60% will survive to adulthood, but of these only 50–60% will be fertile. Of the fertile crosses, the rate of trangenic events is between 10 and 20%. Thus, by injecting 300 embryos, it should possible to generate between 5 and 10 independent lines for each construct.

Analysis of lacZ mRNA Expression Patterns by *in Situ* Hybridization

Many methods are used for analyzing the expression patterns driven by *lacZ* fusion genes in *Drosophila*. Some of these are designed to detect the β-galactosidase protein product, either by activity assays or by immunohistochemistry (see protocols in reference 10). However, for experiments designed to study transcription mechanisms or mRNA localization, it is best to detect the mRNA directly by *in situ* hybridization. The earliest method for *in situ* hybridization to *Drosophila* embryos involved hybridizing radioactive DNA probes to tissue sections.[17] This procedure has been largely replaced by nonradioactive methods that permit detection of expression patterns in whole mount preparations of embryos.[18] These experiments rely on the incorporation of modified nucleotides into the probe sequences. The most common modification is the addition of a digoxigenin-11 epitope (DIG) to either 2'-deoxyuridine-5-triphosphate (dUTP) for DNA probes or uridine-5'-triphosphate (UTP) for RNA probes. The DIG moiety is then recognized by a specific antibody, which is conjugated to an enzyme such as alkaline phosphatase (AP). Detection occurs via an enzymatic reaction that yields a colored precipitate.

As suggested earlier, these experiments can be performed using either DNA or RNA probes. DNA probes are easier to make and are acceptable for many experiments. However, we have found that RNA probes provide significantly better results, with stronger signal intensity and extremely

---

[17] E. Hafen, M. Levine, R. Garber, and W. Gehring, *EMBO J.* **2,** 617 (1983).
[18] D. Tautz and C. Pfeifle, *Chromosoma* **98,** 81 (1989).

low background. This is probably a result of the increased stability of RNA–RNA hybrids compared to DNA–RNA hybrids, which permits one to increase the temperature of hybridization and washes when using RNA probes. For example, in our normal hybridization conditions (50% formamide, 5× SSC), the hybridization temperature for DNA probes is 42°, which can be increased to 55° for RNA probes.

Preparation of Digoxigenin-UTP-Labeled Antisense RNA Probes

Solutions

3 M sodium acetate: pH to 5.2 with glacial acetic acid

Tris-saturated phenol: Solid phenol is melted in a hot water bath and extracted three times with equal volumes of 0.2 M Tris, pH 7.5, and stored at 4° for several weeks

Chloroform-isoamyl alcohol: This is a mixture of 24 parts chloroform with 1 part isoamyl alcohol

Note: The following solutions should be prepared using RNase-free reagents and DEPC-treated H_2O.[15]

10× transcription buffer: 400 mM Tris, pH 7.5, 60 mM $MgCl_2$, 100 mM NaCl, 20 mM spermidine–HCl; store in small aliquots at −20°

10× dithiothreitol (DTT): 50 mM DTT in H_2O, stored in aliquots at −20°

2× carbonate buffer: 120 mM Na_2CO_3, 80 mM $NaHCO_3$, pH to 10.2

4 M LiCl

Stop solution: 0.2 M sodium acetate, pH to 6.0 with glacial acetic acid

Hybridization solution: 50% deionized formamide, 5× SSC, 100 μg/ml boiled, sonicated salmon sperm DNA, 50 μg/ml heparin, 0.1% Tween 80; this solution can be stored indefinitely at 4°

Materials

pBluescript (KSII⁺)-*lacZ:* This clone contains a 2.4-kb fragment of the *lacZ* gene inserted into the polylinker of the cloning vector pBluescript (KSII⁺). This clone contains promoters for two bacteriophage polymerases (T3 and T7) that flank the polylinker region. In this particular clone, the T3 promoter flanks the 3′ end of lacZ sequences and is used to make an antisense probe.

10× DIG-UTP NTP mix: 10 mM ATP, 10 mM GTP, 10 mM CTP, 6 mM UTP, 4 mM DIG-UTP; this mix is available commercially from Boehringer Mannheim (Mannheim, Germany)

RNase inhibitor: 50 U/μl (Roche Molecular Biochemicals)

RNA polymerase (T3, T7, or SP6 polymerase, depending on the particular clone used as template)

Carrier RNA: 20 mg/ml tRNA. Dessicated tRNA from a commercial source (Roche Molecular Biochemicals) is dissolved in 10 mM Tris, pH 7.5, 1 mM NaEDTA, 0.3 M sodium acetate, pH 5.2, extracted twice with Tris-saturated phenol, extracted once with chloroform : isoamyl alcohol, precipitated with 2 volumes of EtOH, washed with RNase-free 70% EtOH, and dissolved at 20 mg/ml in DEPC-treated H$_2$O.

Procedure. There are two major concerns in preparing the template for the transcription reaction. First, the DNA vector must be linearized by restriction digestion using an enzyme that cuts at or near the 5′ end of the *lacZ* sequences. Second, the linearized template must be RNase free for the transcription reaction. For the pBluescript (KSII$^+$)-*lacZ* clone, we digest 20 μg of plasmid DNA with the enzyme *PstI* in a total volume of 50 μl. After digestion is complete, we add 40 μl H$_2$O and 10 μl of 3 M sodium acetate, pH 5.2, and mix well. This is followed by two extractions with 100 μl Tris-saturated phenol, one extraction with 100 μl chloroform : isoamyl alcohol, and precipitation with 200 μl 100% EtOH. The pellet is then washed with RNase-free 70% EtOH and dissolved in 20 μl DEPC-treated H$_2$O. The template concentration should be checked by running a small aliquot on an agarose gel with a known standard.

Approximately 1 μg should be used in the transcription reaction, which also includes 1 μl 10× transcription buffer, 1 μl 10× DIG-UTP NTP mix, 1 μl 10× DTT, 1 μl RNase inhibitor, 10 units T3 polymerase, and DEPC–H$_2$O to a total volume of 10 μl. The reaction is incubated for 2 hr at 37°. At the end of this period, 15 μl H$_2$O is added, along with 25 μl 2× carbonate buffer, and the solution is incubated at 65° for 40 min. The carbonate treatment reduces the size of the individual DIG-labeled RNA fragments for better penetration into the embryonic tissues. This reaction is stopped by the addition of 50 μl of stop solution. The probe is then precipitated as follows: 10 μl of 4 M LiCl and 5 μl 20 mg/ml tRNA (a carrier) are added to the probe reaction and mixed together by vortexing briefly. Then 300 μl 100% EtOH is added, and the sample is mixed well and frozen at −20° for 15 min. The precipitated RNA is then pelleted by centrifugation at high speed for 20 min, and the pellet is washed with 70% EtOH, dried briefly, and dissolved in 150 μl hybridization solution. RNA probes made in this way and stored at −20° are extremely stable, with some probes retaining excellent activity for several years.

In Situ Hybridization to Whole Embryos Using Antisense RNA Probes

Most reporter gene studies are focused on the patterning potential of particular *cis*-regulatory sequences that are included in the design of the construct. However, expression patterns observed in individual transgenic lines can vary significantly, depending on their positions of insertion in the genome. Thus it is extremely important to generate multiple independent lines for each construct and to analyze them in parallel. Our general rule is to analyze at least three independent lines as a first test. If the expression patterns are not consistent within those lines, then more lines must be analyzed until a consensus is reached. For each line, 300–500 transgenic embryos are collected and stained in each *in situ* hybridization experiment. In most cases, these embryos must be within a specific age range, as many genes are expressed only at particular times in development. To collect sufficient embryos in a reasonable time, we set up one collection bottle with ~100 males and 200–300 females for each line.

Solutions

10× phosphate-buffered saline (PBS): 1.3 M NaCl, 70 mM Na$_2$HPO$_4$, 30 mM NaH$_2$PO$_4$, adjust pH to 7.4 with HCl

Fixation buffer: 1.3× PBS, 67 mM [ethylenebis(oxyethylenenitrilo)]tetraacetic acid, pH to 8.0 using NaOH

PBT: 1× PBS, 0.1% Tween 80

Hybridization solution: 50% deionized formamide, 5× SSC, 100 μg/ml boiled, sonicated salmon sperm DNA, 50 μg/ml heparin, 0.1% Tween 80; this solution can be stored indefinitely at 4°

AP staining buffer: 100 mM Tris, pH 9.5, 100 mM NaCl, 50 mM MgCl$_2$, 0.1% Tween 80, stored at room temperature

Materials

Proteinase K: 10 mg/ml in H$_2$O stored at −20° in small aliquots

DIG-UTP-labeled *lacZ* RNA probe: Preparation has been outlined previously. For each experiment, we use 0.2–0.5 μl of probe, which is diluted in 50 μl hybridization mix. Immediately before use, the probe is heated to 90° for 3 min and stored on ice before adding it to the embryos.

Alkaline phosphatase [AP-conjugated anti-DIG antibody (Fab fragments)]: This antibody is available commercially (Boehringer Mannheim). We routinely preabsorb it against fixed *Drosophila* embryos to remove nonspecific-binding activities. This is accomplished by diluting the antibody 1:10 in PBT and incubating with an equal volume of fixed embryos for 2 hr at room temperature. The preabsorbed antibody is then stored in aliquots at 4°, where it is stable for up to 6 months. Immediately before use, the antibody is diluted 1:200 in PBT.

NBT (4-nitroblue tetrazolium chloride): available from Boehringer
   Mannheim, stored in aliquots at −20°
BCIP (5-bromo-4-chloro-3-indolyl-phosphate): available from Boehringer Mannheim, stored in aliquots at −20°
AP staining solution: For each experiment, we use 3.6 μl NBT and 2.8
   μl BCIP, which is diluted in 400 μl AP staining buffer immediately
   before use.
Aquapolymount: multipurpose aqueous mounting medium available
   from Polysciences, Inc. (Warrington, PA). This is stored at 4°, but
   should be warmed to room temperature before use.

Procedure. Embryos are collected on apple juice plates, aged for an
appropriate time, transferred to steel mesh baskets, dechorionated, and
washed with distilled H_2O as outlined previously. The embryos are then
transferred using a dry artist's brush to a 20-ml glass scintillation vial
that contains 3 ml fixation buffer. After adding 1 ml 37% formaldehyde
and 4 ml heptane, the vials are capped tightly and shaken vigorously
on an orbital shaker for 25 min at room temperature. This is the first
fixation step. Fixed dechorionated embryos will be at the interface
between the lower aqueous phase and the upper heptane phase after
shaking. Embryos that retain their chorions sink to the bottom of the
aqueous phase. These embryos and the whole bottom phase are then
removed and discarded using a Pasteur pipette. Embryos that were at
the interface will now be at the bottom of the vial. The vitelline
membranes of these embryos are then removed by adding 8 ml methanol
and shaking hard for 1 min. Devitellinized embryos will sink to the
bottom of the vial, whereas those that retain their vitelline membranes
will stay at the interface. These embryos and the rest of the top phase
are removed by aspiration, and embryos at the bottom are transferred
to a 1.5-ml microcentrifuge tube using a Pasteur pipette. They are then
rinsed twice in methanol. For this, approximately 1 ml is added to the
tube, which is inverted to suspend all the embryos. Once the embryos
sink to the bottom, the supernatant is removed and replaced by the
next rinse. They are then dehydrated by rinsing them six times in 100%
ethanol, after which they can be stored idefinitely at −20°. As mentioned
earlier, the optimum number of embryos for each experiment is between
300 and 500. This represents a volume of approximately 25–40 μl of
fixed dehydrated embryos. Several collections may be required to accumulate this quantity. However, it should be noted that working with too
many embryos in a single tube can also cause problems, such as the
formation of large clumps of embryos during hybridization.

To prepare the embryos for hybridization, they are rinsed twice more
in 100% ethanol and incubated in a 1:1 mixture of ethanol:xylene for 30

min at room temperature. All washes and longer incubations at room temperature are performed on a nutating platform (available from Becton-Dickinson, Sparks, MD). The xylenes step helps clear the yolk from center parts of the embryo, making it easier to visualize internal structures. The embryos are then rinsed three times in 100% ethanol to remove the xylenes, rinsed twice in methanol, and washed for 5 min in a 1:1 mixture of methanol:PBT with 5% formaldehyde (freshly made). This is followed by a 25-min postfixation step in PBT with 5% formaldehyde and three 5-min washes in PBT. The embryos are then incubated for 5–10 min in PBT with 4 μg/ml proteinase K. This step increases the ability of the probe to penetrate embryonic tissues. However, too much treatment with proteinase K causes the embryos to disintegrate during later steps, so each batch should be tested before use. The embryos are then washed three times for 5 min each in PBT and postfixed again by incubating for 25 min in PBT with 5% formaldehyde. The formaldehyde is removed by washing four times for 5 min each in PBT, and the embryos are incubated for 5 min in a 1:1 mixture of PBT:hybridization mix. This mixture is then replaced by hybridization mix, and the embryos are prehybridized for 1–2 hr in a 55° water bath. During the prehybridization period, the hybridization solution should be changed at least once. The prehybridization solution is then removed and replaced by 50 μl of DIG-UTP-labeled *lacZ* RNA probe. The embryos are mixed gently by repeatedly tapping the bottom of the tube and are incubated overnight at 55°.

The next day, the probe is removed by three 20-min washes in 1 ml hybridization solution at 55°. The final wash is replaced by a 1:1 mixture of hybridization solution:PBT and is incubated for 10 min at room temperature. The embryos are then washed four times for 10 min each in PBT to remove all traces of the hybridization solution and are incubated in PBT containing the AP-anti-DIG antibody for 1–2 hr at room temperature. This is followed by four 15-min washes in PBT at room temperature and two 5-min washes in AP staining buffer. The final wash is removed, and 400 μl of AP staining solution is added to the embryos. A small aliquot is pipetted onto a staining dish, and the reaction is monitored under a dissecting scope. The reaction may take from 20 min to more than an hour, depending on the specific activity of the probe and the abundance of *lacZ* transcripts. The reaction is stopped by washing the embryos twice for 5 min each in PBT. At this point, the embryos can be mounted onto microscope slides. This is accomplished by removing as much PBT as possible, adding ~100 μl of aquapolymount, and pipetting the embryo mixture onto a microscope slide. A coverslip is then added, and the embryos are kept overnight so that the aquapolymount can set up. However, we have found that embryos mounted directly after the staining procedure tend to be quite fragile and may flatten extensively over time. To prevent this, we routinely refix the embryos in PBT with 5% formaldehyde and wash twice in PBT before mounting.

Acknowledgments

The injection protocol presented here was developed in Michael Levine's laboratory and refined in my laboratory. I thank Rachel Kraut, who first taught me to inject, and David Kosman, Manfred Frasch, Miki Fujioka, Vikram Vasisht, and Katerina Theodosopoulou for suggestions that improved our technique. I thank Jessica Treisman for the Bluescript *lacZ* plasmid used here and for suggestions on making RNA probes. The *in situ* protocol using RNA probes is adapted from the protocol for DNA probes developed by Tautz and Pfeifle, but has been substantially revised due to suggestions by David Kosman, Tony Ip, Xue Lin Wu, and Vikram Vasisht. Finally, I thank Xuelin Wu and Vikram Vasisht for comments on this manuscript.

This work was supported by a research grant from the National Institutes of Health (GM51946).

[11] Utility of the Secreted Placental Alkaline Phosphatase Reporter Enzyme

By BRYAN R. CULLEN

Introduction

Analyses of the mechanisms underlying the transcriptional and posttranscriptional regulation of gene expression in eukaryotic cells have been facilitated greatly by the use of reporter enzymes. The critical characteristics of a useful reporter enzyme are (1) the availability of a sensitive and relatively simple quantitative assay and (2) the absence of an equivalent endogenous activity. The three most commonly used reporter enzymes, i.e., chloramphenicol acetyltransferase (CAT), β-galactosidase, and luciferase, each fully satisfy these two criteria, at least in part because they are all of nonmammalian origin. These enzymes are also similar in that each is expressed intracellularly and assay therefore requires lysis of the expressing cells. In contrast, the secreted placental alkaline phosphatase (SEAP) reporter is secreted efficiently into the culture medium by all higher eukaryotic cells tested thus far, including *Xenopus* oocytes and insect cells. The secreted SEAP protein is extremely stable and can be quantified readily using a simple colorimetric assay or a highly sensitive chemiluminescent assay. The fact that SEAP is secreted provides several potential advantages, and no obvious disadvantages, for the use of SEAP as a reporter enzyme including the following:

1. No preparation of cell lysates is required.
2. Because the transfected cells are not required for measurement of SEAP activity, they can be used for other purposes, such as mRNA or nuclear run-on analysis.
3. The rate of change of SEAP expression in response to, for example, drug treatment can be measured by sequential sampling of the media.
4. Sample collection and analysis can be automated readily using 96-well microtiter plates.

Placental alkaline phosphatase (PLAP), the wild-type precursor of the artificial SEAP indicator, is normally expressed on the cell surface where it is retained by a phosphatidylinositol-glycan (PI-G) linkage to the cell membrane. Attachment of PI-G modification occurs at an aspartic acid residue located at position 484 in the 513 amino acid PLAP protein and requires sequences located carboxy-terminal to 484.[1] The SEAP gene as originally described is truncated at residue 489 and is, as a result, efficiently secreted rather than PI-G modified.[2] PLAP is only found in higher primates, including humans, and can be readily distinguished from the more prevalent bone and intestinal forms of alkaline phosphatase based on its resistance to inactivation by incubation at 65° and to the drug L-homoarginine. These other forms of alkaline phosphatase do not, therefore, interfere with measurement of SEAP activity even if they are present. However, PLAP is expressed in certain human tissues, most obviously in the placenta, but also at low levels in the liver and in testes. Of more concern is that PLAP is expressed ectopically in a small percentage of human tumors and can be released in an active, soluble form into the medium used to culture these tumor cells after proteolytic removal of the PI-G tail. This phenomenon, while relatively rare, should obviously be checked before a particular human cell line is transfected with a SEAP indicator construct. Because PLAP is only found in higher primates, cell lines from nonprimates such as mice never exhibit any endogenous PLAP activity. Despite this latter, minor caveat, the evident advantages of SEAP as an indicator enzyme have led to an increasing use of SEAP in published research papers and in screening assays designed to identify drugs that affect gene expression pathways relevant to disease pathogenesis.

[1] J. Berger, A. D. Howard, L. Brink, L. Gerber, J. Hauber, B. R. Cullen, and S. Udenfriend, *J. Biol. Chem.* **263,** 10016 (1988).
[2] J. Berger, J. Hauber, R. Hauber, R. Geiger, and B. R. Cullen, *Gene* **66,** 1 (1988).

Principle of Method

We have previously described a SEAP-based indicator plasmid, termed pBC12/PL/SEAP, that encodes a promoterless SEAP gene.[3] Specifically, pBC12/PL/SEAP contains the SEAP gene flanked 5' by a polylinker sequence and 3' by an intron and polyadenylation signals derived from the genomic rat preproinsulin II gene. The ability to propagate this vector in bacteria is conferred by an ampicillin resistance maker and origin of DNA replication derived from pBR322. In addition, pBC12/PL/SEAP contains a functional but transcriptionally inert origin of replication from simian virus 40 (SV40), thus permitting plasmid replication in primate cells that express the SV40 T antigen, such as COS and 293T. The polylinker sequence is convenient for the introduction of regulatory sequences of interest, and derivatives containing the long terminal repeat (LTR) promoters of Rous sarcoma virus (RSV) and human immunodeficiency virus (HIV) have been described, as has a construct bearing the immediate early promoter of human cytomegalovirus (CMV).[2] These plasmids are all available from the author. Equivalent promoterless and control SEAP plasmids are available commercially from Clontech (Palo Alto, CA) and Tropix (Bedford, MA).

After insertion of the promoter of interest into pBC12/PL/SEAP, the resultant expression plasmid can be introduced into tissue culture cell lines using an appropriate transfection protocol. Vectors expressing *trans*-acting factors that modulate the activity of the introduced promoter can be contransfected if desired. Maximal levels of SEAP are generally secreted between 48 and 72 hr after transfection. In our hands, SEAP activity is essentially stable in tissue culture media at 37°.

Colorimetric Assay for SEAP

Materials and Reagents

Diethanolamine (Fisher Scientific, Springfield, NJ)
L-Homoarginine and *p*-nitrophenol phosphate disodium (Sigma Chemical Co., St. Louis, MO)
2× SEAP buffer (for 50 ml):

Amount	Stock	Final concentration
10.51 g	Diethanolamine (100% solution)	2 M
50 μl	1 M MgCl$_2$	1 mM
226 mg	L-Homoarginine	20 mM

[3] B. R. Cullen and M. H. Malim, *Methods Enzymol.* **216,** 362 (1992).

Make up in water and store at 4° (no need to autoclave). This buffer should be at pH 9.8 without requiring further addition, adjust using HCl if necessary.

120 mM p-nitrophenol phosphate: dissolve 158 mg in 5 ml of 1× SEAP buffer (make fresh)

Procedure

1. Transfect cells using your favorite procedure and include a negative control (e.g., a transfection cocktail with no SEAP expression vector in it).
2. Culture at 37° for 48–72 hr.
3. Remove 250 μl of each culture supernatant and transfer to Eppendorf tubes. To be on the safe side, maintain the cultures at 37° until "good" SEAP data have been obtained.
4. Spin for 2 min at room temperature in an Eppendorf microfuge at full speed to pellet any detached cells. Transfer 200 μl to a fresh tube.
5. Heat tubes at 65° for 10 min to inactivate endogenous phosphatases (SEAP is relatively heat stable). Store at −20°, if you wish, after this step.
6. In Eppendorf tubes, add 100 μl of 2× SEAP buffer to 100 μl of the supernatant medium (or empirically determined dilutions thereof). As a zero standard make one sample substituting unused tissue culture medium. Mix gently.
7. Transfer the contents of each tube to a well of a flat-bottomed microtiter plate. Avoid creating air bubbles.
8. Incubate the plate at 37° for 10 min.
9. During this incubation, make the p-nitrophenol phosphate (substrate) solution and prewarm to 37°.
10. Add 20 μl of the substrate solution to each well, preferably using a multipipetter.
11. Using an enzyme-linked immunosorbent assay (ELISA) plate reader, measure the light absorbance at 405 mm (A_{405}) at regular intervals (e.g., every 5 min) over the next 30 min while continuing to incubate the plate at 37°.
12. Calculate the levels of SEAP activity using data points that yield a rate of change in light absorbance that is linear with respect to time. (If available, a computer-linked kinetic ELISA plate reader is most convenient, as this will precisely calculate the linear change in A_{405} in all 96 wells of a microtiter plate over time.)
13. SEAP activity can be expressed simply as the change in light absorbance per minute per sample. More formally, SEAP activ-

ity can be given in milliunits per milliliter, where 1 mU is defined as the amount of SEAP that will hydrolyze 1.0 pmol of p-nitrophenyl phosphate per minute. This equals an increase of 0.04 A_{405} units/min. The specific activity of SEAP is 2000 mU/μg protein.

The colorimetric assay for SEAP just described is very simple, extremely inexpensive, and is able to detect SEAP levels of as little as ~10 pg/100-μl sample, which is sufficient for most uses. However, a significant disadvantage of this assay is its narrow linear dynamic range, which requires careful attention, as noted in step 12, if data obtained are to be meaningful. To overcome this problem, and to increase the sensitivity of the SEAP assay to levels as low as 0.1 pg/100 μl, several companies have developed similar chemiluminescent assays for SEAP based on the use of the phosphatase substrate disodium 3-(4-methoxyspiro[1,2-dioxetane-3,2'(5'-chloro)-tricyclo-[3.3.1.1.]decan]-4-yl)phenyl phosphate (CSPD). The dephosphorylation of CSPD by SEAP has been reported[4] to result in a sustained luminescence that remains constant for up to an hour and that can be measured readily using a luminometer.

Chemiluminescent Assay for SEAP

The chemiluminescent assay for SEAP is performed most readily using one of the similar commercial kits available from Tropix (Bedford, MA), Clontech (Palo Alto, CA), or Roche Molecular Biochemicals (Indianapolis, IN). The first six steps of the colorimetric and chemiluminescent assays are broadly identical. However, the next step in the latter assay requires the addition of a 1.25 mM solution of the CSPD substrate prepared in a proprietary chemiluminescent enhancer solution, both of which are provided as part of the kit. Emitted light can then be measured using a standard tube luminometer, such as a Turner Model TD-20e, with a 5- to 15-sec integration time. Alternately, if a large number of samples are to be assayed, a plate luminometer can be used. Detailed procedures for chemiluminescent detection of SEAP are provided by each kit manufacturer and have also been published.[4] The chemiluminescent assay for SEAP has been reported to be linear over a range of 0.1 pg to 1 ng of SEAP per sample and is at least as sensitive as commercially available assays for luciferase.

Two Novel Uses of the SEAP Indicator Gene

While the majority of published studies that involved SEAP used it in a relatively typical way, i.e., to map functional elements in promoters or enhancers, to analyze promoter response to specific signaling molecules,

---

[4] T.-T. Yang, P. Sinai, P. A. Kitts, and S. R. Kain, *Biotechniques* **23**, 1110 (1997).

or to serve as an internal control in transfection experiments, it is worth highlighting two papers that used SEAP in especially innovative ways.

Flanagan and Leder[5] used SEAP as a tag to facilitate identification of the ligand for the *kit* cell surface receptor. Specifically, these scientists genetically fused the extracellular domain of *kit* to the amino terminus of the mature (i.e., signal peptide deleted) SEAP protein. This resulted in the synthesis of a secreted, soluble affinity reagent, bearing a fully active SEAP enzyme tag, that could be traced easily and sensitively. Using this reagent, these authors were able to demonstrate specific binding of *kit* to certain cell lines by *in situ* staining for alkaline phosphatase activity. The approach of generating fusion proteins consisting of SEAP fused to ligands, or to the extracellular domains of receptors or viral envelope proteins, would appear to be generally applicable to the analysis and quantitation of target protein expression patterns.

Means *et al.*[6] used SEAP to design cell lines that permit the rapid, sensitive, and quantitative assay of human immunodeficiency virus type 1 (HIV-1) and simian immunodeficiency virus (SIV) infectivity. Specifically, these authors derived CD4+ T-cell lines that bear the SEAP indicator gene under the control of either the HIV-1 or SIV LTR promoter element. These LTRs do not give rise to significant levels of expression of linked genes in the absence of the cognate virally encoded Tat transcription factor. However, Tat is provided when these indicator cells are infected with either HIV-1 or SIV, thus resulting in a dramatic activation in SEAP expression, some 48 to 72 hr after infection, that the authors show is directly proportional to the level of infecting virus. These cells therefore permit the rapid titering of viral stocks and have also proved ideal for measuring the neutralizing titer of human or simian sera for HIV-1 or SIV, respectively. The heat resistance of SEAP is particularly useful in this system, as any residual infectious virus present in the supernatant media will be destroyed by the 65° incubation that forms part of the standard SEAP assay.

Although the SEAP indicator enzyme has yet to achieve the widespread popularity of, for example, CAT, the secreted nature of SEAP offers a number of obvious practical advantages. When taken together with the ease and sensitivity of available assays for SEAP, this suggests that SEAP will be increasingly widely used by scientists interested in the analysis of gene expression or in screening for drugs that can modulate this critical and highly complex process. The two examples, given earlier, of novel uses of SEAP are indicative of the many unusual, even unique, applications that this indicator system has the potential to support.

[5] J. G. Flanagan and P. Leder, *Cell* **63,** 185 (1990).
[6] R. E. Means, T. Greenough, and R. C. Desrosiers, *J. Virol.* **71,** 7895 (1997).

[12] Fusions to Imidazopyrazinone-Type Luciferases and Aequorin as Reporters

By Satoshi Inouye

Introduction

Light emission in certain organisms occurs as a result of the luciferin–luciferase reaction[1]:

$$\text{Luciferin} + O_2 \xrightarrow{\text{Luciferase}} \text{Oxyluciferin} + \text{Light}$$

In bioluminescent marine organisms, an imidazopyrazinone (3,7-dihydroimidazopyrazin-3-one) compound is used for the luciferase reaction. Two imidazopyrazinone-type luciferins, coelenterazine[2] and *Cypridina* luciferin[3] (also called *Vargula* luciferin; *Cypridina* is used in this article), have been isolated. These two luciferins are similar with a central imidazolepyrazine nucleus in their structures (Fig. 1).

Coelenterazine is widely distributed in luminous coelenterates, fishes, shrimps, and squids[4,5] and is also known as *Watasenia* preluciferin,[2] *Oplophorus* luciferin, and *Renilla* luciferin. Among coelenterazine-type luciferases, *Renilla* luciferase[6] and *Oplophorus* luciferase[7] have been isolated and their biochemical properties characterized. Further, coelenterazine serves as the chromogenic compound of photoproteins[8] such as aequorin, phialidin, halistaurin, and obelin. Luciferin and luciferase of *Cypridina* have been isolated from the sea firefly, *Cypridina hilgendorfii* (presently *Vargula hilgendorfii*), and its luminescence system has been well investigated.[9] *Cypridina* luciferin is also present in certain kinds of fishes.[9]

In the field of molecular biology, several reporter genes are used for monitoring the gene expression in whole organisms, as well as in single

[1] E. N. Harvey, *in* "Bioluminescence." Academic Press, New York, 1952.
[2] S. Inoue, S. Sugiura, H. Kakoi, K. Hashizume, T. Goto, and H. Iio, *Chem. Lett.,* 141 (1975).
[3] Y. Kishi, T. Goto, Y. Hirata, O. Shimomura, and F. H. Johnson, *Tetrahedron Lett.,* 3427 (1966).
[4] O. Shimomura, S. Inoue, F. H. Johnson, and Y. Haneda, *Comp. Biochem. Physiol.* **65B,** 435 (1980).
[5] C. M. Thompson, P. J. Herring, and A. K. Campbell, *J. Biolumin. Chemilumin.* **12,** 87 (1997).
[6] J. C. Matthews, K. Hori, and M. J. Cormier, *Biochemistry* **16,** 85 (1977).
[7] O. Shimomura, T. Masugi, F. H. Johnson, and Y. Haneda, *Biochemistry* **17,** 994 (1978).
[8] O. Shimomura and F. H. Johnson, *Nature* **256,** 236 (1975).
[9] F. H. Johnson and O. Shimomura, *Methods Enzymol.* **57,** 331 (1978).

METHODS IN ENZYMOLOGY, VOL. 326

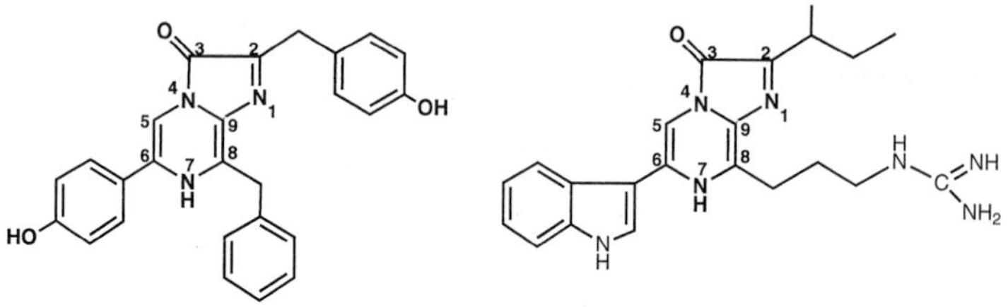

Coelenterazine
(=*Oplophorus* luciferin,
Renilla luciferin,
Watasenia preluciferin)

Cypridina luciferin
(=*Vargula* luciferin)

FIG. 1. Chemical structures of coelenterazine and *Cypridina* luciferin.

cells.[10] Luciferases, including firefly luciferase and photoproteins, are useful reporter molecules with various advantages for the bioluminescent system: they are sensitive, rapid, and harmless. Moreover, the detection of photon produced by luciferase permits real-time imaging of gene expression and of the dynamic changes of target protein in living cells.[10] The luminescence reaction of imidazopyrazinone-type luciferases is simpler in required components than that of firefly luciferase, as the imidazopyrazinone-type luciferase systems do not need any cofactors such as ATP and Mg^{2+}.[10]

Genes for Imidazopyrazinone-Type Luciferases and Photoproteins

cDNAs for *Renilla* luciferase,[11] *Cypridina* luciferase,[12] and photoproteins including aequorin,[13] phialidin (=clytin),[14] halistaurin (=mitrocomin),[15] and obelin[16] have been cloned and expressed in prokaryotic and/ or eukaryotic cells. These have been used as reporter proteins/genes, as summarized in Table I.

---

[10] I. Bronstein, J. Fortin, P. E. Stanley, G. S. A. B. Stewart, and L. J. Kricka, *Anal. Biochem.* **219**, 169 (1994).

[11] W. W. Lorenz, R. O. McCann, M. Longiaru, and M. J. Cormier, *Proc. Natl. Acad. Sci. U.S.A.* **88**, 4438 (1991).

[12] E. M. Thompson, S. Nagata, and F. I. Tsuji, *Proc. Natl. Acad. Sci. U.S.A.* **86**, 6576 (1989).

[13] S. Inouye, M. Noguchi, Y. Sakaki, Y. Takagi, T. Miyata, S. Iwanaga, T. Miyata, and F. I. Tsuji, *Proc. Natl. Acad. Sci. U.S.A.* **82**, 3154 (1985).

[14] S. Inouye and F. I. Tsuji, *FEBS Lett.* **315**, 343 (1993).

[15] T. F. Fagan, Y. Ohmiya, J. R. Blinks, S. Inouye, and F. I. Tsuji, *FEBS Lett.* **333**, 301 (1993).

[16] B. A. Illarionov, V. S. Bondar, V. A. Illarionova, and E. S. Vysotski, *Gene* **153**, 273 (1995).

TABLE I
Expression of Luciferase and Apoaequorin Genes in Bacterial and Mammalian Cells

cDNAs (GenBank Acc. No)	MW (kDa)	Normal location	E. coli		Mammalian cells		Activity of fusion protein
			Cytoplasm	Periplasm	Cytoplasm	Secretion	
Renilla luciferase (M63501)	36 (311)	Cytosol	Yes	N.R.[a]	Cytosol	Yes[b]	Yes (N-/C-)[c]
Apoaequorin (L29571)	21.5 (189)	Cytosol	Yes[d]	Yes[e]	Cytosol Mitochondria[f]	Yes[g]	Yes (N-)
Cypridina luciferase (M25666)	58.5 (527)	Secreted	No[h]	No[i]	Cytosol[j]	Yes	Yes (N-/C-)
Firefly luciferase (M15077)	61 (550)	Peroxisome	Yes	Yes[k]	Peroxisome[l] Cytosol	No[m]	Yes (N-)

[a] Not reported.

[b] With the signal peptide sequence of human interleukin-2 added [J. Liu, D. J. O'Kane, and A. Escher, *Gene* **203**, 141 (1997)].

[c] N-, N-terminal fusion; C-, C-terminal fusion.

[d] >90% inclusion bodies.

[e] With the *OmpA* signal peptide sequence added [S. Inouye, S. Aoyama, T. Miyata, F. I. Tsuji, and Y. Sakaki, *J. Biochem.* (Tokyo) **105**, 473 (1989)].

[f] With the signal peptide sequence of human cytochrome c oxidase added [R. Rizzuto, A. W. M. Simpson, M. Brini, and T. Pozzan, *Nature* **358**, 325 (1992)].

[g] With the signal peptide sequence of human follistatin added [S. Inouye and F. I. Tsuji, *Anal. Biochem.* **201**, 114 (1992)].

[h] Inclusion bodies.

[i] With *pelB* signal peptide sequence added [Y. Maeda, H. Ueda, T. Hara, J. Kazami, G. Kawano, E. Suzuki, and T. Nagamune, *BioTechniques* **20**, 116 (1996)].

[j] With the signal peptide sequence deleted at position 1 to 28 amino acid residues in N terminus.

[k] With *pelB* signal peptide sequence added [P. Billiald, M. Mousli, M. Goyffon, and D. Vaux, *Biotechnol. Lett.* **19**, 1037 (1997)].

[l] With the peroxisomal targeting sequence (-SKL) deleted at the C terminus [S. J. Gould, G.-A. Keller, N. Hosken, J. Wilkinson, and S. Subramani, *J. Cell Biol.* **108**, 1657 (1989)].

[m] With the signal peptide sequence of human follistatin added and the peroxisomal targeting sequence (-SKL) deleted at the C terminus (S. Inouye and K. Umesono, unpublished result).

Renilla Luciferase

Renilla luciferase (EC 1.13.12.5), found in the bioluminescent sea pansy *Renilla reniformis,* is a monomeric protein (36 kDa) that catalyzes the oxidation of coelenterazine to produce light (λ_{max} = 475 nm), coelenteramide (oxyluciferin) and CO_2.[6] The *Renilla* luciferase gene has been cloned, and the protein is made up of 311 amino acid residues.[11] The gene has been tested as a reporter enzyme in various cells.[17–19] *Renilla* luciferase with the signal peptide sequence could be secreted from mammalian cells, but the expression level of the luciferase found in the cultured medium was lower than in the case of cytoplasmic expression.[19] The substrate specificity of purified recombinant *Renilla* luciferase was examined, as summarized in Table II.[20,21] Of various coelenterazine analogs, *e*-coelenterazine and *v*-coelenterazine (Fig. 2) are more efficient than coelenterazine in the luminescent reaction; the initial light intensities with *e*- and *v*-coelenterazine were 7.5 and 6.4 times higher than coelenterazine, respectively. With regard to the spectra of light emission with *Renilla* luciferase, *e*-coelenterazine gave a bimodel spectrum with its peaks at 418 and 475 nm, whereas *v*-coelenterazine showed a red-shifted spectrum with a peak at 512 nm.[20]

Oplophorus Luciferase

Oplophorus luciferase from the deep-sea shrimp, *Oplophorus gracilorostris,* was partially purified and characterized,[4] but its primary structure is still unknown. The luciferase is a secretional enzyme such as *Cypridina* luciferase and catalyzes the oxidation of coelenterazine (=*Oplophorus* luciferin) to produce blue light at 455 nm. *Oplophorus* luciferase differs significantly from *Renilla* luciferase and is highly valuable for use as a reporter protein, e.g., the high quantum yield (0.34 at 22°) and the high optimum temperature for light emission (40°). The light intensity measured with coelenterazine was of the order of *Oplophorus* luciferase (100%) > *Renilla* luciferase (14.1%) > apoaequorin (0.01%). Bisdeoxycoelenterazine

[17] W. W. Lorenz, M. J. Cormier, D. J. O'Kane, D. Hau, A. A. Escher, and A. A. Szalay, *J. Biolumin. Chemilumin.* **11,** 31 (1996).
[18] R. Myerhofer, W. H. R. Langridge, M. J. Cormier, and A. A. Szalay, *Plant J.* **7,** 1031 (1995).
[19] J. Liu, D. J. O'Kane, and A. Escher, *Gene* **203,** 141 (1997).
[20] S. Inouye and O. Shimomura, *Biochem. Biophys. Res. Commun.* **233,** 349 (1997).
[21] H. Nakmura, C. Wu, A. Murai, S. Inouye, and O. Shimomura, *Tetrahedron Lett.* **38,** 6405 (1997).

TABLE II
SUBSTRATE SPECIFICITY OF *Renilla* LUCIFERASE, *Oplophorus* LUCIFERASE, APOAEQUORIN, AND AEQUORIN[a]

Coelenterazine analogs		*Renilla* luciferase			*Oplophorus* luciferase			Apoaequorin	Aequorin	
Prefix and compound no.	Substitution	Initial intensity (%)	Total light (%)	Emission peak (nm)	Initial intensity (%)	Total light (%)	Emission peak (nm)	Initial intensity (%)	Total light (%)	Emission peak (nm)
None (**1**)	None	100[b]	100[c]	475	100[d]	100[e]	454	100[f]	100[g]	465
Bis	2 & 6:-$CH_2C_6H_5$	0.32	0.45	400/440	79	66	448	1.5	0.02	—
h (**2**)	2:-$CH_2C_6H_5$	57	41	475	97	75	452	220	82	464
f (**3**)	2:-$CH_2C_6H_4F(p)$	58	28	472	26	23	446	270	80	473
cl (**6**)	2:-$CH_2C_6H_4Cl(p)$	8	16	478	7	24	450	27	85	463
n (**9**)	2: β-Naphthylmethyl	68	47	475	45	25	449	3.4	26	467
	2:-$CH_2C_6H_4C_6H_5(p)$	5	—		5	—		0.3	1	
(**33**)	2:-$CH_2CH(CH_3)_2$	0.5	0.9	473	21	32	457	3.7		
(**34**)	2:-$CH_2C_6H_{11}(c)$		—		16	32	462	1.8		
	2:-$CH_2CH_2C_6H_5$		—		55	63	460	2	9.5	
	2:-$CH_2CH_2CH_2C_6H_5$		—		33	27		4	30	
m(5) (**23**)	5:-CH_3	5			33	54	440	53	28	438
a	6:-$C_6H_4NH_2(p)$	21	113	468	13	75	460	0.2	8	465
	6:-$C_6H_4NHCH_3(p)$	41	45	475	0.2	0.6	456	0.2		
	6:-$C_6H_4OCH_3(p)$	2	1		0.07			0.1		
	8:-CH_3		—		0.03			0.3	—	
b (**10**)	8:-$(CH_2)_4CH_3$	8	38	469	2	—		100	79	448
cp (**12**)	8:-$CH_2C_5H_9(c)$	16	23	470	48	62	443	600	95	442
ch (**13**)	8:-$CH_2C_6H_{11}(c)$	40	54	470	23	57	444	500	100	452
(**14**)	8:-$CH_2CH_2C_6H_{11}(c)$	3	22	468	0.08	0.14		22	77	442
m(8) (**15**)	8:-$CH_2CH_2C_6H_5$	0.4	—	—	—	—		43	60	444
m	X:-CH_2-	9	35	475	0.02	—		0.5	1	
e (**24**)	X:-CH_2CH_2-	750	137	418/475	82	54	459	0.8	50	405/465
(**38**)	X:-$CH_2CH_2CH_2$-	7	27	480	33	41	442	0.1	6	
v	X:-$CH=CH$-	640	73	512	4	4	480	0.05	3.5	

[a] The prefix represents the structural modification; the prefix and compound number correspond to those used previously [O. Shimomura, Y. Kishi, and S. Inouye, *Biochem. J.* **296**, 549 (1993)]. An em dash indicates insignificant light emission; a (c) denotes "cyclo."
[b] 1.9×10^{11} quanta/sec.
[c] 1.06×10^{13} quanta.
[d] 1.34×10^{12} quanta/sec.
[e] 3.5×10^{13} quanta.
[f] 1.6×10^{9} quanta/sec.
[g] 5.0×10^{13} quanta.

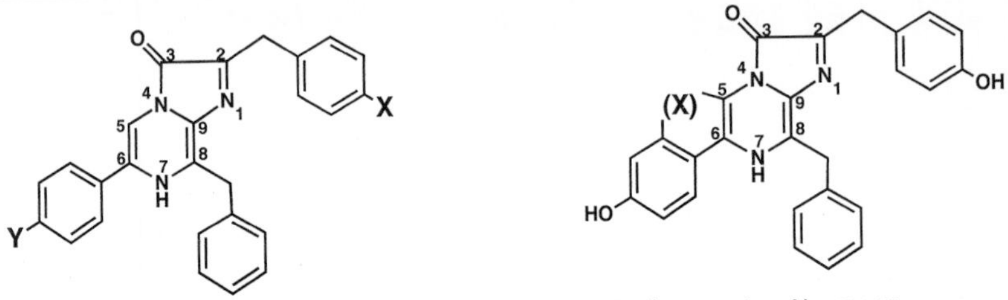

Coelenterazine: X = OH, Y = OH
2-Benzyl coelenterazine *h*(2): X = H, Y = OH
Bisdeoxy coelenterazine *Bis*: X = H, Y = H

e-Coelenterazine: X = CH$_2$CH$_2$
v-Coelenterazine: X = CH=CH

FIG. 2. Chemical structures of coelenterazine analogs.

(Fig. 2) shows efficient luminescence with *Oplophorus* luciferase, but not with *Renilla* luciferase or apoaequorin (Table I).[21] *Oplophorus* luciferase can be used as a sensitive reporter protein when its gene is obtained.

Aequorin (Apoaequorin)

Aequorin, a Ca^{2+}-binding photoprotein isolated from the bioluminescent jellyfish *Aequorea aequorea,* is made from apoaequorin (apoprotein), molecular oxygen, and coelenterazine.[22] When a trace of Ca^{2+} (or Sr^{2+}) is added, aequorin emits a flash of blue light (λ_{max} = 465 nm) by an intramolecular reaction, decomposing into apoaequorin, coelenteramide, and CO$_2$. Thus, aequorin has been used as a calcium indicator. Aequorin can be regenerated from apoaequorin by incubation with coelenterazine, dissolved oxygen, EDTA, and reducing reagents.[8,22] After light emitting by calcium ions, the coelenteramide–apoaequorin complex shows the blue fluorescence, and coelenteramide is released from the protein moiety. The calcium-bound form of apoaequorin catalyzes the oxidation of coelenterazine, resulting in a weak, continuous luminescence, thus apparently functioning as a luciferase that requires Ca^{2+} as a cofactor.[8,20] Sequence analysis of cDNA showed that apoaequorin consists of 189 amino acid residues in a single polypeptide chain (21.5 kDa) with three EF hand structures (Ca^{2+}-binding

---

[22] O. Shimomura, *in* "Natural Products and Biological Activities" (H. Imura, T. Goto, T. Murachi, and T. Nakajima, eds.), p. 33. University of Tokyo Press, Tokyo, 1986.

sites).[13] Apoaequorin cDNA can be expressed in both eukaryotic and pro-karyotic cells.[23-26] After conversion of apoaequorin into aequorin by incuba-tion with coelenterazine, reducing reagents, and molecular oxygen, lumines-cence activity is determined by adding Ca^{2+}. Secretion or targeting signal sequences fused to the amino terminus of apoaequorin can work,[25,26] but those fused to the carboxyl terminus do not show significant activities. Real-time imaging analysis of calcium signaling in a living cell by the expression of apoaequorin cDNA has been tried, but some problems were found, such as slow membrane permeability of coelenterazine and inefficiency in aequorin regeneration; physical effects on the apoaequorin expression in cells are still unclear.[27] In the assay *in vitro,* the regeneration time to aequorin from apoaequorin with coelenterazine and oxygen takes a few hours (at least 2 hr) under calcium-free and reducing conditions by adding excess EDTA and 2-mercaptoethanol at low temperatures (4°).[8] Using *in vivo* systems such as living cells, the regeneration conditions of the expressed apoaequorin did not establish well. Critical problems in the assay *in vivo,* including real-time imaging of calcium ions, come from the regeneration process from the expressed apoaequorin to aequorin in living cells, as follows. First, apoaequorin itself can bind calcium ions without coelen-terazine the same as aequorin *in vitro,* but the calcium-bound form of apoaequorin cannot regenerate to aequorin.[8,20] Thus, the apoaequorin ex-pressed in living cells may bind calcium ions but cannot regenerate to aequorin efficiently. Second, when regeneration was carried out *in vitro* at 37°, regeneration to aequorin failed. For retaining efficient regeneration of expressed apoaequorin in living cells, the cultured cells can be shifted to lower temperature (4°), but the cells will suffer some stress and there will be an effect on calcium signaling.

Various coelenterazine analogs have been synthesized, and correspond-ing preparations of semisynthetic aequorin made. Their light-emitting prop-erties, such as calcium sensitivity, emission spectra, luminescence capacity, and regeneration time, have been characterized.[28,29]

[23] S. Inouye, S. Aoyama, T. Miyata, F. I. Tsuji, and Y. Sakaki, *J. Biochem.* (*Tokyo*) **105,** 473 (1989).

[24] H. Tanahashi, T. Ito, S. Inouye, F. I. Tsuji, and Y. Sakaki, *Gene* **96,** 249 (1990).

[25] S. Inouye and F. I. Tsuji, *Anal. Biochem.* **201,** 114 (1992).

[26] R. Rizzuto, A. W. M. Simpson, M. Brini, and T. Pozzan, *Nature* **358,** 325 (1992).

[27] R. Creton, M. E. Steele, and L. F. Jaffe, *Cell Calcium* **22,** 439 (1997).

[28] O. Shimomura, B. Musicki, Y. Kishi, and S. Inouye, *Cell Calcium* **14,** 373 (1993).

[29] O. Shimomura, Y. Kishi, and S. Inouye, *Biochem. J.* **296,** 549 (1993).

Cypridina (=Vargula) Luciferase

The marine ostracod crustacean *Cypridina hilgendorfii* (presently *Vargula hilgendorfii*) secretes luciferase (EC 1.13.12.6) and *Cypridina* luciferin in sea water, where they mix and react to produce light at 460 nm.[9] The gene for *Cypridina* luciferase has been cloned,[30] and the primary structure of *Cypridina* luciferase consists of 555 amino acid residues possessing the signal peptide sequence for secretion. The cDNA as a reporter gene has been expressed in various mammalian cells.[30–33] *Cypridina* luciferase secretion was used for studying real-time imaging of the protein trafficking pathway in a living cell, using an image intensifier coupled to a charged-coupled device camera.[31–33]

Assay Methods for Luciferases and Aequorin

Luciferins (Coelenterazine and Cypridina Luciferin)

Commercial Source. The commercial suppliers of luciferins are listed with the Internet Web page, as follows.
a. Coelenterazine and its analogs
   Argus Fine Chemicals Limited
      Science Park Square, Flamer, Brighton, East Sussex BN1 9SB, UK
      http://www.net-escape.co.uk/business/argus/
   Biosynth International, Inc.
      1665 W. Quincy Ave. Suite 155, Naperville, IL 60540
      http://www.biosynth.com.
   Molecular Probe, Inc.
      4849 Pitchford Ave. Eugene, OR 97402
      http://www.probes.com.
   Prolume, Ltd.
      1085 William Pitt Way, Pittsburgh, PA 15238
      http://www.prolume.com.
b. *Cypridina (=Vargula)* luciferin
   Prolume, Ltd.
      1085 William Pitt Way, Pittsburgh, PA 15238
      http://www.prolume.com.

[30] E. M. Thompson, S. Nagata, and F. I. Tsuji, *Gene* **96,** 257 (1990).
[31] S. Inouye, Y. Ohmiya, Y. Toya, and F. I. Tsuji, *Proc. Natl. Acad. Sci. U.S.A.* **89,** 9584 (1992).
[32] E. M. Thompson, P. Adenot, F. I. Tsuji, and J.-P. Renard, *Proc. Natl. Acad. Sci. U.S.A.* **92,** 1317 (1995).
[33] G. Miesenböck and J. E. Rothman, *Proc. Natl. Acad. Sci. U.S.A.* **94,** 3402 (1997).

Luciferin Stock Solution

Coelenterazine (MW = 423, ε = 9800 at 435 nm in methanol, 0.1 mg/ml = 236 μM) dissolved in methanol and stored at −80° (for a few months)

Cypridina luciferin (MW = 478, ε = 8800 at 435 nm in methanol, 0.1 mg/ml = 209 μM) dissolved in methanol or *n*-butanol and stored at −80°

Notes for Luciferin

a. The stock solution of luciferins at −20° degrades slowly. For long-term storage, lyophilized luciferin should be protected from air oxidation and photodegradation by keeping it under argon or nitrogen gas in the dark at −80° or −20°.

b. Luciferins in aqueous solution are unstable and subject to rapid oxidation at room temperature. When luciferins are to be dissolved in a buffer, only small aliquots sufficient for each assay should be made.

c. Luciferins luminesce in aprotic solvents such as dimethyl sulfoxide and glycol dimethyl ester (diglyme) without luciferase.

d. The high concentration of protein solution (>1%), including serum albumin and crude extracts of cells, catalyzes the oxidation of luciferins and gives a weak continuous luminescence.

e. Dehydrocoelenterazine, a contaminant of coelenterazine produced during certain chemical synthetic processes, shows competitive inhibition in *Renilla* and *Oplophorus* luciferase reactions.

f. *Cypridina* luciferin did not use *Renilla* luciferase, *Oplophorus* luciferase, or aequorin.

Assay Procedures

Renilla luciferase (quantum yield; 0.10–0.11 at 22–23°, specific activity of recombinant protein; 1.9×10^{13} quanta/sec mg protein)[20]: *Renilla* luciferase expressed in various cells is extracted with 25 mM Tris–HCl, pH 7.5, containing 10 mM EDTA, and is used as the source of *Renilla* luciferase. The reaction mixture contains *Renilla* luciferase in 25 mM Tris–HCl, pH 7.5, containing 100 mM NaCl. The reaction is initiated by adding coelenterazine solution (final concentration 100 nM) and is recorded with a lumiphotometer.

Oplophorus luciferase (quantum yield; 0.34 at 22°, specific activity of native protein; 1.34×10^{14} quanta/sec mg protein)[7,20]: The reaction mixture contains *Oplophorus* luciferase in 15 mM Tris–HCl, pH 8.3, containing 50 mM NaCl. The reaction is started by adding coelenterazine (final concentration 100 nM) and the luminescence is measured with a lumiphotometer.

Aequorin (quantum yield; 0.15 at 25°, specific activity of recombinant protein; 5.0×10^{15} quanta/mg protein)[22]: Apoaequorin expressed in various cells is extracted with 25 mM Tris–HCl, pH 7.4, containing 10 mM EDTA. Apoaequorin is regenerated into aequorin by incubation in 25 mM Tris–HCl, pH 7.4, containing 10 mM EDTA, 5 mM 2-mercaptoethanol, and 100 nM coelenterazine on ice (4°) for 2 hr. 2-Mercaptoethanol used in the regeneration step can be replaced by dithiothreitol or dithioerythritol. The aequorin regenerated is measured by injecting an equal volume of a 30 mM CaCl$_2$ solution made with the buffer; the flash of emitted light is recorded with a lumiphotometer.

Cypridina luciferase (quantum yield; 0.28 at 4°, specific activity of native protein; 8×10^{16} quanta/sec mg protein)[9,34]: *Cypridina* luciferase secreted into the culture medium from mammalian cells is collected and used for the assay. The reaction mixture contains *Cypridina* luciferase in 100 mM Tris–HCl, pH 7.2; the reaction is started by injecting 100 nM *Cypridina* luciferin and then the luminescence produced is recorded.

Notes for Assay

a. Assay conditions for total volume and enzyme concentrations in the reaction mixture are dependent on the cuvette size and the sensitivity of photon detection (type of photomultiplier) in a lumiphotometer.

b. Before starting the assay of samples, the linearity of the light production by expressed luciferase should be check in same assay volume.

c. In the case of transient expression in mammalian cells, cells (2×10^5) were grown in a 35-mm well plate for 24 hr and were transfected with 2 μg of expression plasmid using a transfection reagent. After a 36-hr incubation, the cultured medium was collected and cells were washed with phosphate-buffered saline and suspended in 0.5 ml of extraction buffers. The samples were stored at −80° before use and cells were disrupted by thawing and freezing. For assay, 1–50 μl of cell extracts or cultured medium was used in 100–1000 μl of reaction mixtures.

Acknowledgments

I thank Drs. O. Shimomura and H. Nakamura for their review and comments.

[34] O. Shimomura and F. H. Johnson, *Science* **164,** 1299 (1969).

[13] Novel Methods for Chemiluminescent Detection of Reporter Enzymes

By Corinne E. M. Olesen, Yu-Xin Yan, Betty Liu, Dina Martin, Brian D'Eon, Ray Judware, Chris Martin, John C. Voyta, and Irena Bronstein

Introduction

Reporter gene assays are widely used for studying gene regulation and function.[1] Reporter quantitation enables mapping of promoter and enhancer regions, and identification of factors, mechanisms, or compounds that alter gene expression levels. Chemiluminescent and bioluminescent reporter enzyme quantitation methods are highly sensitive and provide convenient assay formats. We have developed chemiluminescent 1,2-dioxetane substrates and detection protocols for several different reporter enzymes, including β-galactosidase, β-glucuronidase, and alkaline phosphatase, which provide highly sensitive alternatives to colorimetric and fluorescent substrates.[2] In addition, 1,2-dioxetane substrates are used in combination with luciferase detection for dual measurement of two reporter enzymes in a single sample. We also describe a new extended glow bioluminescent luciferase assay system.

We have developed adamantyl 1,2-dioxetanes that are direct chemiluminescent substrates for alkaline phosphatase and other hydrolytic enzymes. Hydrolytic cleavage of adamantyl-1,2-dioxetane substrates, such as CSPD by alkaline phosphatase and Galacton-*Star* by β-galactosidase, results in the formation of a metastable anion, which further fragments to form an excited state anion that emits light (Fig. 1). The enzymatic cleavage reaction produces an electron-rich dioxetane phenolate anion, which initiates a decomposition mechanism called chemically initiated electron exchange luminescence (CIEEL).[3,4] Charge transfer from the phenolate to the dioxetane ring promotes a concerted cleavage of two bonds of the cyclic peroxide, releasing about 100 kcal to chemiexcite one of the resulting carbonyl fragments to a singlet electronic state. The excited species emits light at approximately 480 nm as it converts back to the ground state.

[1] J. Alam and J. L. Cook, *Anal. Biochem.* **188,** 245 (1990).
[2] I. Bronstein, J. Fortin, P. E. Stanley, G. S. A. B. Stewart, and L. Kricka, *Anal. Biochem.* **219,** 169 (1994).
[3] J. Y. Koo and G. B. Schuster, *J. Am. Chem. Soc.* **99,** 6107 (1977).
[4] J. Y. Koo and G. B. Schuster, *J. Am. Chem. Soc.* **100,** 4496 (1978).

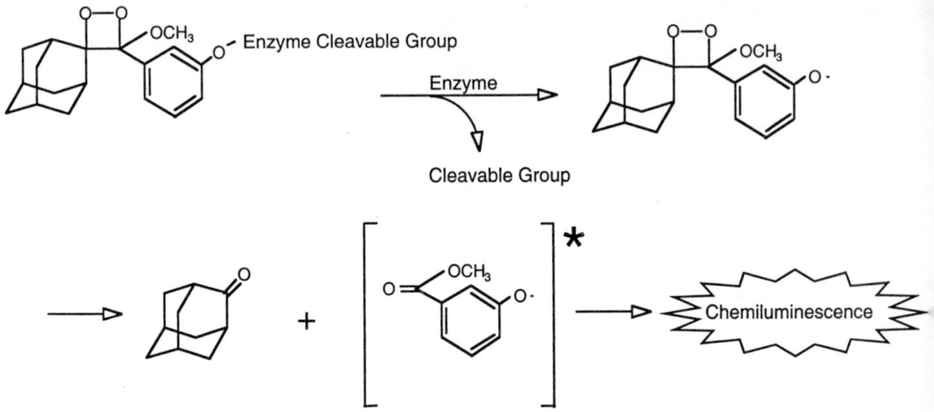

FIG. 1. Light emission mechanism of 1,2-dioxetane substrates.

The chemiluminescent signal obtained from the enzyme-catalyzed diox-etane decomposition reaction is typically a steady-state glow that can be measured over many hours, depending on the enzyme and particular sub-strate used. In the presence of excess substrate, the chemiluminescent signal rises as the metastable dioxetane anion accumulates and production of anion exceeds decomposition. When the rate of anion production equals the rate of decomposition, a steady-state plateau is reached. Prolonged glow emission kinetics simplify measurement of the light signal.

The use of macromolecular enhancers provides a significant improve-ment in the intensity of luminescence from 1,2-dioxetanes.[5-11] In aqueous solution, proton transfer events result in a 1000-fold decrease in chemilumi-nescence intensity compared to that obtained in organic solvents. We have developed proprietary water-soluble quaternary polymers, such as Sapphire and Sapphire-II, which enhance light emission approximately 100-fold in aqueous solution. These polymers provide hydrophobic microdomains that

[5] I. Bronstein, R. R. Juo, J. C. Voyta, and B. Edwards, in "Bioluminescence and Chemilumines-cence: Current Status" (P. Stanley and L. J. Kricka, eds.), p. 73. Wiley, Chichester, 1991.
[6] A. P. Schaap, H. Akhavan, and L. J. Romano, Clin. Chem. **35,** 1863 (1989).
[7] I. Bronstein, J. C. Voyta, Y. Vant Erve, and L. J. Kricka, Clin. Chem. **37,** 1526 (1991).
[8] I. Bronstein, J. Fortin, J. C. Voyta, and L. J. Kricka, BioTechniques **12**(4), 500 (1992).
[9] Y. Vant Erve, J. C. Voyta, B. Edwards, L. J. Kricka, and I. Bronstein, in "Bioluminescence and Chemiluminescence: Status Report" (A. Szalay, L. J. Kricka, and P. Stanley, eds.), p. 306. Wiley, Chichester, 1993.
[10] I. Bronstein, J. C. Voyta, B. Edwards, A. Sparks, and R. R. Juo, Spectrum **7**(2), 10 (1994).
[11] B. Edwards, A. Sparks, J. C. Voyta, and I. Bronstein, in "Bioluminescence and Chemilumi-nescence: Fundamentals and Applied Aspects" (A. K. Campbell, L. J. Kricka, and P. E. Stanley, eds.), p. 56. Wiley, Chichester, 1994.

sequester the dephosphorylated dioxetane anion and protect the excited state emitter from aqueous proton transfer. Additional polymer formulations, Emerald and Emerald-II, contain fluorescein as an energy transfer acceptor, further increasing the intensity of light and shifting the maximum wavelength of emission to 530 nm.

This article describes chemiluminescent reporter gene assays incorporating 1,2-dioxetane substrates, including the Galacto-*Star* and Gal-Screen assays for β-galactosidase, the GUS-Light assay for β-glucuronidase, the Phospha-Light assay for placental alkaline phosphatase, dual reporter assays for luciferase/β-galactosidase (Dual-Light assay) and luciferase/β-glucuronidase, and the Luc-Screen system, an extended glow bioluminescent luciferase assay. The Gal-Screen and Luc-Screen systems are performed in the presence of cell culture media, with a mix and read assay format, and thus are particularly suitable for use in high throughput, automated screening applications.

A. β-Galactosidase Reporter Gene Assays

Background

The bacterial β-galactosidase gene is widely employed as a reporter enzyme in many organisms. Chemiluminescent 1,2-dioxetane substrates for β-galactosidase, including Galacton, Galacton-Plus, and Galacton-*Star* substrates, provide highly sensitive enzyme detection with a wide dynamic range, from 2 fg to 20 ng of purified enzyme.[12–15] This sensitivity is approximately 100- to 1000-fold higher than published detection limits with the fluorescent substrate methylumbelliferyl-β-D-galactopyranoside (MUG, 2 pg) and the colorimetric substrate *o*-nitrophenyl-β-D-galactoside (ONPG, 100 pg), respectively.[12]

Chemiluminescent reporter assays with 1,2-dioxetane substrates have been performed extensively with mammalian cell culture and tissue extracts.[16,17] In addition, 1,2-dioxetanes have been used to measure β-galacto-

[12] V. K. Jain and I. T. Magrath, *Anal. Biochem.* **199**, 119 (1991).

[13] I. Bronstein, J. Fortin, J. C. Voyta, C. E. M. Olesen, and L. J. Kricka, *in* "Bioluminescence and Chemiluminescence: Fundamentals and Applied Aspects" (A. K. Campbell, L. J. Kricka, and P. E. Stanley, eds.), p. 20. Wiley, Chichester, 1994.

[14] C. S. Martin, C. E. M. Olesen, B. Liu, J. C. Voyta, J. L. Shumway, R.-R. Juo, and I. Bronstein, *in* "Bioluminescence and Chemiluminescence: Molecular Reporting with Photons" (J. W. Hastings, L. J. Kricka, and P. E. Stanley, eds.), p. 525. Wiley, Chichester, 1997.

[15] I. Bronstein, C. S. Martin, J. J. Fortin, C. E. M. Olesen, and J. C. Voyta, *Clin. Chem.* **42**, 1542 (1996).

[16] K. L. O'Connor and L. A. Culp, *Oncol. Rep.* **1**, 869 (1994).

[17] N. Shaper, A. Harduin-Lepers, and J. H. Shaper, *J. Biol. Chem.* **269**, 25165 (1994).

sidase activity in yeast extracts,[18,19] including the two-hybrid system[20-26] to study protein:protein interactions and a one-hybrid system[27] to study DNA:protein interactions. In addition, extracts from microinjected frog embryos,[28] protozoan parasites,[29] and bacteria[30] have also been assayed with 1,2-dioxetane substrates. A novel mammalian two-hybrid system, based on β-galactosidase peptide complementation to report protein:protein interactions, has been performed with Galacton-Plus substrate.[31,32] β-Galactosidase peptide complementation has also been utilized in chemiluminescent assays for myoblast cell fusion in cell culture.[31,33] A chemiluminescent cytotoxicity assay, measuring release of β-galactosidase reporter enzyme from transfected cells, has been performed with Galacton substrate.[34]

Galacto-Light and Galacto-Light Plus assays, which utilize the Galacton and Galacton-Plus substrates, respectively, have been described.[13] The Galacto-*Star* and Gal-Screen assay systems incorporate Galacton-*Star* substrate. Enzymatic deglycosylation of Galacton-*Star* produces an unstable 1,2-dioxetane anion intermediate, which decomposes with the concomitant production of light. A luminescent reaction with continuous light signal emission is initiated upon the addition of substrate to enzyme with concurrent enzymatic production and decomposition of the anion intermediate.

[18] J. E. Remacle, G. Albrecht, R. Brys, G. H. Braus, and D. Huylebroeck, *EMBO J.* **16**, 5722 (1997).

[19] V. R. Stoldt, A. Sonneborn, C. E. Leuker, and J. F. Ernst, *EMBO J.* **16**, 1982 (1997).

[20] Y. Bourne, M. H. Watson, M. J. Hickey, W. Holmes, W. Rocque, S. I. Reed, and J. A. Tainer, *Cell* **84**, 863 (1996).

[21] L. A. Carver and C. A. Bradfield, *J. Biol. Chem.* **272**, 11452 (1997).

[22] R. Groisman, H. Masutani, M.-P. Leibovitch, P. Robin, I. Soudant, D. Trouche, and A. Harel-Bellan, *J. Biol. Chem.* **271**, 5258 (1996).

[23] J. B. Hogenesch, W. K. Chan, V. H. Jackiw, R. C. Brown, Y.-Z. Gu, M. Pray-Grant, G. H. Perdew, and C. A. Bradfield, *J. Biol. Chem.* **272**, 8581 (1997).

[24] A. N. Hollenberg, T. Monden, J. P. Madura, K. Lee, and F. E. Wondisford, *J. Biol. Chem.* **271**, 28516 (1996).

[25] M.-J. Lee, M. Evans, and T. Hla, *J. Biol. Chem.* **271**, 11272 (1996).

[26] T. Ulmasov, R. M. Larkin, and T. J. Guilfoyle, *J. Biol. Chem.* **271**, 5085 (1996).

[27] S. S. Wolf, K. Roder, and M. Schweizer, *BioTechniques* **20**, 568 (1996).

[28] C. Gove, M. Walmsley, S. Nijjar, D. Bertwistle, M. Guille, G. Partington, A. Bomford, and R. Patient, *EMBO J.* **16**, 355 (1997).

[29] J. K. Beetham, K. S. Myung, J. J. McCoy, M. E. Wilson, and J. E. Donelson, *J. Biol. Chem.* **272**, 17360 (1997).

[30] E. D'Haese, H. J. Nelis, and W. Reybroeck, *Appl. Environ. Microbiol.* **63**, 4116 (1997).

[31] W. A. Mohler and H. M. Blau, *Proc. Natl. Acad. Sci. U.S.A.* **93**, 12423 (1996).

[32] F. Rossi, C. A. Charlton, and H. M. Blau, *Proc. Natl. Acad. Sci. U.S.A.* **94**, 8405 (1997).

[33] C. A. Charlton, W. A. Mohler, G. L. Radice, R. O. Hynes, and H. M. Blau, *J. Cell Biol.* **138**, 331 (1997).

[34] H. Schäfer, A. Schäfer, A. F. Kiderlen, K. N. Masihi, and R. Burger, *J. Immunol. Methods* **204**, 89 (1997).

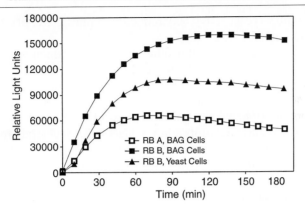

FIG. 2. Kinetic comparison of Gal-Screen reaction buffer A (RB A) and reaction buffer B (RB B) with mammalian and yeast cells. Gal-Screen assays were performed on Ψ2BAGα (CRL-9560, ATCC, Rockville, MD), an NIH/3T3 derivative that constitutively expresses bacterial β-galactosidase from a stably inserted retroviral construct (10,000 cells/well in DMEM/10% calf serum), or SFY526 yeast cells containing the *lacZ* gene (23,000 cells/well in YPD) in a 96-well microplate. Light signal intensity was measured in repeat mode at 27° for 1 sec/well in a TR717 microplate luminometer.

With the Galacto-*Star* assay, cells or cell lysate is incubated with Galacton-*Star* substrate and luminescence enhancer until maximum intensity light emission is reached (approximately 60–90 min). Light emission intensity remains constant for nearly 1 hr. With the Gal-Screen assay, a single reagent containing substrate, luminescence enhancer and lysis agents, is added directly to cells in the presence of cell culture medium, providing a simple mix and read protocol. The kinetics of light emission are similar to the Galacto-*Star* assay. The Gal-Screen system may be used with adherent and nonadherent cell lines, and in the presence or absence of phenol red. In addition, the Gal-Screen system is also used for the direct lysis and assay of yeast cells in the presence of culture medium. Light emission kinetics of the Gal-Screen assay system, performed with both mammalian cells and yeast cells, are demonstrated in Fig. 2.

Reagents

Reagents are components of either the Galacto-*Star* or Gal-Screen reporter gene assay systems (Tropix, Inc., Bedford, MA).

Lysis solution (mammalian cells): 100 m*M* potassium phosphate, pH 7.8, 0.2% Triton X-100

5× Z buffer (yeast cells): 500 mM sodium phosphate, pH 7.1, 50 mM KCl, 5 mM MgSO$_4$

Galacto-*Star* reaction buffer diluent: 100 mM sodium phosphate, pH 7.5, 1 mM MgCl$_2$, 5% Sapphire-II enhancer

Galacton-*Star* substrate: 10 mM Galacton-*Star* substrate (3-chloro-5-(4-methoxyspiro{1,2-dioxetane-3,2'-(4'-chloro)-tricyclo-[3.3.1.1^{3,7}]-decan}-4-yl)phenyl β-D-galactopyranoside)

Gal-Screen substrate: proprietary Galacton-*Star* substrate formulation

Gal-Screen buffer A or Gal-Screen buffer B: contain luminescence enhancer and lysis reagents

Other reagents:

β-Galactosidase (G-5635, Sigma Chemical Co., St. Louis, MO): positive control enzyme (optional)

Phenylmethyl sulfonyl fluoride (PMSF; P-7626, Sigma) or 4-(2-aminoethyl)-benzenesulfonyl fluoride (AEBSF) (A-8456, Sigma): protease inhibitor (optional)

Leupeptin (L-2884, Sigma): protease inhibitor (optional)

Phosphate-buffered saline (PBS): 75 mM sodium phosphate, pH 7.2, 68 mM NaCl

Reconstitution buffer: 100 mM sodium phosphate, pH 7.0, 0.1% bovine serum albumin (BSA; fraction V)

Protocols

Protocol A.I: Preparation of Cell Extracts from Mammalian Tissue Culture Cells

Prepare extracts at 4°.

1. Add fresh dithiothreitol (DTT) to 0.5 mM to the required volume (250 μl per 60-mm culture plate) of lysis solution (optional, see Note 1A). For tissue extracts, add DTT (optional) and protease inhibitors to lysis solution (see Note 2A).
2. Rinse cells twice with PBS.
3. Add lysis solution to cover cells (250 μl per 60-mm culture plate).
4. Detach cells from plate using a cell scraper. Nonadherent cells should be pelleted and sufficient lysis solution added to cover cells. Resuspend and lyse cells by pipetting.
5. Transfer lysate to a microcentrifuge tube and centrifuge for 2 min to pellet any debris.
6. Transfer supernatant to a fresh tube. Extracts may be used immediately or stored at −70°. Perform heat inactivation (Protocol A.VII) or proceed to detection protocol (Protocol A.III).

Protocol A.II: Preparation of Yeast Cell Extracts (Freeze/Thaw)

1. Dilute 5× Z buffer 1 : 5 in MilliQ H_2O or equivalent. Add fresh DTT to 0.5 mM (optional, see Note 1A). (One sample typically requires 1.8–2 ml of 1× Z buffer.)
2. Transfer 1.5 ml of a freshly grown yeast culture (OD_{600} = 0.4–0.6) to a microcentrifuge tube. Centrifuge at 12,000g for 30 sec. Note: If OD_{600} is less than 0.4, use more than 1.5 ml culture.
3. Remove supernatant and resuspend yeast pellet in 1.5 ml of 1× Z buffer.
4. Centrifuge at 12,000g for 30 sec.
5. Remove supernatant and resuspend pellet in 300 μl of 1× Z buffer.
6. Vortex cell suspension and transfer 100 μl to a fresh microcentrifuge tube.
7. Perform freeze/thaw cycle twice: Place tube in liquid N_2 until frozen. Thaw in a 37° water bath.
8. Centrifuge at 12,000g for 5 min at 4° in a microcentrifuge.
9. Transfer supernatant to a fresh tube and store on ice. Proceed to detection protocol (Protocol A.III).

Protocol A.III: Galacto-Star Detection Protocol

Perform assays at room temperature.

1. Dilute Galacton-*Star* substrate 1 : 50 with reaction buffer diluent to make the required volume of reaction buffer (100 μl/well) and equilibrate to room temperature.
2. Add 2–10 μl of cell extracts into microplate wells. Lysis solution or 1× Z buffer should be added to give the same total volume in each well (see Note 3A).
3. Add 100 μl/well of reaction buffer. Incubate for 60–90 min at 26–28° until maximum intensity light emission is reached (see Notes 4A and 5A).
4. Measure intensity of light emission in a microplate luminometer for 0.1–1 sec/well or with a charge-coupled device (CCD) imaging system.

Protocol A.IV: Galacto-Star Direct Lysis Assay Protocol for
Microplate Cultures of Adherent Cells

Perform all steps at room temperature.

1. Dilute Galacton-*Star* substrate 1 : 50 with reaction buffer diluent to make the required volume of reaction buffer (100 μl/well) and equilibrate to room temperature.

2. Remove medium from cells cultured in a 96-well microplate and rinse once with PBS.
3. Add 10 μl/well of lysis solution and incubate for 10 min (see Note 6A).
4. Add 100 μl/well of reaction buffer and incubate for 60–90 min at 26–28° until maximum intensity light emission is reached (see Notes 4A and 5A).
5. Measure intensity of light emission in a microplate luminometer for 0.1–1 sec/well or with a CCD imaging system.

Protocol A.V: Gal-Screen Assay Protocol for Mammalian or Yeast Cells in Culture Medium

1. Dilute Gal-Screen substrate 1:25 with Gal-Screen buffer A or B (i.e., mix 40 μl Gal-Screen substrate + 960 μl Gal-Screen buffer A or B, 100 μl/well required, see Note 7A) to make reaction buffer A or B and equilibrate to room temperature. Gal-Screen substrate should be freshly diluted each time.
2. Add 100 μl/well or reaction buffer A or B to a 96-well microplate containing 100 μl/well of cells in culture medium (see Notes 6A–11A). Incubate at 26–28° for approximately 60–90 min or until constant intensity light emission is reached (see Notes 4A and 5A).
3. Measure intensity of light emission in a microplate luminometer for 0.1–1 sec/well or with a CCD imaging system.

Protocol A.VI: Preparation of Controls

POSITIVE CONTROL. Reconstitute lyophilized β-galactosidase to 1 mg/ml in reconstitution buffer. Generate a standard curve by serially diluting stock enzyme in lysis buffer or 1× Z buffer containing 0.1% BSA. Two to 20 ng of enzyme should be used for the high end amount.

Negative Control. Assay an equivalent volume of mock-transfected cell extract (see Note 12A).

Protocol A.VII: Heat Inactivation of Endogenous β-Galactosidase Activity in Cell and Tissue Extracts

Some cell lines may exhibit relatively high levels of endogenous mammalian β-galactosidase activity. This may contribute to higher background, which will decrease the assay sensitivity by lowering the signal-to-noise ratio. The bacterial β-galactosidase reporter enzyme is more thermostable than the mammalian enzyme activity, and thus incubation at an elevated temperature can be performed to preferentially inactivate the endogenous mammalian background activity. Protocols for heat inactivation of endogenous mammalian β-galactosidase activity in cell extracts[17,35] and tissue extracts[17] have been described. Incubation of cell extracts for 1 hr at 50° has been shown to reduce the activity of mammalian β-galactosidase, ranging from a 3- to 4-fold reduction in cell lines with low levels of endogenous activity to a 20- to 40-fold reduction in cell lines with higher levels of endogenous activity.[35]

1. Following extract preparation, heat extract at 48° for 50 min (for cultured cell extracts) or 60 min (for tissue extracts).
2. Proceed with detection protocol (Protocol A.III).

Notes/Troubleshooting

1A. DTT may be added to stabilize β-galactosidase activity. However, concentrations of reducing agents higher than 0.5 mM will decrease the half-life of light emission of 1,2-dioxetanes. If the half-life of light emission is critical, reducing agents should be omitted from the lysis solution. If a lysis buffer containing excess DTT has been used, the addition of hydrogen peroxide to a final concentration of 10 mM (add 1 μl of 30% H_2O_2 per 1 ml) to the Galacto-*Star* reaction buffer will prevent rapid decay of signal half-life.

2A. For preparation of tissue extracts, add PMSF or AEBSF protease inhibitor to a final concentration of 0.2 mM and leupeptin to 5 μg/ml to the lysis solution. AEBSF is a water-soluble serine protease inhibitor similar to PMSF.

3A. The amount of cell extract required may vary depending on the expression level and instrumentation. Use 2–5 μl of extract for positive controls and 10 μl of extract for potentially low levels of enzyme.

4A. Measurements may be performed in as short as 20–30 min if the time between reaction buffer addition and light signal measurement is identical for all samples.

5A. Assays are ideally performed at 26–28°. If the incubation temperature is significantly different, the kinetics of light emission will vary, as the rate of an enzyme reaction is dependent on temperature.

6A. Heat inactivation has not been found to be effective performed on cells directly in the culture plate and thus is not recommended following lysis solution addition with this protocol or with the Gal-Screen protocol.

7A. Gal-Screen buffer A is optimized for mammalian cell cultures and has been used with transfected adherent and nonadherent cell lines, includ-

[35] D. C. Young, S. D. Kingsley, K. A. Ryan, and F. J. Dutko, *Anal. Biochem.* **215,** 24 (1993).

ing NIH/3T3, CHO-K1, and K562. Gal-Screen buffer B is optimized for yeast cultures and has also been shown to work with mammalian cell cultures. The alternative buffer formulations result in differing light emission kinetics. For mammalian cells, reaction buffer A provides faster kinetics, which may be advantageous at lower incubation temperatures. An approximately twofold higher signal intensity is obtained with reaction buffer B, but assay sensitivity with both buffers is equivalent. The choice of reagents depends on the particular assay requirements and desired kinetic performance.

8A. It is also possible to use 384-well plates with the Gal-Screen assay. Culture and assay reagent volumes should be reduced to 25 μl (or as desired, maintaining a 1:1 ratio of culture:reagent volumes without overfilling wells).

9A. For mammalian cell cultures, a typical cell density is 10,000–50,000 cells/well in 100 μl for 96-well plates or 1000–10,000 cells/well in 25 μl for 384-well plates. For yeast cell cultures, a typical cell density is 10,000–75,000 cells/well in 100 μl for 96-well plates or 5000–40,000 cells/well in 25 μl for 384-well plates.

10A. Cell culture medium (mammalian cells) containing phenol red indicator may be used without affecting assay sensitivity. The presence of phenol red results in signal reduction due to light absorbance, but the signal/noise is not affected. The type of culture medium and presence/absence of serum may contribute to slight variability in signal intensity and assay kinetics. It is recommended to initially perform a kinetic analysis of the particular cell line/culture medium system to determine desired signal measurement time. It is not necessary to measure light signal at the time of peak light emission, as long as all samples are measured in a consistent time frame.

11A. It has been observed that the use of different types of yeast media may affect the light emission kinetics of the Gal-Screen assay. The reaction kinetics should be evaluated initially to determine the optimum measurement time.

12A. Many mammalian cells and tissues have some amount of endogenous mammalian β-galactosidase activity, which contributes to assay background. The chemiluminescent β-galactosidase assays are performed at a pH that is more favorable for the activity of the bacterial enzyme, which has a higher pH optimum than the endogenous mammalian enzyme. It is important to determine the level of endogenous enzyme in nontransfected cells to establish the assay background. In addition, significant reduction of endogenous activity may be achieved by heat inactivation with cell or tissue extracts.[17,35]

13A. It has been observed that cell extracts from primary porcine and murine smooth muscle cells (artery tissue) contain an inhibitor of transfected β-galactosidase reporter enzyme activity.[36] Removal of the inhibitory activity from artery extracts is accomplished by treatment of the extract with Chelex-100, which removes divalent cations.

B. β-Glucuronidase Reporter Gene Assay

Background

The bacterial β-glucuronidase (GUS) gene[37] has become widely used as a reporter for the analysis of gene expression in plants and may also be used in mammalian cells.[38] Glucuron 1,2-dioxetane substrate enables highly sensitive chemiluminescent detection of β-glucuronidase activity.[13,39,40] The GUS-Light reporter gene assay incorporates Glucuron substrate and Emerald luminescence enhancer and provides a wide dynamic range, enabling detection of 60 fg to 2 ng of purified enzyme. At least 100-fold higher sensitivity is achieved with chemiluminescence compared with the fluorescent assay detection limit of 0.02 ng.[41] The GUS-Light assay shows good correlation to results obtained with the fluorescent substrate 4-methylumbelliferyl β-D-glucuronide (4-MUG).[39]

Cell lysate is incubated with Glucuron substrate for 1 hr, during which enzymatic deglycosylation forms a metastable dioxetane anion intermediate, which leads to generation of an excited state species. The sample is then placed in a luminometer and accelerator is added, which raises the pH and introduces a chemiluminescence enhancer, terminating the β-glucuronidase activity and accelerating breakdown of the dioxetane with concomitant emission of light. Glucuron substrate has a half-life of light emission of approximately 4.5 min after the addition of accelerator and should be used with instrumentation with automatic injectors.

[36] H. Oswald, F. Heinemann, S. Nikol, B. Salmons, and W. H. Günzburg, *BioTechniques* **21,** 78 (1997).
[37] R. A. Jefferson, T. A. Kavanagh, and M. W. Bevan, *EMBO J.* **6,** 3901 (1987).
[38] D. R. Gallie, J. N. Feder, and V. Walbot, *in* "GUS Protocols: Using the GUS Gene as a Reporter of Gene Expression" (S. R. Gallagher, ed.), p. 181. Academic Press, San Diego, 1992.
[39] I. Bronstein, J. J. Fortin, J. C. Voyta, R.-R. Juo, B. Edwards, C. E. M. Olesen, N. Lijam, and L. J. Kricka, *BioTechniques* **17,** 172 (1994).
[40] G. Hansen and M.-D. Chilton, *Proc. Natl. Acad. Sci. U.S.A.* **93,** 14978 (1996).
[41] A. G. Rao and P. Flynn, *in* "GUS Protocols: Using the GUS Gene as a Reporter of Gene Expression" (S. R. Gallagher, ed.), p. 89. Academic Press, San Diego, 1992.

Reagents

Reagents are components of the GUS-Light reporter gene assay system (Tropix).

Lysis solution (plant cells or tissue): 50 mM sodium phosphate, pH 7.0, 10 mM EDTA, 0.1% sodium lauryl sarcosine, 0.1% Triton X-100

Lysis solution (mammalian cells): 100 mM potassium phosphate, pH 7.8, 0.2% Triton X-100

Glucuron substrate: 10 mM Glucuron substrate (sodium 3-(4-methoxyspiro{1,2-dioxetane-3,2'-(5'-chloro)-tricyclo-[3.3.1.1^{3,7}]decan}-4-yl)phenyl β-D-glucuronate)

Reaction buffer diluent: 0.1 M sodium phosphate, pH 7.0, 10 mM EDTA

Light emission accelerator

Other reagents:

β-Glucuronidase (G-7896, Sigma): Positive control enzyme (optional)

PBS: 75 mM sodium phosphate, pH 7.2, 68 mM NaCl

Reconstitution buffer: 0.1 M sodium phosphate, pH 7.0, 0.1% BSA (fraction V)

Protocols

Protocol B.I: Preparation of Cell Extracts

Prepare extracts at 4°.

1. Add 2-mercaptoethanol (to 10 mM) or DTT (to 0.5 mM) to the required volume (250 μl per 60-mm culture plate) of lysis solution (optional, see Note 1B).

2. Prepare sample: rinse cells with PBS to remove culture media or prepare plant material.

3. Add lysis solution to cover cells (250 μl of lysis solution per 60-mm culture plate) or plant material (250 μl of lysis solution per 25 mg of plant material).

4. Detach cells from culture plate using a cell scraper. For plant material, homogenize cells or tissue.

5. Transfer lysate to a microcentrifuge tube and centrifuge for 2 min to pellet debris.

6. Transfer supernatant to a fresh tube. Extracts may be used immediately or stored at −70°. Proceed to detection protocol (Protocol B.II).

Protocol B.II: GUS-Light Detection Protocol

Perform all steps at room temperature.

1. Dilute Glucuron substrate 1 : 100 with reaction buffer diluent to make

reaction bufer (70 μl/well required) and equilibrate reaction buffer and light emission accelerator to room temperature.

2. Add 2–20 μl of cell extracts to microplate wells. Lysis solution should be added to give a consistent final volume in each well (see Notes 2B and 3B).

3. Add 70 μl/well of reaction buffer and incubate for 60 min (see Note 4B). Incubations may be as short as 15 min (especially if high levels of expression are anticipated), but the linear range of the assay may decrease.

4. Place microplate in luminometer. Inject 100 μl/well of light emission accelerator (see Note 5B). After a 1- to 2-sec delay following injection, measure each sample for 1–5 sec. Alternatively, the NorthStar HTS workstation CCD-based imaging system can be used for rapid reagent injection and immediate whole-plate imaging for measurement.

Protocol B.III: Preparation of Controls

POSITIVE CONTROL. Reconstitute lyophilized β-glucuronidase to 1 mg/ml in reconstitution buffer. Store at 4°. Generate a standard curve by serially diluting stock enzyme in lysis solution containing 0.1% BSA. Two nanograms of enzyme should be used for the high end amount.

NEGATIVE CONTROL. Assay an equivalent volume of nontransformed extract (see Note 6B).

Notes/Troubleshooting

1B. Reducing agents may be added to increase enzyme stability, particularly if extracts are not used immediately. They are not required for the chemiluminescent reaction.

2B. The amount of cell extract required may vary depending on the expression level and instrumentation. Use 5 μl of extract for positive controls and 10–20 μl of extract for potentially low levels of enzyme.

3B. Chlorophyll in concentrated samples may interfere with the chemiluminescent signal intensity. Therefore, if high levels of chlorophyll are present, several extract dilutions should be assayed.

4B. Light intensities are time dependent. Reaction buffer should be added to samples in the same time frame as they are measured. For example, if it takes 1 sec to complete measurement of each well, then reaction buffer should be added to wells every 1 sec.

5B. If manual injection is used, the accelerator should be added in the same consistent time frame as the reaction buffer is added (i.e., the incubation time in reaction buffer should be identical for each well).

6B. Bacterial contamination of plant material may cause assay background, as β-glucuronidase activity is found in several different soil bacteria.

Care should be taken to avoid the presence of bacteria in plant samples. In addition, some plant tissues contain endogenous β-glucuronidase and virtually all mammalian tissues contain some level of endogenous mammalian β-glucuronidase activity. Mammalian β-glucuronidase is primarily lysosomal and has a pH optimum of pH 3.5–5.0, thus interference with the GUS-Light assay is minimal. However, with any extract, it is important to determine the assay background with nontransformed or nontransfected extract.

C. Placental Alkaline Phosphatase Reporter Gene Assay

Background

Secreted placental alkaline phosphatase (SEAP) has been increasingly utilized as a eukaryotic reporter gene. A gene construct bearing a mutation in the membrane localization domain of human placental alkaline phosphatase causes the normally membrane-anchored protein to be secreted from the cell.[42] Thus, SEAP detection is performed with a sample of cell culture medium while the cell population remains intact. CSPD 1,2-dioxetane alkaline phosphatase substrate may be used for highly sensitive detection of SEAP or nonsecreted placental alkaline phosphatase (PLAP). The Phospha-Light reporter gene assay system incorporates CSPD substrate and Emerald luminescence enhancer for high sensitivity and a dynamic range covering five orders of magnitude of purified enzyme concentration.[13,15,39] With CSPD substrate, approximately 3 fg of purified placental alkaline phosphatase can be detected, which is 1000-fold greater sensitivity than that reported for the colorimetric substrate p-nitrophenyl phosphate (pNPP, 10 pg).[1] In the presence of culture media, sensitivity with chemiluminescent detection is reduced approximately 10-fold due to increased background contributed by serum alkaline phosphatase. A comparison of detection with CSPD to fluorescent detection with 4-methylumbelliferyl phosphate (MUP) demonstrates over 100-fold higher sensitivity with chemiluminescent detection (unpublished data). CSPD substrate has been used for quantitation of both secreted placental alkaline phosphatase reporter enzyme in cell culture media[43,44] and nonsecreted placental alkaline phosphatase in cell[45] and tissue extracts.[16]

[42] J. Berger, J. Hauber, R. Hauber, R. Geiger, and B. R. Cullen, *Gene* 66, 1 (1988).
[43] R. E. Jones, D. Defeo-Jones, E. M. McAvoy, G. A. Vuocolo, R. J. Wegrzyn, K. M. Haskell, and A. Oliff, *Oncogene* 6, 745 (1991).
[44] R. E. Means, T. Greenough, and R. C. Desrosiers, *J. Virol.* 71, 7895 (1997).
[45] K. Guo and K. Walsh, *J. Biol. Chem.* 272, 791 (1997).

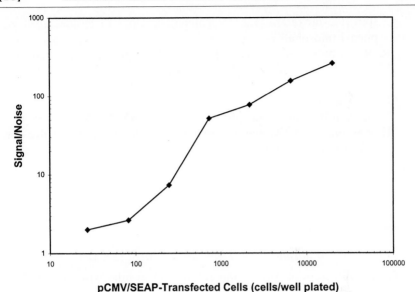

pCMV/SEAP-Transfected Cells (cells/well plated)

FIG. 3. Phospha-Light assay: Measurement of secreted placental alkaline phosphatase reporter enzyme with pCMV/SEAP-transfected NIH/3T3 cells. NIH/3T3 cells were transfected with pCMV/SEAP (Tropix) and allowed to recover for 2 hr. Cells were trypsinized and seeded in a 96-well plate at the indicated densities and incubated for 18 hr. Fifty microliters of culture medium was removed from each well and assayed with Phospha-Light exactly according to the protocol. Light emission was measured in the NorthStar HTS workstation CCD-based imaging system.

Secreted placental alkaline phosphatase is measured from 24 to 72 hr after cell transfection.[46] Figure 3 demonstrates Phospha-Light quantitation of SEAP activity from pCMV/SEAP-transfected NIH/3T3 cells, performed 24 hr after transfection. Highly sensitive detection of SEAP is obtained, from less than 100 cells, with a dynamic range of three orders of magnitude of cell concentration.

Reagents

Reagents are components of the Phospha-Light reporter gene assay system (Tropix).

5× dilution buffer

Assay buffer: contains endogenous alkaline phosphatase inhibitors

CSPD substrate: 25 mM CSPD substrate (disodium 3-(4-methoxy-

[46] B. Cullen and M. Malim, *Methods Enzymol.* **216,** 362 (1992).

spiro{1,2-dioxetane-3,2'-(5'-chloro)tricyclo[3.3.1.1^{3,7}]decan}-4-yl)-phenyl phosphate)

Reaction buffer diluent: contains Emerald luminescence enhancer

Human placental alkaline phosphatase: 3 μg/ml (0.75 U/ml), positive control enzyme, store at −20°.

Other reagents:

PBS: 75 mM sodium phosphate, pH 7.2, 68 mM NaCl

Human placental alkaline phosphatase (P-3895, Sigma): positive control enzyme (optional)

Reconstitution buffer: 1× dilution buffer, 0.1% BSA (fraction V), 50% glycerol (optional)

Protocols

Protocol C.I: Phospha-Light Detection Protocol (SEAP)

Perform all steps at room temperature, unless otherwise indicated.

1. Dilute CSPD substrate 1:20 with reaction buffer diluent to make the required volume of reaction buffer (50 μl/well) and dilute 5× dilution buffer to 1× with H$_2$O (150 μl/sample). Equilibrate all reagents to room temperature.
2. Aliquot 50 μl of culture medium into a microfuge tube, add 150 μl of 1× dilution buffer (sufficient for triplicates), and heat at 65° for 30 min (see Note 1C).
3. Cool to room temperature by placing on ice briefly.
4. Add 50 μl of diluted, heat-treated culture media to three microplate wells.
5. Add 50 μl/well of assay buffer and incubate for 5 min (see Note 2C).
6. Add 50 μl/well of reaction buffer and incubate for 20 min (see Note 3C).
7. Measure intensity of light emission in a microplate luminometer for 0.1–1 sec/well or with a CCD imaging system.

Protocol C.II: Extract Preparation (PLAP)

Prepare extracts at 4°.

1. Dilute 5× dilution buffer to 1× with H$_2$O. Add Triton X-100 to a final concentration of 0.2% (v/v). (One 60-mm culture plate typically requires 250 μl of 1× dilution buffer.)
2. Rinse cells twice with PBS.
3. Add 1× dilution buffer/0.2% Triton X-100 to cover cells.

4. Detach cells from culture plate using cell scraper. Prepare extract by trituration and transfer to a microcentrifuge tube. Cell extracts may be used immediately or stored at $-70°$. Nonadherent cells should be pelleted and sufficient $1\times$ dilution buffer/0.2% Triton X-100 added to cover cells. Resuspend and lyse cells by trituration.

5. Mix 15 μl of cell extract and 185 μl of $1\times$ dilution buffer and heat at $65°$ for 30 min.

6. Proceed with detection protocol (Protocol C.I), starting at step 3.

Protocol C.III: Phospha-Light Direct Assay Protocol for Microplate Cultures of Adherent Cells (PLAP)

Perform all steps at room temperature. Heat inactivation should not be performed with this protocol.

1. Dilute CSPD substrate 1:20 with reaction buffer diluent to make reaction buffer (50 μl/well required) and dilute $5\times$ dilution buffer to $1\times$ with H_2O (50 μl/well required). Equilibrate all reagents to room temperature.
2. Rinse wells once with PBS.
3. Add 50 μl/well of $1\times$ dilution buffer and incubate for 5 min.
4. Add 50 μl/well of assay buffer and incubate for 5 min.
5. Add 50 μl/well of reaction buffer and incubate for 20 min.
6. Measure intensity of light emission in a microplate luminometer for 0.1–1 sec/well or with a CCD imaging system.

Protocol C.IV: Preparation of Controls

POSITIVE CONTROL. Generate a standard curve by serially diluting stock enzyme in $1\times$ dilution buffer or mock-transfected cell culture media. A 10-μl aliquot of stock enzyme should be used for the high end amount. Alternatively, stock enzyme may be prepared by reconstituting lyophilized human placental alkaline phosphatase to 1 mg/ml in $1\times$ dilution buffer containing 0.1% BSA and 50% glycerol. Store at $-20°$.

NEGATIVE CONTROL. Assay an equivalent volume of culture media from mock-transfected cells (see Notes 4C–6C).

Notes/Troubleshooting

1C. Chemiluminescent reporter assays for placental alkaline phosphatase may be conducted in cells that have endogenous nonplacental alkaline phosphatase activity. Human placental alkaline phosphatase is highly thermostable compared to other mammalian nonplacental alkaline phosphatase isozymes. Incubation at an elevated temperature results in selective inactivation of endogenous nonplacental alkaline phosphatase from both cells and from the serum component of the culture medium. We have found that heat inactivation performed directly on cells in the culture dish is not effective, and thus is not recommended with the direct PLAP assay (Protocol C.III).

2C. Do not add assay buffer to culture media until it has cooled.

3C. Light intensities are time dependent. The time between reaction buffer addition and light signal measurement should be identical for all wells.

4C. Endogenous nonplacental enzyme activity is reduced significantly using a combination of heat inactivation and differential inhibitors that do not significantly inhibit the transfected placental isozyme. However, it is important to determine the level of endogenous enzyme in media or extracts from nontransfected cells to establish assay background.

5C. The most significant contribution of nonplacental alkaline phosphatase activity is from serum. Therefore, the use of serum-free medium may improve background considerably. In addition, we have found that using heat-inactivated serum (56° for 1 hr) to prepare media provides a significant reduction in background.

6C. Certain cell lines, such as HeLa and others derived from cervical cancers, and some other tumor cell lines may express high levels of endogenous placental alkaline phosphatase, resulting in high assay backgrounds.[47] Therefore, the use of PLAP or SEAP as a reporter system in these cell lines is generally not recommended.

D. Luc-Screen Luciferase Reporter Gene Assay

Background

Traditional luciferase assays generate a fast peak-decay pattern of light emission; to measure the "flash" light signal that immediately follows substrate addition, an instrument with a reagent injector is needed. Each well of a plate is injected and measured immediately due to fast kinetics of the flash assay. Such a format is not amenable to automated, high throughput screening applications. The Luc-Screen reporter gene assay system provides extended glow light emission that is generated directly in microplate cultures in the presence of cell culture medium with a mix and read protocol.

---

[47] F. J. Benham, J. Fogh, and H. Harris, *Int. J. Cancer* **27,** 637 (1981).

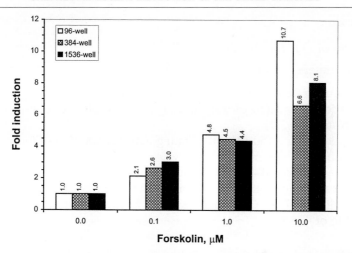

Fig. 4. Luc-Screen assay: Forskolin induction of pCRE-luc-transfected NIH/3T3 cells. NIH/3T3 cells were transfected with pCRE-luc (Stratagene, La Jolla, CA) and incubated overnight. Cells were seeded in 96 (2×10^4 cells/well)-, 384 (4000 cells/well)-, and 1536 (1180 cells/well)-well microplates in the presence of varying concentrations of forskolin (Calbiochem, San Diego, CA), incubated for 17 hr, and assayed with Luc-Screen reagents. Light emission was measured with the NorthStar HTS workstation for 96 (1 min), 384 (2 min), and 1536 (4 min) wells.

Detection of 50 fg of purified luciferase is achieved in the presence of culture medium, with a dynamic range of 50 fg to 100 ng of luciferase. The light emission has a signal half-life of 4 to 5 hr. The Luc-Screen assay is performed with adherent or suspension cell lines and in the presence or absence of phenol red. Figure 4 demonstrates the use of the Luc-Screen assay for quantitation of forskolin-induced luciferase activity in pCRE-luc-transfected NIH/3T3 cells. Transfected cells were plated in the presence of forskolin inducer in 96-, 384-, and 1536-well microplate density formats.

Reagents

Reagents are components of the Luc-Screen reporter gene assay system (Tropix).
Luc-Screen buffer 1
Luc-Screen buffer 2
Other reagents:
Luciferase (L-1759, Sigma): positive control enzyme (optional)

Protocol

Protocol D.I: Luc-Screen Assay Protocol

Perform assay at room temperature.

1. Equilibrate buffers to room temperature.
2. Add buffer 1 (50 μl/well) to cells in 100 μl of culture medium (see Notes 1D–2D).
3. Add buffer 2 (50 μl/well). Timing between addition of buffer 1 and buffer 2 is not critical.
4. Incubate for 10 min (see Note 3D).
5. Measure light emission in a microplate luminometer for 0.1–1 sec/well or with a CCD imaging system. Light signal decays with a half-life of 4–5 hr.

Notes/Troubleshooting

1D. A two- to fourfold decrease in signal intensity is typically observed in culture medium containing phenol red.

2D. If desired, PBS containing 1 mM Mg^{2+} and 1 mM Ca^{2+} may be substituted for culture medium.

3D. For steady light emission, the temperature during assay incubation and measurement should be maintained at 22–27°. When performing kinetic analysis, it is important that the temperature is within this range.

E. Dual Luciferase/β-Galactosidase Reporter Gene Assay

Background

Cotransfection of two reporter gene constructs, an experimental and a constitutively expressed control, is frequently employed to normalize reporter activity for transfection efficiency. Luciferase and β-galactosidase reporter genes are ideal for cotransfections due to the availability of covenient assays with similar sensitivities, dynamic ranges, and luminescent readouts. The Dual-Light luminescent reporter gene assay system combines these luminescent reactions for highly sensitive, sequential detection of luciferase and β-galactosidase in a single extract sample.[15,48–54] High sensi-

[48] I. Bronstein, C. S. Martin, C. E. M. Olesen, and J. C. Voyta, *in* "Bioluminescence and Chemiluminescence: Molecular Reporting with Photons" (J. W. Hastings, L. J. Kricka, and P. E. Stanley, eds.), p. 451. Wiley, Chichester, 1997.

[49] C. S. Martin, P. A. Wight, A. Dobretsova, and I. Bronstein, *BioTechniques* **21**, 520 (1996).

[50] T. Bourcier, G. Sukhova, and P. Libby, *J. Biol. Chem.* **272**, 15817 (1997).

tivity and a wide dynamic range are attained, with detection of 1 fg to 20 ng of purified luciferase and 10 fg to 20 ng of purified β-galactosidase.

The Dual-Light reporter assay incorporates luciferin and Galacton-Plus substrates. The luciferase signal is measured immediately after the addition of a solution containing luciferin and Galacton-Plus substrates. The luciferase reaction produces a light signal that decays with a half-life of approximately 1 min. At this point, the light signal from the β-galactosidase reaction is initially negligible due to the low pH (7.8) and absence of luminescence enhancer. After a 30- to 60-min incubation, the light signal from the accumulated product of the β-galactosidase/Galacton-Plus substrate reaction is initiated by addition of a light emission accelerator, which raises the pH and provides luminescent enhancer to increase light intensity. Light emission from the β-galactosidase reaction exhibits very slow decay with a half-life of up to 180 min. At this stage, residual light emission from the luciferase reaction is minimal due to its rapid kinetic decay and the quenching effect of the accelerator.

Generally, only very high luciferase concentrations (1 ng or greater of enzyme) may interfere with the low-level detection of β-galactosidase (see Fig. 5). The level of interference produced by one enzyme at the low-level detection of the second enzyme was measured by assaying individual purified enzyme dilution series. In the absence of luciferase, the presence of β-Gal in amounts up to 1 ng exhibits no impact on the signal from the luciferase portion of the assay over background. However, in a cell-based assay, we have observed interference of a high level of β-Gal with luciferase (see Note 3E). When β-Gal is not present, quantities of luciferase below 1 ng do not increase the background signal in the β-Gal assay. At very high luciferase levels, 1 ng or above, residual luciferase signal is measurable; however, this level of signal interferes only in an extremely low-level detection of β-Gal. A longer delay after the addition of accelerator prior to the measurement of light intensity results in decreased levels of residual luciferase signal when extremely high levels of luciferase are present.

A dual assay for quantitation of both luciferase and β-glucuronidase activities in a single extract is performed by substituting the Glucuron substrate for the Galacton-Plus substrate (Protocol E.II). Figure 6 demonstrates the use of Dual-Light for quantitation of both luciferase and β-

[51] Y. V. Fedorov, N. C. Jones, and B. B. Olwin, *Mol. Cell. Biol.* **18**(10), 5780 (1998).
[52] T. Iwata, S. Sato, J. Jimenez, M. McGowan, M. Moroni, A. Dey, N. Ibaraki, V. N. Reddy, and D. Carper, *J. Biol. Chem.* **274**(12), 7993 (1999).
[53] U. Laufs and J. K. Liao, *J. Biol. Chem.* **273**(37), 24266 (1998).
[54] R. F. McClure, C. J. Heppelman, and C. V. Paya, *J. Biol. Chem.* **274**(12), 7756 (1999).

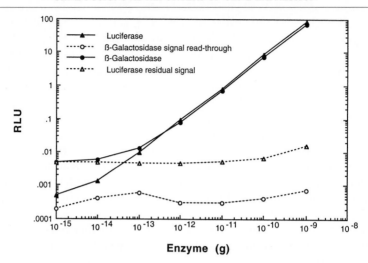

Enzyme (g)

FIG. 5. Detection of luciferase and β-Gal with Dual-Light and levels of background signal. Purified luciferase and β-galactosidase were diluted in lysis solution. Each dilution set was assayed with Dual-Light as described, using 10 μl of each dilution. Light emission was measured in a Dynatech Laboratories ML2250 Microlite microplate luminometer. Luciferase—Signal intensity following injection of buffer B into diluted luciferase. Luciferase residual signal—Signal intensity following injection of Accelerator-II into diluted luciferase. β-Galactosidase—Signal intensity following injection of Accelerator-II into diluted β-galactosidase. β-Galactosidase signal read through—Signal intensity following injection of buffer B into diluted β-galactosidase.

galactosidase reporter enzymes from NIH/3T3 cells cotransfected with pGL3, which constitutively expresses luciferase from the SV40 promoter/ enhancer, and pCMVβ, which constitutively expresses β-Gal from the CMV promoter/enhancer. For both enzymes, detection over a wide range of cell concentration is achieved and can be performed in sample wells containing less than 100 (β-Gal) to several hundred cells (luciferase).

Reagents

Reagents are components of the Dual-Light reporter gene assay system (Tropix).
Lysis solution: 100 mM potassium phosphate, pH 7.8, 0.2% Triton X-100
Buffer A: Lyophilized reaction buffer, pH 7.8. Reconstitute in 5 ml of sterile deionized or Milli-Q H$_2$O.
Buffer B: Lyophilized luciferin. Reconstitute in 22 ml of sterile deionized or Milli-Q H$_2$O. Add Galacton-Plus substrate (or Glucuron

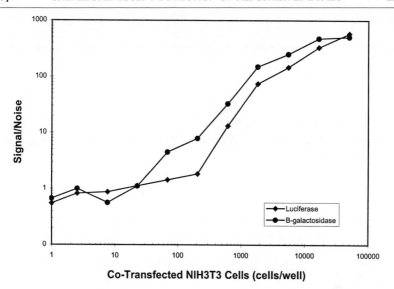

Co-Transfected NIH3T3 Cells (cells/well)

Fig. 6. Dual-Light assay: Measurement of luciferase and β-Gal reporter enzyme activities in pGL3 (Promega, Madison, WI) and pCMVβ (Tropix) cotransfected cells. NIH 3T3 cells were cotransfected and allowed to recover for 18 hr. Cells were trypsinized and seeded in a 96-well plate at the indicated densities and incubated for 6 hr. Lysis solution (10 μl/well) was added directly to each well and then the Dual-Light assay was performed according to the protocol. Light emission was measured in the NorthStar HTS workstation.

substrate) diluted 1 : 100 immediately before use. Do not store Buffer B and Galacton-Plus substrate (Glucuron substrate) mixtures for longer than 24 hr.

Galacton-Plus substrate: 10 mM Galacton-Plus substrate (3-chloro-5-(4-methoxyspiro{1,2-dioxetane-3,2'-(5'-chloro)tricyclo[3.3.1.1^{3,7}] decan}-4-yl)phenyl-β-D-galactopyranoside)

Glucuron substrate: 10 mM Glucuron substrate (available separately)

Accelerator-II: luminescence accelerator containing Sapphire-II enhancer

Other reagents:

β-Galactosidase (G-5635, Sigma): positive control enzyme (optional)

Luciferase (L-1759, Sigma): positive control enzyme (optional)

β-Glucuronidase (G-7896, Sigma): positive control enzyme (optional)

PBS: 75 mM sodium phosphate, pH 7.2, 68 mM NaCl

Reconstitution buffer: 100 mM sodium phosphate, pH 7.0, 0.1% BSA (fraction V)

Protocols

Protocol E.I: Preparation of Cell Extracts from Tissue Culture Cells

Prepare extracts at 4°.

1. Add fresh DTT to 0.5 mM to the required volume of lysis solution (optional, see Notes 1E and 2E).
2. Rinse cells twice with PBS.
3. Add lysis solution to cover cells (250 μl of lysis buffer for a 60-mm culture plate).
4. Detach cells from culture plate using a cell scraper.
5. Transfer lysate to a microcentrifuge tube and centrifuge for 2 min to pellet any debris.
6. Transfer supernatant to a fresh tube. Extracts may be used immediately or stored at −70°. Proceed to detection protocol (Protocol E.II).

Protocol E.II: Dual-Light Luciferase/β-Galactosidase Assay Protocol

Perform assay at room temperature.

1. Dilute sufficient Galacton-Plus substrate 1 : 100 with buffer B and equilibrate all buffers to room temperature.
2. Add 2–10 μl of cell extracts into microplate wells (see Notes 3E–6E).
3. Add 25 μl/well of buffer A.
4. Within 10 min, inject 100 μl of buffer B. After a 2-sec delay, measure the luciferase signal for 0.1–1 sec/well (see Note 7E).
5. Incubate samples for 30–60 min.
6. Inject 100 μl of Accelerator-II. After a 2-sec delay, measure the β-galactosidase signal for 0.1–1 sec/well or follow the same delay and measurement time used for the addition of buffer B (see Notes 8E and 9E). Alternatively, the NorthStar HTS workstation CCD-based imaging system can be used for rapid reagent injection and immediate whole-plate imaging for measurement.

Protocol E.III: Preparation of Controls

POSITIVE CONTROL. Reconstitute lyophilized β-galactosidase (β-glucuronidase) to 1 mg/ml in reconstitution buffer. Store at 4°. Generate a standard curve by serially diluting the enzyme in lysis solution containing 0.1% BSA. Reconstitute lyophilized luciferase to 1 mg/ml in reconstitution buffer. Small aliquots should be stored at −70° and used only once after thawing. Prepare serial dilutions as described earlier. Two to 20 ng of β-galactosidase (β-glucuronidase) and 1–10 ng of luciferase should be used for the high end amounts.

NEGATIVE CONTROL. Assay an equivalent volume of mock-transfected cell extract (see Notes 10E and 11E).

Notes/Troubleshooting

1E. DTT may be added fresh prior to use to a final concentration of 0.5 mM to stabilize luciferase activity. However, higher concentrations of reducing agents such as 2-mercaptoethanol and DTT will decrease the half-life of light emission of the Galacton-Plus substrate. If the extended half-life of light emission from Galacton-Plus substrate is desired (i.e., with manual accelerator addition), reducing agents should be omitted from the lysis solution. If a lysis solution containing excess DTT has been used, the addition of hydrogen peroxide to Accelerator-II to a final concentration of 10 mM (add 1 μl of 30% H_2O_2 per 1 ml of Accelerator-II) will prevent rapid decay of signal half-life.

2E. The lysis solution may be substituted with alternative lysis solutions and protocols. However, high levels of reducing agents may interfere with the 1,2-dioxetane substrate, causing rapid signal decay. Alternative lysis solutions should be tested first to ensure the desired kinetic performance.

3E. Because the volume of cell extract assayed for each enzyme is identical, the ratio of control reporter to experimental vector used in a transfection should be adjusted to ensure that individual enzyme signal intensities are within the detection range of the instrument used. We have observed interference with the luciferase signal in cells transfected with the pCMVβ vector, which constitutively expresses β-Gal at a very high level. In addition to contributing a high background during the luciferase measurement, the presence of such a high level of β-Gal enzyme reduces the expression level of luciferase. Thus, with such a high level of expression from the CMV promoter, it may be necessary to adjust the amount of control plasmid used.

4E. If using less than 10 μl of extract, the lysis solution should be added to bring the total volume up to 10 μl to ensure that the concentration of reducing agent and other components is identical in each well.

5E. The amount of cell extract required may vary depending on the level of expression and instrumentation used. It is important to adjust the amount of extract to keep the signal intensity within the linear range of the instrument.

6E. For direct assay of adherent cells cultured in a 96-well microplate, remove culture medium and wash wells once with PBS. Then add 10 μl of lysis solution directly to each well and proceed to step 3.

7E. Signal intensities are time dependent. The time between the addition of buffer B and measurement should be identical for each well. Instruments with automatic injectors will eliminate this concern. Longer or shorter measurements and delay times may be utilized but the same timing should be used when measuring the β-galactosidase signal after the addition of Accelerator-II.

8E. If manual addition is used, then Accelerator-II should be added in the same consistent time frame as buffer B.

9E. The Dual-Light system is suitable for use with luminometers with automatic injectors. If only a single injector is available, rinse the injector thoroughly between injection of buffer B and accelerator. Manual injection may be performed if samples are measured at the same interval after adding accelerator.

10E. High levels of endogenous β-galactosidase activity in cells or tissues may interfere with measurement of the β-galactosidase reporter enzyme. Endogenous enzyme activity is minimized at the pH at which the Dual-Light assay reaction is performed[12]; however, it is important to assay the level of endogenous enzyme with nontransfected cell extracts.

11E. Heat inactivation to reduce endogenous β-galactosidase activity should not be performed due to the detrimental effect on luciferase activity. When high endogenous β-galactosidase activity necessitates heat inactivation, assays for luciferase and β-galactosidase should be performed individually.

F. Microplates / Instrumentation

Microplates

The use of opaque white microplates is recommended for microplate-based assays. Black microplates may also be used, although the signal intensity will be much lower due to absorbance of the emitted light. For cell-based assays, the use of clear-bottom, white, tissue culture-treated microplates is recommended to allow microscopic examination of cultures. Well-to-well crosstalk increases with the use of clear-bottom plates, due to light piping, and thus plates should be set up such that potentially very bright wells are not adjacent to low signal wells. White backing sheets may be applied to the plate bottom prior to signal measurement. The absolute signal intensity will be higher (approximately twofold), as the white backing reflects light toward the light emission detector and eliminates light absorption by the black plate platform, but relative signal levels are unaffected.

Luminometers

Chemiluminescent reporter gene assays are performed in a variety of commercially available luminometers. For optimum results, a standard curve should be initially generated with a purified enzyme to determine the linear assay range that is achievable with the particular instrument. These luminometers are normally photomultiplier tube (PMT)-based instruments that move each well of a microplate directly below the PMT detector or lens/fiber-optic light collection interface. Most of these instruments exhibit a measurement dynamic range of five to six orders of magnitude. This is necessary because many 1,2-dioxetane chemiluminescent assays exhibit linear signal generation over six orders of magnitude. The TR717 microplate luminometer (Tropix Division, PE Biosystems) provides 96- and 384-well measurement capability and automated reagent injection.

CCD-Based Imaging Systems

Sensitive detection of chemiluminescent signals in 96-well and higher density microplates is also possible with charge-coupled device-based camera instrumentation for rapid and accurate data acquisition.[55–57] The North-Star HTS workstation (Tropix) is a fully integrated luminescence detection system comprising CCD camera-based whole plate imaging for both "glow" and "flash" luminescence chemistries, reagent injection, data analysis, wavelength discrimination, and multiformat compatibility, including 96-, 384-, and 1536-well microplate densities.

Scintillation Counters

A liquid scintillation counter may be used as a substitute for a luminometer; however, sensitivity may be lower.[58–60] When using a scintillation counter, it is necessary to turn off the coincident circuit to measure chemiluminescence directly (single photon-counting mode). The instrument manufacturer should be contacted to determine how this is accomplished. If it is not possible to turn off the coincident circuit, a linear relationship is

[55] C. S. Martin and I. Bronstein, *J. Biolumin. Chemilumin.* **9,** 145 (1994).
[56] A. Dzgoev, M. Mecklenburg, P. Larsson, and B. Danielsson, *Anal. Chem.* **68,** 3364 (1996).
[57] L. J. Kricka, X. Ji, O. Nozaki, and P. Wilding, *J. Biolumin. Chemilumin.* **9,** 135 (1994).
[58] R. Fulton and B. Van Ness, *BioTechniques* **14,** 762 (1993).
[59] V. T. Nguyen, M. Morange, and O. Bensaude, *Anal. Biochem.* **171,** 404 (1988).
[60] G. Erkel, U. Becker, and T. Anke, *J. Antibiot.* **49,** 1189 (1996).

established by taking the square root of the counts per minute measured minus the instrument background.

$$\text{Actual} = (\text{measured-background})^{1/2}$$

Summary

Chemiluminescent reporter gene assays provide highly sensitive, quantitative detection in simple, rapid assay formats for detection of reporter enzymes that are widely employed in gene expression studies. Chemiluminescent detection methodologies typically provide up to $100-1000\times$ higher sensitivities than may be achieved with fluorescent or colorimetric enzyme substrates. The variety of chemiluminescent 1,2-dioxetane substrates available enable assay versatility, allowing optimization of assay formats with the available instrumentation, and are ideal for use in gene expression assays performed in both biomedical and pharmaceutical research. In addition, 1,2,-dioxetane chemistries can be multiplexed with luciferase detection reagents for dual detection of multiple enzymes in a single sample. These assays are amenable to automation with a broad range of instrumentation for high throughput compound screening.

[14] Fusions to Chloramphenicol Acetyltransferase as a Reporter

By CLAYTON BULLOCK and CORNELIA GORMAN

Introduction

The bacterial gene chloramphenicol acetyltransferase, or *cat*, encodes an enzyme that inactivates the antibiotic chloramphenicol (which binds to the bacterial and mitochondrial 50S ribosome, thereby blocking peptidyltransferase) by acetylation to 3-acetyl and 1,3-diacetyl derivatives according to the reactions[1] shown in Fig. 1. One step of this is a pH-dependent and enzyme-independent acyl migration. The K_m of CAT for acetyl-CoA is around 50 μM, and the K_m for chloramphenicol is around 10 μM.[2] Since the initial discovery and characterization of the chloramphenicol resistance

[1] W. V. Shaw, *J. Biol. Chem.* **242,** 687 (1967).
[2] W. V. Shaw, *Methods Enzymol.* **43,** 737 (1975).

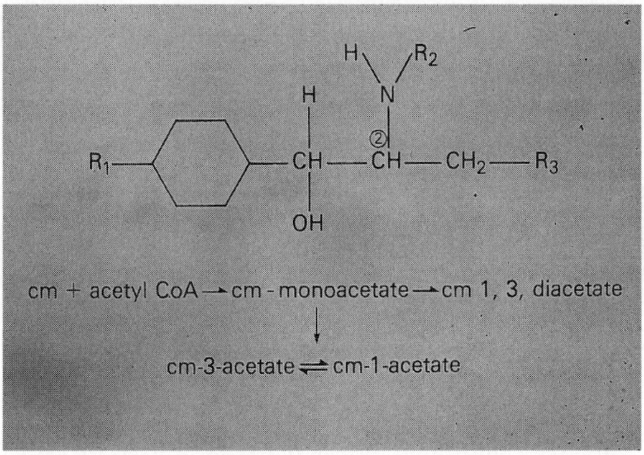

FIG. 1. The structure of chloramphenicol is shown. The acetylation reaction catalyzed by the CAT enzyme is also shown. Chloramphenicol can be acetylated in either the 1 carbon or the 3 carbon position. Two distinct forms of the monoacetate form of chloramphenicol are possible as well as a diacetylated form.

factor in *Escherichia coli* by W. V. Shaw, the enzyme has been exploited for several purposes. These include its use in assays to monitor serum chloramphenicol levels in patients being treated for bacterial infections[3-6] and assays used to characterize resistant organisms.[2] Considerably more interest in the CAT gene developed, however, after it was shown that, by virtue of the absence of any similar endogenous activity in mammalian and other nonbacterial cells, CAT can be exploited as a reporter gene to characterize promoters and other transcriptional control elements by linking these elements to the CAT gene in recombinant plasmids and transfecting these into cells.[7,8] Since the discovery by Gorman *et al.*[7] of CAT as a gene expression tool, the system has been widely employed to characterize transfection techniques for tissues and cell lines, for characterizing DNA mutagenesis and repair systems,[9-11] for genetic and biochemical analyses of transcription factors and transcription initiation complexes, for characterization of signaling pathways, for characterizing the tissue specificity of

[3] A. L. Smith and D. H. Smith, *Clin. Chem.* **24,** 1452 (1978).
[4] P. S. Letiman, T. J. White, and W. V. Shaw, *Antimicrob. Agents Chemother.* **10,** 347 (1976).
[5] R. Daigneault and M. Guitard, *J. Infect. Dis.* **133,** 515(1976).
[6] L. Robison, R. Seligsohn, and S, Lerner, *Antimicrob. Agents Chemother.* **13,** 25 (1978).
[7] C. Gorman *et al., Mol. Cell Biol.* **2,** 1044 (1982).
[8] C. Gorman *et al., Proc. Natl. Acad. Sci. U.S.A.* **79,** 6777 (1982).
[9] A. Lehman and A. Oomen, *Nucleic Acid Res.* **13,** 2087 (1985).
[10] J. Burke and A. Mogg, *Nucleic Acid Res.* **13,** 1317 (1985).
[11] H. Mitani, J. Komura, and A. Shima, *Mutat. Res.* **236,** 77 (1990).

promoters in transgenic animals,[12] and for evaluating the efficacy of delivery systems for gene therapy in intact animals.

Since 1985, CAT has been the overwhelming choice for many of the experiments using reporter genes. Even though there are now many other useful reporter gene systems, over 16,000 citations use a form of the CAT assay. Given the popularity of this approach and of the CAT gene in particular, it is reasonable for investigators to want a quick, easy, reproducible, safe, and cost-effective assay for measuring CAT activity, hence the plethora of CAT assays developed over the years. Following a discussion of tissue and cell extract preparation, this discussion considers three types of CAT assay methods, a thin-layer chromatography (TLC) assay,[13] a mixed phase assay, and a high-performance liquid chromatography (HPLC) assay. Parameters that can influence the sensitivity and reproducibility of the assays include the concentration and stability of the reagents, the temperature of the reaction, the reaction time, the extraction efficiency of reaction products in solvents, the presence of other acetylases in the reaction, the reaction volume, the type of cell transfected, the preparation of cell extracts, and the amount of extracts used in the reaction.

Preparation of Cell Extracts

Extracts are made from cell cultures 48 hr post-transfection with a CAT expression vector by resuspending washed, pelleted cells in 0.25 M Tris, pH 7.4–8, and either sonicating them or subjecting them to three rounds of alternately freezing in dry ice followed by thawing at 37°. In a typical experiment, cells from a 10 cm dish are resuspended in 100 μ and 20 μl of the extract used per reaction. Organs or tissues from transgenic animals must be homogenized with a homogenizer in a larger volume. Following lysis, the preparation is centrifuged at high speed (e.g., 12,000g for 10 min) to remove nonsoluble cellular debris, and the supernatant, containing the soluble CAT enzyme, is removed for assay. It may be desirable to quantify the CAT activity/mg of protein, in which case EDTA (to 5 mM) and protease inhibitors (PMSF to 1 mM) and aprotinin (to 2 μg/ml) may be added and an aliquot of the supernatant reserved for protein quantification. It is desirable to heat the remaining supernatant to 60° to inactivate endogenous deactylase activity present in many cell extracts.[14] This last step may increase the sensitivity of the assay in tissues where CAT expression is low.

[12] H. Westphal, P. Overbeek, J. Khillan, A. Cheplinksy, A. Schmidt, K. Mahon, K. Bernstein, J. Piatigorsky, and B. DeCrombrugghe, *Cold Spring Harb. Symp. Quant. Biol.* **50** (1985).
[13] J. Cohen *et al., Proc. Natl. Acad. Sci. U.S.A.* **77,** 1078 (1980).
[14] D. Crabb and J. Dixon, *Anal. Biochem.* **163,** 88 (1987).

CAT is a stable enzyme and samples may be stored frozen for future assays, both before and after cell extracts are prepared. In an alternative protocol for preparing cell extracts, Seed and Sheen[15] first incubate the cells for 5 min at 22° with a buffer made of 20 mM Tris–HCl, pH 7.5, and 2 mM MgCl$_2$; once this has been removed the cells are solubolized in a solution of the same buffer that also contains 0.1% Triton X-100. The cells are solublized for 5 min at 22° before the lysed cells are centrifuged in a microfuge for 2 min to remove cell debris as well as nuclei. A slightly different method of solubilizing the cells is used by Chireux et al.[16] Here the cells are first washed in Tris-buffered saline solution and solubilized in a 0.1 M Tris–HCl buffer at pH 7.8. containing 0.7% Nonidet-P40.

Enzymatic Assay from Intact Cells and Animals

In addition to the just described methods of sample preparation, CAT enzyme activity can also be assayed from the medium of CAT expressing cells in culture[17] and from urine of transgenic animals.[18] Although all the methods described earlier assay the CAT following solubilization of the host cells expressing the protein, the traditional assays have been modified to allow measurement of CAT activity from intact cells and animals. These assays rely on the ability of chloramphenicol or CAT protein to pass though the cell membrane.[17] An added advantage is the fact that this assay relies on acetyl-CoA present in the cells being assayed and therefore does not require the addition of exogenous acetyl-CoA. For this assay,[14] CM is added directly to the cell culture medium. At various time points after the addition of chloramphenicol, media are removed and extracted twice with ethyl acetate. Following the extraction steps, the amounts of the various forms of acetyl-chloramphenicol can be determined by any of the methods described earlier. The authors use the traditional TLC method to quantitate the amount of acetylated chloramphenicol present in the cell culture media. Typically a final concentration of 50 to 200 μM chloramphenicol is added to between 1 and 10 ml of media. As with any of the CAT assays, the percentage acetylation is a function of incubation time, number of cells transfected, and the chloramphenicol concentration used.

In a parallel type of experiment, CAT activity has been measured from intact animals by quantifying the amount of CAT activity present in urine transformed with CAT expressing bone marrow cells[18] or in serum in

[15] B. Seed and J. Sheen, Gene 67, 271 (1988).
[16] M. Chireux, J. Raynal, and M. Weber, Anal. Biochem. 219, 147 (1994).
[17] D. Alter and K. Subramanian. Biotech. 6, 526 (1988).
[18] R. Narayanan et al., Exp. Cell. Res. 174, 297 (1988).

mice following tail vein injection of lipid DNA complexes containing CAT expression vector (C. M. Gorman, personal communication).

Thin-Layer Chromatography

This is the method first described by Shaw,[1,2] Cohen,[13] and Gorman et al.[7,8] in which the acetylated derivatives of [14C]chloramphenicol are separated from [14C]chloramphenicol by silica gel thin-layer chromatography and quantified as percentage acetylation. Although time-consuming, labor-intensive, and expensive, this technique is also sensitive (as little as 0.0005 U of CAT activity can be detected, which can be extended by increasing the concentration of labeled chloramphenicol), specific, and quantitative.

The reaction is performed in a volume of 180 μl containing 100 μl 0.25 Tris (pH 7.4–8), 20 μl cell extract (about 5×10^6 cells), 1 μCi [14C]chloramphenicol (50 mCi/mM, 111 μM). These are equilibrated to 37°, and the reaction is started with the addition of 20 μl 4 mM acetyl-CoA (444 μM). Control reactions contain extracts from either mock-transfected cells or purified CAT. The reaction is carried out for 30 min at 37° and is stopped with the addition of 2 ml cold ethylacetate, which extracts the chloramphenicol and its acetylated derivatives. The organic layer is removed, dried overnight (a Speed-Vac works well), and resuspended in 30 μl ethyl acetate, spotted on a silica gel thin-layer plate, and run with chloroform–methanol (95:5). Autoradiography is used to identify the spots of chloramphenicol and its mono and diacetyl derivatives (Fig. 2). These are then scraped and counted in a scintillation cocktail. Results can be expressed as the percentage of chloramphenicol acetylated by volume or milligrams of extract or the units of enzyme activity calculated. For an example of how this type of data is often presented, see Fig. 3.

Variations

Several variations on this technique have been developed that obviate the need for the autoradiography step, the scintillation counting step, and either reduce or replace the radioisotope. For example, because chloramphenicol and acetylated chloramphenicols fluoresce under UV light (254–304 nm), the use of a TLC plate with fluorescent additives (e.g., Merck) allows the substrate and products of the reaction to be visualized directly, a photograph of the TLC plate made with appropriate photographic equipment, and sparing use of radioisotope (0.01 μCi 14C-labeled chloramphenicol + 7.5 nM unlabeled).[19] As an alternative to scraping off the spots and

---

[19] R. Aubin et al., Nucleic Acid Res. **15** 23, 10069 (1987).

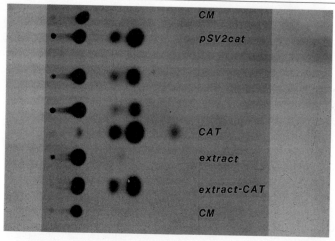

CM

pSV2cat

CAT

extract

extract-CAT

CM

Fig. 2. A autoradiograph of a TLC of the products of a CAT assay are shown. The silica gel TLC was run with chloroform–methanol (95:5, ascending). The chloramphenicol (CM) indicates the position of the origin. The acetylated forms of chloramphenicol separate due to differential mobility within the solvent used. The predominate form of acetylated-CM is the 3-acetate form of the monoacetylated CM. This position is preferentially acetylated. When the reaction is run to completion, as seen in the lane with the control CAT enzyme, the diacetylated form of acetylated-CM is also present.

counting them, a thin-layer scanner may be used that detects and quantitates β particles.[20]

Finally, the development of fluorescent substrates abolishes the need for radioisotope completely.[21–24] Several fluorophore derivatives of chloramphenicol have been described (e.g., BODIPY chloramphenicol and BODIPY 1-deoxychloramphenicol, Molecular Probes, Eugene, OR) with K_m, values for CAT similar to those of chloramphenicol. The procedure is the same as for the other TLC assays except that after running the TLC, the substrate and products are visualized under UV light (although they can be seen under ambient light), the plate is photographed under UV light, and the spots are cut out and extracted with methanol. Following extraction, the sample is quantitated by excitation emission flourometry at 490 and 512 nm, respectively.

[20] A. Berthold, *Biomed. Biochim. Acta* **49** 12, 1243 (1990).
[21] D. Hruby *et al.*, *Biotech.* **8,** 170 (1990).
[22] S. Young *et al.*, *Anal. Biochem.* **197,** 401 (1991).
[23] D. Hruby and E. Wilson, *Methods Enzymol.* **216,** 369 (1992).
[24] C. Lefevre *et al.*, *Biotech.* **19,** 488 (1995).

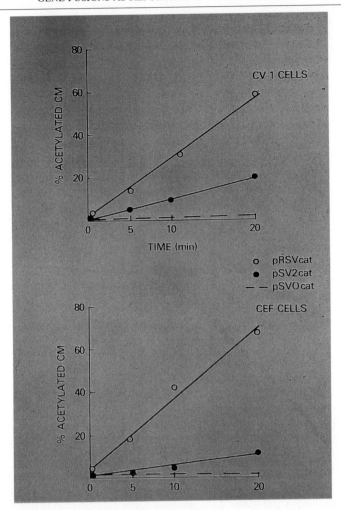

FIG. 3. Expression levels from three CAT expression vectors are compared following transfection into two different cell lines: CV-1 cells are monkey kidney cells and CEF cells are chick embryo fibroblast cells. A time course for each reaction was performed, and data are expressed as the percentage of acetylated-CM present over time.

Phase Extraction and Mixed Phase Assays

Two kinds of assays take advantage of the differential solubility between nonpolar acetylated chloramphenicol products and polar, acetyl-CoA and chloramphenicol reactants. In phase extraction assays, the reaction is stopped and extracted with an organic solvent, and the extracted acetylated chloramphenicols are counted in scintillation fluid. In mixed phase assays, the reaction is overlaid with scintillation fluid, into which the acetylated chloramphenicols diffuse, and the samples are counted at various time points of the reaction in progress. Both kinds of procedures avoid the time-consuming autoradiography and scraping of radioactive spots involved in the TLC assay, and some of these assays obviate the overnight drying step as well. Furthermore, the use of [³H]acetyl-CoA in place of [¹⁴C]chloramphenicol in some of these assays makes them considerably less expensive than the TLC assay, and because [³H]acetyl-CoA is available at higher specific activity than [¹⁴C]chloramphenicol, the sensitivity of these assays can be increased by using a higher specific activity in the reaction. Indeed, the sensitivity of some of these assays is comparable to or exceeds that of the TLC assay. The instability of acetyl-CoA, as well as the presence of acetylases and thioesterases in many cell extracts, is, however, a drawback to the use of labeled acetyl-CoA. The use of benzene in some of these assays is another drawback as benzene is a potent carcinogen and thus difficult to dispose of.

Phase Extraction Assays

Four such CAT assays have been described[25–28] (see Table I). Many of these assays have the advantage of having a level of sensitivity that compares well to the TLC assay, and the sensitivity can be increased with the use of higher specific activity. Additional attractive features of the Sleigh[24] assay include limiting concentration of labeled acetyl-CoA, which makes the assay economical, the avoidance of an overnight drying step, making it possible to obtain quantifiable data the same day the assay is performed, and the avoidance of benzene. As acetyl-CoA is limiting, however, care must be taken both in handling the reactant and in heat treating the cell extracts to abolish endogenous deacetylase and thoiesterase activities. This assay is unsuitable for assays of CAT activity from tissues harvested from transgenic animals.

The assay by Nordeen *et al.*[25] compensates for the instability of acetyl-CoA by using a system that generates ³H-labeled acetyl-CoA from ³H-labeled sodium acetate, CoA, yeast-CoA synthetase, and ATP. The use of [³H]sodium acetate instead of [¹⁴C]chloramphenicol makes this assay still more economical. The first step of the reaction is to generate [³H]acetyl-CoA. The reaction is then started with the addition of 50 μl cell extract.

[25] M. Sleigh. *Anal. Biochem.* **156,** 251 (1986).
[26] S. Nordeen. P. Green, and D. Fowlkes, *DNA* **6,** 173 (1987).
[27] D. Nielsen. T. Chang. and D. Shapiro, *Anal. Biochem.* **179,** 19 (1989).
[28] L. Sankaran, *Anal. Biochem.* **200,** 180 (1992).

TABLE I
PHASE EXTRACTION ASSAYS

Author	[CM]	[Acetyl donor]	Reaction volume (μl)	Temperature	Extraction	Fluor	Sensitivity
Sleigh[25]	1.6 mM	90 μM acetyl-CoA, 0.1 μCi	100	37°	2× 100 μl ethyl acetate	1 ml	10^{-5} to 10^{-4} U
Nordeen et al.[26]	1.33 mM	400 μM acetyl-CoA, 7.8 μCi, synthesized from ^{3}H-labeled sodium acetate	250	37°	1 ml benzene, two time points, 100 μl each	3 ml	10^{-7} to 10^{-5}U
Nielsen et al.[27]	1 mM	30 μM acetyl-CoA, 0.4 μCi	100	37°	1 ml 7 M urea	10 ml 0.8% PPO in toluene	10^{-5} to 10^{-2} U
Sankaran[28]	1 mM	100 μM acetyl-CoA, 0.1 μCi	100	Variable	500 μl borat–5 M NaCl	0.4% PPO–0.005% POPOP in toluene	10^{-7} to 10^{-5} U
Seed and Sheen[15]	0.1 mM, 0.2 μCi either [^{14}C] or [^{3}H]	250 μM butyryl-CoA	100	37°	200 μl TMPD:mixed xylenes 2:1		10^{-7} to 10^{-4} U

At two different time points, 100 μl of the reaction is removed and 1 ml cold benzene is added to stop the reaction and extract the acetylated chloramphenicol. Following vigorous mixing and separation of the phases with centrifugation, a 750-μl aliquot of the benzene phase is removed to a 6-ml scintillation vial and the sample is dried overnight in a Speed-Vac. The sample is then resuspended in 3 ml scintillation fluor and the samples are counted. The sensitivity of this assay (10^{-7} U) again is comparable to that of the TLC assay, and as with the assay by Sleigh,[25] the sensitivity can be increased with higher specific activity. This assay also offers the advantage of generating labeled substrate so that neither breakdown nor depletion of substrate is an issue. Indeed, this assay can be allowed to run for 24 hr, and the assay is linear over a wide range of CAT activities, up to at least 10^{-5} U. Furthermore, the use of [^{3}H]sodium acetate makes this assay extremely economical. However, the use of benzene and the need for an overnight drying step are drawbacks. The evaporation is necessary due to the partitioning of a small amount of labeled volatile material into the benzene phase, and additional background is observed, necessitating the use of (-)chloramphenicol controls. Care must be taken also not to overdry the samples as quenching is observed when the sample is baked onto the scintillation vial.

The assay by Nielsen et al.[27] also employs the inexpensive [^{3}H]acetyl-CoA, avoids the use of benzene, and increase the sensitivity of the assay by optimizing the concentration of acetyl-CoA and reducing the background. Neilsen et al.[27] found that by reducing the concentration of acetyl-CoA from 100 to 30 μM with a consequent increase in specific activity, they increased the sensitivity of detection from 0.001 U of CAT to 0.00001 U in their assay. Some of this increase is also due to the use of 7 M urea to terminate the reaction, which decreases the partitioning of the [^{3}H]acetyl-CoA out of the aqueous phase. The reaction is terminated at 2 hr by removing 90 μl of the reaction mixture to 1 ml 7 M urea in a 20-ml scintillation vial. Ten milliliters of 0.8% PPO (2,5-diphenyloxazole) in toluene is then added, the phases are mixed by shaking and are allowed to separate, and the samples are counted. The assay is linear over 3 orders of magnitude (0.00001 to 0.01), and the increased sensitivity allows CAT activity even from a promoterless construct, SVOCAT, to be detected over background in transfected *Xenopus laevis* fibroblasts. As with the assay by Sleigh,[25] the limiting concentration and instability of the acetyl-CoA in the reaction can cause quenching. Also, as noted by Sankaran[28] (see later), when using conditions in which the enzyme is in excess of substrate (the K_m of acetyl-CoA for CAT is in the 100 μM range), the assay cannot accurately measure enzyme activities directly but requires the use of standards.

Further improvements in the speed and sensitivity of the phase extraction method are found in the assay by Sankaran.[28] These improvements are achieved by use of a more efficient extraction procedure and by the discovery that the assay can be carried out at 60°, which obviates the need for a heat inactivation step and increases the rate of the reaction. This assays requires that (-)chloramphenicol and (-)enzyme controls are included. Following the addition of the benzene/toluene mixture, the vial is shaken and centrifuged, and the sample is counted. The extraction efficiency of the acetylated chloramphenicol into the toluene phase is almost 100%, as compared with an efficiency of only 70% when 7 M urea was used in this assay or an efficiency of only 58% when Econofluor II was substituted for the toluene scintillant (see Mixed Phase section later). Background counts are low, around 0.25%, but will increase over time with the hydrolysis of acetyl-CoA. Furthermore, when enzyme activities were calculated directly with this assay, the results are close to the value of those provided by the suppliers of purified enzymes and with those of the TLC method, making the inclusion of standard enzymes unnecessary. Sankaran's experiments[28] with the temperature effect showed that compared with the assay at 23°, enzyme activity increases 1.9-, 2.2-, and 3.4-fold at 37°, 45°, and 60°, respectively. Performing the assay at 60° obviates the need for heat inactivation, increases the specific activity of some cell types, and allows for a shortening of the reaction time (even to 5 min). Sankaran[28] also found that when the concentration of acetyl-CoA was increased from 0.1 μCi and 100 μM to 0.5 μCi and 1 mM, the enzyme activities were significantly higher, with enzyme activities at the lower concentration of acetyl-CoA being only 76% of those obtained with saturating levels of substrate. Using the lower concentration, however, balances the need for a sensitive assay with economy. The amount of label in this assay is lower than that of Nielsen et al.[27] (0.1 vs 0.4 μCi), and the assay is still more sensitive, with as little as 0.0000001 units detectable. The only drawback to this assay is the use of benzene in the scintillation fluid.

The final phase extraction assay considered here is the one by Seed and Sheen.[15] Exploiting the relatively low specificity of CAT for acetyl donors, butyryl-CoA is substituted for acetyl-CoA in the reaction, and the greater hydrophobicity of the butyrylated chloramphenicol as compared with acetylated chloramphenicols enhances the discrimination between reactants and products in the phase extraction procedure. Either [³H]chloramphenicol or the more expensive [¹⁴C]chloramphenicol can be used in this assay. Experimenting with different substituted benzenes for their ability to extract products and discriminate products from substrates, a 2:1 mixture of tetramethylpentadecanes (TMPD) and mixed xylene isomers was found to extract 0.15% [¹⁴C]chloramphenicol and 40% of the butyrylated [¹⁴C]chloramphenicol, for a discrimination of 270-fold. To stop the reaction, 2 volumes of the TMPD : xylene mixture is added to and mixed vigorously with the reaction. Following centrifugation, 90% of the organic phase is removed to a scintillation vial and counted. The [³H]chloramphenicol gives a higher background, which can be reduced by preextraction of the antibiotic with solvent. The background can be reduced and sensitivity increased by back extracting the organic phase with 0.5 volumes TE (10 mM Tris, pH 7.8, 1 mM EDTA). The assay is linear over 2 to 3 orders of magnitude of enzyme concentration, and when the tritiated compound is used, the assay is less than $1 per sample. The cost of using [¹⁴C]chloramphenicol is less than $2 per sample.

Mixed Phase Assays

In these assays (see Table II), the reaction mixture is gently overlaid with scintillation fluid, into which the products of the reaction diffuse while the reaction is in progress. The advantage of these assays is that they allow the course of the reaction to be monitored and quantitated instead of quantitating a reaction end point. The reaction can even be performed at room temperature in a scintillation counter. As with phase extraction assays, the sensitivity of the reaction increases with higher specific activity of substrate, and the need for sensitivity can be balanced with the need for economy. For example, the original version of this assay by Neuman[29] uses 5 μCi [³H]acetyl-CoA (200 mCi/mmol) diluted with 100 μM unlabeled acetyl-CoA, 1 mM chloramphenicol, total volume 250 μl overlaid with 5 ml scintillation fluor (Econofluor, Du Pont NEN). Eastman[30] found that by using 0.1 μCi undiluted [³H]acetyl-CoA (0.28 μM final) but keeping everything else the same, a 4- to 6-fold increase in sensitivity is obtained, with a concomitant 10-fold decrease in background and a large reduction in cost as well. Eastman[29] also observed that this assay was linear with enzyme concentration only when less than 50% of the tritium entered the scintillation cocktail, a condition satisfied in the range of 0.001 to 0.01 CAT units. As noted by Sankaran,[28] the extraction efficiency of Econofluor II is <58% (as compared with a toluene-based scintillant), thus requiring the use of enzyme standards in the reaction to calculate enzyme activity. Quenching is observed with this assay due to the hydrolysis of acetyl-CoA, particularly at low concentrations. The relatively low sensitivity of these assays makes them unsuitable for cell lines that transfect poorly or for promoters with low levels of transcription.

As with phase extraction assays that employ limiting concentrations of [³H]acetyl-CoA, the problem of acetyl-CoA instability can be overcome by using yeast-CoA synthetase and ³H-labeled sodium acetate, which gener-

[29] J. Neumann, C. Morency, and K. Russian, *Biotech.* **5**, 444 (1987).
[30] A. Eastman, *Biotech.* **5**, 730 (1987).

TABLE II
MIXED PHASE ASSAYS

Author	[CAM]	[Acetyl donor]	Reaction volume (μl)	Temperature	Fluor	Sensitivity
Neumann et al.[29]	1 mM	100 μM acetyl-CoA, 5 μCi	250	Room	5 ml Econofluor II	0.002 to 0.01 U
Eastman[30]	1 mM	0.28 μM acetyl-CoA, 0.1 μCi	250	Room	5 ml Econofluor II	0.001 to 0.01 U
Purschke and Muller[31]	1.26 mM	185 μM acetyl-CoA, 6 μCi synthesized from ^{3}H-labeled sodium acetate	250	Room	5 ml Econofluor II	0.0001 to 0.06 U
Chireux, et al.[16]	0.625 mM	3.75 μM acetyl-CoA, 0.15 Ci	200	Room	3 ml toluene +5 g/liter PPO	4 ng protein (100 transfected cells) to 0.01 U

ates labeled [³H]acetyl-CoA with high specific activity throughout the reaction, increasing the sensitivity of the assay and decreasing the cost. This technique is employed in the assay by Purschke and Mueller.[31] As with the assay by Nordeen et al.[26] on which it is based, the reaction to generate [³H]acetyl-CoA is first set up and then 60-μl aliquots of this are added to the CAT, yielding a final volume of 250 μl. The reaction is overlaid with 5 ml Econofluor II. The reactions are started at 1-min intervals with the addition of the reference CAT or samples. The limit of sensitivity of this assay is 0.0001 U and is linear to 0.06 U, which compares favorably to the sensitivity and range of the TLC assay. Again, reference standards need to be used in this assay as the scintillation fluor extracts <58% of the acetylated chloramphenicols.

As with the phase extraction assays by Nielsen et al.[27] and Sankaran,[28] the sensitivity of the mixed phase assay can be increased either by optimizing the concentration of [³H]acetyl-CoA or by using a toluene-based scintillant that extracts the acetylated chloramphenicol more efficiently than Econofluor II. Both parameters are considered in the assay by Chireux et al.[16] Assays performed with reactions containing 625 μM chloramphenicol, 3.75 μM [³H]acetyl-CoA (200 mCi/mmol), and with increasing final concentrations of cold acetyl-CoA from 0 to 205 μM [total volume to 200 μl made up with 0.1 M Tris, pH 7.8, overlaid with 3 ml toluene with 5 g/liter diphenyloxazole (PPO) and counted at timed intervals for up to 8 hr] showed decreasing signal-to-noise ratios and decreasing slopes with increasing concentrations of the cold acetyl-CoA. In the absence of cold acetyl-CoA, CAT activity can be detected in the range of 4–40 ng protein, with the lower limit corresponding to about 50–100 PC-12 rat pheochromocytoma cells electroporated with CMV-CAT plasmid. In contrast, the same extract assayed by the TLC method (with 3 μM [¹⁴C]chloramphenicol (56 mCi/mmol) and 530 μM acetyl-CoA) required at least 100 ng of protein to detect CAT activity. The background in this assay corresponds to about 1% of the total radioactivity. The upper limit of detection in this assay corresponds to 0.25% substrate conversion per minute and a 20-min incubation or, assuming an extraction efficiency of 100%, about 0.01 U. Beyond this enzyme activity, actual data are less than the value predicted from linear regression analysis of lower activities. Departure from linearity beyond 0.01 U is due both to acetyl-CoA instability and the slow rate of partitioning of [³H]acetylchloramphenicols into the organic phase. After a preincubation of 20 hr at 37°, reaction rates are only about 75% of reactions done with fresh assay mixtures. Although the diacetylated chloramphenicol partitions into the toluene phase in about 20 sec, monoacetylated chloramphenicols

---

[31] W. Purschke and P. Mueller, Biotech. **16,** 264 (1994).

partition out more slowly, with a time constant of $T = 10.5$ min. Thus, at rates higher than 0.01 U, the reaction rate exceeds the rate of partitioning of acetylchloramphenicol into the toluene phase.

High-Performance Liquid Chromatography

The first assays using CAT and HPLC techniques were designed for determining levels of chloramphenicol in patients being treated with the drug for bacterial infections,[32,33] but the technique has also been adapted for CAT as a reporter gene. The reactions and variations on the HPLC techniques are similar to those described previously (see Table III). For example, some of the assays described employ a time-consuming extraction and drying step, which allows more of the sample to be quantitated than those procedures where extraction is avoided. Where extraction is avoided, the speed of the assay is offset by a loss in sensitivity. This loss in sensitivity can be compensated by using a radioactive substrate and an HPLC setup that measures counts instead of UV absorbance. Alternatively, a fluorescent substrate (BODIPY chloramphenicol, Molecular Probes) can be used and the assay quantitated by excitation emission spectroscopy. As with the TLC technique, HPLC monitors the consumption of the reactants as well as the appearance of products over time. The main advantage of HPLC is automation, reducing the manual labor required to perform the assay in laboratories where HPLC equipment is available.

In the assay by Young et al.,[34] the reaction is terminated and the chloramphenicols are extracted with 1 ml ethyl acetate, dried, and resuspended in 20 μl 25% acetonitrile (in water). This is then applied to the HPLC apparatus and eluted with a 25–45% acetonitrile gradient over 2 min. The amount of each species is quantitated by integrating the absorbance at 278 nm, and the results can be expressed either as percentage acetylation or enzyme units quantitated from moles acetylated per unit time. By maintaining the pH at 7.4 or lower, the amount of 1.3-diacetyl- and 1 acetylcholoramphenicol formed can be minimized, thus simplifying the calculations. The sensitivity of this assay compares well to that of the TLC and is much faster.

In the assay by Siegert et al.,[35] the extraction and drying steps are avoided by terminating the reaction with an equal volume of acetonitrile, which precipitates the CAT. Twenty microliters of the reaction is then applied to the HPLC apparatus, and the species are eluted with 50% metha-

[32] A. Weber, K. Opheim, J. Koup, and A. Smith, *Antimicrob. Agent Chemother.* **19,** 323 (1981).
[33] A. Lovering, L. White, and D. Reeves, *J. Antimicrob. Chemother.* **17,** 821 (1986).
[34] S. Young, A. E. Jackson, D. Puett, and M. H. Melner, *DNA* **4,** 469 (1985).
[35] H. Siegert, B. Wittig, and S. Wolfl, *J. Clin. Chem. Clin. Biochem.* **28,** 217 (1990).

TABLE III
HPLC CAT Assays

Author	[CAM]	[Acetyl donor]	Reaction volume (μl)	pH	Extraction	Eluent	Detection	Sensitivity
Young, et al.[34]	188.6 μg/ml	Acetyl-CoA 3.56 mg/ml	150	7.4	1 ml ethyl acetate	25–45% acetonitrile in water	Absorbance at 278 nm	0.00005 to 0.001 U
Siegert, et al.[35]	27.6 μM	4 nmol, 552 μM	145	Not defined	None	50% methanol	Absorbance at 275 nm	No data
Zajac et al.[35]	11 μM, 0.1 μCi	750 μM butyryl-CoA	146	3.8	None	30–100% acetonitrile in water	Scintillation counting	2.5×10^{-8} to 7.5×10^{-6} U
Waldon et al.[36]	2.27 μM BODIPY CAM	363 μM	220	Not defined	Ethyl acetate	0.001% trifluoroacetic acid : 27% acetonitrile	490-nm excitation; 512-nm emission	4×10^{-8} to 4×10^{-5}

nol and quantitated by integrating absorbance at 275 nm. Although the limit of detection was not determined, this assay appears much less sensitive than the one by Young et al.[34] and is linear only to 0.004 units. The range on linearity can be extended to higher CAT units by increasing the concentration of chloramphenicol in the reaction.

In the assay by Zajac et al.[36] the extraction and drying steps are also avoided but the sensitivity of the assay is actually better than the one by Young et al.[34] due to the use of [^{14}C]chloramphenicol and quantitation by (automated) scintillation counting. The resolution between substrate and products is improved by using butyryl-CoA instead of acetyl-CoA, and quantitation is made easier by the coelution of the various butyrylated chloramphenicols.

Finally, an HPLC technique has been described by Waldon et al.[37] that uses the fluorescent chloramphenicol derivative, BODIPY chloramphenicol. In comparison to the same assay performed with chloramphenicol and quantitated by UV absorbance, the assay employing BODIPY chloramphenicol and quantitated with excitation emission spectroscopy at 490 and 512 nm, respectively, was 1000-fold more sensitive and able to assay CAT activity from a single hair follicle of a mouse transgenic for cat under the control of an ultrahigh sulfur keratin gene promoter.

Detection of CAT Expression Levels Using Immunological Techniques

In addition to the various enzymatic assays discussed previously, several authors have adapted standard immunological methods to measure CAT expression using anti-CAT antibodies. These methods range from detection of CAT expression in single cells both in vitro[38–40] and in vivo[40] to quantitation of CAT protein through the use of an ELISA and finally to determination of CAT expression levels via Western blot analysis.[41] The central difference between measuring CAT expression with immunological techniques versus the enzymatic reaction is that the immunological approaches measure the absolute amount of CAT protein produced rather than estimate relative expression levels based on the enzymatic reaction.

These immunological techniques have been shown to be particularly

[36] J. Zajac, J. M. Gerardi, A. K. Kearns, and F. R. Bringhurs, DNA 7, 509 (1988).
[37] D. Waldon, M. F. Kubicek, G. A. Johnson, and A. E. Buhl, Eur. J. Clin. Chem. Clin. Biochem. 31, 42 (1993).
[38] G. H. Smith, P. J. Doherty, R. B. Stead, C. M. Gorman, D. E. Graham, and B. H. Howard, Anal. Biochem. 156, 17 (1986).
[39] P. B. Antin, J. H. Mar, and C. P. Ordahl, BioTechniques 6, 640 (1988).
[40] R. E. Harlan, D. Mondal, R. Coleman, and Om Prakash, BioTechniques 10, 304 (1991).
[41] D. K. Burns and R. M. Crowl, Anal. Biochem. 162, 399 (1987).

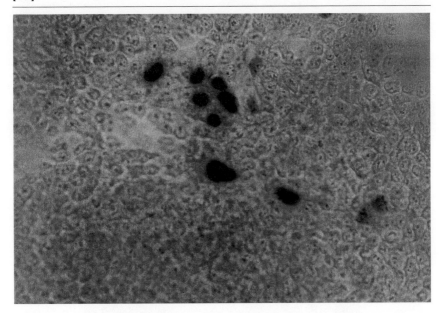

FIG. 4. Immunocytochemical detection of CAT expression in teratocarcinoma cells with a CAT monoclonal antibody. F9 cells were transfected with pSV2CAT[7] and stained with an antibody for CAT.

useful when the percentage or type of cells expressing the gene or anatomical detail is of interest. These types of analysis of CAT expression are especially valuable with *in vivo* expression systems such as transgenic animals or in gene therapy experiments.[42] For the cellular visualization of CAT expressing cells, a number of variations have been used, ranging from the direct conjugation of an anti-CAT antibody[8,38] to the use of techniques that amplify the signal such as biotinlyated or fluorescent secondary antibodies[39,40] or anti-CAT digoxigenin[42] (Fig. 4). In some cases it may be desirable to monitor protein levels directly for studies exploring translation functions. One immunological assay that is adapted easily for this kind of analysis is a Western dot blot.[41] Another immunological-based assay that has become quite popular is the CAT ELISA.

CAT Gene Expression Measured by RNA Analysis

For many questions addressing the various points of control of gene expression (transcriptional, posttranscriptional, and translation) it may also

[42] C. M. Gorman, M. Aikawa, B. Fox, E. Fox, C. Lapuz, B. Michaud, H. Nguyen, E. Roche, T. Sawa, and J. P. Wiener-Kronish, *Gene Ther.* **4,** 983 (1997).

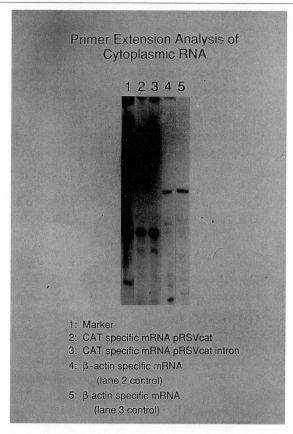

Fig. 5. Human embryonic kidney cells (293) were transfected with the two pRSVcat[8] plasmids. Following transfection, cytoplasmic mRNA was isolated. Primer extension was performed on RNA samples using either a CAT-specific primer[43] or a primer specific for human β-actin. Primer-specific extension products were then separated by acrylamide gel electrophoresis and detected by autoradiography.

be desirable to quantitate the level of CAT RNA present in a cell or organism. Such assays must be able to distinguish between CAT DNA used to transfect the cell or organism and the CAT-specific RNA to be measured. Huang and Gorman[43] measured CAT-specific RNA using traditional primer extension methods and used the presence of an intron in the expression vector to distinguish between the input transfecting DNA and the CAT-specific transcripts (Fig. 5). Because the CAT transcripts have the intron

[43] M. Huang and C. M. Gorman, *Nucleic Acids Res.* **18,** 937 (1990).

sequences removed, the RNA-specific primer extension product will be shorter than any product that results from priming of the transfecting DNA. Knuchel et al.[44] have developed a CAT-specific RT-PCR assay that also allows for the quantitation of CAT mRNA but not CAT DNA. In this study, no intron was included in the CAT expression vector so a "tailed" RT-PCR approach was developed to distinguish between CAT mRNA and CAT DNA. This approach uses a tailed antisense primer for the RT step; 20 bp are complementary to the CAT mRNA and 27 bp are not complementary to the CAT gene. Optimal conditions for this RT-PCR reaction were 2.0 Mg^{2+} for the PCR buffer, an annealing temperature of 55°, and 35 cycles.

CAT-TR tailed RT primer
5′CATCGATGACAAGCTTAGGTATCGATACCATTCATCCGCTTATC3′
TAIL-R antisense primer specific for the tail sequence
5′CATCGATGACAAAGCTTAGGTATCGATA3′

Underlined sequences are not complementary to CAT sequences. Based on this assay, CAT RNA produced a linear signal in between the range of 2×10^{-10} to 0.1×10^{-16} pmol. The sensitivity of this assay was able to measure the equivalent of eight copies of CAT mRNA. The mRNA for CAT could be measured easily for 72 hr, and the stability of the CAT mRNA was determined to be at least 24 hr.

[44] M. Knuchel, D. P. Bednarik, N. Chikkala, F. Villinger, T. M. Folks, and A. A. Anasri, *J. Virol. Methods* **48**, 325 (1994).

[15] Fusions to β-Lactamase as a Reporter for Gene Expression in Live Mammalian Cells

By GREGOR ZLOKARNIK

Introduction

β-Lactamase has been developed as a reporter enzyme for monitoring gene expression in mammalian cells.[1,2] The enzyme is of bacterial origin without any mammalian homologues and permits quantitative analysis of

[1] J. T. Moore, S. T. Davis, and I. K. Dev, *Anal. Biochem.* **247**, 203 (1997).
[2] G. Zlokarnik, P. A. Negulescu, T. E. Knapp, L. Mere, N. Burres, L. Feng, M. Whitney, K. Roemer, and R. Y. Tsien, *Science* **279**, 84 (1998).

reporter expression in single mammalian cells without interference from an endogenous background. It can be fused to the N and C termini of proteins without loss of activity, which allows monitoring the expression levels of the tagged protein.[1,3] It is distinguished from popular luciferase and β-galactosidase reporter enzymes[4,5] in that it can be detected sensitively in individual living cells under standard culture conditions. This is accomplished with a fluorogenic membrane-permeable substrate ester that diffuses into cells and provides high intracellular concentrations of accumulated fluorescent substrate. The reporter is detected sensitively by its catalysis of substrate hydrolysis, which results in a highly amplified fluorescent signal, allowing analysis of expression of proteins that are present in cells at low copy numbers. The fluorescent signal from live cells permits facile functional selection of reporter-expressing cells with flow cytometry or fluorescence microscopy. Selected cells usually remain viable and can be used for subsequent analyses that require intact cells.

The Reporter Enzyme

The β-lactamase enzyme used as a reporter in mammalian cells is a member of a large and structurally diverse family of enzymes that cleave β-lactam antibiotics such as penicillins and cephalosporins. The enzyme that was chosen as reporter in mammalian cells corresponds in part to the mature *Escherichia coli* enzyme encoded by the ampicillin resistance gene (Amp[r]) of the pUC18 and pBS plasmids. It differs from the published TEM-1 β-lactamase sequence[6] by conservative mutations V82I and A184V that are present in pUC18 and pBS plasmids and introduction of a new N terminus. The new N terminus was created by deletion of the first 23 amino acids, which comprise the signal sequence, and introduction of a methionine and mutation H24D, which provide an optimal Kozak sequence[7] for mammalian expression. This enzyme is preferred as a reporter over its more distant cousins because it exhibits fast and simple kinetics with cephalosporin substrates, allowing its ready quantitation. The structural gene encoding the reporter β-lactamase is referred to as *bla*.

Functional reporter fusion proteins may be constructed by replacing either the ATG initiator codon or the termination codon with the appro-

[3] M. Whitney, E. Rockenstein, G. Cantin, T. Knapp, G. Zlokarnik, P. Sanders, K. Durick, F. F. Craig, and P. A. Negulescu, *Nature Biotechnol.* **16,** 1329 (1998).

[4] J. R. de Wet, K. V. Wood, M. DeLuca, D. R. Helinski, and S. Subramani, *Mol. Cell. Biol.* **7,** 725 (1987).

[5] P. A. Norton and J. M. Coffin, *Mol. Cell. Biol.* **5,** 281 (1985).

[6] J. G. Sutcliffe, *Proc. Natl. Acad. Sci. U.S.A.* **75,** 3737 (1978).

[7] M. Kozak, *Nucleic Acids Res.* **12,** 857 (1984).

priate sequence encoding the protein polypeptide chain.[1] The unmodified reporter enzyme has a half-life of approximately 3 hr in living mammalian cells,[2] permitting the monitoring of events that lead to a decrease in reporter gene expression and result in lower cellular reporter enzyme levels.

The Substrate and Its Ester

β-lactamase is detected sensitively within living cells with the fluorescent substrate coumarin cephalosporin fluorescein No. 2 (CCF2).[2] CCF2 is a cephalosporin derivative, which is labeled with a fluorescent donor 7-hydroxycoumarin derivative and an acceptor fluorescein fluorophore. In the intact substrate, excitation of the donor leads to efficient resonance energy transfer to the acceptor fluorescein, which emits green fluorescence. β-Lactamase catalyzes the hydrolysis of the cephalosporin, which results in the separation of the donor fluorophore from the acceptor. After hydrolysis, excitation of the donor 7-hydroxycoumarin fluorophore results in emission of blue fluorescence. The substrate is delivered into live cells via its membrane-permeable substrate ester derivative, CCF2/AM (AM is the acronym for an acetoxymethylester). In this substrate derivative the moieties in CCF2 are modified as a butyrate ester (hydroxycoumarin), acetoxymethylester (cephalosporin), and diacetate ester (fluorescein). This substrate ester diffuses into cells and is converted to the free substrate, CCF2, by endogenous mammalian cytoplasmic esterases. The free substrate carries several negative charges at physiological pH, which help retain it in the cytoplasm of the cell.

Data Analysis

Cells loaded with the β-lactamase substrate CCF2 fluoresce green in the absence of the reporter. In cells expressing the reporter, β-lactamase converts the green fluorescent substrate to its blue fluorescent hydrolysis product. The substrate, which exhibits an easily visible color change on hydrolysis, has some distinct advantages over intensity-based substrates. The original fluorescence allows verification of whether the substrate has indeed been delivered into the cell. The change in the color of fluorescence provides a simple and sensitive way of visualizing whether a reporter construct is expressed and whether a cell population is homogeneous. The dual wavelength readout allows for ratiometric analysis, which is preferred, as it is less affected by experimental fluctuations than single wavelength measurements. Ratiometric analysis of CCF2 hydrolysis uses the fluorescence emission intensity value measured for the donor fluorophore and divides it by the fluorescence intensity value measured for the acceptor emission to give an "intensity ratio." This ratio is a measure of the progression of the substrate hydrolysis reaction, increasing in value as donor fluorescence increases with hydrolysis of the substrate. Ratiometric readouts (also referred to as "ratioing") reduce the noise in assay measurements substantially, as factors that affect fluorescence intensities at both wavelengths cancel out in the ratio. Experimental variables whose impact on the assay results are reduced substantially by ratioing include variations in cell size, cell number, cellular probe concentration, and fluctuations in excitation intensity.

Typical Experimental Conditions for Live Cell Assays

In its simplest form, the β-lactamase reporter assay is performed as an end point assay. First, the *bla* reporter gene is introduced into cells either transiently or stably, followed by an experiment in which expression of the reporter is modulated over several hours or even days under physiological conditions. Then, at the end of the experiment, cellular β-lactamase activity is determined by fluorescence measurement under separate optimized conditions outlined later.

Cells are typically loaded with a serum-free loading buffer containing 1 μM CCF2/AM at ambient temperature (18–22°) for 60–90 min. Intracellular fluorescence is usually first visible about 15 min after loading and increases steadily for about 60 minutes. Longer incubation times lead to little further increase in cellular fluorescence as equilibrium between dye uptake and leakage from cells is reached. This equilibrium may be reached at times and fluorescence intensities that differ from the ones depicted in Fig. 1 and will depend on the cell line, the specific experimental conditions, and the

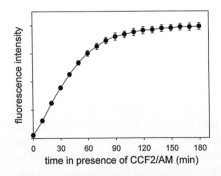

FIG. 1. Time course of cellular substrate loading at 22° measured by fluorescence intensity. Jurkat cells were kept in the presence of 1 μM CCF2/AM, and fluorescence intensity was measured at 535 ± 12.5 nm with excitation at 395 ± 12.5 nm on a Cytofluor 4000 fluorescence microtiter plate reader at 10-min intervals. Each data point represents the average of four sample wells and each well contained 10^5 cells. Error bars represent standard error.

concentration of CCF2/AM in the loading medium. Results of loading Jurkat cells with 1 μM CCF2/AM at ambient temperature (22°) are shown in Fig. 1.

Commonly used assay cell lines were also loaded with 1 μM CCF2/AM at ambient temperature for 1 hr. The following increases in cellular fluorescence relative to endogenous background were measured on an epifluorescence microscope equipped for quantitative image analysis: Jurkat (85-fold), HeLa (120-fold), HEK 293 (75-fold), GH3 (56-fold), COS-7 (44-fold), CHO (38-fold), and CV-1 (28-fold). Suspension cells were loaded at cell densities up to $1–2 \times 10^6$ cells/ml. Adherent cells were loaded at 60–80% confluence, as loading was significantly less efficient when cells were completely confluent. Loading cells at higher temperatures, such as 37°, led to more rapid loading of the substrate with the equilibrium being reached earlier, but for all cell lines tested, equilibrium concentrations of intracellular substrate were significantly lower at 37° than at ambient temperatures (data not shown). This is explained with a disproportionately larger effect of temperature on increase in dye export than on the processes involved in dye loading.

Substrate-Loading Protocols

The substrate ester CCF2/AM is dissolved to 1 mM in dry dimethyl sulfoxide (DMSO). This stock solution can be frozen and be kept protected from light for future use (for about 1 month). To prepare 1 ml of 1 μM CCF2/AM loading buffer, 1 μl of the 1 mM CCF2/AM stock solution in DMSO is diluted with 9 μl of a DMSO solution containing 100 mg/ml of the dispersant Pluronic-F127 and 0.1% acetic acid. The resulting DMSO solution is added with vigorous agitation to 1 ml of serum-free medium containing HEPES (25 mM, pH 7.3) or a suitable buffer such as HBS (pH 7.3). The resulting medium is referred to as the loading buffer. Because CCF2/AM has limited stability in aqueous solutions, the loading buffer should be used within 1 hr of its preparation.

For loading adherent cells the medium is aspirated from cells and re-placed with loading buffer. After the desired loading period (typically 1 hr), the loading buffer is aspirated and replaced with fresh medium.

For loading of suspension cells, cells are spun gently to a pellet by centrifugation, the supernatant removed, and the cells resuspended in load-ing buffer. After the desired loading period, cells are again spun gently into a pellet by centrifugation, the loading buffer aspirated, and the cells resuspended in fresh medium.

Adherent cells typically load best when 50–75% confluent. Most cell lines will accumulate more substrate when loaded at room temperature and kept at room temperature thereafter, as substrate loss in many cell lines increases significantly at 37°. Removing the solution containing substrate ester CCF2/AM after the desired loading period and replacing it with solution or media without dye ester reduces (removes) solution fluorescence. β-Lactamase nonexpressing cells will appear more greenish if left to convert the ester for another 15–30 min after loading and washout of the substrate ester.

Aqueous Substrate Ester Solutions

It is common for researchers to prepare concentrated aqueous reagent solutions for use in their experiments. This is to avoid adding too high a concentration of organic solvent to the preparation and to allow addition of accurate volumes of reagents, as the error in accuracy increases as addition volumes decrease. Unfortunately, this practice can cause problems when applied to CCF2/AM, which is a large hydrophobic molecule that has relatively low solubility in water. Making 100×, 10×, or even 5× aqueous loading solutions will usually lead to precipitation of CCF2/AM from solution (although this may not be visible to the unaided eye), resulting in inefficient loading.

To allow preparation of aqueous reagent stock solutions (CCF2/AM loading solutions), organic cosolvents were investigated for their ability to prevent precipitation of CCF2/AM from aqueous solution and their lack of interference with the cells' viability and β-lactamase reporter activity. Polyethylene glycol with an average molecular weight of 400 g/mol (PEG-400) is such a cosolvent. It will prevent CCF2/AM precipitation from aqueous 6× or 11× loading solutions when present at a concentration of 24% (v/v). Many cell lines tolerate PEG-400 in the solution (at least for the purpose of determining intracellular β-lactamase concentrations at the end of an experiment). For instance, Jurkat, HeLa, HEK 293, GH3, COS-7, CHO, and CV-1 will tolerate up to 5% PEG-400 in the final medium during substrate loading.

Table I gives the volumes of reagents needed to prepare 6× or 11×

TABLE I
VOLUMES OF REAGENTS FOR 6 AND 11× CCF2/AM LOADING SOLUTIONS

Loading solution	6×	11×
1 mM CCF2/AM in DMSO	6 μl	12 μl
DMSO, 100 mg/ml Pluronic F127, 0.1% acetic acid	54 μl	48 μl
24% (v/v) PEG-400 in serum-free medium or buffer	1 ml	1 ml
To 100 μl cell medium, add:	20 μl	10 μl

aqueous CCF2/AM loading solutions. (*Note*: Aqueous CCF2/AM loading solutions deteriorate quickly at ambient temperatures and should be used within 1 hr of preparation and unused portions discarded.) An appropriate volume of 1 mM CCF2/AM stock solution in DMSO is added to DMSO containing 100 mg/ml of the dispersant Pluronic-F127 and 0.1% acetic acid. The resulting DMSO solution is added with vigorous agitation to 1 ml of 24% (v/v) PEG-400 in serum-free medium containing HEPES (25 mM) or in a suitable buffer such as HBS (pH 7.3). The resulting solution is added to cells in their culture medium. After loading, the CCF2/AM containing medium is aspirated and replaced with fresh medium or, to avoid hypotonic shock, replaced with medium containing 4% PEG-400 (after loading with 6× solution) or medium containing 2% PEG-400 (after loading with 11× solution). (*Note:* Care should be taken with addition of the PEG-400 containing stock solutions and their removal. PEG-400 containing media and buffer solutions may be hypertonic and may cause cells to shrink initially. After a period of time in the presence of PEG-400, cells may regain original volume but at an increased intracellular tonicity. At this time, changing cells from PEG-400 containing solutions into nominally isotonic solutions may cause cells to swell. Cell lines not mentioned earlier, including primary cells, neurons, and so on, may not tolerate the change in tonicity or exposure to PEG-400.)

Blocking Active Substrate Export

If cells appear only dimly fluorescent by visual inspection on an epifluorescence microscope after following the substrate loading procedure just described, this may be because of active transport of the substrate CCF2 out of the cells. Active transport of CCF2 out of cells via anion transporters can be reduced by loading and keeping cells in the presence of a standard inhibitor of nonspecific anion transport, probenecid,[8] at a final concentration of 2.5 mM. Probenecid (Sigma), an organic acid, is dissolved in aqueous solution with the addition of one equivalent of an aqueous base, such as aqueous sodium hydroxide. A 250 mM probenecid stock solution (100×) is prepared by the addition of 2.5 ml 1 N sodium hydroxide solution to a suspension of 2.5 mmol probenecid in 7.5 ml water and vigorous agitation to dissolve the organic acid. Adding this solution to the loading buffer and the medium used to wash the cells at a final concentration of 1% aids in retaining CCF2 and its enzymatic hydrolysis product.

[8] F. Di Virgilio, T. H. Steinberg, J. A. Swanson, and S. C. Silverstein, *J. Immunol.* **140**, 915 (1988).

Discussion of Cell Loading

The fluorogenic β-lactamase substrate ester, CCF2/AM, is designed to be hydrophobic to diffuse into mammalian cells where it is converted by mammalian cytoplasmic esterases to free the polar β-lactamase substrate, CCF2. Ester protection groups used in CCF2/AM (butyrate, acetate, and acetoxymethyl ester) provide the necessary lability in the cytoplasmic environment for loading of the cells. These groups are well established for delivery of prodrugs as well as fluorescent probes into cells. Ester hydrolysis liberates CCF2, which is the substrate for β-lactamase, and its hydrolysis gives rise to blue cytoplasmic fluorescence in cells (Fig. 2, path A).

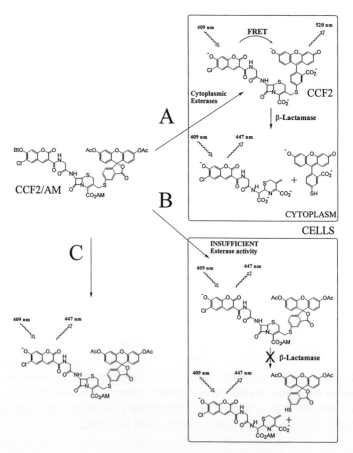

Fig. 2. Schematic of the various fates of CCF2/AM during cell loading. (A) Loading into healthy, esterase expressing cells. (B) Loading of compromised cells or cells lacking esterase. (C) Hydrolysis in medium.

The reason for the bluish appearance of compromised or overgrown cells is not entirely apparent. One can speculate that these cells may have decreased esterase activity, which results in inefficient removal of the acetate esters of fluorescein (Fig. 2, path B). Cells in which the donor is deprotected much more readily than the fluorescein acceptor will appear bluish. This is because, unlike the free fluorescein in CCF2, the esterified fluorescein acceptor in CCF2/AM is not a quencher of the free donor fluorophore. A suggested procedure is to load cells over extended periods of time (up to several hours), after which compromised cells often will have lost their dye and cells low in esterase activity will have had time to process the substrate ester. Cells that are not viable and do not exclude trypan blue may fluoresce blue during the loading procedure. Loading in the presence of trypan blue, with accumulation of the dye inside nonviable cells, efficiently reduces fluorescence from partially hydrolyzed CCF2/AM and stained cells appear nonfluorescent.

Although much more stable in aqueous medium than in the cytoplasm of cells, ester groups will hydrolyze to some degree in aqueous buffer at neutral pH. This occurs over the course of many hours for the butyrate ester or hours to days for fluorescein acetates and AM-ester. Buffer components such as albumin or serum speed this hydrolysis and are therefore not recommended during cell loading. The butyrate protection group of the hydroxycoumarin donor fluorophore is the most prone to hydrolysis in aqueous solution (Fig. 2, path C). A solution of CCF2/AM in which the butyrate protection group has been lost will have some blue fluorescence. In the cell, where fluorescein acetates are removed by cellular esterases, fluorescein quenches the donor efficiently. The partial hydrolysis of the butyrate ester is usually not a concern for loading of the substrate into cells, as under the recommended loading conditions (30–90 min, room temperature, absence of albumin or serum), the predominant amount of CCF2/AM remains esterified and available for loading.

Quantitative Analyses

Typically, reporter gene expression is reported relative to a set level, such as the fold induction or repression of reporter activity compared to background activity. Rarely, the number of reporter enzymes present per cell is determined quantitatively.[9] This is somewhat unfortunate because it does not allow quantitative comparisons of reporter expression levels between unrelated experiments or between different biological back-

[9] G. P. Nolan, S. Fiering, J. F. Nicolas, and L. A. Herzenberg, *Proc. Natl. Acad. Sci. U.S.A.* **85,** 2603 (1988).

grounds. The lack of quantitative information is mainly due to the fact that for most reporter enzyme–substrate combinations the kinetics have not been determined to allow quantitative determinations in the sample, the enzyme kinetics are complicated, or the signal is difficult to calibrate. The alternative, which is to prepare a standard curve with purified enzyme and assess the amounts present in the sample by comparison, requires commercial availability of the reporter enzyme with known purity and activity.[10] This is not the case in most instances.

The kinetics of the β-lactamase reporter enzyme with its fluorescent substrate CCF2 have been determined.[2] Kinetic data allow quantitative determination of the reporter activity in a sample and the experimental procedures are outlined.

β-Lactamase Activity in Live Cells

As the concentration of substrate in individual cells is typically unknown, the progress of the β-lactamase-catalyzed substrate hydrolysis reaction is determined from the intensity ratio of donor-to-acceptor fluorescence emissions. This intensity ratio is less affected by variations in intracellular substrate concentrations than measurements at the individual donor and acceptor wavelengths and is therefore better suited to assess the progression of the enzyme reaction in cells. Because the increase in blue fluorescence and the drop in green fluorescence with substrate hydrolysis are mostly linear with the amount of substrate hydrolyzed, a plot of the intensity ratio against substrate conversion has a hyperbolic shape. This is due to the nonlinear contribution of the numerator and denominator to the ratio. The two procedures that follow allow the progression of the substrate hydrolysis reaction to be related to the observed intensity ratios.

A linear relationship between the amount of substrate hydrolyzed and the ratio can be achieved experimentally by measuring the fluorescence intensity at the crossover point of substrate and product emission spectra (around 500 nm), which is independent of substrate conversion. When this value is used as the denominator, instead of acceptor fluorescence intensity at its 520-nm peak, a linear relationship of the progression of substrate hydrolysis and ratio value is obtained. This experimental modification provides the benefit of reducing noise in the assay, as perturbations that affect emission intensities at both wavelengths to similar degrees cancel out in the ratio, although it does result in a smaller numerical range of the ratio values.

Alternatively, one can calculate the fraction (f) at time (t) of hydrolyzed substrate from experimentally determined ratio values using the following

---

[10] V. K. Jain and I. T. Magrath, *Anal. Biochem.* **199**, 119 (1991).

equation. Additional calibration values needed are obtained readily by measuring the fluorescence intensities with excitation of the hydroxycoumarin donor of unhydrolyzed and fully hydrolyzed substrate at the blue and green emission wavelengths.

$$f_{(t)} = \frac{R_{(t)}F_{SG} - F_{SB}}{(F_{PB} - F_{SB}) + R_{(t)}(F_{SG} - F_{PG})} \tag{1}$$

Progression of CCF2 hydrolysis reaction $f_{(t)}$ derived from ratio values $R_{(t)}$ is shown in Eq. (1) in which $f_{(t)}$ is fraction of substrate converted to product at time (t), $R_{(t)}$ is ratio value of blue/green fluorescence intensities at time (t), F_{SG} is fluorescence emission intensity of substrate in the green channel, F_{SB} is fluorescence emission intensity of substrate in the blue channel, F_{PB} is fluorescence emission intensity of product in the blue channel, and F_{PG} is fluorescence emission intensity of product in the green channel.

In vitro, the product is obtained by the incubation of substrate with a suitable concentration of the β-lactamase enzyme to achieve full substrate hydrolysis. In experiments with cells, fluorescence intensity values for CCF2-loaded cells with no β-lactamase activity (e.g., wild-type cells) and cells with fully hydrolyzed CCF2 [e.g., clonal cells expressing *bla* under viral (e.g., CMV, SV40) promoter control] need to be determined.

β-Lactamase Expressing Cells in Populations

Epifluorescence microscopes and flow cytometers display single-cell resolution and permit direct quantitative determination of the fraction of reporter enzyme or reporter-fusion expressing cells in a population with high fidelity. However, many other instruments for fluorescence analysis do not have single-cell resolution capabilities, such as plate readers and fluorescence spectrophotometers. To determine the fraction of reporter enzyme-expressing cells in a population with an instrument lacking single-cell resolution, intensity ratios for donor and acceptor emission intensities can be determined for substrate-loaded cells in a cell population. Typically, donor and acceptor fluorescence intensity values determined in the measurement show a linear dependence with the fraction of blue cells in the population. The plot of a simple blue/green ratio has a hyperbolic shape due to the nonlinear contribution of the numerator and denominator. In general, one can accurately discriminate 5–10% changes in the percentages of blue cells in a population-based assay without spatial resolution (such as in a fluorescence microtiter plate reader) (Fig. 3).

To determine the fraction of β-lactamase expressing blue cells [$f_{(B)}$] in the population accurately, this value can be computed from the intensity

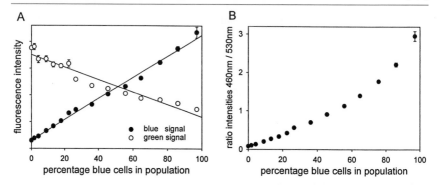

Fig. 3. (A) Change in blue and green emission intensities as a function of percentage blue cells in a 96-well plate. Wild-type and CMV-*bla* Jurkat cells (clonal Jurkat cell line with *bla* under CMV promoter control, expressing approximately 1.5×10^4 enzymes/cell) were mixed and the percentage of blue cells present was verified by flow cytometry. Each data point represents the average of six sample wells. Each well contained 10^5 cells. Data were obtained on a Cytofluor 4000 plate reader with 395 ± 12.5 nm excitation and 460 ± 20-nm (blue) and 535 ± 12.5-nm (green) emission filters. (B) Numerical ratio of blue and green intensity values graphed against the percentage of blue cells in the population.

ratio using Eq. (2). It permits determination of the fraction of blue cells in a population from experimentally determined ratio values. For calibration purposes, this computation also requires fluorescence intensity measurements of a reporter negative cell population (e.g., wild-type cells) and a cell population in which all cells express the enzyme (e.g., positive control such as clonal cells with CMV-*bla* or SV40-*bla*). Also, for accuracy, cells should be kept with substrate for sufficient time before fluorescence measurement to allow for complete conversion of intracellular substrate to product in cells expressing the reporter.

$$f_{(B)} = \frac{R_{(P)}F_{NG} - F_{NB}}{(F_{PB} - F_{NB}) + R_{(P)}(F_{NG} - F_{PG})} \tag{2}$$

Fraction of blue cells $f_{(B)}$ in a population derived for ratio values $R_{(P)}$ is shown in Eq. (2) in which f_B is fraction of blue cells in the population, $R_{(P)}$ is ratio value of blue/green fluorescence intensities of cell population, F_{NG} is green signal from a reporter negative cell population (e.g., wild type), F_{NB} is blue signal from a reporter negative cell population (e.g., wild type), F_{PB} is blue signal from a reporter positive control (e.g., clonal CMV-*bla* population), and F_{PG} is green signal from a reporter positive control (e.g., clonal CMV-*bla* population).

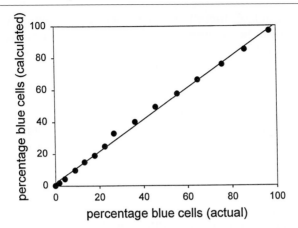

Fig. 4. Percentage of blue cells derived from ratio values plotted against the percentage of blue cells determined by flow cytometry for the same samples.

Figure 4 demonstrates the advantages of dual wavelength analysis in reducing noise stemming from variations in cell number and cellular substrate loading. The percentages of blue cells determined mathematically from the ratio values plotted in Fig. 1B are graphed against the percentages of blue cells confirmed for the same cell populations using flow cytometry. As can be seen, the scatter around the regression line found for individual wavelength readings in Fig. 3A is substantially smaller after ratioing and mathematical conversion of ratios back to apparent percentages of blue cells present. Equation (2) computes the fraction of blue cells, a variable whose value is not readily obtained from individual emission intensity readings of cell populations alone.

Quantitation in Lysates

Concentrations of active β-lactamase can be determined accurately in buffer solutions. This permits the average intracellular β-lactamase concentration in reporter positive cells to be determined from the β-lactamase activity in lysates from a known numbers of cells.

Suspension cells are harvested by centrifugation, and adherent cells are collected by dissociation with a EGTA containing buffer. If proteases need to be used to dissociate cells, great care should be taken to ensure complete removal of protease activity with multiple cell washes with protease-free buffer (or addition of a protease inhibitor). Cells are resuspended in isotonic phosphate buffer, pH 7.3, to a cell density of 10^7 cells/ml in a small tube with cap. A 10-μl sample of cell suspension is removed and diluted into

10 ml of buffer suitable for cell counting on a Coulter counter. Alternatively, the cell count can be determined using a hemocytometer. The remainder of cells are lysed by shock freezing. The tube is exposed to liquid nitrogen until the content freezes and is warmed to 30° in a water bath. After three freeze/thaw cycles, the cell debris is spun into a pellet by centrifugation in a cooled (4°) tabletop microcentrifuge at maximum permissible speed. The lysate solution (50 μl) is placed into a well of a black clear-bottom microtiter plate. The average activity in wells from two twofold serial dilutions of 20 nM purified β-lactamase derived from pUC18[11] (a kind gift of S. Mobashery, Wayne State University, Detroit, MI) in phosphate buffer (50 μl) serves as a standard in the calculation of β-lactamase activity for cell lysis samples. Wells with 50 μl buffer alone and wells with lysate of untransfected cells serve as baseline controls.

As an alternative to the freeze/thaw method, cells may be lysed using a final concentration of 0.5% CHAPS, which apparently has little effect on β-lactamase activity at this concentration.

Quantitation with Standard Curve. To the samples (50 μl volume), one adds 50 μl of 20 μM CCF2 in phosphate buffer, pH 7.3, for a final concentration of 10 μM CCF2. The microtiter plate is transferred to a fluorescence plate reader, and fluorescence intensity measurement (447 nm, product channel, e.g., 460 ± 20 nm) readings are taken, initially at 3- to 5-min intervals for the first hour and then hourly for another 4 hr to allow assessment of very low enzyme concentrations. Hydrolysis of CCF2 is complete within 1 hr at the highest β-lactamase concentrations in the standard (10 nM enzyme), which gives the signal for 100% conversion. This value, subtracted by the value for blue fluorescence reading from a well containing CCF2 in buffer alone, gives the total fluorescence change associated with 1 nmol CCF2 (100 μl of 10 μM) converted. It is used to calculate the amount of CCF2 hydrolyzed as the magnitude of the blue fluorescence signal is proportional to blue fluorescence from the sample wells. The rate of increase in blue well fluorescence, for less than 5% conversion of CCF2 in each well, is a measure of β-lactamase activity in that well. The amount of β-lactamase present in wells containing cell lysate can be determined from the standard curve calculated for the purified enzyme. This activity is then normalized to the number of cells in the lysate, which was determined by cell count. In addition, the percentage of β-lactamase expressing cells in the population can be determined in cells before lysis by loading with CCF2/AM and analysis by visual inspection on a fluorescence microscope

[11] G. Zafaralla, E. K. Manavathu, S. A. Lerner, and S. Mobashery, *Biochemistry* **31**, 3847 (1992).

or by flow cytometry and the cell count modified to represent reporter-expressing cells only.

Note: Care should be taken to ensure that the detector has a linear response in the intensity range chosen for the experiment.

Quantitation without Standard Curve. A simple estimate of cellular β-lactamase concentration relies on fluorescence intensity values for the substrate and the fully hydrolyzed CCF2 sample and the published kinetic values for CCF2 hydrolysis by TEM-1 β-lactamase. From the published values (K_{cat} = 29 ± 1 sec^{-1} and K_m = 23 ± 1 μM),[2] the expected hydrolysis rate for 10 μM CCF2 is about 9 sec^{-1} (mol substrate/mol enzyme). Steady-state values for hydrolysis of 10 μM CCF2 solutions differ only slightly from initial rates of hydrolysis (steady state rates of 7.3 and 6.7 sec^{-1} after 5 hr were determined in two independent experiments). Reliance on these data allows analysis without preparation of a standard curve with purified enzyme. Commercially available preparations of enzyme (i.e., Sigma, Penicillinase *E. coli* 205 RTEM) are suitable to catalyze hydrolysis of the substrate in order to obtain the fluorescence intensity values for fully hydrolyzed CCF2.

Maximizing the Performance of the Reporter Assay

In a cell-based assay it is desirable to be able to stop or at least slow the enzymatic reaction of the reporter enzyme to avoid exhaustion of substrate in cells that have high concentrations of reporter enzyme. Phenylethylthio-β-galactopyranoside is a competitive inhibitor of the reporter enzyme β-galactosidase[12] and has been used successfully to stop β-galactosidase activity in cells prior to flow cytometry.[13] It would be just as desirable to be able to inhibit the reporter enzyme with an irreversible or suicide inhibitor because, in addition to using it as a stopping agent, it would also allow inhibition of background reporter activity present prior to an assay. To abolish background activity only, cells would be treated with the inhibitor, followed by washout and quantitative analysis of reporter enzyme synthesized *de novo,* after the treatment with the inhibitor.

Irreversible Inhibition of Intracellular β-Lactamase

The reporter enzyme β-lactamase can be inactivated irreversibly in living cells with clavulanic acid. Although clavulanic acid is a substrate for β-lactamase, it also acts as an inhibitor because 1 out of 115 turnovers of the substrate results in cross-linking of residues in the active site of the

[12] C. K. De Bruyne and M. Yde, *Carbohydr. Res.* **56,** 153 (1977).
[13] S. N. Fiering, M. Roederer, G. P. Nolan, D. R. Micklem, D. R. Parks, and L. A. Herzenberg, *Cytometry* **12,** 291 (1991).

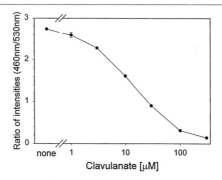

Fig. 5. Inhibition of intracellular β-lactamase reporter activity as a function of clavulanic acid concentration. CMV-*bla* Jurkat cells (expressing approximately 1.5×10^4 enzymes/cell) were treated for 16 hr with clavulanic acid at 37°. Cells were then loaded with 1 μM CCF2/AM for 1 hr, and population fluorescence intensity ratios were analyzed in a clear-bottom, 96-well plate using a Cytofluor 4000 plate reader with excitation at 395 ± 12.5 nm (bandpass filter) and 460 ± 20-nm and 535 ± 12.5-nm emission filters. Each data point represents the average of four sample wells and each well contained 10^5 cells. Error bars represent standard error. A population of cells lacking the reporter (wt Jurkat) gives a ratio of 0.15 under similar experimental conditions (not shown).

enzyme.[14] Cross-linking leads to irreversible inhibition of the enzyme, which allows the inhibitor to be removed from the medium without reappearance of activity. The apparent IC_{50} for inhibition of intracellular β-lactamase in the cytoplasm of cells for overnight exposures to inhibitor is about 10 μM (Fig. 5), although *in vitro* it has a K_i for the enzyme of 0.4 μM.[15] This apparent lower potency of clavulanic acid in the cell-based assay is likely due to the limited access of the inhibitor to the highly concentrated intracellular reporter enzyme.

The protein synthesis inhibitor cycloheximide was used to arrest *de novo* reporter synthesis and to assess how long the reporter enzyme remains inhibited after inhibitor washout (not shown). The inhibition persisted throughout the entire length of the experiment (5 hr), which is a longer period than the half-life of the enzyme in cells. This period is sufficient to allow clavulanic acid to be used to remove β-lactamase activity that may be present in cells prior to performing an assay.

The lithium salt of clavulanic acid is available from U.S. Pharmacopeia (USP, Maryland).

[14] J. Fisher, R. L. Charnas, and J. R. Knowles, *Biochemistry* **17**, 2180 (1978).
[15] U. Imtiaz, E. Billings, J. R. Knox, E. K. Manavathu, S. A. Lerner, and S. Mobashery, *J. Am. Chem. Soc.* **115**, 4435 (1993).

Experimental Procedure to Reduce Background

One day prior to the assay, CHO cells in which β-lactamase reporter expression is dependent on the activation status of a G-protein-coupled receptor (GPCR) are plated in six clear-bottom 96-well microtiter plates at 75% confluency in the presence or absence of 300 μM clavulanic acid. After 16 hr, all cells are washed once with clavulanic acid-free medium containing 0.1% bovine serum albumin. Cells are then exposed to various concentrations of agonist for the GPCR. Cells are incubated for 3 hr at 37° for reporter gene expression to occur, after which β-lactamase activity in the cells is determined using CCF2/AM. Figure 6 shows the increase of β-lactamase reporter activity in the cells with agonist concentration for cells with and without clavulanic acid pretreatment. The curve shapes of the plotted ratio values are very similar, with similar EC_{50} concentrations for the agonist. However, background β-lactamase activity is much lower in the assay using cells that are pretreated with clavulanic acid. The agonist-induced ratio change is about 2.5-fold for untreated cells, whereas clavulanic acid pretreatment gives a ratio change of 7-fold in the assay.

In cases where stimulation of cells occurs over much longer periods (10 hr or longer), background activity may not only be from a preexisting reporter enzyme, but also from a reporter enzyme synthesized *de novo* in the absence of stimulus. Clavulanic acid pretreatment may only reduce the former and not the latter, and therefore may not be as effective as in

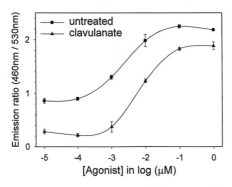

Fig. 6. Effect of clavulanic acid pretreatment on the dynamic range (fold ratio change) in an assay with background reporter activity. CHO cells expressing the β-lactamase reporter gene under control of a G-protein-coupled receptor were incubated with agonist for 3 hr with or without prior overnight exposure to 300 μM clavulanic acid. β-Lactamase reporter activity was determined by 1-hr loading with 1 μM CCF2/AM, and population fluorescence intensity ratios were analyzed in a clear-bottom, 96-well plate using a Cytofluor 4000 plate reader. Each data point represents the average of four sample wells and each well contained 10^5 cells. Error bars represent standard error.

the depicted example (Fig. 6) in curbing total background signal under those conditions.

Modulating the Dynamic Range

Cells that express high concentrations of the β-lactamase reporter turn blue fluorescent with the loading of substrate, as substrate loading becomes rate limiting. One can use the inhibitor to shift the dynamic range of the cell-based reporter assay to higher reporter concentrations by inhibiting a fraction of the expressed reporter enzyme. In this application, the inhibitor clavulanic acid is added to cells after the gene expression experiment but prior to or concomitant with the addition of CCF2/AM. The inhibitor concentration is adjusted for partial inhibition of the intracellular reporter activity. With this treatment, substrate loading is no longer rate limiting and the assay can be used to monitor changes in reporter expression at high expression levels.

Instrumentation

Fluorescence Microscopy

An inverted microscope equipped for epifluorescence using either a xenon or mercury excitation lamp is suitable for use in viewing fluorescence from CCF2 and its β-lactamase-catalyzed hydrolysis product in cells. For visual inspection of cells, a long-pass filter passing blue and green fluorescence light is needed to allow visual inspection to determine whether the cells fluoresce green or blue. For epifluorescence microscopes, a long-pass dichroic mirror is needed to separate excitation and emission light and should be matched to the excitation filter (i.e., maximally reflect the excitation light around 405 nm, yet allow good transmission of the emitted light).

A suitable filter set for observing β-lactamase activity is

Excitation filter: 405 ± 10 nm
Emission filter: 435 nm long pass
Dichroic mirror: 425 nm

Filters for microscopes are available through Chroma Technology Corp. and Omega Optical, Inc., both in Vermont.

As with many fluorescent dyes, care should be taken to avoid photobleaching of dye-loaded cells. The fluorescent substrate is particularly sensitive to continuous illumination with UV, or any other wavelength of light suitable to excite the dye, through a high magnification, high numerical aperture objective. In the case of CCF2, intense illumination can cause the acceptor fluorophore to be bleached (destroyed) with loss of fluorescence resonance energy transfer (FRET) and appearance of donor fluorescence.

This effect is progressive and nonreversible with an increase in light exposure leading to more acceptor destruction.

Photobleaching is reduced by limiting the exposure of cells to excitation light for brief fluorescence analyses lasting a few seconds at a time. Alternatively, the light exposure of the substrate can be reduced by use of a lower magnification objective, which spreads the excitation beam over a larger area. For photography, use of high-sensitivity film, such as 400 ASA or greater, is recommended.

Ratio-Imaging Microscopy

For microscope-based ratiometric analysis, blue and green fluorescence emissions are analyzed separately by filtering emitted light separately through two emission filters, passing either blue or green fluorescence. The ratio of blue fluorescence and green fluorescence intensities allows for numerical analysis of data (color is expressed as a numeric ratio of the intensities at the two wavelengths after background subtraction). Imaging microscopy and analysis can be used to monitor gene expression in single cells in real time.[2]

A suitable filter set for ratiometric analysis is

Excitation filter:	405 ± 10 nm
Dichroic mirror:	425 nm
Emission filter (blue):	460 ± 25 nm
Emission filter (green):	535 ± 25 nm

Fluorescence Microtiter Plate Readers

Most fluorescence microtiter plate readers are suitable for measuring β-lactamase reporter activity with the substrate CCF2. Although plate readers with top reading geometry permit analysis of β-lactamase reporter expression, live-cell experiments are best performed in black-walled, 96-well microtiter clear-bottom plates with low fluorescence background and the signal read on a bottom-count instrument, as cells usually settle to or grow on the well bottom.

A suitable filter set for plate readers is

Excitation filter:	405 ± 10 nm	(superior to 395 ± 12.5 nm)
Emission filter (blue):	460 ± 20 nm	(or 470 ± 30 nm)
Emission filter (green):	535 ± 12.5 nm	(or 535 ± 15 nm)

Filters for plate readers are available through Chroma Technology Corp., Vermont.

After background subtraction, ratios can be calculated by dividing the 460-nm emission (blue channel) reading by the 535-nm emission (green channel) reading. True background values should be determined for each read, as these are highly dependent on instrument-specific factors and on the length of time the lamp in the instrument has been lit.

Flow Cytometers

Flow cytometers permitting UV or violet excitation and collection of two emission wavelengths are suitable for analysis of β-lactamase expression in cells. Suitable excitation sources include an argon laser delivering UV excitation (351–364 nm), a krypton laser with excitation lines at 407 or 413 nm or multiline violet at 407–415 nm, or a mercury arc lamp with excitation through a filter centered at 405 nm (e.g., 405 nm with a 10-nm bandwidth). UV excitation (351–364 nm) from an argon laser is more than adequate, although for maximum signal over noise, excitation in the violet with a krypton laser is better (we have not compared these results with excitation from a mercury arc lamp). Optical filters and a dichroic beam splitter suitable for use with CCF2 on a flow cytometer are:

Emission filter (blue): 460 ± 25 nm
Emission filter (green): 535 ± 20 nm
Dichroic mirror: 490 nm long pass

Filters are available through Chroma Technology Corp. and Omega Optical, Inc., both in Vermont.

Discussion

Reporter genes are often used as markers of the expression of a particular protein. It is important to realize that in the process of replacing or modifying the structural gene of the protein of study with the reporter gene, many processing events, including splicing of mRNA and secondary protein modifications, are likely to differ from those experienced by the native protein. Conversely, fusion of a protein to the reporter may affect its kinetics with substrates. Barring these complications, which affect all reporter proteins to varying degrees, the β-lactamase reporter permits tracking of gene expression events, including expression of low abundance proteins in individual living cells. Together with the finding that the reporter can be fused to N and C termini of proteins without loss of function, these properties make it ideally suited for gene-tagging and gene-trapping applications.[3]

The current fluorogenic substrate, CCF2, is freely diffusable in the cytoplasm and does not permit spatial localization of the enzyme to specific cytoplasmic locations. However, the simple Michaelis–Menten type enzyme kinetics of this reporter–substrate pair allow quantitative determination of reporter activity and permit comparison of reporter levels across different experiments and biological backgrounds. The dual wavelength readout reduces susceptibility of the reporter assay to variations in cell number, substrate concentration, light intensity, and sample volume.

The β-lactamase reporter enzyme can be inhibited irreversibly inside living cells, allowing β-lactamase background activity (i.e., from "leaky" promoters) to be removed before an experiment. Inhibitor treatment may also allow β-lactamase to be used as a live cell reporter of high protein expression levels, a domain currently occupied by the GFP reporter. This is achieved by partial inhibition of intracellular β-lactamase activity until it falls within the linear range of the assay.

The β-lactamase reporter assay is easy to use, with simple reagent incubations and nondestructive optical detection in live cells. This permits facile preparation of reporter cell lines through functional selection of clones by their optical properties and subsequent propagation of selected cells. It facilitates the transition from molecular biology, with the tagging of genes and elucidation of their regulation, to cell biology with its study of intracellular signals, cell differentiation pathways, and proliferation events in living cells.

Acknowledgments

I thank Tom Knapp for performing experiments involving flow cytometry and Susanne Bayat for performing β-lactamase inhibition experiments.

Section IV

Functional Tags for Biochemical Purification

[16] Purification of Proteins Using Polyhistidine Affinity Tags

By Joshua A. Bornhorst and Joseph J. Falke

The expression and subsequent purification of recombinant proteins are widely employed in biochemical studies. A powerful purification method involves the use of peptide affinity tags, which are fused to the protein of interest and used to expedite protein purification via affinity chromatography.[1,2] A widely employed method utilizes immobilized metal-affinity chromatography (IMAC) to purify recombinant proteins containing a short affinity tag consisting of polyhistidine residues. IMAC is based on the interactions between a transition metal ion (Co^{2+}, Ni^{2+}, Cu^{2+}, Zn^{2+}) immobilized on a matrix and specific amino acid side chains. Histidine is the amino acid that exhibits the strongest interaction with immobilized metal ion matrices, as electron donor groups on the histidine imidazole ring readily form coordination bonds with the immobilized transition metal. Peptides containing sequences of consecutive histidine residues are efficiently retained on IMAC column matrices. Following washing of the matrix material, peptides containing polyhistidine sequences can be easily eluted by either adjusting the pH of the column buffer or adding free imidazole to the column buffer.[3]

IMAC is a versatile method that can be utilized to rapidly purify polyhistidine affinity-tagged proteins, resulting in 100-fold enrichments in a single purification step.[4] Affinity-tagged protein purities can be achieved at up to 95% purity by IMAC in high yield.[5,6] Purification using polyhistidine tags has been carried out successfully using a number of expression systems, including *Escherichia coli*,[7] *Saccharomyces cerevisiae*,[8] mammalian cells,[9]

[1] J. Nilsson, S. Ståhl, J. Lundeberg, M. Uhlén, and P. Å. Nygren, *Protein Expr. Purif.* **11**, 1 (1997).
[2] J. W. Jarvik and C. A. Telmer, *Annu. Rev. Genet.* **32**, 601 (1998).
[3] J. Porath, *Protein Expr. Purif.* **3**, 263 (1992).
[4] J. Schmitt, H. Hess, and H. G. Stunnenberg, *Mol. Biol. Rep.* **18**, 223 (1993).
[5] E. Hochuli, W. Bannwarth, H. Döbeli, R. Gentz, and D. Stüber, *Bio/Technology* **6**, 1321 (1988).
[6] R. Janknecht, G. de Martynoff, J. Lou, R. A. Hipskind, A. Nordheim, and H. G. Stunnenberg, *Proc. Natl. Acad. Sci. U.S.A.* **88**, 8972 (1991).
[7] M. W. Van Dyke, M. Sirito, and M. Sawadogo, *Gene* **111**, 99 (1992).
[8] D. C. Kaslow and J. Shiloach, *Bio/Technology* **12**, 494 (1994).
[9] R. Janknecht and A. Nordheim, *Gene* **121**, 321 (1992).

and baculovirus-infected insect cells.[10] However, this purification method may not be sufficient for tagged proteins expressed at low levels that require significantly greater than 100-fold enrichment or for the preparation of highly homogeneous protein samples. In such cases, either the use of a different affinity tag or the use of the polyhistidine tag in conjunction with additional purification techniques should be considered.

General Considerations

Incorporation of the Polyhistidine Affinity Tag

Affinity tags consisting of six polyhistidine residues are commonly used in IMAC. Whereas tags of six histidine residues are generally long enough to yield high-affinity interactions with the matrix, both shorter and longer affinity tags have been used successfully. In some cases the use of longer polyhistidine tags has resulted in increased purity due to the ability to use more stringent washing steps.[11] Still, it is advisable to use the smallest number of histidine residues as required for efficient purification to minimize possible perturbation of protein function. In general, a six histidine tag is an appropriate choice for the first trial when adding a novel polyhistidine tag to a protein.

Polyhistidine affinity tags are commonly placed on either the N or the C terminus of recombinant proteins. Optimal placement of the tag is protein specific. A potential problem is inaccessibility of the protein tag to the immobilized metal due to occlusion of the tag in the folded protein. Moving the affinity tag to the opposite terminus of the protein or carrying out the purification under denaturing conditions often resolves this problem. In principle, it is possible that the affinity tag may interfere with protein activity, although the relatively small size and charge of the polyhistidine affinity tag ensure that protein activity is rarely affected. Thus, the affinity tag usually does not need to be removed following protein purification.[12] If necessary, the affinity tag can be removed by use of a protease cleavage site inserted between the tag and the protein.[13]

Polyhistidine affinity tags are small enough to be incorporated easily into any expression vector. These tags can be added onto target genes by

[10] A. Kuusinen, M. Arvola, C. Oker-Blom, and K. Keinänen, *Eur. J. Biochem.* **233,** 720 (1995).

[11] R. Grisshammer and J. Tucker, *Protein Expr. Purif.* **11,** 53 (1997).

[12] J. Crowe, H. Döbeli, R. Gentz, E. Hochuli, D. Stüber, and K. Henco, *Methods Mol. Biol.* **31,** 371 (1994).

[13] D. B. Nikolov, S. H. Hu, J. Lin, A. Gasch, A. Hoffmann, M. Horikoshi, N. H. Chua, R. G. Roeder, and S. K. Burley, *Nature* **360,** 40 (1992).

site-directed mutagenesis or by polymerase chain reaction methods. DNA fragments coding for the polyhistidine affinity tag can also be created from synthetic oligonucleotides and cloned into an appropriate location in the desired plasmid.[14] Alternatively, there are a wide variety of commercially available cloning vectors for the generation and expression of polyhistidine-tagged recombinant proteins in different expression systems. Vectors for secreted polyhistidine affinity tagged proteins have also been developed for *E. coli.*[15]

Affinity Matrices

A variety of immobilized metal matrices are available for use in IMAC. Initial IMAC reports used iminodiacetic acid (IDA) as a matrix to chelate transition metals through three coordination sites.[16] A problem with the use of IDA matrices is that the metal ion is only weakly bound to such a three-coordinate matrix. Metal leaching from the matrix during purification causes lowered yields and impure products.[12] More recently, purification of polyhistidine affinity-tagged proteins has been facilitated by the development of the commercially available matrices nickel–nitrilotriacetic acid (Ni^{2+}-NTA)[17] and Co^{2+}–carboxylmethylaspartate (Co^{2+}-CMA),[18] which are coupled to a solid support resin. These matrices securely coordinate metal ions through four coordination sites while leaving two of the transition metal coordination sites exposed to interact with histidine residues in the affinity tag. Molecular models of these interactions are shown in Fig. 1. Much of the versatility of IMAC stems from the ability of these resins to tolerate a wide range of conditions, including the presence of protein denaturants and detergents. The stability of metal binding in these resins also allows the resins to be regenerated and reused several times. The Ni^{2+}–NTA matrix (available from Qiagen) has a binding capacity of 5–10 mg protein/ml of matrix resin and a high binding affinity ($K_d = 10^{-13}$ M) for the six residue polyhistidine tag at pH 8.0.[4] The Co^{2+}–CMA matrix (Talon resin, available from Clontech) has a somewhat lower affinity for the polyhistidine affinity tag than the Ni^{2+}–NTA resin, resulting in elution of the tagged proteins under milder conditions. The Co^{2+}–CMA also has been reported to exhibit less nonspecific protein binding than the Ni^{2+}–NTA

[14] N. E. David, M. Gee, B. Andersen, F. Naider, J. Thorner, and R. C. Stevens, *J. Biol, Chem.* **272,** 15553 (1997).
[15] A. Skerra, *Gene* **141,** 79 (1994).
[16] J. Porath, J. Carlsson, I. Olsson, and G. Belfrage, *Nature* **258,** 598 (1975).
[17] E. Hochuli, H. Döbeli, and A. Schacher, *J. Chromatogr.* **411,** 177 (1987).
[18] G. Chaga, J. Hopp, and P. Nelson, *Biotechnol. Appl. Biochem.* **29,** 19 (1999).

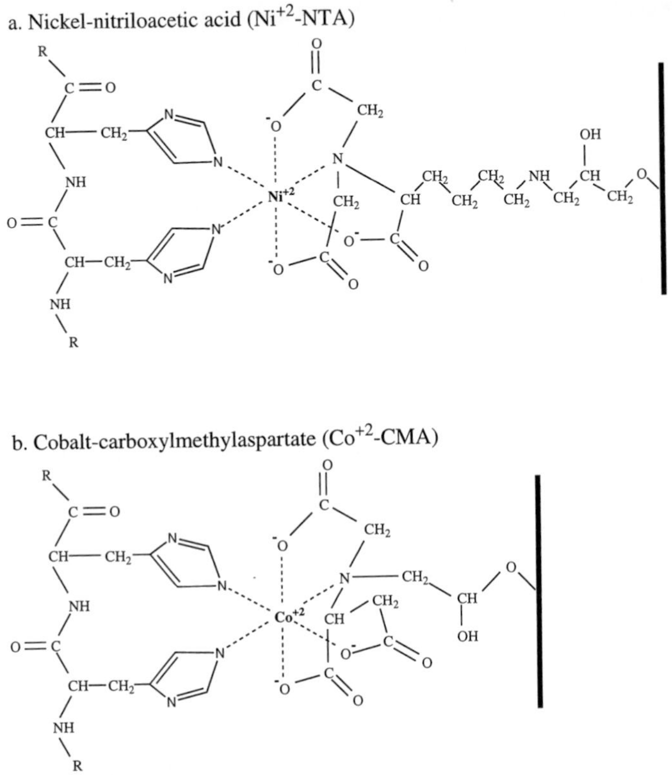

a. Nickel-nitriloacetic acid (Ni^{+2}-NTA)

b. Cobalt-carboxylmethylaspartate (Co^{+2}-CMA)

FIG. 1. Models of the interactions between the polyhistidine affinity tag and two immobilized metal affinity chromatography matrices. (a) The nickel–nitrilotriacetic acid matrix (Ni^{+2}–NTA) [from J. Crowe, H. Döbeli, R. Gentz, E. Hochuli, D. Stüber, and K. Henco, *Methods Mol. Biol.* **31**, 371 (1994)]. (b) The cobalt–carboxylmethylaspartate matrix (Co^{2+}–CMA) (from G. Tchaga, Clontech, personal communication). In both cases, the metal ion exhibits octahedral coordination by four matrix ligands and two histidine side chains, the latter provided by the polyhistidine affinity tag.

resin, resulting in higher elution product purity.[19] The binding capacity of the Co^{2+}–CMA resin is also about 5–10 mg of protein/ml of resin.

Purification under Native and Denaturing Conditions

Purification using the polyhistidine tag can be performed under either native or denaturing conditions by IMAC. The use of mild buffer conditions

---

[19] T.-T. Yang, P. S. Nelson, G. L. Bush, D. I. Meyer, and S. R. Kain, *Am. Biotechnol. Lab.* **I**, 12 (1997).

and imidazole as the elutant often yields biologically active purification products. Proteins that remain soluble in the cytoplasm, or that are secreted, usually can be purified using these native conditions. However, purification under native conditions may be hindered if the target protein is insoluble, aggregates in inclusion bodies, or possesses a tertiary structure that occludes the polyhistidine affinity tag. In such cases, proteins can be purified by the use of denaturing conditions such as 6 M guanidinium hydrochloride or 8 M urea during the purification process. Interaction of the resin with the polyhistidine tag does not require a specific conformation of the peptide tag, which makes effective purification with the use of denaturing conditions possible. Purification under denaturing conditions can depress the activity of phosphatases and proteolytic enzymes.[6] The use of urea as a denaturant is often preferable as 6 M guanidinium hydrochloride precipitates in the presence of SDS, interfering with subsequent SDS–PAGE analysis. Proteins purified under denaturing conditions can then be refolded into their active states by dialyzing away the denaturants.[20] In some cases, proteins can be refolded while bound to the resin.[21]

Purification of Membrane Proteins

Polyhistidine-tagged membrane proteins can be purified by IMAC using detergent-containing buffers to solubilize the proteins during the chromatographic process.[22,23] IMAC of membrane proteins has been carried out successfully in a variety of ionic and nonionic detergents. It is difficult to predict which detergent will be most suitable for IMAC in a given membrane protein system.[11] Although caution should be used, the Ni^{2+}–NTA and Co^{2+}–CMA matrices are generally able to tolerate limited amounts of nonionic and ionic detergents. Following IMAC, it is possible to restore the activity of purified polyhistidine-tagged membrane proteins by reconstitution into membrane vesicles.[14]

Nonspecific Binding

A problem with the use of polyhistidine affinity tags is nonspecific binding of untagged proteins. Although histidine occurs relatively infrequently (2% of all protein residues are histidine), some cellular proteins contain two or more adjacent histidine residues.[4] These proteins have an

[20] P. T. Wingfield, in "Current Protocols in Protein Science" (J. E. Coligan, B. M. Dunn, H. L. Ploegh, D. W. Speicher, and P. T. Wingfield, eds.), p. 6.1.1. Wiley, New York, 1995.
[21] D. Sinha, M. Bakhshi, and R. Vora, *Biotechniques* **17**, 509 (1994).
[22] R. Flachmann and W. Kühlbrandt, *Proc. Natl. Acad. Sci. U.S.A.* **93**, 14966 (1996).
[23] J. J. Janssen, P. H. Bovee-Geurts, M. Merkx, and W. J. DeGrip, *J. Biol. Chem.* **270**, 11222 (1995).

affinity for the IMAC matrix and may coelute with the protein of interest, resulting in significant contamination of the final product. This problem is generally more pronounced in systems other than *E. coli*. Mammalian systems, for example, have a higher natural abundance of proteins containing consecutive histidine residues.[12] Disulfide bond formation between the protein of interest and other proteins can also lead to contamination. The use of 10 mM 2-mercaptoethanol in the loading, wash, and elution buffers generally eliminates this potential problem. Nonspecific hydrophobic interactions can also cause some copurification with the desired protein. Including low levels (up to 1%) of the nonionic detergent Triton X-100 or Tween 20 in the protein buffers can reduce these interactions without substantially affecting the binding of the tagged protein to the Ni^{2+}–NTA or the Co^{2+}–CMA matrices. The addition of salt (up to 500 mM NaCl), glycerol (up to 20%), or low levels of ethanol (up to 20%) can also reduce nonspecific hydrophobic protein interactions with these matrices. Optimum levels of these buffer components should be determined experimentally for individual proteins.

Purification Procedure

Design of Protein Binding, Washing, and Elution Steps

Binding of the polyhistidine-tagged proteins can be performed using either a column or a batch procedure. Cell lysis should be done in buffered solution adjusted to pH 8.0. When the column procedure is utilized, the resin is packed into a column and the cell lysate is slowly loaded (3 to 4 column volumes per hour) onto the column. The batch procedure involves incubating the affinity matrix resin in the cell lysate solution and then packing the resin into a column. During incubation at 4°, the resin can be suspended in the cell lysate solution by shaking or stirring. Use of the batch procedure often results in more efficient binding of the tagged protein. With either method, the use of the minimum amount of resin needed to bind the tagged protein is recommended. The tagged protein usually has a higher binding affinity than other proteins that bind nonspecifically to the resin. Thus, when the minimum amount of resin is used, the tagged protein will fill most of the available binding sites, reducing the number of nonspecific proteins that bind. Sodium chloride (up to 500 mM) and low levels of imidazole (up to 20 mM) can also be included in the binding buffer to reduce the number of proteins that bind nonspecifically to the resin. Most protease inhibitors, with the exception of metal-chelating agents such as EDTA, can be included in all buffers. There are commercially available protease inhibitor cocktails specifically designed for use in IMAC purifications (Sigma).

Following binding of the tagged protein, the column can be washed to remove nonspecific proteins that bind weakly to the column. If desired, the inclusion of imidazole (10–50 mM for Ni^{2+}–NTA, 10 mM for Co^{2+}–CMA) in the wash buffer will increase the stringency of the wash and elute nonspecifically bound proteins more effectively. Alternatively, a wash buffer with a pH lower than that of the binding buffer (pH 6.3 for Ni^{2+}–NTA, pH 7.0 for Co^{2+}–CMA) can be employed to remove nonspecifically bound proteins. Agents such as detergents, 2-mercaptoethanol, and sodium chloride are often included in the wash and binding buffer to reduce nonspecific protein binding.

Three different methods can be used to elute the tagged protein of interest. Lowering the pH (to 5.3–4.5 for Ni^{2+}–NTA, 6.0 for Co^{2+}–CMA) protonates the imidazole nitrogen atom of the histidine residue (pK_a 6.0) and disrupts the coordination bond between the histidine and the transition metal. The histidine analog imidazole can also be used to competitively elute the bound polyhistidine residues (concentrations of 100 mM or higher for Ni^{2+}–NTA, 50 mM or higher for Co^{2+}–CMA). If the tagged protein forms oligomers, more stringent conditions, such as lower pH or higher concentrations of imidazole, may be required to elute the protein. While both of these elution methods are effective, the use of imidazole is often preferable as exposure to low pH may damage the protein of interest. Note, however, that heating a sample that contains imidazole to boiling prior to SDS–PAGE can cause acid-labile bonds to hydrolyze. Instead, it is recommended to heat the sample to no more than 37° for 5 min in the SDS loading buffer prior to analysis.[12] Including chelating agents, such as EDTA (100 mM) in the elution buffer, can facilitate maximal elution of proteins from the resin. This treatment will strip the metal atoms away from the matrix, resulting in contamination of the elute. The presence of the chelating agent or the metal in the eluate may interfere with enzyme activity. Moreover, the matrix cannot be reused following the use of a chelating agent without recharging.

Protocol: Native Purification of a Soluble Polyhistidine-Tagged Protein

The following protocol is one that is designed for the purification of the soluble ERK2 protein tagged with six N-terminal histidine residues from *E. coli* utilizing a Ni^{2+}–NTA resin under nondenaturing conditions (see Fig. 2).[24] While this protocol may need minor optimization for other

[24] D. J. Robbins, E. Zhen, H. Owaki, C. A. Vanderbilt, D. Ebert, T. D. Geppert, and M. H. Cobb, *J. Biol. Chem.* **268,** 5097 (1993).

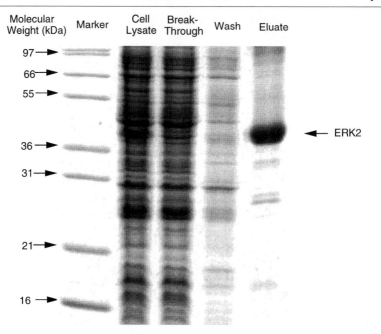

FIG. 2. SDS–PAGE analysis of a representative polyhistidine-tagged protein purification using a nickel–nitrilotriacetic acid (Ni^{2+}–NTA) matrix. The 42-kDa MAP-kinase protein ERK2 was affinity tagged at its N terminus with six histidines [D. J. Robbins, E. Zhen, H. Owaki, C. A. Vanderbuilt, D. Ebert, T. D. Geppert, and M. H. Cobb, *J. Biol. Chem.* **268,** 5097 (1993)], and was isolated using the procedure detailed in the text. Shown are equal volumes of the cell lysate, the breakthrough material that failed to bind to the resin during the batch step, the wash material obtained after loading the resin into the column, and the eluate from the column. The 15% Laemmli gel was visualized by Coomassie staining.

polyhistidine-tagged proteins, it should serve as a good starting point for the purification of soluble proteins in the native state. Urea or guanidinium hydrochloride can be added to all buffers to perform this procedure under denaturing conditions. Additional agents, including detergents, 2-mercaptoethanol, and various protease inhibitors, can also be added to the buffers to prevent nonspecific binding or proteolysis.

1. Lyse cells expressing the tagged protein by sonication on ice in loading buffer (300 mM NaCl, 50 mM NaH$_2$PO$_4$, pH to 8.0 with NaOH). Approximately 3–5 ml of loading buffer should be used per gram (wet weight) of cells. Keep the lysate as cold as possible to minimize possible proteolysis.

2. Immediately centrifuge lysate by spinning at 30,000g for 30 min at 4°.

3. Add 50% Ni^{2+}–NTA slurry (Qiagen) preequilibrated in ice-cold loading buffer to the supernatant. Add a sufficient amount of 50% Ni^{2+}–NTA slurry to bind the polyhistidine-tagged protein (5–10 mg/ml resin). Stir or rotate at 4° for 1 hr.

4. Load the resin onto a column. Wash the resin with 20 column volumes of loading buffer at 4°.

5. Wash the resin with 20 column volumes of wash buffer at 4° (same as loading buffer but also containing 10 mM imidazole, pH to 8.0 with HCl).

6. Elute with a 20 column volume gradient of 10 to 250 mM imidazole in loading buffer (pH to 8.0 with HCl). Collect 1-ml fractions and assay for fractions containing the desired protein using SDS–PAGE. Pool the desired fractions.

Comparison of IMAC to Other Affinity Chromatographic Techniques

Because a wide number of other affinity tag systems are available, it is useful to consider the advantages and disadvantages of the polyhistidine affinity tag and IMAC. The nonspecific binding of proteins to the IMAC column is a major disadvantage, especially when the tagged protein is not expressed at high levels. In some cases, the use of other affinity tag systems results in higher degrees of enrichment of the affinity-tagged protein than can be attained when polyhistidine affinity tags are employed. For example, biotinylation-accepting domain affinity tags have been shown to provide protein samples of higher yield and purity than polyhistidine affinity tags.[25,26]

Still, the polyhistidine affinity tag also possesses several important advantages. This tag is easily added to the protein of interest and, for IMAC purification of highly expressed proteins, can readily provide purities of up to 95% with 90% recovery of the tagged protein in a single purification step. The relatively small size and charge of the histidine affinity tag ensure that it rarely affects protein function. Polyhistidine-tagged proteins can be eluted under mild conditions from the IMAC resin, which allows them to retain their biological activity. Purification can be carried out readily under denaturing conditions that allow for the purification of insoluble proteins and proteins in which the affinity tag is occluded by the native tertiary structure of the protein. The IMAC affinity matrix resin is not affected by protease or nuclease activities in the extract; thus unlike many other biological affinity procedures, IMAC can be used effectively as an initial purification step with crude cell lysates. Following IMAC, proteins can often be readily purified further using other chromatographic methods. The immobi-

[25] J. Tucker and R. Grisshammer, *Biochem. J.* **317**, 891 (1996).
[26] Y. Pouny, C. Weitzman, and H. R. Kaback, *Biochemistry* **37**, 15713 (1998).

lization of polyhistidine-tagged proteins on IMAC matrices can also be used for protein–protein interaction studies.[27] Overall, the use of IMAC to purify tagged proteins provides a rapid and inexpensive purification method in comparison to other affinity protein purification methods.

The advantages of the polyhistidine affinity tag can be combined with the advantages of other affinity tags through the use of multiaffinity fusion systems wherein two or more affinity tags are attached to the same protein. These multifusion systems provide great flexibility during the purification process by allowing for protein immobilization and elution to be performed under a variety of conditions.[1,28] In addition, the use of multiaffinity tag purification procedures can result in products of higher purity than can be achieved using each individual affinity domain alone.[29] A number of affinity tag domains have been coupled successfully with the polyhistidine tag on recombinant proteins, including the GST affinity tag,[30] a modified S-peptide of ribonuclease A,[29] and both the albumin-binding protein fusion domain and a biotinylation accepting domain.[28]

Acknowledgments

The authors thank Natalie Ahn and Melanie Cobb for providing the plasmid NpT7-5, which expresses six histidine-tagged ERK2. The authors thank Irene Ota, Eric Nalefski, Mark Benson, and Julie Poelchau for providing valuable comments regarding the manuscript. We also gratefully acknowledge NIH Grants GM 40731 and GM 48203.

[27] T. Lu, M. Van Dyke, and M. Sawadogo, *Anal. Biochem.* **213**, 318 (1993).
[28] J. Nilsson, M. Larsson, S. Ståhl, P. Å. Nygren, and M. Uhlén, *J. Mol. Recognit.* **9**, 585 (1996).
[29] J. S. Kim and R. T. Raines, *Anal. Biochem.* **219**, 165 (1994).
[30] C. A. Panagiotidis and S. J. Silverstein, *Gene* **164**, 45 (1995).

[17] Generating Fusions to Glutathione S-Transferase for Protein Studies

By Donald B. Smith

History of pGEX Vectors

pGEX vectors and the abbreviation of GST for glutathione S-transferase have become well known to those attempting to express polypeptides in heterologous systems, but the serendipitous history of this expression system is less familiar. Studies at the Walter and Eliza Hall Institute (WEHI) of Medical Research in Melbourne, Australia, of the resistance of 129/J mice to infection with the parasitic worm *Schistosoma japonicum* indicated that resistance was associated with the presence of antibodies that recognized a 26-kDa parasite protein (Sj26).[1,2] Screening of a *S. japonicum* cDNA expression library with antibodies eluted from the 20- to 30-kDa region of a Western blot of parasite proteins probed with rabbit antiserum led to the identification of several clones[3] encoding an open reading frame that showed 42% homology with mammalian mu class glutathione *S*-transferases.[4] When this gene was expressed in *Escherichia coli*, GST activity could be detected in the supernatant of lysed bacteria, and activity could be purified after passage over a column of immobilized-reduced glutathione.[5] However, when material purified in this way was used to immunize susceptible mice, only limited protection against infection was observed.[6] Subsequent work suggests that the innate resistance of 129/J mice may have more to do with innate differences in the portal system than with the response to Sj26,[7] although the possibility of developing a vaccine based on GST is still being pursued for *Schistosoma*[8–10] and *Fasciola hepatica.*[11]

The idea of using GST as an affinity tag for fusion proteins came about because the author's fellow British post-doc, Kevin Johnson, was having difficulty expressing antigens of the parasite *Taenia ovis* in a form that could protect against challenge infections. Our standard method at this time was to purify insoluble β-galactosidase fusion proteins by size-exclusion chromatography of denatured inclusion bodies. After hearing the author's seminar on the expression of Sj26 in *E. coli* and its affinity purification as

[1] G. F. Mitchell, K. M. Cruise, E. G. Garcia, and W. U. Tiu, *J. Parasitol.* **70,** 983 (1984).

[2] G. F. Mitchell, J. A. Beall, K. M. Cruise, W. U. Tiu, and E. G. Garcia, *Parasite Immunol.* **7,** 165. (1985).

[3] J. A. Beall and G. F. Mitchell, *J. Immunol. Methods* **86,** 217 (1986).

[4] D. B. Smith, K. M. Davern, P. G. Board, W. U. Tiu, E. G. Garcia, and G. F. Mitchell, *Proc. Natl. Acad. Sci. U.S.A.* **83,** 8703 (1986).

[5] D. B. Smith, M. R. Rubira, R. J. Simpson, K. M. Davern, W. U. Tiu, P. G. Board, and G. F. Mitchell, *Mol. Biochem. Parasitol.* **27,** 249 (1988).

[6] G. F. Mitchell, E. G. Garcia, K. M. Davern, W. U. Tiu, and D. B. Smith, *Trans. R. Soc. Trop. Med. Hyg.* **82,** 885 (1988).

[7] G. F. Mitchell, *Parasite Immunol.* **11,** 713 (1989).

[8] M. G. Taylor, M. C. Huggins, F. H. Shi, J. J. Lin, E. Tian, P. Ye, W. Shen, C. G. Qian, B. F. Lin, and Q. D. Bickle, *Vaccine* **16,** 1290 (1998).

[9] J. M. Grzych, J. DeBont, J. L. Liu, J. L. Neyrinck, J. Fontaine, J. Vercruysse, and A. Capron, *Infect. Immun.* **66,** 1142 (1998).

[10] S. X. Liu, Y. K. He, G. C. Song, X. S. Luo, Y. X. Xu, and D. P. McManus, *Vet. Parasitol.* **69,** 39 (1997).

[11] C. A. Morrison, T. Colin, J. L. Sexton, F. Bowen, J. Wicker, T. Friedel, and T. W. Spithill, *Vaccine* **14,** 1603 (1996).

a soluble protein on glutathione-agarose, he suggested that it might be possible to turn the plasmid into a general expression vector that would allow affinity purification of fusion proteins under nondenaturing condition. Vectors of this type had previously been designed using the IgG-binding domains of *Staphylococcus aureus* protein A,[12] but such fusions were obviously unsuitable for use as immunological reagents. A few months later the author had constructed the pGEX-1, -2T, and -3X plasmids in which the GST open reading frame was followed by a protease cleavage site, multiple cloning sites, and termination codons.[13] A high level of expression was achieved using the inducible *tac* promoter, with tight control of expression achieved by including the *lacI*q gene in the plasmid. Luckily, many GST fusion proteins, including Johnson's *T. ovis* antigens, remained soluble and could be still be purified under nondenaturing conditions by affinity chromatography on glutathione-agarose beads.

Coincident with the completion of this work, the company AMRAD was set up by a group of Australian research institutes in order to capitalize on their scientific discoveries, and the pGEX vectors became their first product. Another lucky break at this time was that the author's friend, Lynn Corcoran (who had suggested using the *tac* promoter), took the vectors with her when she moved from WEHI to David Baltimore's laboratory at the Whitehead Institute in Boston. She used the pGEX vectors there to express the B-cell transcription factor *Oct-2* in order to raise antisera and also to localize DNA-binding domains within the protein. Like a virus infection, use of the vectors spread through the Whitehead and was then spread to other American laboratories by Cornelis Murre, Margie Oettinger, and Bruce Mayer, and GST fusion proteins began to be used increasingly as tools to study protein–DNA interactions[14] and protein–protein interactions.[15] A range of pGEX vectors and reagents are supplied by Amersham Pharmacia. Licenses for commercial applications are available from AMRAD Corporation Limited (Kew, Victoria 3101, Australia).

The pGEX Family

Most of the prokaryotic pGEX vectors have essentially the same structure. A plasmid of about 5 kb contains the β-lactamase gene conferring

[12] B. Nilsson, L. Abrahmsen, and M. Uhlen, *EMBO J.* **4,** 1075 (1985).
[13] D. B. Smith and K. S. Johnson, *Gene* **67,** 31 (1988).
[14] A. B. Lassar, J. N. Buskin, D. Lockshon, R. L. Davis, S. Apone, S. D. Hauschka, and H. Weintraub, *Cell* **58,** 823 (1989).
[15] B. J. Mayer, P. K. Jackson, and D. Baltimore, *Proc. Natl. Acad. Sci. U.S.A.* **88,** 627 (1991).

resistance to ampicillin, the pBR322 origin of replication, and the over-expressing $lacI^q$ allele that efficiently represses basal transcription from the *lac* promoter. The entire open reading frame of the *S. japonicum* 26-kDa GST is placed downstream of the strong *tac* promoter and is followed by multiple cloning sites (*Bam*HI, *Sma*I, and *Eco*RI) and translation termination codons in all three frames. The first set of vectors (pGEX-1, -2T, and -3X) were constructed so that the multiple cloning sites occur in different codon positions so that the reading frame of the insert can be matched with that of the vector.[13] The vector pGEX-2T contains the recognition site of thrombin upstream of the multiple cloning sites, whereas pGEX-3X contains that of blood coagulation factor Xa. Most pGEX vectors are named according to the convention that suffixes of T and X refer to the presence of thrombin and factor Xa cleavage sites, whereas terminal numbers refer to the reading frame of the multiple cloning sites.

Reading Frame/Cloning Site Alterations

A variety of modified vectors have been constructed over the last decade. First are vectors that provide multiple cloning sites in the two remaining reading frames in combination with factor Xa (pGEX-A and -B[16]) or thrombin protease cleavage sites [pGEX-1λT (Amersham Pharmacia), pGEX-2T-1,[17] and -3T[18]]. Additional cloning sites (*Sal*I, *Xho*I, and *Not*I) are present in all three frames in the vectors pGEX-4T-1, 2, and 3 (thrombin cleavage, Amersham Pharmacia) and pGEX-5X-1, 2, and 3 (factor Xa, Amersham Pharmacia), whereas the factor Xa cleavage site vectors pMI101, 102, and 103 have *Bgl*II and *Eco*RI sites in all three frames.[19]

Protease Site Modifications

A second class of modified vectors are designed to increase the efficiency with which site-specific proteases cleave fusion proteins. Enhanced cleavage with thrombin can be achieved using vectors in which a polyglycine stretch is introduced either before (pGEX-KT,[20] pGEX-5G/LIC[21]) or after (pGEX-

[16] D. Valle, J. Kun, J. Linss, E. D. Garcia, and S. Goldenberg, *Insect Bio. Mol. Biol.* **23,** 457 (1993).

[17] E. Olsen and S. S. Mohapatra, *Int. Arch. Allergy Immunol.* **98,** 343 (1992).

[18] B. Frorath, C. C. Abney, H. Berthold, M. Scanarini, and W. Northemann, *BioTechniques* **12,** 558 (1992).

[19] M. Ikeda, K. Arai, and H. Masai, *Gene* **181,** 167 (1996).

[20] D. J. Hakes and J. E. Dixon, *Anal. Biochem.* **202,** 293 (1992).

[21] R. S. Haun and J. Moss, *Gene* **112,** 37 (1992).

KG,[22] pGEX-His-2,[23] and pGSTag,[24] all containing additional cloning sites) the thrombin recognition site. A related vector, pGEX-KN, contains a polyglycine stretch upstream of a unique NotI site so that inserts can be engineered to contain a thrombin cleavage site at the exact N terminus of the exogenous polypeptide.[20] A similar effect is achieved through a different strategy using the vector pGEX-5G/LIC in which polymerase chain reaction (PCR) fragments generated using specially designed PCR primers are converted to overhanging ends using T4 DNA polymerase and are then annealed with similarly treated vector without the need for ligation.[21] The efficiency of factor Xa cleavage can also be increased by introducing a polyglycine stretch after the cleavage site in the maltose-binding protein expression vector system,[25] but similar vectors have yet to be produced for GST fusions. More drastically, the vectors pGEX-6P1, 2, and 3 (Amersham Pharmacia) include a peptide sequence that is optimally cleaved by the human rhinovirus 3C protease at 5°. This protease is itself available as a GST fusion protein (PreScission protease, Amersham Pharmacia) so that both the GST carrier and the protease can be removed after proteolysis by affinity chromatography on glutathione-agarose. An enterokinase cleavage site is present in the vectors pGEX-His-2[23] and pGEX-6T (derived from pGEX-3T[26]), but this protease appears to be less efficient and relatively non-specific.

Secondary Affinity Tags

Another type of modification to the pGEX vectors has been the inclusion of a sequence encoding a second affinity tag so that the protein becomes a tripartite fusion. The rationale behind this approach is that affinity purification using different tags situated at either end of the fusion protein allows partially degraded or incomplete fusion proteins to be eliminated. The vectors pGEX-His-2 (thrombin and enterokinase cleavage sites[23]) and pALEX (factor Xa cleavage[27]) combine expanded multiple cloning sites with a COOH-terminal hexahistidine region that binds to Ni^{2+}–agarose, whereas pGEX-ST (thrombin[28]) includes a 10 amino acid sequence that binds to streptavidin. These vectors also provide a simple way of eliminating

[22] K. Guan and J. E. Dixon, *Anal. Biochem.* **192**, 262 (1992).
[23] S. A. Strugnell, B. A. Wiefling, and H. F. DeLuca, *Anal. Biochem.* **254**, 147 (1997).
[24] D. Ron and H. Dressler, *BioTechniques* **13**, 866 (1992).
[25] P. L. Rodriguez and L. Carrasco, *BioTechniques* **18**, 238 (1995).
[26] H. Berg, M. Walter, L. Mauch, J. Seissler, and W. Northemann, *J. Immunol Methods* **164**, 221 (1993).
[27] C. A. Panagiotidis and S. J. Silverstein, *Gene* **164**, (1995) 45.
[28] G. Sun and R. J. A. Budde, *Anal. Biochem.* **231**, 458 (1995).

proteases after cleavage of GST from the purified fusion protein by affinity purification using the COOH-terminal tag. NH_2-terminal histidines are present in the vector pGEX-6T,[26] facilitating the purification of insoluble GST fusions.

NH_2-Terminal Fusions

A more radical rearrangement of the pGEX vectors is provided by vectors such as pETGEXCT[29] that allow the expression of foreign polypeptides as NH_2-terminal fusions with GST. Limited results using a variety of vectors suggest that these fusions are at least as stable as COOH-terminal fusions,[29–32] although they may often be insoluble (A. Sharrocks, personal communication), and subcloning is slightly complicated by the need to ensure that both ends of the DNA insert are in the appropriate reading frame. The three-dimensional structure of the *S. japonicum* 26-kDa GST encoded by pGEX-3X has been solved and reveals that both NH_2 and COOH termini are relatively unstructured and surface accessible, both in the subunit and as a dimer, although the glutathione-binding site appears to be NH_2-terminal.[33]

Additional Functions

Another series of modified pGEX vectors contain polypeptide domains that simplify the detection of GST fusion proteins so that they can be used more readily as probes for the screening of cDNA libraries or in the analysis of protein–protein interactions. The vector pAGEX-2T encodes two copies of the protein A IgG-binding domain between a thrombin cleavage site and the multiple cloning sites and so can be detected directly with standard enzyme-linked immunoglobulins.[34] The vectors pGEX-2TK[35] (Amersham Pharmacia) and pGSTag[24] contain thrombin cleavage sites followed by peptide sequences that can be phosphorylated *in vitro* by cAMP-dependent heart muscle kinase and protein kinase A, respectively. A further refinement is introduced into the vectors pGEX-GTH and pET-HTG[30] in which

[29] A. D. Sharrocks, *Gene* **138,** 105 (1994).
[30] T. H. Jensen, A. Jensen, and J. Kjems, *Gene* **162,** 235 (1995).
[31] M. Harris and K. Coates, *J. Gen. Virol.* **74,** 1581 (1993).
[32] K. J. Airenne and M. S. Kulomaa, *Gene* **167,** 63 (1995).
[33] M. A. McTigue, D. R. Williams, and J. A. Tainer, *J. Mol. Biol.* **246,** 21 (1995).
[34] D. B. Smith, L. C. Berger, and A. G. Wildeman, *Nucleic Acids Res.* **21,** 359 (1993).
[35] W. G. Kaelin, W. Krek, W. R. Sellers, J. A. DeCaprio, F. Ajchenbaum, C. S. Fuchs, T. Chittenden, Y. Li, P. J. Farnham, M. A. Blanar, D. M. Livingston, and E. K. Flemington, *Cell* **70,** 351 (1992).

the phosphorylation site is positioned so that it is at the opposite end of the fusion protein from the GST moiety, which is either NH_2- or COOH-terminal, ensuring that only full-length polypeptides will be purified and labeled.

Note that some of the vectors described earlier (pGEX-HTG,[30] pETGEXCT,[29] pALEX[27]) are hybrids between pGEX and pET so that expression of GST fusion proteins is dependent on inducible expression of the T7 RNA polymerase. These vectors have been designed to reduce the level of expression in the absence of induction, but no direct comparisons have been reported. GST fusions can also be expressed from a bacteriophage λ vector (λGEX5[36]) or from a phagemid [pGHX (-)[37]].

Eukaryotic Vectors

Finally, mention should be made of the variety of expression vectors that have been constructed for the expression of GST fusion proteins in eukaryotic systems. Inducible vectors for *Saccharomyces cerevisiae* include pEG-KT and -KG[38] in which a thrombin cleavage site is preceded or followed by a stretch of five glycine residues and pYGEX-4T-1[39] derived from pGEX-4T-1, whereas for *Schizosaccharomyces pombe* the vectors pESP1, 2, and 4 include both thrombin and enterokinase cleavage sites.[40,41] Several transfer vectors have been designed that allow the expression of GST fusion proteins in insect cells using recombinant baculoviruses. The transfer vectors pAcG1, 2T, and 3X correspond to pGEX-1, -2T, and -3X,[42] whereas pVLGST is derived from pGEX-2T but has an expanded multiple cloning site and lacks the multiple termination codons.[43] Another baculovirus transfer plasmid, pAcSG2T-tag, contains a signal sequence at the NH_2 terminus that directs the secretion of fusion proteins into the tissue culture medium, whereas the COOH terminus carries an epitope tag.[44] Stable expression of GST fusion proteins in mammalian cells can be

[36] R. Fukunaga and T. Hunter, *EMBO J.* **16,** 1921 (1997).

[37] T. Hunter and G. J. Hunter, *BioTechniques* **24,** 194 (1998).

[38] D. A. Mitchell, T. K. Marshall, and R. J. Deschenes, *Yeast* **9,** 715 (1993).

[39] M. A. Romanos, F. J. Hughes, S. A. Comerford, and C. A. Scorer, *Gene* **152,** 137 (1995).

[40] Q. Lu, J. C. Bauer, and A. Greener, *Gene* **200,** 135 (1997).

[41] T. Hosfield and Q. Lu, *Gene* **269,** 10 (1999).

[42] A. H. Davies, J. B. M. Jowett, and I. M. Jones, *Bio/Technology* **11,** 933 (1993).

[43] J. M. Beekman, A. J. Cooney, J. F. Elliston, S. Y. Tsai, and M. Tsai, *Gene* **146,** 285 (1994).

[44] Y. Wang, A. H. Davies, and I. M. Jones, *Virology* **208,** 142 (1995).

achieved using the vector pLEF[45] or the vectors pCMGT and pGMGX.[46] The first two vectors contain the thrombin cleavage site of pGEX-2T, whereas the third contains that of pGEX-3X. Another vector contains COOH-terminal histidine residues together with a protein kinase A phosphorylation site.[47] All of these mammalian expression vectors include the T7 promoter so that fusion proteins can be generated *in vitro* by coupled transcription/translation. Such *in vitro* transcripts generated using the vectors pXen-1, -2, and -3 include *Xenopus* β-globin untranslated regions so that GST fusion proteins can be expressed in *Xenopus* oocytes following microinjection.[48]

General Considerations

The major reason for using GST as a fusion partner is that many fusions remain soluble and stable even at high levels of expression and can be purified by affinity chromatography under nondenaturing conditions. However, it is well to remember that as Robert Burns wrote, "The best laid schemes o' mice an' men Gang aft a-gley," things do not always work out simply. For each fusion protein it will be necessary to optimize the region of the polypeptide to be expressed, the time course of expression, the factors that influence solubility, stablility, and the efficiency of purification, and the conditions required for proteolytic removal of the GST carrier. In addition, it should be noted that conditions for expression and purification of GST fusion proteins in eukaryotic systems are relatively poorly defined, and some ingenuity may be required in order to maximize yields in these systems. For example, the high endogenous level of glutathione in *S. cerevisiae* means that cell extracts should be dialyzed before absorption of GST fusions to glutathione agarose (personal communication.) The methods described in this article refer to GST fusions expressed in *E. coli* and may require adaptation for eukaryotic systems.

Some studies require the expression of a foreign polypeptide in its entirety, and in this case, problems with instability, insolubility, or the level of expression can only be addressed once the appropriate DNA fragment has been cloned into the chosen vector. Where it is not necessary to produce the fusion protein as a single polypeptide it may be prudent to produce a

[45] F. Rudert, E. Visser, G. Gradl, P. Grandison, L. Shemshedini, Y. Wang, A. Grierson, and J. Watson, *Gene* **169,** 281 (1996).
[46] A. Chumakov and H. P. Koeffler, *Gene* **131,** 231 (1993).
[47] B. Chatton, A. Bahr, J. Acker, and C. Kedinger, *BioTechniques* **18,** 142 (1995).
[48] M. C. MacNicol, D. Pot, and A. M. MacNicol, *Gene* **196,** 25 (1997).

variety of constructs containing different portions of the protein as a set of overlapping fragments. Although fusions as large as a 100 kDa have been expressed and purified, the level of expression generally decreases with increasing size, as also does the affinity of GST fusions for glutathione-agarose.[49] Another important factor that affects the solubility of fusion proteins is the presence of strongly hydrophobic regions, and there are several examples where removing hydrophobic signal sequences or membrane anchor regions from constructs has increased the solubility of the fusion protein.[13,50] Another study suggests that the influence of hydrophobic sequences on the level of expression is greatest when they are at the extreme COOH terminus.[51]

The NH_2-terminal structure of polypeptides expressed using standard pGEX vectors depends on whether cleavage with a site-specific protease is used to remove the GST moiety. However, it is worth remembering that even after cleavage, polypeptides will contain 2–16 additional vector-derived NH_2-terminal amino acids, depending on the vector and the site used for subcloning. Some fusion proteins will also retain polyglycine, kinase sites, or protein A domains. The only exceptions are the vectors pGEX-KN[20] and pGEX-5G/LIC[21] that allow fragments to be cloned immediately following a thrombin cleavage site. Similar considerations apply to the COOH terminus of fusion proteins. In most pGEX vectors the multiple cloning sites are followed by TGA termination codons in all three frames. While this ensures that translation does not continue into plasmid sequences, it means that 3–16 amino acids encoded by the multiple cloning sites will be present at the COOH terminus. This problem can be averted by including a TGA termination codon in the DNA fragment insert; many common bacterial strains encode suppressor tRNAs that allow readthrough of UAA and/or UAG codons, resulting in the expression of multiple products with different sites of termination.[52]

Recombinant pGEX plasmids can be transformed into a wide variety of E. coli strains, as most pGEX plasmids carry the overexpressed $lacI^q$ allele of the lac repressor that prevents fusion protein expression. Transformants are plated on Luria plates containing 50 μg ml^{-1} ampicillin and are grown overnight at 37°.

[49] J. V. Frangioni and B. G. Neel, *Anal. Biochem.* **210,** 179 (1993).
[50] S. Franke, F. Gunzer, L. H. Wieler, G. Baljer, and H. Karch, *Vet. Microbiol.* **43,** 41 (1995).
[51] S. Y. Sheu and S. J. Lo, *Gene* **160,** 179 (1995).
[52] M. J. Raftery, C. A. Harrison, and C. L. Geczy, *Rapid Comm. Mass Spectr.* **11,** 405 (1997).

Screening for Expression of Fusion Proteins

Transformants expressing the desired polypeptide can be conveniently identified as follows.

1. Pick isolated colonies of potential transformants and of the parental vector into 15-ml tubes containing 2 ml L broth, 100 μg ml^{-1} ampicillin and grow in a shaking incubator at 37° until cloudy (3–5 hr).

2. Induce expression by adding isopropyl-β-D-thiogalactopyranoside (IPTG) to 0.1 mM and grow for a further 1–3 hr.

3. Analyze 10 μl of the culture by electrophoresis through a 10% polyacrylamide–SDS gel (SDS–PAGE). Cell pellets can be collected by centrifugation at 13,000g for 30 sec and stored at −20°.

4. After Coomassie Blue staining of the gel, transformants will normally be identified by the absence of the 26-kDa protein observed for the parental vector and by the presence of one or more novel bands of higher molecular weight.

Some fusion proteins may be difficult to detect against the background of host proteins. A more sensitive way of detecting GST fusion proteins is to carry out a small-scale purification.

1. Prepare a 50% slurry of glutathione-agarose in phosphate buffered saline (PBS: 150 mM NaCl, 16 mM Na$_2$HPO$_4$, 4 mM H$_2$PO$_4$, pH 7.3) by swelling powdered glutathione-agarose beads (Sigma) in PBS for 30 min. These swollen beads, or beads supplied as a 75% slurry (glutathione Sepharose 4B, Amersham Pharmacia), should be collected by centrifugation at 1000g for 10 sec and washed three times with a 10× volume of PBS before making up to 50% in PBS. Beads can be stored at 4° in PBS containing 0.05% sodium azide or protease inhibitors for 1 month[24] or indefinitely at less than −20° in PBS containing 10% glycerol.[49]

2. Resuspend cell pellets from induced cultures (step 3 above) in 300 μl PBS.

3. Gently sonicate resuspended cells on ice using a small-diameter probe until clear (about 10 sec). Alternatively, cells can be resuspended in buffer containing 100 μg/ml lysozyme and 10 μg/ml DNase I, and lysed by 3–10 cycles of freezing in a dry-ice ethanol bath followed by thawing in water at 25°.

4. Pellet insoluble material by centrifugation at 13,000g for 5 min.

5. After transferring the supernatant to a fresh tube, add 50 μl of a 50% solution of glutathione-agarose beads. An alternative is to use prepacked glutathione agarose MicroSpin columns (Amersham Pharmacia).

6. Incubate at room temperature for 2 min. Collect beads by centrifugation at 1000g for 10 sec and wash three times with 1 ml PBS.

7. Add 20 μl 1× SDS loading buffer to the washed beads and analyze by SDS–PAGE and Coomassie Blue staining. Fusion proteins will be larger than the 26- to 28-kDa GST protein purified from cells transformed with the parent vector.

These methods work well for soluble fusion proteins, but direct screening for expression may be unsatisfactory in cases where the proportion of transformants to parental vector is very low or when the fusion protein is insoluble or expressed at very low levels. A different screening method can be used where a specific serum or antibody is available for the protein to be expressed. Cell lysates prepared from different clones as described in steps 2–4 earlier are incubated for 1 hr in individual wells of an ELISA plate precoated with a monoclonal antibody specific for GST (Amersham Pharmacia). After washing, binding of a second antibody specific for the non-GST portion of the fusion protein is detected using an appropriate enzyme-linked anti-immunoglobulin.

Where necessary, the structure of recombinant plasmids can be checked by restriction analysis of miniprep plasmid DNA. Digestion with the enzymes AvaII, HincII, or XmnI yields DNA fragments containing the insert together with 1000, 1400, or 880 nucleotides, respectively, of the flanking region sequence, with the exact size depending on the choice of parental pGEX vector. Alternatively, oligonucleotide primers flanking the multiple cloning site can be used to check the presence of an insert of the expected size by PCR or by nucleotide sequence analysis. Suitable primers are 5'-GCATGGCCTTTGCAGGG-3' located about 60 nucleotides upstream of the multiple cloning site or 5'-GTCTCCGGGAGCTGCAT-3' located about 90 nucleotides downstream. GenBank accession numbers of pGEX vectors are M21676 (pGEX-1), M97937 (pGEX5G/LIC), U13849-58 (pGEX-1λT, -2T, -2TK, -3X, -4T-1, -2, and -3, 5X-1, -2, and -3), U78872-4 (pGEX-6P-1, -2, and -3), AB014641 (pGEX-pUC-3T), and AJ223813 [pGHX(-)].

In cases where it proves impossible to isolate subclones of a particular DNA fragment, it may be that expression of the fusion protein is toxic to E. coli. Although expression from the tac promoter is efficiently repressed by the overexpressed lacI^q allele encoded by the pGEX vectors, some residual expression is sometimes observed in the absence of induction.[53] This background expression can be reduced by plating cells on media containing 0.2–2% glucose,[49,53] suggesting that it results from readthrough

---

[53] M. Saluta and P. A. Bell, Life Sci. News 1, 15 (1998).

from the *lacZ* promoter that follows the *lacI*[q] gene.[13] Alternatively, the fusion protein could be expressed using one of the pET-based vectors (pGEX-HTG,[30] pETGEXCT,[29] pALEX[27]) in which expression is dependent on inducible expression of the T7 RNA.

Purification of Fusion Proteins

No single method can be given for the purification of GST fusion proteins, as each fusion protein will have slightly different properties. Therefore, in all cases the first step is to investigate how that particular fusion protein is expressed.

1. Pick a single colony into 10 ml L broth containing 100 μg ml⁻¹ of ampicillin and grow overnight in a shaking incubator at 37°.

2. Add 1 ml of the overnight culture to 10 ml fresh broth and grow for 1 hr before adding IPTG to 0.1 m*M*. Remove a 10-μl sample and a 1.5-ml sample. Pellet cells from the 1.5-ml sample and store at −20°.

3. Grow for a further 5 hr, sampling every hour.

4. Analyze the 10-μl aliquots by SDS–PAGE in order to characterize the time course of fusion protein expression.

5. Using the cell pellet from the time point with peak fusion protein expression, repeat the small-scale purification described in steps 7–9 above, analyzing equivalent quantities of the cell pellet, insoluble material, soluble material, unwashed beads after absorption, and washed beads.

The simplest case is if the fusion protein is efficiently removed from the supernatant after incubation with glutathione-agarose beads and can be recovered from the washed beads. In this case the fusion is soluble and stable and can be purified by scaling up the small-scale purification.

1. Dilute 100 ml of an overnight culture into 1 liter of fresh L broth containing 100 μg ml⁻¹ of ampicillin. Grow and induce according to time course results.

2. Pellet cells by centrifugation at 5000g and resuspend in 20 ml ice-cold PBS containing reducing agents such as 1–5 m*M* dithiothreitol (DTT) or 0.1% 2-mercaptoethanol (J. Goding, personal communication).

3. Sonicate on ice using a 5-mm-diameter probe tip. Adjust power setting so that the suspension does not froth and the solution changes to a dull gray color within about 5 min. Lysis occurs very readily with cells transformed with the parental pGEX vectors, but may take much longer for some fusion proteins. More efficient lysis may be obtained by pretreating cells with lysozyme and/or by including a number of freeze/thaw cycles prior to sonication.

4. Add Triton X-100 to 1% and centrifuge at 10,000g for 5 min at 4°. Add the supernatant to a 50-ml tube and add 1 ml of preswollen 50% glutathione-agarose beads and, inverting occasionally, incubate at 4° for up to 30 min.

5. Collect beads by centrifugation at 500g for 30 sec and wash three times with 50 ml ice-cold PBS.

6. Elute fusion protein by incubating beads with an equal volume of freshly made 50 mM Tris–HCl (pH 8) containing 10 mM reduced glutathione at room temperature for 5 min with occasional gentle agitation.

7. Remove the supernatant obtained after centrifugation at 500g for 30 sec and store aliquots at −80° after adding glycerol to 10% (v/v).

Problems with Expression and Purification

Reported yields of GST fusion proteins vary from 0.1 to 50 mg liter^{-1} with a median of 3.5. Where the yield obtained is insufficient the level of expression and the efficiency of recovery can be improved by several means.

Low Level of Expression

The level at which a particular fusion protein is expressed can vary greatly between different *E. coli* strains[18,26,54] and so it is often useful to compare expression in a variety of host cells. Expression can also be enhanced by using an enriched growth medium[54] or by resuspending cells in fresh medium prior to induction (K. Johnson, personal communication). It may also be worthwhile investigating the effect of varying the period and timing of induction or using higher concentrations of IPTG.

Instability

When a fusion protein is unstable, a ladder of breakdown products may be observed, often including a prominent 26- to 28-kDa product corresponding to GST. The stability of fusions during cell growth can be increased by reducing the temperature of cell growth or by delaying and shortening the induction period.[53] Another approach it to express fusions in protease-deficient *E. coli* strains such as BL21,[53] although the yield and stability can also differ between different nonmutant strains.[18,55] Fusion proteins may

[54] G. Y. Lee, J. Zhu, L. Yu, and C. Yu, *Biochim. Biophys. Acta* **1363**, 35 (1998).
[55] D. J. Kemp, D. B. Smith, S. J. Foote, N. Samaras, and M. G. Peterson, *Proc. Natl. Acad. Sci. U.S.A.* **86**, 2423 (1989).

also become unstable during purification, and in these cases it may be possible to increase yields by ensuring that all lysis solutions remain ice cold during sonication. Lysis buffers can also be supplemented with protease inhibitors, such as 1 mM EDTA, 1 mM phenylmethylsulfonyl fluoride, 10 μg ml^{-1} leupeptin or aprotinin, 5 mM benzamidine, or 15 mM trethanolamine, or by the addition of 1–2% sarkosyl.[49] An alternative strategy for eliminating fusion protein degradation products is to use the vectors pGEX-His-2,[23] pALEX,[27] or pGEX-ST[28] that encode a second affinity tag at the COOH terminus.

Insolubility

Low yields of fusion proteins will be observed if the fusion is largely insoluble, and in this case a variety of alternative approaches can be followed. Solubility may differ between different *E. coli* strains, whereas some fusion proteins become soluble when the induced cultures are grown for 10–15 hr at temperatures of 10–25° or if cells are resuspended in a buffer other than PBS. Excessive sonication can lead to insolubility and better results can sometimes be achieved if lysis is instead achieved using a sonication bath,[49] cycles of freeze/thaw, or a French press. If the fusion protein remains insoluble under these conditions, then cells can be treated before lysis with an empirically determined concentration of sarkosyl in the presence of 1 mM EDTA, followed by the addition of a molar equivalent of Triton X-100 prior to affinity purification in order to sequester the sarkosyl.[49] Similar results have been reported by adding the nonionic detergent palmityl sulfobetaine (Zwittergen) to 2%.[56] Other ways of increasing the proportion of fusion protein that is soluble are to grow cells in the presence of 2.5 mM betaine and 440 mM sorbitol[57] or to use thioredoxin reductase mutant cells in which the formation of disulfide bond formation is enhanced.[58] If none of these measures work it may be necessary to abandon attempts to recover the fusion protein under nondenaturing conditions. Persistently insoluble fusion proteins can be purified from the insoluble pellet after resolubilization in 0.5% sarkosyl, 8 M urea, and/or 1% SDS and recovered by size fractionation, after electrophoresis by SDS–PAGE. Fusions containing a hexa-histidine tag can be purified by Ni^{2+} affinity chromatography, although the GST moiety would then be redundant.

[56] H. F. Seow, M. J. Mucha, L. Hurst, J. S. Rothel, and P. R. Wood, *Vet. Immunol. Immunpathol.* **56,** 107 (1997).

[57] L. Yu, K. Deng, and C. Yu, *J. Biol. Chem.* **270,** 25634 (1995).

[58] Y. Bobovnikova, P. N. Graves, H. Vlase, and T. F. Davies, *Endocrinology* **138,** 588 (1997).

Inefficient Affinity Purification

The efficiency with which fusion proteins bind to glutathione-agarose can be reduced if cells are subjected to excessive sonication.[49] Evidence also shows that large fusion proteins bind less efficiently than smaller ones,[13,49] but the generality of these observations is unknown; ideally, such comparisons should be made between fusions containing different numbers of repeats of the same polypeptide domain. This problem can be circumvented by simply increasing the amount of glutathione-agarose. Binding may also be enhanced by including 5 mM DTT or 2-mercaptoethanol in all storage and binding buffers so that the glutathione remains in the reduced state. One report suggests that the binding capacity of glutathione-agarose differs between different commercial products.[49]

Inefficient Elution from Glutathione-Agarose

Where the elution of fusion-bound fusion protein is inefficient, even using freshly prepared elution buffer containing 5 mM DTT, it may be possible to increase yields by carrying out the protease cleavage *in situ* while the fusion protein is bound to the glutathione-agarose[38] or by supplementing the elution buffer with small quantities of detergents such as 0.1% SDS,[55] 0.1% Triton X-100, 2% N-ocytlglucoside,[49] Tween 1%, or 0.2% NP-40.[59]

Copurification with E. coli Proteins

In some instances, material eluted from glutathione-agarose contains an additional protein species of about 60 kDa that is unrelated to the GST fusion protein. This corresponds to one of the *E. coli* heat-shock-induced chaperonins and this molecule can be removed by treatment with 5 mM MgCl$_2$ and 5 mM ATP followed by affinity purification on glutathione-agarose.[60]

Protease Cleavage of Fusion Proteins

Most pGEX vectors encode GST fusion proteins that include a polypeptide sequence recognized by a site-specific protease so that the GST moiety can be removed and the foreign polypeptide studied in its own right. How-

[59] L. Seroude and D. L. Cribbs, *Nucleic Acids Res.* **22,** 4356 (1994).
[60] Z. Keresztessy, J. Hughes, L. Kiss, and M. Hughes, *Biochem. J.* **314,** 41 (1996).

ever, many studies in which proteins have been expressed using the pGEX vectors have failed to find significant differences in antigenicity, enzyme activity, or binding properties between the GST fusion protein and the cleaved product. Because cleavage of fusion proteins is not always efficient or accurate, it may be prudent only to use proteolysis when the intact fusion protein proves unsatisfactory.

The most straightforward way of carrying out the cleavage reaction is to add protease to fusion protein while it is absorbed to glutathione-agarose beads (step 5 in the purification protocol). Because the efficiency of cleavage varies greatly for different fusion proteins, it is important to determine the optimal ratio of protease to fusion protein over the range 0.001–0.05 (w/w). Details of the cleavage reaction vary for different proteases.

Thrombin

Thrombin cleavage can be carried out at temperatures between 20 and 37° for 0.3–16 hr in 50 mM Tris–HCl (pH 8), 10–150 mM NaCl, 2.5 mM CaCl$_2$. Nonspecific lysis by bacterial proteases during prolonged incubations can be reduced by including the protease inhibitor aprotinin.[17] A variety of modified pGEX vectors have been constructed (pGEX-KT, pGEX-5G/LIC, pGEX-KG, pGEX-His-2, or pGSTag) in which the efficiency of thrombin cleavage is enhanced because of the presence of an adjacent stretch of polyglycine. Thrombin can be removed from the cleaved product by affinity purification on *p*-aminobenzamidine agarose [1 hr at 20° in 50 mM Tris–HCl (pH 8), 0.5 M NaCl].[57,61]

Factor Xa

Although factor Xa has been less popular than thrombin for cleaving GST fusion proteins, perhaps because cleavage requires longer incubation periods (up to 18 hr) and higher ratios of enzyme to fusion protein are necessary (0.005–0.1), there are several examples of its successful use. Cleavage can be carried out at temperatures ranging from 4 to 25° in the same cleavage buffer as used for thrombin.

PreScission

This protease comprises the 3C protease of human rhinovirus fused at its NH$_2$ terminus to GST (Amersham Pharmacia) and is used in conjunction

[61] M. G. Kazanietz, J. J. Barchi, J. G. Omichinski, and P. M. Blumberg, *J. Biol. Chem.* **270,** 14679 (1995).

with the vectors pGEX-6P-1, -2, and -3. Cleavage occurs at the optimum temperature of 5° (range 5–15°) in 50 mM Tris–HCl (pH 7.0), 150 mM NaCl, 1 mM EDTA, and 1 mM DTT. Because the protease is itself a GST fusion protein, both it and the cleaved GST will be retained on gluthatione-agarose while the cleaved product is released into the aqueous phase. How well this new system will perform has yet to be established.

Enterokinase

Vectors pGEX-6T and pGEX-His-2 both contain peptide sequences recognized by enterokinase. The efficiency of cleavage is little affected by the amino acid residue following the cleavage site,[41] although cleavage in 50 mM phosphate buffer appears to be relatively nonspecific.[23]

Applications of GST Fusion Proteins

Most of the publications that cite pGEX vectors have used GST fusion proteins either as antigens for immunological or vaccination studies or as a way of producing functionally active enzymes for biochemical or structural studies. A particularly fruitful application has been the use of GST fusion proteins in the analysis of protein–DNA and protein–protein interactions. Typically, a GST fusion protein acts as bait to purify specifically interacting proteins that are then isolated by affinity purification with glutathione-agarose beads. A related application has been to use the pGEX plasmids to construct a cDNA expression library, which was then probed with DNA or protein probes in order to identify novel DNA-binding or protein-binding domains.[19] Larger cDNA expression libraries can be constructed using the bacteriophage expression vector λGEX5.[36] Another important advance has been the development of a variety of eukaryotic versions of the pGEX vectors that allow proteins to be produced outside of the reducing environment of the bacterial cell so that proteins can fold in a more native fashion, as well as permit posttranslational modifications such as glycosylation and phosphorylation.

The pGEX vectors and GST fusion proteins have become a basic tool of the molecular biologist with more than 4000 citations to the original paper. However, GST was chosen as the fusion partner only by chance, and it would be surprising if a systematic search of bacterial and eukaryotic proteins did not produce even better fusion partners.

[18] Use of the *Strep*-Tag and Streptavidin for Detection and Purification of Recombinant Proteins

By ARNE SKERRA and THOMAS G. M. SCHMIDT

Introduction

Today the sequencing of entire microbial genomes is no longer a formidable challenge and even the human genome will be elucidated in a foreseeable time. Understanding the so-called proteome is thus becoming recognized as the next major biochemical task. Consequently, methods are needed that permit the isolation of diverse gene products in their functional state, under standardized conditions, using inexpensive and simple but reliable procedures. The *Strep*-tag method[1] provides a promising solution for this demand.

The *Strep*-tag constitutes a nine amino acid peptide with the sequence Ala-Trp-Arg-His-Pro-Gln-Phe-Gly-Gly, which can be fused easily to a recombinant protein during subcloning of its cDNA or gene. This peptide stretch confers reversible binding activity toward the well-known protein reagent streptavidin. Hence, it can be employed for the one-step purification of a corresponding fusion protein via streptavidin affinity chromatography. In addition, the *Strep*-tag can be used for detection, on Western blots or ELISAs, using streptavidin-enzyme conjugates.

The *Strep*-tag was originally developed as a generic affinity tag for the rapid isolation of bacterially produced antibody fragments[1] without the need for a specific antigen affinity column. Its amino acid sequence was selected in a specialized filter sandwich colony screen from a plasmid-encoded library of random peptides, which were displayed at the C terminus of an immunoglobulin V_H domain as part of a recombinant antilysozyme F_V fragment. The F_V fragment was secreted in *Escherichia coli,* released from the colonies, and captured to an antigen-coated membrane so that the attached peptides could be probed for binding activity with a streptavidin-alkaline phosphatase conjugate. In repeated rounds of screening the *Strep*-tag was identified as a nonhydrophobic amino acid sequence with sufficient intrinsic affinity toward streptavidin, as finally judged according to its practical performance in detection and purification experiments.

Notably, the *Strep*-tag has the property of binding to streptavidin competitively with biotin, the natural ligand of this protein. This behavior

[1] T. G. M. Schmidt and A. Skerra, *Protein Eng.* **6,** 109 (1993).

permits the use of very gentle conditions for the elution of a bound *Strep*-tag fusion protein from the streptavidin affinity column, just by applying a diluted solution of biotin or one of its chemical derivatives. Thus, the *Strep*-tag enabled the purification of a fully functional F_v fragment when attached to the V_H domain,[1] even though these small antibody fragments are known for their fragility due to the weak association between V_H and V_L.

In subsequent studies it was demonstrated that the *Strep*-tag cannot only be used for the purification of F_v fragments (when either fused to V_H or V_L), but also in a totally different protein context.[2] As a first step toward generic application, a standardized purification protocol for recombinant proteins produced in *E. coli* was established. For this purpose the heterologous production of a well-defined truncated version of streptavidin—comprising residues Glu^{14} to Ser^{139} plus an N-terminal methionine—turned out to be critical.[2] Furthermore, with growing use of the *Strep*-tag the question arose whether this module can be displayed functionally in other places of a recombinant protein than just at the C terminus. Depending on the application, it might be desirable to fix the tag to the N terminus or even internally in a polypeptide chain, e.g., in case of fusion proteins that are composed of different domains. Therefore, the *Strep*-tag II with the modified sequence Asn-Trp-Ser-His-Pro-Gln-Phe-Glu-Lys was developed as a better suitable variant,[3] although with slightly diminished streptavidin-binding activity. In its sequence the penultimate Glu residue functionally substituted the free terminal carboxylate group of the *Strep*-tag, which on complex formation participates in a salt bridge with streptavidin and is hence critical for binding. This peculiar interaction was confirmed by crystallographic analysis of the complexes between recombinant "core" streptavidin and the two corresponding peptides.[3]

The structural investigation not only provided insight into the mechanism of molecular mimicry between the *Strep*-tag and biotin, but it also permitted the improvement of the artificial peptide–protein interaction by engineering streptavidin itself. Following targeted random mutagenesis of a flexible loop adjoining the binding site, streptavidin mutants could be selected with enhanced affinity both for the *Strep*-tag and for the *Strep*-tag II.[4] Streptavidin mutant No. 1, which had the amino acid sequence of the residues 44 to 47 changed from Glu-Ser-Ala-Val to Val-Thr-Ala-Arg, exhibited significantly improved performance, especially in the purification of *Strep*-tag II fusion proteins, when coupled to a chromatographic support.[4] The mutant streptavidin was therefore later termed StrepTactin.

[2] T. G. M. Schmidt and A. Skerra, *J. Chromatogr. A* **676**, 337 (1994).
[3] T. G. M. Schmidt, J. Koepke, R. Frank, and A. Skerra, *J. Mol. Biol.* **255**, 753 (1996).
[4] S. Voss and A. Skerra, *Protein Eng.* **10**, 975 (1997).

Generally, recombinant fusion proteins carrying either the *Strep*-tag or the *Strep*-tag II get more tightly bound to such StrepTactin affinity columns. These columns can be prepared with matrices that are amenable to HPLC applications.[5] StrepTactin proved to be particularly helpful in critical situations, e.g., when sticky components in a cell extract necessitate extended washing or when the affinity tag is partially obscured by some part of the recombinant polypeptide chain. In addition, novel assay formats are now enabled. For example, StrepTactin-coated ELISA plates can be used for the functional capturing and subsequent detection of *Strep*-tag (II) fusion proteins.[4]

Thus, after a series of biomolecular engineering steps, we present a technique that has become a simple and powerful tool for the purification and detection of proteins in a reproducible manner (Fig. 1). The following sections describe the current methodology for the application of the *Strep*-tag. Most of the procedures have been tested for a variety of recombinant gene products and constitute standard protocols. First, we outline the aspects of bacterial production strategies for *Strep*-tag fusion proteins. We describe protocols for expression cloning in *E. coli,* synthesis, and purification via StrepTactin affinity chromatography. Then we deal with the use of the *Strep*-tag (II) for the purpose of detection. Finally, we summarize and discuss some illustrative *Strep*-tag (II) applications.

Production of *Strep*-Tag Fusion Proteins in *E. coli*

General Considerations

Most of the experience in the production of *Strep*-tag fusion proteins has so far been gathered with *E. coli* as a host organism. Due to its facile and rapid genetic manipulation and its inexpensive handling, even at the fermenter scale, *E. coli* represents an ideal expression host, especially for highly parallelled use. As discussed further later, this gram-negative bacterium offers two general routes for expression: (i) intracellularly in the cytosol and (ii) secretory in the periplasmic space. For both of these strategies, but especially in the case of secretion, it must be considered that the production of a heterologous protein can be accompanied by a toxic effect on the bacterium, which may impair cell growth and plasmid stability.

Consequently, it is recommended to stringently regulate the biosynthesis of the recombinant gene product. When the foreign protein is cytotoxic, even the production of minute quantities can result in a drastic selection

---

[5] A. Skerra and T. G. M. Schmidt, *Biomol. Eng.* **16,** 79 (1999).

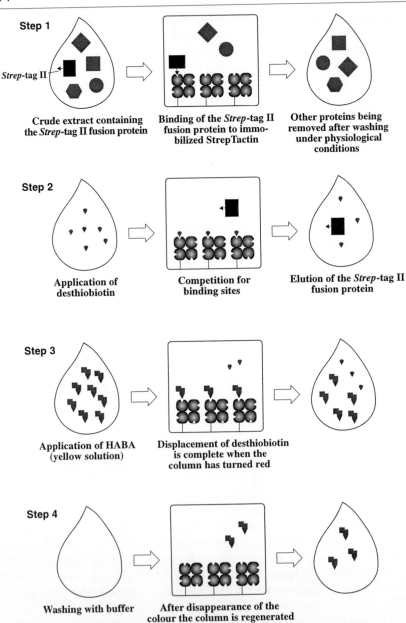

Step 1

Strep-tag II

Crude extract containing
the *Strep*-tag II fusion protein

Binding of the *Strep*-tag II
fusion protein to immo-
bilized StrepTactin

Other proteins being
removed after washing
under physiological
conditions

Step 2

Application of
desthiobiotin

Competition for
binding sites

Elution of the *Strep*-tag II
fusion protein

Step 3

Application of HABA
(yellow solution)

Displacement of desthiobiotin
is complete when the
column has turned red

Step 4

Washing with buffer

After disappearance of the
colour the column is regenerated

FIG. 1. Generic purification scheme for *Strep*-tag (II) fusion proteins.

TABLE I
GENERAL FEATURES OF pASK-IBA VECTORS CARRYING THE *tet*$^{p/o}$

Plasmid	*Strep*-tag II	Secretion	Cleavage
pASK-IBA2	C-terminal	Yes	No
pASK-IBA3	C-terminal	No	No
pASK-IBA4	N-terminal	Yes	No
pASK-IBA5	N-terminal	No	No
pASK-IBA6	N-terminal	Yes	Yes
pASK-IBA7	N-terminal	No	Yes

against those *E. coli* cells that harbor a functional expression plasmid. In such cases, tight repression of the promoter is required. Our preferred solution for inducible expression is the recently developed tetracycline promoter system,[6] which has been implemented in all current *Strep*-tag II expression vectors (Table I). The strength of the *tet*A promoter is comparable to that of the *lac*UV5 promoter. It can be fully induced by the addition of anhydrotetracycline at a concentration that is not antibiotically effective.[6] The constitutive expression of the *tet* repressor gene, which is also encoded on the expression vectors (Fig. 2), guarantees tight repression of this system in the absence of inducer.[6–8]

The *lac* promoter, for comparison, is susceptible to catabolite repression (i.e., it is dependent on the cAMP level and metabolic state of the bacterial cell) and depends strongly on the number of repressor molecules, which may vary due to the presence of chromosomally and/or plasmid-encoded gene copies. In contrast, the *tet*A promoter/operator is not coupled to ordinary cellular regulation mechanisms. Therefore, when using the *tet* system, no principal restriction exists regarding the choice of the culture medium or of a particular expression strain.[6] The *tet* promoter system proved to be useful for the production of recombinant proteins both in the shaker flask with a conventional rich broth and at the fermenter scale using synthetic glucose minimal medium.[9,10]

Cytosolic versus Periplasmic Expression

Generally, a periplasmic secretion strategy is necessary for the functional production of proteins possessing structural disulfide bonds.[11,12] This is often

[6] A. Skerra, *Gene* **151**, 131 (1994).
[7] H. Loferer, M. Hammar, and S. Normark, *Mol. Microbiol.* **26**, 11 (1997).
[8] M. T. Korpela, J. S. Kurittu, J. T. Karvinen, and M. T. Karp, *Anal. Chem.* **70**, 4457 (1998).
[9] W. Schiweck and A. Skerra, *Proteins Struct. Funct. Genet.* **23**, 561 (1995).
[10] C. Bandtlow, W. Schiweck, H.-H. Tai, M. E. Schwab, and A. Skerra, *Eur. J. Biochem.* **241**, 468 (1996).
[11] A. Skerra and A. Plückthun, *Science* **240**, 1038 (1988).
[12] R. Glockshuber, T. Schmidt, and A. Plückthun, *Biochemistry* **31**, 1270 (1992).

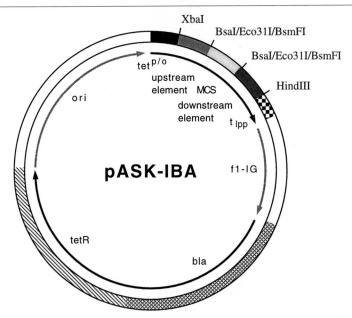

FIG. 2. Structure of pASK-IBA expression vectors. MCS, f1-IG, bla, ori, t$_{lpp}$, and *tet*R denote the multiple cloning site, intergenic region of phage f1, β-lactamase gene (ampicillin resistance), origin of replication from the pUC family of plasmids, *lpp* terminator of transcription, and *tet* repressor gene, respectively. The expression cassette is under transcriptional control of the *tet*A promoter/operator. The *tet* repressor gene has been placed as a second cistron immediately behind the β-lactamase gene and is thus under transcriptional control of its constitutive promoter. Cutting the vector with the type IIS restriction enzymes *Bsa*I, *Eco*31I, or *Bsm*FI leads to the generation of defined 5' overhangs immediately adjacent to the upstream and downstream elements, respectively, which are not mutually compatible and can therefore not religate. Upstream elements may comprise DNA sequences encoding the OmpA signal peptide, an initiator methionine, the *Strep*-tag II, and/or the factor X$_a$ recognition sequence, whereas downstream elements may be the *Strep*-tag II or just a stop codon (cf. Fig. 3). The gene to be inserted should be equipped with compatible overhangs, e.g., via PCR (see Fig. 3 for primer design).

the case for eukaryotic secretory biomolecules. The reducing conditions in the cytosol of *E. coli* do not allow efficient disulfide bond formation so that either aggregation and deposition as inclusion bodies or degradation of the unfolded polypeptide is observed.

Secretion of a recombinant protein into the periplasm of *E. coli* can be effected by N-terminal fusion with a bacterial leader peptide, which becomes selectively cleaved off after membrane translocation by the *E. coli* signal peptidase I. Our vectors encode the OmpA signal peptide,[13] but other

---

[13] N. R. Movva, K. Nakamura, and M. Inouye, *J. Biol. Chem.* **255,** 27 (1980).

bacterial leader sequences are equally established. Periplasmic secretion has an additional advantage insofar as the recombinant protein is separated from host cell proteases, which are mostly resident in the bacterial cytosol. Furthermore, because the *E. coli* outer membrane can be selectively disintegrated by mild treatment with EDTA, lysozyme, and so on (see later), the cytosolic components are easily removed by centrifugation as a whole, together with cell wall and membrane constituents. Finally, the periplasmic space is accessible for molecules with masses up to 600 Da so that folding or stability of the recombinant protein can be influenced during expression[14] by adding appropriate substances, e.g., thiol compounds[15] or nonmetabolizable sugars,[16] to the culture medium.

Protein secretion is, however, often accompanied by aggregate formation in the periplasm. The yield of correctly folded, soluble protein can then be raised by lowering the growth temperature of the cells.[17] Periplasmic secretion on a preparative scale is not necessarily restricted to proteins of secretory nature. The strategy can also be extended to cytosolic proteins or enzymes, so long as they do not contain too many cysteine residues or stop-transfer sequences.[18]

Concerning the efficient production of nonsecretory proteins in the general case, however, cytosolic expression is likely to be the strategy of choice. The yield of recombinant protein in the bacterial cell is usually higher, although quite a large fraction of it may be deposited as inclusion bodies. The soluble part that is liberated on breaking up the cells constitutes in most cases functional and correctly folded protein. Solubilization of aggregates and refolding can raise the yields, but such steps are laborious and need optimization for each particular protein. Therefore, especially with respect to standardized and parallelled protein production, it is clearly more feasible to purify just that fraction of the protein that is soluble. Fortunately, the formation of inclusion bodies can again be diminished and the yield of directly purifiable protein frequently raised by lowering the growth temperature of *E. coli*.[19] Therefore, the culturing of *E. coli* for expression purposes at temperatures below 37° is usually good strategy.

Despite the higher content of recombinant protein within the cell under conditions of cytosolic expression, the amount of contaminating host cell proteins is certainly higher and, even more importantly, their composition is significantly more complex. Nevertheless, *Strep*-tag purification works efficiently with crude *E. coli* extracts so that bioactive proteins are normally

[14] C. Wülfing and A. Plückthun, *Mol. Microbiol.* **12,** 685 (1994).
[15] H. N. Müller and A. Skerra, *J. Mol. Biol.* **230,** 725 (1993).
[16] G. A. Bowden and G. Georgiou, *Biotechnol. Progr.* **4,** 97 (1988).
[17] A. Skerra and A. Plückthun, *Protein Eng.* **4,** 971 (1991).
[18] T. Tudyka and A. Skerra, *Protein Sci.* **6,** 2180 (1997).
[19] C. H. Schein, *Bio/Technology* **7,** 1141 (1989).

isolated from the cytosol in high purity and yield. The low nonspecific binding activity of the streptavidin or StrepTactin affinity matrix is of particular advantage in this regard.

There is just one aspect that should be considered when working up whole cell extracts of *E. coli* or other host organisms. The cytosol contains biotinylated proteins that bind strongly to streptavidin. *Escherichia coli,* for example, possesses the biotin carboxyl carrier protein (BCCP, 22.5 kDa)[20] as part of the fatty acid synthesis machinery. Cross-reaction with such proteins can be avoided, both in purification and detection applications, by masking the biotin groups with avidin from hen egg white. Similarly as the bacterial protein streptavidin, avidin forms a kinetically stable complex with biotin, but contrasting, it does not recognize the *Strep*-tag.[1]

Strep-Tag II Expression Vectors and Cloning Strategy

Initially we constructed *E. coli* expression vectors that encoded the *Strep*-tag[1] and carried either the *lac* promoter[1] or, more recently, the *tet* promoter.[4,6] This section describes a series of current *tet* promoter vectors, which were developed for producing both N- and C-terminal fusion proteins with the *Strep*-tag II. In addition, a choice is offered between producing the recombinant protein directly in the cytosol or, via the vector-encoded OmpA signal sequence, in the periplasm of *E. coli*. Still another route, which finally leads to the production of authentic protein (i.e., without the tag), is possible when using the vectors that encode an N-terminal *Strep*-tag II (with or without OmpA leader) and a subsequent factor X_a cleavage site. Thus, at least six pASK-IBA vectors are available for the bacterial synthesis of *Strep*-tag II fusion proteins (Table I).

All of these pASK-IBA vectors (Fig. 2) share a backbone with the same genetic elements, which were derived from the plasmid pASK75.[6] Their expression cassettes, whose sequences are summarized in Fig. 3, are under transcriptional control of the tightly regulated *tet*A promoter/operator and are flanked by singular *Xba*I and *Hin*dIII restriction sites. Transcription ends at the lipoprotein terminator. The *tet* repressor is encoded on the same vector under constitutive expression of the β-lactamase promoter as an artificial second cistron.[6] This arrangement ensures a balanced stochiometry between repressor molecules in the cell and plasmid copy number.

Efficient initiation of translation of the cloned foreign gene is provoked by a tandem ribosome-binding site (RBS). If present, the OmpA signal sequence mediates secretion of the recombinant protein into the periplas-

[20] M. R. Sutton, R. R. Fall, A. M. Nervi, A. W. Alberts, P. R. Vagelos, and R. A. Bradshaw, *J. Biol. Chem.* **252,** 3934 (1977).

mic space. By action of the signal peptidase, the natural N terminus of the protein should be liberated, as long as the signal sequence is attached precisely and the *Strep*-tag II is fused to the C terminus. Finally, the intergenic region of the bacteriophage f1 facilitates the preparation of single-stranded DNA for site-directed mutagenesis.[21,22]

Expression cloning of foreign genes can be achieved in a standardized fashion by means of type IIS restriction enzymes *Eco*31I, *Bsa*I, or *Bsm*FI, which recognize an asymmetric nucleotide sequence and cleave the DNA at a defined distance to one side of it. Cleavage sites for these enzymes are placed in the multiple cloning region in a way that, after digest, the whole segment is cut out and overhangs are generated in the neighboring functional DNA elements (see Fig. 3). These elements include, depending on the chosen vector, an ATG start codon, coding regions for the OmpA signal peptide, as well as for the *Strep*-tag II, and a translational stop codon. The resulting 5' protruding ends are noncompatible with each other so that religation of the cleaved vector is prohibited and oriented gene insertion is possible.

Compatible overhangs can be generated at both ends of the cDNA or gene to be cloned via polymerase chain reaction (PCR) with appropriately hybridizing primers that contribute the necessary nucleotide stretches (see Fig. 3). Again, type IIS restriction sites can be used for cutting the PCR product (*Bsa*I is the enzyme of choice in this case whereas *Eco*31I is recommended for cutting the vector) whereby the recognition sequences are preferentially oriented outward the amplified fragment so that the encoded amino acid sequence is not affected and the recognition sites finally become lost on cleavage. This cloning strategy allows precise one-step fusion of the cDNA or gene with the vector and its functional parts, especially the *Strep*-tag II-encoding sequence.

The reading frame is not interrupted by stop codons throughout the multiple cloning regions of the listed vectors so that a conventional insertion strategy is possible as well. The incorporation of extraneous amino acids into the recombinant protein should, however, be considered in this case. It should furthermore be noted that the reading frame differs between expression vectors pASK-IBA 2, 3, 6, and 7 on one hand and pASK-IBA 4 and 5 on the other. In addition, if the foreign gene is cloned in one of the vectors with the N-terminal *Strep*-tag II, a change from cytosolic to periplasmic expression (or vice versa) can be achieved easily by subcloning using the pair of *Hin*dIII (or *Eco*RV) and *Nhe*I restriction sites.

[21] T. A. Kunkel, J. D. Roberts, and R. A. Zakour, *Methods Enzymol.* **154**, 367 (1987).
[22] J. Geisselsoder, F. Witney, and P. Yuckenberg, *BioTechniques* **5**, 786 (1987).

pASK-IBA2:

```
                        Eco3II  PshAI                                              PshAI Eco3II
XbaI            RBS     BsaI NcoI BsmFI EcoRI SstI KpnI SmaI BamHI XhoI SalI PstI BsmFI NcoI BsaI Eco47III   BstBI        HindIII
TCTAGATAACGAGGGCAAAAAATGAAAAAG......GCGCAagccGGATCCCGAATTCGAGCTCGAGGTGCAACCTGCAGGGGACCATGGTCTCAgcgcTTGAGCCACCCGCAGTTCGAAAAATAATAAGCTT
                        MetLysLys   OmpA         AlaGlnAlaGlySerTrpSerHisProGlnPheGluLysSerLeuVal...AspLeuGlyIleGlySerAla1aTrpSerHisProGlnPheGlyGluLysEnd
                                                                                                      Link              Strep-tag II

Cloning primers: Forward: 5'-nnnnnnGGTCTCNGCGCC(N)20; Reverse: 5'-nnnnnnGGTCTCNCGCGCT(N)17
```

pASK-IBA3:

```
            Eco3II PshAI                                                         PshAI Eco3II
XbaI        BsaI SacII BsmFI EcoRI SstI KpnI SmaI BamHI XhoI SalI PstI BsmFI NcoI BsaI Eco47III   BstBI        HindIII
TCTAGATAACgtGGAGACCGCGGTCCCGAATtCGAGCTCGGTACCCGGGGATCCTCGAGGTGCAACCTGCAGGGGACCATGGTCTCAgcgcTTGAGCCACCCGCAGTTCGAAAAATAATAAGCTT
            MetGlyAspArgGlyProGlnPheGluLysSerLeuVal...AspLeuGlyIleGlySerAla1aTrpSerHisProGlnPheGlyGluLysEnd
                                                                   Link                Strep-tag II

Cloning primers: Forward: 5'-nnnnnnGGTCTCNGAATG(N)17; Reverse: 5'-nnnnnnGGTCTCNCGCGCT(N)20
```

pASK-IBA4:

```
                                                         BbeI
                                                         EheI
                                                         NarI
                                              Eco3II PshAI KasI
XbaI            RBS     NheI                   BsmFI SacII BsmFI EcoRI SstI KpnI SmaI BamHI XhoI SalI PstI BsmFI NcoI BsaI EcoRV    HindIII
TCTAGATAACGAGGGCAAAAAAT......GCGCAgccGGAGCTAGCtGCAGACGCGGTCCCGAATTCGAGCTCGGTACCCGGGGATCCTCGAGGTGCAGATCTGCAGGGGACCATGGTCTCTgatatCTAACTAAGCTT
                MetLysLys  OmpA  AlaGlnAlaIleAlaSerTrpSerHisProGlnPheGluLysGlyArgGlyAlaGluGluThrAlaValProAsnSerSerSerVal...ProGlyIleProArgGlyArgProAlaGlyGlyArgProTrpSerLeuIleSerSerAsnEnd
                            Link              Strep-tag II                                       AspArgGlyProGlnPheGluLysGlyArgGlyAlaGlyAlaAsnGlyThrArgGlySerValGlyValAlaAspLeuGlyIleGlySerAla...AspHisGlyLysEnd

Cloning primers: Forward: 5'-nnnnnnGGTCTCNGTGCC(N)20; Reverse: 5'-nnnnnnGGTCTCNTATCA(N)20
```

pASK-IBA5:

```
                                 BbeI
                                 EheI
                                 NarI
                      Eco3II PshAI KasI
XbaI            RBS   BsmFI SacII BsmFI EcoRI SstI KpnI SmaI BamHI XhoI SalI PstI BsmFI NcoI BsaI EcoRV    HindIII
TCTAGATAACgtgGAGAGCACCCGCAGTTCGAAAAAGGCCGAGCGCGGTCCCGAATTCGAGCTCGGTACCCGGGGATCCTCGAGGTGCAGATCTGCAGGGGACCATGGTCTCTgatatCTAACTAAGCTT
                MetAlaSerTrpSerHisProGlnPheGlyGlyArgGlyAlaGluThrAlaValProAsnSerSerSerVal...ProGlyIleProArgGlyArgProAlaGlyGlyArgProTrpSerLeuIleSerSerAsnEnd
                Link    Strep-tag II                                   AspArgGlyProGlnPheGluLysGlyArgGlyAlaGlyAlaAsnGlyThrArgGlySerValGlyValAlaAspLeuGlyIleGlySerAla...AspHisGlyLysEnd

Cloning primers: Forward: 5'-nnnnnnGGTCTCNGTGCC(N)20; Reverse: 5'-nnnnnnGGTCTCNTATCA(N)20
```

pASK-IBA6:

```
                                                               BbeI
                                                               EheI
                                                               NarI
                                                    Eco3II PshAI KasI
XbaI            RBS     NheI              Eco3II BsmFI SacII BsmFI EcoRI SstI KpnI SmaI BamHI XhoI SalI PstI BsmFI NcoI BsaI EcoRV    HindIII
TCTAGATAACGAGGGCAAAAAAATCGAAGGgcgcCGAGACCGCGGTCCCGAATTCGAGCTCGGTACCCGGGGATCCTCGAGGTGCAGATCTGCAGGGGACCATGGTCTCTgatatCTAACTAAGCTT
                MetLysLys  OmpA  AlaGlnAlaIleAlaSerTrpSerHisProGlnPheGluLysGlyArgGlyAlaGluThrAlaValProAsnSerSerSerVal...ProGlyIleProArgGlyArgProAlaGlyGlyArgProTrpSerLeuIleSerSerAsnEnd
                            Link              Strep-tag II                      GluThrAlaValProAsnSerSerSerVal...ProGlyIleProArgGlyArgProAlaGlyGlyArgProTrpSerLeuIleSerSerAsnEnd
                                          Factor Xa                            ArgProArgSerArgIleAsnSerSerSerVal...ProGlyIleProArgGlyArgProAlaGlyGlyArgProTrpSerLeuIleSerSerAsnEnd

Cloning primers: Forward: 5'-nnnnnnGGTCTCNATATCA(N)20; Reverse: 5'-nnnnnnGGTCTCNTATCA(N)20
```

pASK-IBA7:

```
                                 BbeI
                                 EheI
                                 NarI
                      Eco3II PshAI KasI
XbaI            RBS   NheI  BsmFI SacII BsmFI EcoRI SstI KpnI SmaI BamHI XhoI SalI PstI BsmFI NcoI BsaI EcoRV    HindIII
TCTAGATAACgtGAGGGCAAAAAAATCGAAGGgcgcCGAGACCGCGGTCCCGAATTCGAGCTCGGTACCCGGGGATCCTCGAGGTGCAGATCTGCAGGGGACCATGGTCTCTgatatCTAACTAAGCTT
                MetAlaSerTrpSerHisProGlnPheGlyGlyArgGlyAlaSerArgArgGluGlyArgAspArgArgGlnProTrpIleGluGlyGlyAlaGluThrAlaValProAsnSerSerSerVal...AspLeuGlyIleGlySerAla...AspLeuGlyLysEnd
                Link    Strep-tag II     Factor Xa                    GluThrAlaValProAsnSerSerSerMetValSerArgIleAspEnd
                                                                      ArgProArgSerArgIleAsnSerSerSerVal...ProGlyIleProArgGlyArgProAlaGlyGlyArgProTrpSerLeuIleSerSerAsnEnd

Cloning primers: Forward: 5'-nnnnnnGGTCTCNGCGCC(N)20; Reverse: 5'-nnnnnnGGTCTCNTATCA(N)20
```

Cell Growth, Gene Expression, and Preparation of Bacterial Protein Extract

Host Strains

With the *tet* promoter system, recombinant gene expression is basically independent of the chosen *E. coli* host strain, although differences in the protein secretion and/or folding behavior may be observed. JM83[23] is our preferred strain for periplasmic secretion, whereas BL21[24] gives rise to good levels of soluble protein in the case of cytosolic synthesis. A series of other strains have been tested successfully for recombinant gene expression with the *tet* promoter[6] and to date there are no *E. coli* hosts known to be incompatible with this system.

Culture and Cell Harvest

Procedure. Two milliliters of LB medium[25] containing 100 μg/ml ampicillin (LB/Amp) is inoculated with a fresh colony of the *E. coli* strain harboring the expression plasmid and shaken overnight at 37° (200 rpm).

[23] C. Yanisch-Perron, J. Vieira, and J. Messing, *Gene* **33**, 103 (1985).
[24] F. W. Studier and B. A. Moffatt, *J. Mol. Biol.* **189**, 113 (1986).
[25] J. Sambrook, E. F. Fritsch, and T. Maniatis, "Molecular Cloning: A Laboratory Manual," 2nd Ed. Cold Spring Harbor Laboratory Press, Cold Spring Harbor, NY, 1989.

FIG. 3. Expression cassettes on pASK-IBA vectors. Differences between individual pASK-IBA plasmids are confined to the region between *Xba*I and *Hin*dIII restriction sites. Corresponding sequences are shown for each plasmid. The 5'-protruding ends, which are generated after cutting with *Bsa*I, *Eco*31I, or *Bsm*FI, are indicated with lowercase characters. Generic forward and reverse PCR primer sequences are proposed for every vector in order to obtain appropriate boundaries in the amplified sequence. It is recommended to use at least 20 hybridizing nucleotides, $(N)_{20}$, at each of the 5' or 3' ends, respectively. The forward primer for pASK-IBA3 already includes the ATG start codon so that the hybridizing sequence with the target gene may be reduced. User-defined DNA bases at the 5' end of each primer are denoted with "n". In cases where other restriction sites might be chosen for cloning, care must be taken in correctly fusing the reading frame of the structural gene with the reading frames of the vector-encoded elements. In those vectors with an N-terminal *Strep*-tag II (pASK-IBA4 to pASK-IBA7), the tag is followed by the linker sequence 5'-GGCGCC. This sequence is recognized by four different restriction enzymes (*Kas*I, *Nar*I, *Ehe*I, *Bbe*I). Cleavage with a suitable enzyme and subsequent filling in enables the production of a blunt end in every reading frame. Therefore, any restriction site that may be present at the 5' end of an open reading frame can be used for fusing a target sequence with the *Strep*-tag II by conventional cloning. Stop codons are present on the vectors in all reading frames so that any restriction site within the multiple cloning region can be used for ligation with the 3' end of the gene.

The preculture is transferred to 100 ml LB/Amp medium in a 250-ml flask and is incubated again with agitation (200 rpm). The flask is typically thermostatted at 22° for periplasmic secretion or between 22 and 37° for cytosolic expression. The optical density (OD) of the culture is monitored at 550 nm. When an $OD_{550} = 0.5$ is reached, gene expression is induced by adding 10 μl of an anhydrotetracycline solution (aTc; 2 mg/ml in dimethylformamide, DMF) and shaking is continued for 3 hr. The cells are then harvested by centrifugation (4200g, 12 min, 4°). The supernatant is discarded and the remaining medium is carefully pipetted off. Cells are resuspended in 1 ml of a suitable buffer for the preparation of the extract (see later). The procedure can be carried out at a different scale (e.g., with 2 liters of culture in a 5-liter flask) by keeping volume ratios constant.

Comments. Generally, the optimal temperature for growth of *E. coli* is 37°. To keep culture periods short, expression of the recombinant protein at this temperature is often preferred. However, especially in the case of protein secretion in *E. coli*, it is often observed that the foreign gene product predominantly aggregates in the periplasm if induction is performed at 37°. At the same time cell lysis can occur.[26] The yield of soluble, functional protein can often be increased substantially by lowering the growth temperature to between 22° and 30°. It is recommended that the main culture be incubated at this temperature from the time of inoculation. Lowering the growth temperature just on induction may give rise to less satisfactory results.

Likewise, the induction period can be optimized. A suitable time point for harvest is reached at the beginning of a plateau in the growth curve, which typically happens in the case of protein secretion after 2.5 to 3 hr of induction at 22°. In the case of cytosolic expression at 37°, we also recommend harvest after 2.5 to 3 hr of induction, although some recombinant proteins tend to accumulate for a much longer time. The *tet*A promoter is maximally induced at an aTc concentration of 200 to 500 μg/liter culture. In some cases it is advantageous to use partial induction. This can be achieved by applying an aTc concentration between 20 and 100 μg/liter.

Preparation of Periplasmic Extract

Procedure. The bacterial cell pellet is resuspended at 4° in 1 ml of prechilled buffer P [100 mM Tris–Cl (pH 8.0), 500 mM sucrose, 1 mM Na$_2$EDTA], transferred to a 1.5-ml Eppendorf reaction tube, and incubated on ice for 30 min. Under these conditions the outer membrane of the transformed *E. coli* cells is in most cases sufficiently permeabilized in

---

[26] A. Plückthun and A. Skerra, *Methods Enzymol.* **178**, 497 (1989).

order to release the soluble components of the periplasm,[27] including the *Strep*-tag (II) fusion protein. If liberation of the recombinant protein is not satisfactory, varying concentrations of lysozyme (between 50 and 200 μg/ml) can be added during this step in order to degradate the peptido-glycane. The spheroplasts are removed by centrifugation (microfuge, 14,000 rpm, 5 min, 4°). The cleared supernatant is carefully pipetted off as the periplasmic extract and transferred to a fresh tube.

This protein solution is directly ready for StrepTactin affinity chroma-tography. If performed at a larger scale, e.g., 2 liters of culture, the sphero-plasts are conveniently pelleted in two steps. After removing the bulk quickly (bench top centrifuge, 5000 rpm, 10 min, 4°), the slightly turbid supernatant is cleared rigorously in a second centrifugation (27,000g, 15 min, 4°).

For SDS–PAGE or Western blot analysis, 10- to 20-μl samples of the extract are used. In addition, 10 μl of the cell suspension can be removed during the preceding 30-min incubation step on ice in order to analyze the total cell protein. The whole cell sample is mixed thoroughly, up to a volume of 80 μl, with a solution of 12.5 U/ml benzonase[28] (Merck, Darmstadt, Germany) in 100 mM Tris–Cl (pH 8.0), 5 mM MgCl$_2$, and 20 μl of reducing 5× SDS–PAGE sample buffer [7.5% (w/v) SDS, 25% (v/v) glycerol, 0.25 M Tris–Cl (pH 8.0), 12.5% (v/v) 2-mercaptoethanol, 0.25 mg/ml bromophenol blue] is added. After a 1-hr incubation on ice for degrading chromosomal DNA, the sample should be frozen until SDS–PAGE is performed. Samples are heated to 95° for 5 min before application to the gel.

Comments. If the purification of active metalloenzymes is intended, the periplasmic extract can be prepared using polymyxin B sulfate (2 mg/ml) instead of EDTA.[4,29] The EDTA should then be omitted from the chromatographic buffers as well. In rare instances it may happen that a recombinant protein, even though apparently translocated to the periplasm, is not sufficiently liberated by one of these cell fractionation procedures. It is then recommended to prepare the soluble part of the total cell protein by Frech Press homogenization, ultrasonic, or chemical treatment (see later).

The choice of reagents used for extracting the recombinant protein does not seem to be critical for the subsequent *Strep*-tag purification. Both Tris–Cl and phosphate buffers are suitable for example. NaCl should be

---

[27] H. J. P. Marvin, M. B. A. Beest, and B. Witholt, *J. Bacteriol.* **171,** 5262 (1989).

[28] Benzonase is an extracellular dsDNA-, ssDNA-, and RNA-cleaving endonuclease from *Serratia marcescens*. The commercially available recombinant protein is free of detectable protease activities, has high enzymatic activity and stability under widely variable reaction conditions, and tolerates even the presence of EDTA and SDS.

[29] R. A. Dixon and I. Chopra, *Antimicrob. Agents Chemother.* **29,** 781 (1986).

added if DNA-binding proteins are to be purified or if the recombinant protein tends to interact nonspecifically with other cell components. The NaCl concentration can be raised up to 1 M. Furthermore, StrepTactin affinity chromatography in the presence of mild detergents [e.g., 0.05% (w/v) dodecyl maltoside, 0.1% (v/v) Tween 20, 0.1% (v/v) Triton X-100, 0.1% (w/v) CHAPS] or reducing agents (1–5 mM DL-dithiothreitol, DTT) is also possible. However, the pH should be adjusted to a value of 7 or above so that the His side chain in the *Strep*-tag (II) sequence is not substantially protonated. In order to maintain defined chromatography conditions, the extract may be dialyzed against buffer W [100 mM Tris–Cl (pH 8.0), 1 mM Na$_2$EDTA], possibly with the addition of 150 mM NaCl, prior to purification.

Preparation of Cytosolic Extract

Procedure. The bacterial cell sediment from a 100-ml culture (cf. earlier discussion) is resuspended at 4° in 1 ml of prechilled buffer W. For analysis of the total cell protein by SDS–PAGE, 10 μl from this suspension should be removed and treated with benzonase and sample buffer as described previously. The cell suspension is subjected to ultrasonication under ice cooling using a microtip (Labsonic 1510 operated at 100 W, B. Braun Melsungen, Germany) in repeated periods of 1 to several minutes. Lysis is complete after significant reduction of the optical density at 590 nm [% lysis = $(1 - A_{590}^{\text{sonicate}}/A_{590}^{\text{suspension}}) \times 100$]. Alternatively, especially when working with larger suspension volumes, the cells may be disrupted mechanically in a French Pressure homogenizer (SLM Aminco, Urbana, IL). Complete lysis is achieved by repeatedly passing the suspension through the prechilled pressure cell (typically three times at 15,000 psi) until the homogenate flows smoothly out of the throttle. Insoluble components are removed by rigorous centrifugation (microcentrifuge, 14,000 rpm, 15 min, 4°). The supernatant is carefully pipetted off as the total soluble protein extract and transferred to a clean tube. This protein solution is ready for StrepTactin affinity purification of the recombinant protein.

Bacterial cells can also be broken up by chemical treatment. In this case the cell pellet from a 100-ml culture is resuspended in 2 ml buffer X [100 mM Tris–Cl (pH 8.0), 1 mM Na$_2$EDTA, 0.1% (v/v) Tween 20, 100 μg/ml lysozyme] and incubated on ice for 30 min under gentle shaking. Two millimolar MgCl$_2$ and 10 μg/ml DNase I (4370 U/mg) are added from stock solutions (1 M and 10 mg/ml, respectively) and incubation on ice is continued for another 90 min. EDTA can be omitted in the case of metalloproteins and NaCl or DTT may be added if necessary. The suspension becomes viscous, due to cell lysis, whereas the viscosity disappears again when the chromosomal DNA becomes progressively degraded. The lysate is cleared by centrifugation (microfuge, 14,000 rpm, 10 min, 4°) and the supernatant is transferred to a fresh tube. Storage at $-20°$ is recommended until chromatography is carried out.

In order to avoid the strong binding of intracellular biotinylated proteins, such as BCCP, or of free biotin to the streptavidin or StrepTactin affinity matrix, avidin can be added to the protein extract.[1,2] Egg white avidin masks the biotin groups in a stable complex but has no affinity toward the Strep-tag. This precaution is important in case of low expression levels because larger amounts of cell extract may then be applied to the affinity column. The total biotin content of an E. coli-soluble cell extract from a 1-liter culture with $OD_{550} = 1.0$ is about 1 nmol. Twenty microliters of a 2-mg/ml stock solution of avidin in buffer W should be added per 1 ml of the cell extract (derived from a culture with $OD_{550} \approx 1.0$). After incubation on ice for 30 min, an aggregate usually forms, which is removed by centrifugation. The sample is then ready for StrepTactin affinity chromatography or it can be stored at $-20°$.

Purification of Strep-Tag Fusion Proteins by Streptavidin (or StrepTactin) Affinity Chromatography

General Considerations

In the purification of proteins carrying the original Strep-tag, the use of a streptavidin chromatography matrix usually leads to satisfactory results, as long as a suitable preparation of the core streptavidin is employed.[2] However, if extended washing is necessary, if the affinity tag is not optimally exposed, or if the Strep-tag II is utilized, affinity purification with the StrepTactin matrix provides an advantage, concerning both yield and homogeneity of the desired product. The original Strep-tag is still compatible with the novel chromatography material.

The cell extract containing the Strep-tag (II) fusion protein is preferentially prepared in a mild buffer, as described earlier, and directly applied to the column with the immobilized streptavidin or StrepTactin. Streptavidin, as well as its mutant version, has a low tendency for nonspecific binding so that host cell proteins and other contaminants are rapidly removed by washing. Selective elution of the Strep-tag (II) fusion protein is effected by applying biotin at a low concentration in the same buffer that has been used for washing. If repeated use of the column is intended, an appropriate chemical derivative can be utilized instead of biotin.[5] In the case of immobilized StrepTactin, desthiobiotin is thus the preferred ligand for elution.[4]

Consequently, the method benefits from two steps, binding and competitive displacement, which involve biomolecular specificity. This explains the high purification efficiency, which in most cases permits the isolation of homogeneous protein in one step. In addition, elution is effected under gentle conditions. No pH shift, high salt, detergent, chaotropic reagent, or metal-complexing compound is needed. Hence, the *Strep*-tag technique combines the advantage of quite a small and biochemically inert tag with purification conditions that keep the protein in a functional state.

When using biotin derivatives that bind reversibly to the matrix (such as diaminobiotin in the case of streptavidin or desthiobiotin in the case of StrepTactin), the column can be regenerated simply by extended washing. The regeneration is, however, accelerated by applying a buffer containing the organic dye HABA [2(4'-hydroxyazobenzene)benzoic acid],[30] which becomes weakly complexed by the biotin-binding pocket of streptavidin. Its presence in sufficient excess blocks emerging free binding sites and thus prevents rebinding of desthiobiotin so that this compound is removed more efficiently (cf. Fig. 1). Furthermore, as HABA changes its color from yellow to red on complexation, the regeneration process can be followed visually. When the resin attains a red color, desthiobiotin has been replaced and HABA can then be washed out quickly with buffer alone. Having turned pale again the column is ready for the next purification run.

In fact, the columns can be stored in the presence of HABA so that the status of activity is always obvious. The addition of EDTA prevents bacterial growth. When using Sepharose (Pharmacia, Uppsala, Sweden) as a carrier it can be observed that the buffer flow through the column bed becomes inhomogeneous after a number of purification runs. It is therefore recommended to repack the resin following every fifth use. The column can also be cleaned from time to time. This may be achieved using 5 *M* NaCl, extreme pH (streptavidin tolerates pH from 2 to 13, although some carriers such as Sepharose should be treated with less harsh conditions), or chaotropic salts (6 *M* Gdn-Cl). Despite their nature as a proteinaceous affinity receptor, streptavidin or StrepTactin meet industrial production demands in this respect. The extraordinary stability of streptavidin and its engineered mutant contributes to the long usability of the affinity column and, therefore, to the economy of the whole technique. The only contamination that should be prevented is biotin. Biotin cannot be readily removed from streptavidin or StrepTactin, although the latter version seems to have somewhat diminished affinity for this compound. Therefore, protein extracts with a significant biotin content should be supplemented with avidin prior to the chromatography as described earlier.

Contrasting with the *Strep*-tag, which is functionally confined to the C terminus of a recombinant protein, the *Strep*-tag II can be fused to

---

[30] P. C. Weber, J. J. Wendoloski, M. W. Pantoliano, and F. R. Salemme, *J. Am. Chem. Soc.* **114,** 3197 (1992).

either the N- or the C-terminal end. As both termini can usually be expected to be situated at the surface of the folded protein, steric accessibility is normally not a reason for the choice of the fusion site. Of course it should be kept in mind whether the N or C terminus of a protein is involved in its biochemical activity. However, other aspects, such as length and sequence of the coding region and possible problems with premature termination of protein synthesis, should be considered as well. If a large protein is to be produced, C-terminal fusion of the *Strep*-tag (II) is therefore preferred. However, when there is an internal Shine-Dalgarno sequence together with an in-frame ATG codon in the gene, secondary translational initiation may lead to nondesired protein fragments. The copurification of such side products can be avoided when using an N-terminal *Strep*-tag II.

Finally, the complete proteolytic removal of the affinity tag is only possible when it is attached to the N terminus of the recombinant protein. The explanation is that all commonly used sequence-specific proteases, such as factor X_a or enterokinase, cleave C-terminally with respect to their recognition site. Consequently, the generation of an authentic protein can only be achieved by using an N-terminal tag, followed by the protease recognition site that is located immediately in front of the first amino acid of the mature polypeptide. Such an arrangement is possible when inserting the recombinant gene into pASK-IBA 6 or 7 (Table I, Fig. 3).

Nevertheless, proteolytic cleavage is time-consuming and often not very efficient. Hence, we merely recommend it if indeed no modification of the recombinant protein can be tolerated. This might be the case in conjunction with therapeutic trials. Otherwise, it was shown in many cases that the *Strep*-tag or *Strep*-tag II does not interfere with the folding or bioactivity of a protein so that these tags may be kept during further biochemical use. The attached *Strep*-tag (II) can even be useful in the following experiments, e.g., for the functional and oriented immobilization on StrepTactin-coated ELISA plates (see later).

Analytical Scale: Strep Tactin Affinity Chromatography without Special Equipment

Procedure. This protocol is intended for small ready-to-use colums with 1 ml bed volume of StrepTactin Sepharose (5 mg/ml; IBA GmbH, Göttingen, Germany). The resin is packed between a pair of filter disks so that the column cannot run dry under gravity flow. All operations may be performed at a temperature that is compatible with the stability of the recombinant protein (between 4° and 30°). Depending on expression levels, the volume of the applied cell extract, which is prepared as described earlier, can be varied in order to take full advantage of the column capacity.

A

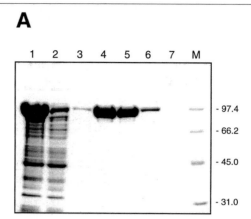

B

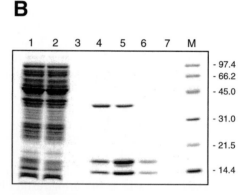

Fig. 4. One-step purification of T7 RNA polymerase and of the human cysteine protease CPP32 under gravity flow. The T7 RNA polymerase gene from *E. coli* BL21(DE3) and the CPP32 cDNA from the human hepatocyte cell line hUH7 were PCR amplified and inserted into pASK-IBA3 in order to achieve C-terminal *Strep*-tag II fusion. Gene expression was performed at 37° in *E. coli* BLR(DE3) or at 25° in XL1-Blue, respectively, and proteins were purified from the cytosolic extract after sonication. As a modification to the standard procedure described in the text, 500 m*M* NaCl and 1 m*M* DTT or 300 m*M* NaCl and 3 m*M* DTT, respectively, were added to buffer W during preparation of the lysate and chromatography. Five hundred microliters or 2 ml of the lysate, respectively, was applied to a 1-ml StrepTactin Sepharose column. (A) Ten percent Coomassie-stained SDS–PAGE with samples from the T7 RNA polymerase purification. Lane 1: cytosolic lysate; lane 2: eluate after the first washing step; lanes 3 to 7: eluates from the second to the sixth elution step; and lane M: molecular size standard (kDa). The procedure yielded 2.5 mg purified T7 RNA polymerase after a single run, corresponding to 50 mg per liter *E. coli* culture. The polymerase is susceptible to proteolytic degradation, which leads to the copurification of small amounts of N-terminally shortened polypeptide [cf. P. Davanloo, A. H. Rosenberg, J. J. Dunn, and F. W. Studier, *Proc. Natl.*

A 1-ml column is sufficient for the purification of 50 to 100 nmol recombinant protein (e.g., 1 mg of a 10-kDa or 5 mg of a 100-kDa protein). After removing excess buffer by draining, the column is equilibrated by washing twice with 4 ml of buffer W (see earlier; in the case of metalloproteins without EDTA).

If the protein extract was frozen, the sample is thawed and centrifuged before application to the column in order to remove aggregates that may have formed (microcentrifuge, 14,000 rpm, 5 min, 4°). When the sample has completely entered the chromatography bed, the column is washed five times with 1 ml of buffer W. The eluate is collected in 1-ml fractions and samples of each fraction may be subjected to analytical SDS–PAGE. Then six portions of 0.5 ml buffer E (buffer W containing 2.5 m*M* desthiobiotin; Sigma, St. Louis, MO) are added and the eluate is collected in similar fractions. Samples of each fraction should be applied to SDS–PAGE. The purified *Strep*-tag II fusion protein usually elutes in the third to fifth fraction obtained with the desthiobiotin solution. The use of this protocol is illustrated with the purification of T7 RNA polymerase and human cysteine protease CPP32, which were produced in *E. coli* with the vector pASK-IBA3 at different portions of cellular protein (Fig. 4).

After pooling the relevant fractions the buffer can be exchanged and desthiobiotin or EDTA can be removed by dialysis or gel filtration, if necessary. For regeneration the column is washed three times with 5 ml buffer R (buffer W containing 1 m*M* HABA; Sigma). The color shift from yellow to red signals the removal of desthiobiotin and the intensity of the red color is an indicator of matrix activity. The column can be stored at 4°, overlayed with 2 ml of this buffer, up to the next purification run.

Preparative Scale: StrepTactin Affinity Chromatography Using Semiautomated Equipment

Procedure. The periplasmic protein fraction (ca. 20 ml) prepared from a 2-liter *E. coli* culture is dialyzed overnight against 2 liters of buffer W.

Acad. Sci. U.S.A. **81,** 2035 (1984)]. (B) Fifteen percent Coomassie-stained SDS–PAGE with samples from the purification of the cysteine protease CPP32. Lane 1: cytosolic lysate; lane 2: eluate after the first washing step; lanes 3 to 7: eluates from the second to the sixth elution step; and lane M: molecular size standard (kDa). The procedure yielded 0.7 mg purified cysteine protease CPP32 after one run, which corresponds to 3.5 mg per liter *E. coli* culture. The 32-kDa proprotease (see lanes 4 and 5) with the C-terminal *Strep*-tag II is processed (by autoproteolysis or via a host protease) to yield p17 and p12, the latter of which still carries the *Strep*-tag II. The resulting active heterotetramer (p17/p12-*Strep*)$_2$ is susceptible to further autoproteolytic degradation (especially p17; see lane 5) [cf. P. R. E. Mittl, S. Di Marco, J. F. Krebs, X. Bai, D. S. Karanewsky, J. P. Priestle, K. J. Tomaselli, and M. G. Grütter, *J. Biol. Chem.* **272,** 6539 (1997)].

If appropriate, i.e., if some cell lysis has occurred, biotin is complexed by adding egg white avidin to a final concentration of 20 μg/ml. After a 30-min incubation on ice the protein solution is centrifuged (bench top centrifuge, 5000 rpm, 20 min, 4°) and cleared by sterile filtration. The chromatography is carried out in a system equipped with a peristaltic pump, a UV detector, and a fraction collector (e.g., Pharmacia Gradifrac). A column with 2 ml StrepTactin Sepharose (5 mg/ml) is equilibrated with buffer W. When a constant baseline is reached (monitored by measuring A_{280}), the protein solution is applied (flow rate 20 ml/hr). The column is washed with buffer W until the A_{280} has diminished to the baseline. Bound protein is then eluted with a solution of 2.5 mM desthiobiotin in buffer W (i.e., buffer E). If nonspecific binding of host cell proteins is observed, the ionic strength of these buffers should be raised, e.g., by adding up to 1 M NaCl (150 mM as a first choice).

For the quick regeneration of the column, excess desthiobiotin is first removed by washing with buffer W until a constant baseline is attained. Then 10 ml of a 5 mM solution of HABA in buffer W is applied (i.e., buffer R). Finally, the column is washed with buffer W again until the red color has disappeared. As an alternative to the application of HABA, the column can be regenerated by washing with buffer W overnight at a flow rate of 5 ml/hr.

Purification of Strep-Tag Fusion Proteins by StrepTactin-HPLC

Procedure. For the purpose of high-throughput preparations or if the recombinant protein is labile, it is desirable to perform chromatography at high speed. In order to demonstrate that the *Strep*-tag purification is applicable to HPLC conditions, StrepTactin is coupled covalently to the POROS 20 perfusion matrix (PE Biosystems, Weiterstadt, Germany). The green fluorescent protein (GFP)[31] carrying the *Strep*-tag II at its C terminus is purified from the cytosol of *E. coli* (Fig. 5). In this case, GFP is produced using the T7 promoter on the plasmid pRSET5a.[32] Bacterial cells are grown at 28° in 2 liter LB/Amp medium and expression is induced at OD$_{550}$ = 0.5 by the addition of 0.5 mM isopropyl-β-D-thiogalactopyranoside (IPTG; Roth, Karlsruhe, Germany). After shaking for 40 hr at 28°, the cells are harvested by centrifugation (4200g, 12 min, 4°), resuspended in 20 ml buffer W, and lysed by sonication. The homogenate is centrifuged (27,000g, 30 min, 4°), and the supernatant is passed through a sterile filter; 1.5 ml of the extract (corresponding to 150 ml of the culture) is injected into a column

[31] A. B. Cubitt, R. Heim, S. R. Adams, A. E. Boyd, L. A. Gross, and R. Y Tsien, *Trends Biochem. Sci.* **20,** 448 (1995).
[32] R. Schoepfer, *Gene* **124,** 83 (1993).

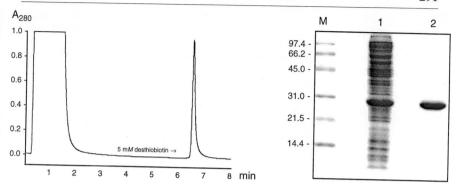

FIG. 5. One-step purification of recombinant GFP with a C-terminal *Strep*-tag II using HPLC. (Left) The elution profile from a StrepTactin column obtained under the chromatographic conditions described in the text was monitored by measuring the absorbance at 280 nm. (Right) Coomassie-stained SDS–PAGE (15%). Lane M: molecular size standard (kDa); lane 1: host cell extract; and lane 2: eluate at 6.7 min.

containing 1.7 ml POROS 20 with 5 mg/ml covalently coupled StrepTactin (IBA GmbH, Göttingen, Germany) using a BioCAD workstation (PE Biosystems, Weiterstadt, Germany). The column is washed with buffer W, and the GFP/*Strep*-tag II fusion protein is eluted with a 5 mM solution of desthiobiotin in the same buffer. Chromatography is performed at a flow rate of 4 ml/min (pressure: 25 bar). After rinsing for 5 min with 5 mM HABA dissolved in buffer W (at 6 ml/min) and a short washing step with the buffer alone, the column is regenerated. This procedure yields ca. 3 mg pure GFP/*Strep*-tag II fusion protein in one purification run (Fig. 5).

Comments. *Strep*-tag fusion proteins expressed in *E. coli* as described in the preceding sections can generally be purified from the sterile-filtered soluble cellular extract at elevated pressure on BioCAD or FPLC workstations using the POROS matrix with covalently coupled StrepTactin (5 mg/ml). The conditions and parameters given in Table II are recommended as a starting point for setting up the equipment.

Proteolytic Clipping of the N-Terminal Strep-Tag II Using Factor X_a

Although removal of the *Strep*-tag II should, in most cases, not be necessary, a convenient protocol for its cleavage was established that yields homogeneous protein by employing the StrepTactin affinity chromatography twice. After purification via *Strep*-tag II, the recombinant protein is digested with biotinylated factor X_a and the reaction mixture is again applied to the affinity column. Correctly cleaved protein lacks *Strep*-tag II and passes through, whereas residual fusion protein, clipped *Strep*-tag II,

TABLE II

PARAMETERS FOR THE PURIFICATION OF *Strep*-TAG II FUSION PROTEINS ON A StrepTactin AFFINITY COLUMN USING HPLC EQUIPMENT

Column dimension:	4.6 × 100 mm (column volume: 1.7 ml)
Starting buffer (W):	100 mM Tris–Cl (pH 8.0)
Eluent (E$_5$):	5 mM desthiobiotin in buffer W
Regeneration (R$_5$):	5 mM HABA in buffer W
Sample:	1-ml cell extract in buffer W containing up to 3 mg of the recombinant *Strep*-tag II fusion protein
Flow rate:	1–4 ml/min (350–1400 cm/hr)
Elution sequence:	3 column volumes 100% W
	Inject sample
	6 column volumes 100% W
	3 column volumes 100% E$_5$
	3 column volumes 100% W
	9 column volumes 100% R$_5$
	5–10 column volumes 100% W
	Inject next sample, etc.

and biotinylated factor X$_a$ are commonly retained by the matrix. *P. aeruginosa* blue copper protein azurin[33] has served here as a model protein.

Azurin is fused precisely with the OmpA signal peptide, *Strep*-tag II, and the factor X$_a$ recognition site in front of its mature N terminus and secreted into the periplasm of *E. coli* using the vector pASK-IBA6. Bacterial cultivation and preparation of the cell extract are performed as described earlier. The *Strep*-tag II/azurin fusion is then purified by chromatography on StrepTactin Sepharose. After buffer exchange and removal of desthiobiotin by gel filtration on Sephadex G25, 100 μg of the purified *Strep*-tag II/azurin fusion protein is cleaved with 20 μg biotinylated factor X$_a$ (Boehringer Mannheim, Germany). The digest is performed in a total volume of 100 μl 100 mM Tris–Cl (pH 8.0), 1 mM Na$_2$EDTA, 5 mM CaCl$_2$ for 48 hr at 37°. About 40% of the *Strep*-tag II/azurin fusion protein is clipped under these conditions. The whole sample is applied to a StrepTactin Sepharose column with a bed volume of 0.5 ml. The column is washed with buffer W and the eluate is collected in 100-μl fractions. The major part of the trimmed azurin (>90%) appears as a homogeneous protein in four successive fractions (Fig. 6). It should be noted that depending on the size of the recombinant protein and on the accessibility of the protease recognition site, the yield of the cleavage reaction can vary from case to case and may in practice be even considerably higher. A Sec[93] → Cys mutant of the human 15 kDa

[33] H. Nar, A. Messerschmidt, R. Huber, M. van de Kamp, and G. W. Canters, *FEBS Lett.* **306**, 119 (1992).

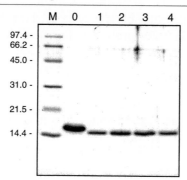

FIG. 6. Preparation of authentic azurin by proteolytic removal of the N-terminal *Strep*-tag II. Coomassie-stained SDS–PAGE (15%) with samples from the experiment described in the text. Lane M: molecular size standard (kDa); lane 0: purified azurin prior to treatment with biotinylated factor X_a; and lanes 1 to 4: samples collected during the second StrepTactin chromatography, after 48 hr proteolysis at 37°, for removal of the cleaved *Strep*-tag II, uncleaved *Strep*-tag II/azurin fusion protein, and the biotinylated protease.

selenoprotein,[34] for example, which was similarly produced as *Strep*-tag II fusion protein, could be processed to more than 90% in the course of 1-hr incubation at 26° with a factor X_a/substrate ratio of 1/250.

Use of *Strep*-Tag for Recombinant Protein Detection

Detection assays for tagged proteins are important tools not only in order to follow their fate when optimizing expression and purification conditions, but to also assist in the elucidation of biochemical function. For example, the localization of the protein to a particular cell compartment can be studied. Practically useful *in vitro* experiments are also made possible if the tag allows for oriented immobilization of the recombinant protein to a derivatized surface. In principle, there are two kinds of detection assays:

i. An assay in the denatured state, e.g., Western blot. Protein blotting techniques are generally utilized for the identification of a recombinant protein after expression and for comparing its amount in crude cell extracts.

ii. An assay under native conditions, e.g., ELISA. If the recombinant protein has a ligand-binding function, ordinary ELISA techniques can be used for verifying its bioactivity. The ligand is immobilized to the microtiter plate and subsequently bound recombinant protein can be detected via its *Strep*-tag (II) in conjunction with enzyme-labeled streptavidin or StrepTactin. However, the order of the molecules can be inverted as well and a

[34] V. N. Gladyshev, K.-T. Jeang, J. C. Wootton, and D. L. Hatfield, *J. Biol. Chem.* **273**, 8910 (1998).

StrepTactin-coated microtiter plate may be used for the functional capturing of the *Strep*-tag (II) fusion protein. In this case, either ligand-binding or
enzymatic activity can be probed in the course of following incubation steps.

Generally, if a surface-bound *Strep*-tag (II) fusion protein is detected,
the apparent sensitivity is raised because of an avidity effect that occurs
when the tetrameric streptavidin—as part of an enzyme conjugate—
interacts with the immobilized biomolecule in a multivalent manner. In
contrast, the inverse assay, which involves the monovalent capture of the
recombinant protein via its *Strep*-tag (II), has only become possible after
the development of StrepTactin with its higher affinity for the peptide.[4] As
demonstrated later, the complex between StrepTactin and a recombinant
Strep-tag (II) fusion protein persists several washing steps without significant leaching.

Detection on Western Blot with Streptavidin-Alkaline
Phosphatase Conjugate

Procedure. After SDS–PAGE[35] and electrotransfer of the proteins to an
appropriate membrane, preferentially nitrocellulose (Biometra, Göttingen,
Germany) or Immobilon P (Millipore, Eschborn, Germany), the membrane
is blocked with 20 ml 3% (w/v) bovine serum albumin (BSA), 0.5% (v/v)
Tween in phosphate-buffered saline (PBS) [4 mM KH$_2$PO$_4$, 16 mM
Na$_2$HPO$_4$, 115 mM NaCl] for 1 hr at room temperature or overnight at 4°.
The use of ordinary milk powder is not recommended, as milk is a rich
source of biotin. The membrane is washed three times for 5 min with PBS
containing 0.1% (v/v) Tween. Prior to detection of the recombinant protein
the membrane is incubated for 10 min in 10 ml PBS/Tween containing 2
μg/ml egg white avidin. In this way, endogenous proteins displaying biotin
groups are specifically masked[1] (cf. Fig. 7); 2.5 μl of streptavidin-alkaline
phosphatase conjugate (SA/AP; Amersham, Braunschweig, Germany) is
then added and incubation is continued for 60 min. Alternatively, the
StrepTactin-alkaline phosphatase conjugate (IBA GmbH, Göttingen,
Germany) can be used, which leads to higher sensitivity. The membrane
is washed three times with PBS/Tween and twice with PBS (ca. 1 min per
washing step; care should also be taken to exchange the liquid underneath
the sheet). The chromogenic reaction is initiated in the presence of 20 ml
100 mM NaCl, 5 mM MgCl$_2$, 100 mM Tris–Cl (pH 8.8) by adding 10 μl
7.5% (w/v) nitroblue tetrazolium (NBT) in 70% (v/v) DMF, and 60 μl 5%
(w/v) 5-bromo-4-chloro-3-indolyl phosphate (BCIP), toluidine salt (Am-
resco, Solon, OH) in DMF.[36] When the pattern has developed the reaction

---

[35] S. P. Fling and D. S. Gregerson, *Anal. Biochem.* **155**, 83 (1986).
[36] M. S. Blake, K. H. Johnston, G. J. Russel-Jones, and E. C. Gotschlich, *Anal. Biochem.* **136**,
175 (1984).

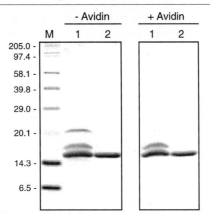

FIG. 7. Western blot detection of the azurin/*Strep*-tag II fusion protein with StrepTactin-HRP conjugate. Azurin with a C-terminal *Strep*-tag II was produced at 30° using the plasmid pASK-IBA2, and the whole cell *E. coli* extract was prepared via sonication. Four microliters of this extract (lane 1) together with 2 μg purified azurin/*Strep*-tag II (lane 2) and a biotinylated molecular size standard (Sigma, lane M; kDa) were separated by 15% SDS–PAGE. Proteins were transferred to nitrocellulose (Biometra, Göttingen, Germany) and, after blocking overnight, the membranes were incubated with StrepTactin/HRP conjugate in the presence or absence of avidin. The addition of egg white avidin essentially eliminated the signal resulting from the biotin carboxyl carrier protein at 22.5 kDa. The signals that were due to the *Strep*-tag II displayed by the mature azurin (15 kDa) or some of its preprotein (17 kDa), with the OmpA signal sequence still attached, remained unchanged.

is stopped by washing the membrane several times with distilled water. The membrane is then air dried and stored in the dark.

Detection on Western Blot with StrepTactin-HRP Conjugate

Procedure. Western blot detection using the StrepTactin-horseradish peroxidase (HRP) conjugate is performed as described earlier with the following variations. The alkaline phosphatase conjugate is replaced by StrepTactin-HRP (IBA GmbH, Göttingen, Germany), again applied at a 1:4000 dilution, and the chromogenic reaction is performed in 20 ml PBS by adding 130 μl 3% (w/v) 4-chloro-1-naphthol in methanol and 20 μl 30% (v/v) H_2O_2 (Fig. 7).

Detection of Recombinant Antibody F_v Fragment Fused with Strep-Tag in ELISA

Procedure. The cavities of a 96-well microtiter plate (Becton Dickinson, Oxnard, CA) are coated overnight at 4° with 100 μl of a lysozyme solution [3 mg/ml in 50 m*M* $NaHCO_3$ (pH 9.6)] as antigen. Control rows are coated with BSA (3 mg/ml in the same buffer). Residual binding sites are blocked at ambient temperature with 2% (w/v) nonfat dry milk (Bio-Rad, Hercules CA) in PBS for 2 hr. After washing three times with PBS/Tween, the periplasmic cell fraction, which was prepared after secretion of the lysozyme-binding antibody D1.3 F_v fragment with the *Strep*-tag at V_H or V_L in *E. coli*,[1] is applied in a total volume of 50 μl PBS/Tween and incubated for 1 hr. After three washing steps, 50 μl SA/AP conjugate is added at a dilution of 1:1000 in PBS/Tween and incubated for 1 hr. Unbound conjugate is removed by washing twice with PBS/Tween and twice with PBS. Finally, 100 μl of a solution of 0.5 mg/ml *p*-nitrophenyl phosphate (pNPP) in 0.9 M diethanolamine (pH 9.6), 1 mM MgCl$_2$ is added. The color signal is developed for 5 to 10 min and stopped by the addition of 100 μl 10 mM EDTA (pH 8.0). Absorption values are determined for the sample wells at 405 nm by means of a microplate reader after subtraction of the blank value for each dilution.

ELISA with StrepTactin-Coated Microtiter Plates for Capturing Strep-Tag II Fusion Proteins

In order to demonstrate that Streptactin-coated ELISA plates can bind *Strep*-tag II fusion proteins in a reproducible and stable manner, alkaline phosphatase from *E. coli* (PhoA), overexpressed with the vector pASK-IBA2 and purified by StrepTactin chromatography, was used as a test system.

Cavities of an *N*-hydroxysuccinimide-activated microtiter plate (Costar, Acton, MA) are covalently coated with 30 μg StrepTactin per well according to the manufacturer's instructions, stabilized with 1% (w/v) BSA, 0.5% (v/v) polyvinyl alcohol, 0.1% (w/v) dextran sulfate in PBS buffer, and lyophilized. Different amounts of the PhoA/*Strep*-tag II fusion protein (0.5 to 4 μg) in 100 μl TBS [50 mM Tris–Cl (pH 7.3), 140 mM NaCl] are incubated in the cavities and subjected to a series of washing steps with 300 μl TBS containing 0.1% (v/v) Tween 20. PhoA that has remained bound is quantified by applying 150 μl 0.5 mg/ml *p*-nitrophenyl phosphate in 100 mM Tris–Cl (pH 8.6), 100 mM NaCl, 1 mM MgCl$_2$, 0.5 mM ZnCl$_2$ and measuring the end point absorbance at 405 nm (690-nm reference) with a microtiter plate reader (SLT, Crailsheim, Germany). As a result the major fraction of PhoA is still bound to the cavities and only a negligible part is lost during washing (Fig. 8).

In a similar assay, two different concentrations of the PhoA/*Strep*-tag II fusion protein (0.25 and 0.5 μg per cavity) are each applied to 48 cavities. The plate is washed three times with 300 μl TBS/Tween per cavity, and the enzymatic activity of the bound PhoA is subsequently measured as described earlier and compared. Variance between single cavities within the assay turns out to be below 2% for each concentration (Fig. 8). These findings show that the *Strep*-tag II is useful for the functional immobilization of a recombinant target protein in highly parallel experiments, e.g., in diagnostic assays or in the screening of antisera against a *Strep*-tag (II) fusion protein.

In order to investigate such an application, the following solid-phase immunoassay is performed: 0.5 or 1 μg purified recombinant *H. pylori* urease with the *Strep*-tag II attached to the C terminus of its heavy chain[37] are each applied to two cavities of a StrepTactin-coated microtiter plate (see earlier) and incubated for 30 min. After three washing steps with TBS/Tween, the cavities are incubated with two different human serum samples diluted 1 : 2000 in PBS containing 0.1% (w/v) BSA. One serum is obtained from a *H. pylori*-infected patient while the other is from a control person. After 30 min the samples are removed by washing as described previously. Bound antibodies are then detected by adding rabbit antihuman IgG/HRP conjugate (DAKO, Hamburg, Germany) diluted 1 : 2000 in TBS/Tween for 30 min. Following three washing steps with TBS/Tween, the chromogenic reaction is started in the presence of 200 μl buffer (24.3 mM citric acid, 51.4 mM Na$_2$HPO$_4$) containing 200 μg/ml 3,3′,5,5′-tetramethylbenzidine and 0.01% (v/v) H$_2$O$_2$. After signal development the reaction is stopped by adding 50 μl 2.5 M H$_2$SO$_4$. Quantification is performed by measuring the absorbance at 450 nm (620-nm reference) in a microplate reader. As a result, the patient's serum yields a pronounced signal in comparison with the serum from the control person (Fig. 8).

Practical Applications

Following purification of a bacterially produced F$_v$ fragment via the *Strep*-tag[1] as a first example, a variety of other polypeptides have been successfully isolated with this method. Table III summarizes published reports. These include different types of proteins, e.g., various antibody fragments and other secretory ligand-binding proteins and enzymes as well as cytosolic polypeptides. Remarkably, heterooligomeric proteins can be purified in a functional state with stoichiometric subunit composition, even if just one of the polypeptide chains is equipped with the *Strep*-tag (II). This was shown not only for F$_v$ fragments with their noncovalently associated V$_H$ and V$_L$ domains, but also for heterodimeric enzymes, such as *H. pylori* urease,[37] and it demonstrates the gentle elution conditions during streptavidin or StrepTactin affinity chromatography so that the quaternary structure of a complex protein is usually retained. An even more striking example is provided by the isolation of several integral membrane protein

[37] T. G. M. Schmidt and A. Skerra, *in* "Protein Engineering in Industrial Biotechnology" (L. Alberghina, ed.), p. 41. Harwood Academic Publishers (2000).

complexes,[38,39] which were purified—in the presence of mild detergents—right from the membrane solubilisate, after capturing with a bacterially produced cognate F_v fragment carrying the *Strep*-tag. Hence, conventionally laborious purification protocols were condensed to one step. Apart from applications for protein purification, Table III provides some examples where the *Strep*-tag (II) was used for detection purposes.

[38] G. Kleymann, C. Ostermeier, B. Ludwig, A. Skerra, and H. Michel, *Bio/Technology* **13**, 155 (1995).
[39] C. Ostermeier, A. Harrenga, U. Ermler, and H. Michel, *Proc. Natl. Acad. Sci. U.S.A.* **94**, 10547 (1997).

A

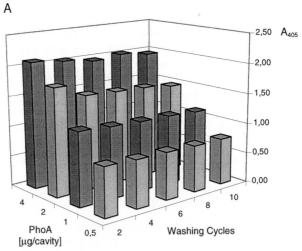

FIG. 8. Capture of *E. coli* alkaline phosphatase (PhoA) and *H. pylori* urease as *Strep*-tag II fusion proteins on StrepTactin-coated microtiter plates. The assays were performed as described in the text. (A) The stability of the complex between PhoA/*Strep*-tag II and StrepTactin, which was covalently coupled to the plastic surface, was characterized depending on the number of washing steps. After eight washing cycles, more than 90% of the *Strep*-tag II fusion protein remained bound to the StrepTactin-coated cavities, irrespective whether the applied amount was 0.5, 1, 2, or 4 µg. (B) The intraassay variance of PhoA/*Strep*-tag II adsorbed to the StrepTactin-coated microtiter plate was determined after three washing steps with applied amounts of 0.5 µg (rows A–D) and 0.25 µg (rows E–H), respectively. The coefficient of variation was 1.56% for 0.5 µg and 1.35% for 0.25 µg. Average signals were 1.1035 and 0.6579, respectively, with standard deviations of 0.0172 and 0.0089. (C) Solid-phase immunoassay for the detection of antibodies against *H. pylori* urease in sera from a patient (serum X) and from a noninfected control person (serum Y). Purified *H. pylori* urease produced with the vector pASK-IBA3 as a *Strep*-tag II fusion protein (1 or 0.5 µg) was immobilized on a StrepTactin-coated microtiter plate. Sera were then applied and bound antibodies were detected with rabbit antihuman IgG/HRP conjugate.

Although most of the recombinant proteins that have been purified with the *Strep*-tag method have been produced in *E. coli,* other host organisms also served for the production of *Strep*-tag (II) fusion proteins. *S. cerevisiae* strain 29A, for example, was employed for the biosynthesis of oat phytochrome A carrying the *Strep*-tag by using the expression vector

B

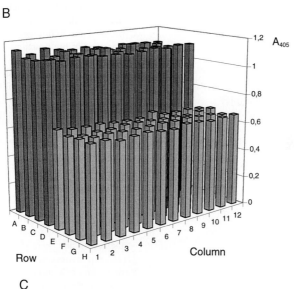

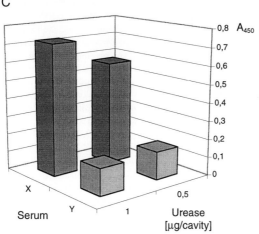

FIG. 8. (*continued*)

TABLE III

EXAMPLES FOR RECOMBINANT PROTEIN PURIFICATION AND DETECTION
WITH THE *Strep*-tag (II)

Strep-tag (II) fusion protein[a]	*Strep*-tag (II) application
V_H domain of the lysozyme-binding D1.3 F_V fragment (mouse hybridoma)	Purification of the intact F_V fragment[b] Detection of the F_V fragment via ELISA[b] and Western blotting[b]
V_L domain of the D1.3 F_V fragment (mouse hybridoma)	Purification of the intact F_V fragment[b] Detection of the F_V fragment via ELISA[b] and Western blotting[b] Capture of the F_V/antigen complex[b]
Azurin (*P. aeruginosa*)	Purification of the functional metalloprotein[c]
Cytochrome b_{562} (*E. coli*)	Purification of the functional heme protein[c]
Cystatin (chicken)	Purification of the functional protein[c]
Retinol-binding protein (pig)	Purification of the functional protein[d]
Bilin-binding protein and mutants (*Pieris brassicae*)	Purification of the functional protein[e] ELISA applications[e]
Fragment of the growth-promoting activity receptor A (chicken)	Purification of the functional protein fragment[f]
Fragment of the ciliary neurotrophic factor receptor (mammalian)	Purification of the functional protein fragment[f]
V_H domain of the cytochrome C oxidase-binding 7E2 F_V fragment (mouse hybridoma)	Purification of the intact F_V fragment[g] Capture of the intact cytochrome C oxidase (4 subunits)[h] Detection of the F_V fragment via immunoelectron microscopy[i] and immunofluorescence microscopy[j]
V_H domains of the ubiquinol:cytochrome C oxidoreductase-binding 7D3 and 2D6 F_V fragments (mouse hybridomas)	Purification of the intact F_V fragment[i] Capture of the intact ubiquinol:cytochrome C oxidoreductase (three subunits)[h] Detection of the F_V fragment via immunoelectron microscopy[i,k] and immunofluorescence microscopy[j]

[a] If not otherwise stated, the proteins were produced in *E. coli*.
[b] T. G. M. Schmidt and A. Skerra, *Protein Eng.* **6**, 109 (1993).
[c] T. G. M. Schmidt and A. Skerra, *J. Chromatogr. A* **676**, 337 (1994).
[d] H. N. Müller and A. Skerra, *Biochemistry* **33**, 14126 (1994).
[e] F. S. Schmidt and A. Skerra, *Eur. J. Biochem.* **219**, 855 (1994); G. Beste, F. S. Schmidt, T. Stibora, and A. Skerra, *Proc. Natl. Acad. Sci. U.S.A.* **96**, 1898 (1999).
[f] S. Heller, T. P. Finn, J. Huber, R. Nishi, M. Geissen, A. W. Püschel, and H. Rohrer, *Development* **121**, 2681 (1995).
[g] C. Ostermeier, L.-O. Essen, and H. Michel, *Proteins: Struct. Funct. Genet.* **21**, 74 (1995).
[h] G. Kleymann, C. Ostermeier, B. Ludwig, A. Skerra, and H. Michel, *Bio/Technology* **13**, 155 (1995).
[i] G. Kleymann, C. Ostermeier, K. Heitmann, W. Haase, and H. Michel, *J. Histochem. Cytochem.* **43**, 607 (1995).
[j] S. Ribrioux, G. Kleymann, W. Haase, K. Heitmann, C. Ostermeier, and H. Michel, *J. Histochem. Cytochem.* **44**, 207 (1996).

TABLE III (*continued*)

Strep-tag (II) fusion protein[a]	*Strep*-tag (II) application
V_H domain of the photosystem I-binding 1E7 F_V fragment (mouse hybridoma)	Purification of the intact F_V fragment[l] Capture of the intact photosystem I (three subunits)[l] Detection of the F_V fragment via Western blotting[l] and immunoelectron microscopy[l]
Phytochrome A (oat)	Purification of the functional, phytochromobilin-reconstituted protein (expressed in yeast)[m]
Alkaline phosphatase (*E. coli*)	Purification of the active metalloenzyme[n,o]
Green fluorescent protein (*A. victoria*)	Purification of the functional protein[p] Detection on Western blot
Subunit B of *H. pylori* urease	Copurification of expressed A and B subunits in stoichiometric amounts[q] Detection on Western blot (subunit B)
Curli assembly factor CsgG (*E. coli* outer membrane)	Purification of the native membrane protein[r]
Cyanobacterial phytochrome Cph1	Purification of the active, light-regulated histidine kinase[s]
Glutathione *S*-transferase and fusion proteins	Purification of the dimeric fusion enzyme[t] Detection in ELISA[t]
V_H domain of the cytochrome *C* oxidase-binding 7E2C50S F_V fragment (mouse hybridoma)	Purification of the intact F_V fragment[u] Capture of the cytochrome *C* oxidase subunits I and II[u]
Hydrogenase precursor (*E. coli*)	Purification of the active precursor (80%) with N-terminal *Strep*-tag II[v]
PdxA and PdxJ (*E. coli*)	Purification of the active enzymes for vitamin B_6 synthesis[w]

[k] G. Kleymann, S. Iwata, H.-H. Wiesmüller, B. Ludwig, and H. Michel, *Eur. J. Biochem.* **230,** 359 (1995).

[l] G. Tsiotis, W. Haase, A. Engel, and H. Michel, *Eur. J. Biochem.* **231,** 823 (1995).

[m] J. T. Murphy and J. C. Lagarias, *Photochem. Photobiol.* **65,** 750 (1997).

[n] M. Hengsakul and A. E. G. Cass, *J. Mol. Biol.* **266,** 621 (1997).

[o] S. Voss and A. Skerra, *Protein Eng.* **10,** 975 (1997).

[p] A. Skerra and T. G. M. Schmidt, *Biomol. Eng.* **16,** 79 (1999).

[q] T. G. M. Schmidt and A. Skerra, *in* "Protein Engineering in Industrial Biotechnology" (L. Alberghina, ed.), p. 41. Harwood Academic Publishers (2000).

[r] H. Loferer, M. Hammar, and S. Normark, *Mol. Microbiol.* **26,** 11 (1997).

[s] K.-C. Yeh, S.-H. Wu, J. T. Murphy, and J. C. Lagarias, *Science* **277,** 1505 (1997).

[t] T. Tudyka and A. Skerra, *Protein Sci.* **6,** 2180 (1997).

[u] C. Ostermeier, A. Harrenga, U. Ermler, and H. Michel, *Proc. Natl. Acad. Sci. U.S.A.* **94,** 10547 (1997).

[v] T. Maier, N. Drapal, M. Thanbichler, and A. Böck, *Anal. Biochem.* **259,** 68 (1998).

[w] B. Laber, W. Maurer, S. Scharf, K. Stepusin, and F. S. Schmidt, *FEBS Lett.* **449,** 45 (1999).

(*continued*)

TABLE III (*continued*)

Strep-tag (II) fusion protein[a]	*Strep*-tag (II) application
Rat epididymal retinoic acid-binding protein	Purification of the functional protein[1]
"Camelized" antibody V_H domains	ELISA detection of the engineered immunoglobulin domains[2]
NADH dehydrogenase fragment of *E. coli* respiratory chain complex I	One-step purification of the multi-subunit protein[3]
bHLHzip domain of human c-Myc protein	Purification of the active leucine zipper domain[4]
Secondary glycine betaine uptake system BetP (*Corynebacterium glutamicum*)	Purification of the functional transmembrane carrier[5]
Human tissue transglutaminase	Purification of active enzyme (expressed in human embryonic kidney cells) and use for ELISA diagnosis of gluten-sensitive enteropathy[6]
Hyaluronate synthase (group A *streptococci*)	Purification of the active, presumably membrane-anchored enzyme[x]
SP1 and SP3 transcription factors	Purification of the soluble proteins (expressed in SF9 cells)[y]
Ligase I, III, IV, and XRCC4 (human)	Purification of the active enzymes (expressed in HeLa cells)[z]
T7 RNA polymerase (*E. coli*)	Purification of the active enzyme (this study)
CPP32 (human)	Purification of the active cysteine protease (this study)

[x] V. Nickel and I. van de Ryn, personal communication (1999).
[y] M. Truß and T. G. M. Schmidt, unpublished (1997).
[z] U. Grawunder, personal communication (1999).
[1] M. Sundaram, A. Sivaprasadarao, D. M. Aalten, and J. B. Findlay, *Biochem. J.* **334**, 155 (1998).
[2] J. Davies and L. Riechmann, *Bio/Technology* **13**, 475 (1995).
[3] S. Bungert, B. Krafft, R. Schlesinger, and T. Friedrich, *FEBS Lett.* **460**, 207 (1999).
[4] N. Zwicker, K. Adelhelm, R. Thiericke, S. Grabley, and F. Hänel, *Biotechniques* **27**, 368 (1999).
[5] R. Rübenhagen, H. Rönsch, H. Jung, R. Krämer, and S. Morbach, *J. Biol. Chem.* **275**, 735 (2000).
[6] M. Sárdy, U. Odenthal, S. Kárpáti, M. Paulsson, and N. Smyth, *Clin. Chem.* **45**, 2142 (1999).

pMAC160.[40] Transcription factors SP1 and SP3 were produced as *Strep*-tag II fusion proteins in SF9 insect cells (suspension culture in TC100 medium with FCS) using the Baculo Gold cloning system (PharMingen, Hamburg, Germany) and the vector pVL1392.

[40] J. A. Wahleitner, L. Li, and J. C. Lagarias, *Proc. Natl. Acad. Sci. U.S.A.* **88**, 10387 (1991).

The *Strep*-tag system has also been utilized for the purification of proteins that were produced in a human cell line (HeLa).[41] For example, cDNAs for the human DNA ligases I,[42] III,[42] and IV[43] and for XRCC4[43] were subcloned into a modified pCDNA3 expression vector (Invitrogen, Groningen, The Netherlands) encoding the *Strep*-tag II for C-terminal fusion. The proteins were synthesized using the transient vaccinia virus overexpression system according to standard protocols.[44] The cells were shock-frozen in liquid nitrogen and lysed by 15 sec sonication in an appropriate buffer [25 mM Na–HEPES (pH 7.9), 300 mM KCl, 10 mM MgCl$_2$, 0.1% (v/v) nonylphenoxypolyethoxyethanol, NP-40, 0.25 % (v/v) Tween-20, supplemented with 1 mM phenylmethylsulfonyl fluoride (PMSF), and 1 μg/ml each of pepstatin, aprotinin A, and leupeptin]. Debris was removed by centrifugation (microfuge, 14,000 rpm, 15 min, 4°). The supernatant was sterile-filtered and applied to a 1-ml StrepTactin column (IBA, Göttingen, Germany) equilibrated with the lysis buffer. The column was washed with 5 ml lysis buffer followed by washing with 5 ml low salt buffer [60 mM Tris–Cl (pH 7.9), 10 mM MgCl$_2$, 10% (v/v) glycerol]. Bound protein was then eluted with a solution of 2.5 mM desthiobiotin in the low salt buffer. Elution fractions were tested for recombinant protein by SDS–PAGE and immunoblotting as well as activity assays.[43] Based on SYPRO-Orange-stained SDS–PAGE, the eluted proteins were >98% pure. Both recombinant DNA ligases were biochemically active and, notably, free of nuclease contamination.[41]

Generally, for the production of *Strep*-tag fusion proteins in eukaryotic host cells, a variety of established expression and vector systems are principally suitable. The coding sequence for the *Strep*-tag (II) can be introduced during the cloning of a cDNA by means of an appropriate PCR primer that adds a corresponding stretch of nucleotides. Two aspects should, however, be taken into consideration for eukaryotic expression. First, the sequence of the *Strep*-tag II was originally described with the amino acid sequence Asn-Trp-Ser- ... at its N-terminal end,[3] which may constitute a signal for N-glycosylation in eukaryotic cells. Because the asparagine residue can be replaced with alanine without any loss of activity regarding the interaction with StrepTactin, the sequence Ala-Trp-Ser-His-Pro-Gln-Phe-Glu-Lys should be used under such circumstances. Second, in contrast with *E. coli,* eukaryotic cells contain several biotinylated proteins. However, the amount of endogenous free or protein-bound biotin is normally not critical

[41] U. Grawunder, personal communication (1999).

[42] A. E. Tomkinson and D. S. Levin, *Bioasssays* **19,** 893 (1997).

[43] U. Grawunder, M. Wilm, X. Wu, P. Kulesza, T. E. Wilson, M. Mann, and M. R. Lieber, *Nature* **388,** 492 (1997).

[44] T. R. Fuerst, E. G. Niles, F. W. Studier, and B. Moss, *Proc. Natl. Acad. Sci. U.S.A.* **83,** 8122 (1986).

when using StrepTactin affinity columns, irrespective of whether the protein was expressed in the cytosol or secreted to the culture medium.[45] If there is doubt about an elevated biotin content in the protein extract a sample can be applied to a small StrepTactin or streptavidin column, followed by rinsing with a solution of HABA. If biotin is present, the upper part of the column will not turn red and from the deduced volume of the pale zone—together with the known immobilized streptavidin concentration—the biotin content can be calculated. Thus one can estimate the minimal amount of egg white avidin that should be added to the extract with the *Strep*-tag fusion protein prior to the chromatography (see earlier discussion).

Conclusions

The *Strep*-tag protein purification system combines the advantage of a well-functionalized streptavidin or StrepTactin interaction matrix, which is comparable in its specificity toward a *Strep*-tag fusion protein to the molecular recognition between antibodies and antigens, with simple chromatography protocols and a remarkable robustness. In respect to possible cleaning conditions for the affinity column, this method meets industrial standards. The frequent reuse of the matrix and the bulk availability of StrepTactin contribute to the economy of this technique both in the research laboratory and at the preparative scale. Because the binding of a *Strep*-tag fusion protein to the affinity matrix is controlled easily and elution is effected under very mild conditions, almost every protein can be isolated in a functional state. The potential of standardizing protocols for the purification of different proteins should be attractive regarding "good manufacturing practice" for the production of medically important polypeptides. Finally, application of the *Strep*-tag for detection or capturing purposes under reproducible conditions, especially in a parallel fashion, opens interesting prospects in proteomics.

Acknowledgments

The authors thank U. Grawunder for his data on the purification of *Strep*-tag II fusion proteins expressed in HeLa cells, G. Holzapfel for data from the assay of the PhoA/*Strep*-tag II fusion protein on StrepTactin-coated microtiter plates, K. Kiem for technical assistance, and L. Rieger for cloning the T7 RNA polymerase gene on pASK-IBA3. T.S. thanks H. Stadler for general support.

---

[45] N. Smyth, U. Odenthal, B. Merkl, and M. Paulsson, *Methods Mol. Biol.* **139**, 49 (2000).

[19] Streptavidin-Containing Chimeric Proteins: Design and Production

By TAKESHI SANO and CHARLES R. CANTOR

Introduction

Avidin and its bacterial equivalent, streptavidin, are proteins that have remarkably high binding affinities for a small water-soluble vitamin, D-biotin (vitamin H).[1,2] The equilibrium dissociation constants (K_d) of the avidin–biotin and streptavidin–biotin complexes are estimated at 10^{-14}–10^{-15} M; these are among the tightest noncovalent interactions known between proteins and their ligands. The extremely tight biotin-binding affinities, along with the ability of biotin to be incorporated easily into various biological materials, allow these two biotin-binding proteins to serve as useful, powerful tools in a variety of biological and medical applications.[3]

Applications of avidin and streptavidin often use their chemical conjugates. For example, chemical conjugates between streptavidin and color-generating enzymes, such as alkaline phosphatase and peroxidase, are commonly used reagents for immuno-detection of target molecules. In such procedures, biotinylated secondary antibodies are used to attach the streptavidin-containing chemical conjugates to the targets via primary antibodies. To make conjugates of natural avidin or streptavidin with partner proteins, covalent chemistry is generally used. Although such chemistry has been used successfully for the production of certain avidin- or streptavidin-containing conjugates, it would be very useful if a partner protein could be fused directly to avidin or streptavidin by recombinant DNA methods. This would allow the production of structurally homogeneous avidin- or streptavidin-containing chimeric proteins, as opposed to chemical avidin- or streptavidin-containing conjugates, which often display structural heterogeneity. It would also facilitate the production of avidin- or streptavidin-containing chimeric proteins in large quantities.

Isolation of the cDNA and the gene for avidin[4] and streptavidin,[5] respec-

[1] N. M. Green, *Adv. Protein Chem.* **29,** 85 (1975).
[2] N. M. Green, *Methods Enzymol.* **184,** 51 (1990).
[3] M. Wilchek and E. A. Bayer, *Methods Enzymol.* **184** (1990).
[4] M. L. Gope, R. A. Keinänen, P. A. Kristo, M. O. Conneley, W. G. Beattie, T. Zarucki-Schultz, B. W. O'Malley, and M. S. Kulomaa, *Nucleic Acids Res.* **15,** 3595 (1987).
[5] C. E. Argaraña, I. D. Kuntz, S. Birken, R. Axel, and C. R. Cantor, *Nucleic Acids Res.* **14,** 1871 (1986).

tively, allowed the production and modification of these proteins by using recombinant DNA technology. A group of such efforts is directed toward the production of avidin- or streptavidin-containing chimeric proteins. The rationale behind these efforts is that, by fusing a protein of interest to avidin or streptavidin, the fused partner protein should acquire highly specific, strong biological recognition capability derived from the avidin or streptavidin moiety. This allows, for example, the labeling, conjugation, and targeting of the fused partner protein to any biological material containing biotin. This article places particular focus on streptavidin-containing chimeric proteins and describes general strategies and considerations for their design and production. Many of these strategies and considerations should also be applicable to the design and production of avidin-containing chimeric proteins, as discussed briefly later.

Streptavidin as a Fusion Partner

When we initiated the design and production of chimeric proteins,[6–8] streptavidin was chosen as a fusion partner over avidin. Streptavidin is a tetrameric protein produced by the bacterium *Streptomyces avidinii*, and it has an estimated biotin-binding affinity (K_d) at 10^{-14} M, similar to that of avidin. Its great similarity to avidin, including the biotin-binding and structural characteristics, resulted in the naming of this protein as the streptomyces equivalent of avidin.

Several properties of streptavidin make this protein more attractive and potentially more useful as a fusion partner than avidin. First, streptavidin has no cysteine residues, whereas each subunit of avidin has two cysteine residues, which make an intrasubunit disulfide bond.[1,2] The lack of sulfhydryl groups should allow easier production and manipulation of the chimeric proteins. Second, streptavidin is free from carbohydrate, whereas avidin is a glycoprotein with one carbohydrate chain per subunit.[1,2] Although the carbohydrate chains of avidin have no apparent contribution to the fundamental properties of avidin, such as biotin binding and subunit association, heterogeneity of the carbohydrate chains is seen.[9] Third, streptavidin is a neutral protein, whereas avidin is a basic protein with an isoelectric point of around 10.[2] The carbohydrate chains also contribute to the high isoelectric point of avidin. The neutral isoelectric point of streptavidin should be

[6] T. Sano and C. R. Cantor, *Biochem. Biophys. Res. Commun.* **176,** 571 (1991).
[7] T. Sano and C. R. Cantor, *Bio/Technology* **9,** 1378 (1991).
[8] T. Sano, A. N. Glazer, and C. R. Cantor, *Proc. Natl Acad. Sci. U.S.A.* **89,** 1534 (1992).
[9] R. C. Bruch and H. B. White III, *Biochemistry* **21,** 5334 (1982).

advantageous in many applications, as it minimizes nonspecific binding to other biological materials. Fourth, the three-dimensional structure of streptavidin determined by X-ray crystallography has been known for some time.[10,11] The availability of the three-dimensional structure was of particular importance in designing and producing streptavidin-containing chimeric proteins.

Recombinant Streptavidins

The gene for streptavidin was cloned in *Escherichia coli* from a genomic library of *S. avidini*.[5] Isolation of the streptavidin gene allowed the production of recombinant streptavidin by using an *E. coli* expression system.[12] The extremely tight biotin-binding affinity makes streptavidin lethal to any cell in which it is expressed because biotin is essential for cell viability. Thus, very tight expression control is essential for streptavidin to be produced in *E. coli* efficiently. This problem was solved by using an enhanced version of the bacteriophage T7 expression system.[13]

In this system, T7 lysozyme, a natural inhibitor of T7 RNA polymerase, is expressed constitutively in host *E. coli* cells. Thus, the activity of any T7 RNA polymerase, produced due to the leakiness of the promoter used for the T7 RNA polymerase gene, can be repressed effectively, allowing a plasmid carrying the cloned streptavidin gene under the control of a strong bacteriophage T7 promoter, *Φ10*, to be maintained stably until the time of induction. Upon induction of the T7 RNA polymerase gene, which results in the overproduction of T7 RNA polymerase, cellular T7 lysozyme is titrated out, and newly synthesized T7 RNA polymerase directs transcription of the streptavidin gene. By using this system, streptavidin is expressed efficiently in host *E. coli,* and it accounts for greater than 30% of total cell protein.[12]

Expressed recombinant streptavidin generally forms inclusion bodies, as seen with many *E. coli* overexpression systems. However, expressed streptavidin can be solubilized easily by treatment with guanidine hydrochloride (or other potent organic denaturants, such as urea), and the solubilized streptavidin can be renatured by slow removal of guanidine

[10] W. A. Hendrickson, A. Pähler, J. L. Smith, Y. Satow, E. A. Merritt, and R. P. Phizackerley, *Proc. Natl. Acad. Sci. U.S.A.* **86,** 2190 (1989).
[11] P. C. Weber, D. H. Ohlendorf, J. J. Wendoloski, and F. R. Salemme, *Science* **243,** 85 (1989).
[12] T. Sano and C. R. Cantor, *Proc. Natl. Acad. Sci. U.S.A.* **87,** 142 (1990).
[13] F. W. Studier, A. H. Rosenberg, J. J. Dunn, and J. W. Dubendorff, *Methods Enzymol.* **185,** 60 (1990).

hydrochloride. Then, renatured, functional streptavidin can be purified by 2-iminobiotin affinity chromatography,[14] in which streptavidin binds to immobilized 2-iminobiotin at basic pH (>pH 9.5) and is released under slightly acidic conditions, such as at pH 4. This relatively simple expression/purification procedure allows the production of purified recombinant streptavidin with full biotin-binding ability at yields of at least a few milligrams per 100 ml of culture.

Streptavidin-Containing Chimeric Proteins

Establishment of the expression/purification procedure allowed the design and production of streptavidin-containing chimeras.[6–8] We have described general protocols for expressing streptavidin-containing chimeras in *E. coli* by using the bacteriophage T7 expression system and procedures for purifying and characterizing expressed streptavidin-containing chimeric proteins.[15] Thus, interested readers should be referred to this article for detail protocols. This section describes several general considerations that should be made when designing and producing streptavidin-containing chimeric proteins.

N-Terminal Fusion versus C-Terminal Fusion

In a streptavidin-containing chimeric protein, the partner protein can, in principle, be fused to either the N terminus or the C terminus of streptavidin. In our limited experience, streptavidin-containing chimeric proteins in which the partner proteins are fused to the N terminus of the streptavidin moiety are problematic. Such streptavidin-containing chimeric proteins can be expressed efficiently in *E. coli* by using the bacteriophage T7 expression system. However, the chimeric proteins, expressed in the form of inclusion bodies, often fail to fold correctly. We do not know why such streptavidin-containing chimeric proteins have limited folding ability. This problem might be solvable by using appropriate linkers between the streptavidin moiety and the partner proteins. In fact, a few streptavidin-containing chimeric proteins, in which partner proteins are fused to the N terminus of the streptavidin moiety, have been produced successfully.[16–18] In contrast,

[14] K. Hofmann, S. W. Wood, C. C. Brinton, J. A. Montibeller, and F. M. Finn, *Proc. Natl. Acad. Sci. U.S.A.* **77,** 4666 (1980).

[15] T. Sano, C. L. Smith, and C. R. Cantor, *Methods Mol. Biol.* **63,** 119 (1997).

[16] S. Dübel, F. Breitling, R. Kontermann, T. Schmidt, A. Skerra, and M. Little, *J. Immunol. Methods* **178,** 201 (1995).

almost all of our attempts to make streptavidin-containing chimeric proteins, in which the partner proteins, including relatively short peptide sequences, are fused to the C terminus of the streptavidin moiety, have been successful without major difficulties at the refolding step.

Tetrameric Nature of Streptavidin

As described earlier, streptavidin is a tetrameric protein. The tetrameric structure is essential for its extremely tight biotin-binding affinity because intersubunit contacts to biotin, made by an adjacent subunit through a subunit–subunit interface, have a significant contribution to the biotin-binding site.[19,20] The three-dimensional structure of nature streptavidin[10,11] suggests that, without significant modifications, streptavidin would not be able to form a stable, functional molecule in a dimeric or monomeric form, although a dimeric streptavidin with reduced biotin-binding affinity and stability has already been produced.[21] When a partner protein is fused to streptavidin, the resulting streptavidin-containing chimeric protein forms a tetramer via its streptavidin moiety, making the fused partner protein also tetrameric. Thus, if the tetrameric structure is undesirable for the partner protein, making a streptavidin-containing chimeric protein is not a choice, unless such molecules are reconstituted with nonfused streptavidin subunits to control the valency of the partner protein (e.g., a chimeric protein consisting of one chimeric subunit and three nonfused subunits). Similarly, if the partner proteins are naturally associated with each other to form multimers, production of a streptavidin-containing chimeric protein in a tetrameric form causes aggregate formation. However, the tetramer formation of streptavidin-containing chimeric proteins is beneficial for certain partner proteins. For example, chimeric proteins between single-chain antibodies and streptavidin showed enhanced antigen-binding affinities over the parental single-chain antibodies because of their tetravalency.[22–24]

---

[17] C. Oker-Blom, A. Orellana, and K. Keinanen, *FEBS Lett.* **389**, 238 (1996).

[18] M. Karp, C. Lindqvist, R. Nissinen, S. Wahlbeck, K. Åkerman, and C. Oker-Blom, *BioTechniques* **20**, 452 (1996).

[19] S. Miyamoto and P. A. Kollman, *Proteins* **16**, 226 (1993).

[20] T. Sano, S. Vajda, G. O. Reznik, C. L. Smith, and C. R. Cantor, *Ann. N.Y. Acad. Sci.* **799**, 383 (1996).

[21] T. Sano, S. Vajda, C. L. Smith, and C. R. Cantor, *Proc. Natl Acad. Sci. U.S.A.* **94**, 6153 (1997).

[22] S. M. Kipriyanov, F. Breitling, M. Little, and S. Dubel, *Hum. Antibodies Hybridomas* **6**, 93 (1995).

Expression Systems

As indicated previously, the original system for the production of streptavidin-containing chimeric proteins is based on the bacteriophage T7 expression system. This expression system offers high yields, facilitating the production of streptavidin-containing chimeric proteins in large quantities. However, expressed streptavidin-containing chimeric proteins generally form inclusion bodies. Thus, the partner proteins, fused to the streptavidin moiety, must be able to refold correctly after solubilization of the inclusion body fraction. This limits the number of partner protein species that can be fused to streptavidin.

Several successful attempts have been made to solve the problem associated with the inclusion body formation of streptavidin-containing chimeric proteins. A simple, yet effective method uses lower incubation temperatures, such as at 20–30°, during expression. This strategy has been used successfully for the expression of other proteins in soluble form; it should also be applicable to the expression of streptavidin[25] and its chimeric proteins[16] in a folded, active form. Other expression systems that should be useful for the production of streptavidin-containing chimeric proteins include other bacterial expression systems, such as those using *Bacillus subtilis* in which expressed recombinant streptavidin is secreted to the culture media,[26] and baculovirus expression systems using insect cells.[17,18] The availability of different expression systems should expand the range of proteins that can be fused to streptavidin.

Use of Avidin as a Fusion Partner

Some of the properties of avidin that were considered potential disadvantages as a fusion partner, described in a previous section, have already been solved. For example, the use of an *E. coli* expression system allows the production of recombinant avidin in a carbohydrate-free form. Avidin mutants with neutral and acidic isoelectric points have been produced by

---

[23] S. M. Kipriyanov, M. Little, H. Krophofer, F. Breitling, S. Gotter, and S. Dubel, *Protein Eng.* **9**, 203 (1996).

[24] L. A. Pearce, G. W. Oddie, G. Coia, A. A. Kortt, P. J. Hudson, and G. G. Lilley, *Biochem. Mol. Biol. Int.* **42**, 1179 (1997).

[25] A. Gallizia, C. de Lalla, E. Nardone, P. Santambrogio, A. Brandazza, A. Sidoli, and P. Arosio, *Protein Expr. Purif.* **14**, 192 (1998).

[26] V. Nagarahan, R. Ramaley, H. Albertson, and M. Chen, *Appl. Environ. Microbiol.* **59**, 3894 (1993).

genetic engineering[27,28] so that electrostatic interactions with acidic molecules, seen with natural avidin, can be minimized. The three-dimensional structure of avidin has also been determined by X-ray crystallography,[29–31] and this structural information is particularly useful in designing and producing avidin-containing chimeric proteins. Production of recombinant avidin and avidin-containing chimeric proteins in soluble, functional form can also be done in *E. coli* by expressing them at low temperatures[32,33] and in insect cells by using the baculovirus expression system,[34] similar to the production of streptavidin and its chimeric proteins. These suggest that avidin can now serve as a useful partner protein in the production of chimeric proteins.

Conclusions

A number of successful efforts have proven the power and usefulness of avidin- and streptavidin-containing chimeric proteins. These and other efforts also suggest the great potential for avidin and streptavidin to become even more useful partner proteins in the production of chimeric proteins. For example, genetic engineering has been used successfully to alter, improve, and enhance the properties of these proteins, including their biotin-binding and structural characteristics. This should expand the range of proteins that can be fused beneficially to avidin and streptavidin, and such avidin- and streptavidin-containing chimeric proteins should facilitate further development of enhanced biological and medical technologies.

Acknowledgments

This work was supported by grants from the National Cancer Institute (CA39782) and the U.S. Department of Energy (DE-FG02-93ER61656).

[27] E. Nardone, C. Rosano, P. Santambrogio, F. Curnis, A. Corti, F. Magni, A. G. Siccardi, G. Paganelli, R. Losso, B. Apreda, M. Bolognesi, A. Sidoli, and P. Arosio, *Eur. J. Biochem.* **256,** 453 (1998).

[28] A. T. Marttila, K. J. Airenne, O. H. Litinen, T. Kulik, E. A. Bayer, M. Wilchek, and M. S. Kulomaa, *FEBS Lett.* **441,** 313 (1998).

[29] O. Libnah, E. A. Bayer, M. Wilchek, and J. L. Sussman, *Proc. Natl. Acad. Sci. U.S.A.* **90,** 5076 (1993).

[30] L. Pugliese, A. Coda, M. Malcovati, and M. Bolognesi, *J. Mol. Biol.* **231,** 698 (1993).

[31] L. Pugliese, M. Malcovati, A. Coda, and M. Bolognesi, *J. Mol. Biol.* **235,** 42 (1994).

[32] K. J. Airenne, P. Sarkkinen, E.-L. Punnonnen, and M. S. Kulomaa, *Gene* **144,** 75 (1994).

[33] K. J. Airnne and M. S. Kulomaa, *Gene* **167,** 63 (1995).

[34] K. J. Airenne, C. Oker-Blom, V. S. Marjomäki, E. A. Bayer, M. Wilchek, and M. S. Kulomaa, *Protein Expr. Purif.* **9,** 100 (1997).

[20] Fusions to Maltose-Binding Protein: Control of Folding and Solubility in Protein Purification

By Deepali Sachdev and John M. Chirgwin

Introduction

Expression in *Escherichia coli* provides an abundant source of recombinant proteins. However, many eukaryotic proteins accumulate in *E. coli* as inclusion bodies.[1,2] Recovery of native protein from inclusion bodies involves isolating and washing the inclusion bodies, followed by their solubilization with a chaotropic agent, such as urea or guanidine,[3] and finally refolding of the protein by removal of the denaturant. For proteins that readily refold *in vitro,* inclusion body formation can be advantageous, as the protein of interest can be 30–90% pure in the inclusion bodies. For many eukaryotic proteins expressed in bacteria, however, the recovery of native protein from inclusion bodies is frequently unsatisfactory, giving low yields of biologically active product.

Despite improvements in the procedures for refolding proteins from inclusion bodies,[4,5] recovery of native protein still remains a common problem. One strategy to assist the recovery of eukaryotic proteins expressed in *E. coli* has been to fuse them to bacterial proteins.[6,7] Such fusions have been used primarily for providing affinity purification tags and secondarily for enhancing solubility and suppressing degradation. Bacterial proteins commonly used as fusion partners include glutathione *S*-transferase (GST), *β*-galactosidase, thioredoxin, and maltose-binding protein (MBP). GST has been used primarily for affinity purification and convenient binding to glutathione-agarose beads in pull-down assays.[8] Some recombinant proteins expressed as fusions to *β*-galactosidase continue to aggregate into inclusion bodies.[9] The large size of oligomeric *β*-galactosidase may make it unsuitable for enhancing soluble expression of eukaryotic proteins. Thioredoxin is

[1] J. F. Kane and D. J. Hartley, *Trends Biotechnol.* **6,** 95 (1988).
[2] R. G. Schoner, L. F. Ellis, and B. E. Schoner, *Bio/Technology* **24,** 349 (1992).
[3] F. A. Marston and D. L. Hartley, *Methods Enzymol.* **182,** 264 (1990).
[4] H. Lilie, E. Schwarz, and R. Rudolph, *Curr. Opin. Biotechnol.* **9,** 497 (1998).
[5] E. D. B. Clark, *Curr. Opin. Biotechnol.* **9,** 157 (1998).
[6] E. R. LaVallie and J. M. McCoy, *Curr. Opin. Biotechnol.* **6,** 501 (1995).
[7] P. Riggs, *in* "Current Protocols in Molecular Biology" (F. M. Ausubel *et al.,* eds.), p. 16.4.1/16.6.1. Greene Associates/Wiley Interscience, New York, 1990.
[8] D. B. Smith and K. S. Johnson, *Gene* **67,** 31 (1988).
[9] F. A. Marston, *Biochem. J.* **240,** 1 (1986).

a small monomer that facilitates the soluble expression of a number of mammalian growth factors and cytokines.[10] Maltose-binding protein has been widely used[11] for the expression of eukaryotic proteins in *E. coli*. MBP is purified easily by chromatography on immobilized amylose resin.[7] Elution is accomplished inexpensively with maltose, which does not interfere with the function of most partner proteins. MBP folding is well characterized both *in vivo* and *in vitro*.[12] It folds efficiently in the bacterial periplasm, its normal site of localization, and in the cytoplasm on deletion of its leader peptide.[13] MBP can assemble *in vivo* from separately expressed fragments,[14] suggesting that its folding is robust.

We have tested MBP as fusion partner for bacterial expression of two mammalian aspartic proteinases, pepsinogen and procathepsin D, with their endoplasmic reticulum (ER) signal peptides deleted. Procathepsin D and pepsinogen show extensive similarity in their polypeptide chain folds and three-dimensional structure.[15] When expressed in *E. coli*, aspartic proteinases such as pepsinogen and procathepsin D are found entirely in inclusion bodies.[16,17] Fusion of these proteins to MBP permits their soluble expression in *E. coli*, suggesting that fusions to MBP may have general utility to control the folding and solubility of recalcitrant eukaryotic proteins. MBP fusions may have particular value in preventing the aggregation and insolubility of nonnative protein folding states, making these accessible to standard methods of biophysical analysis in solution.

Experimental Procedures

The aspartic proteinases are expressed in *E. coli* as maltose-binding protein fusions using the cytoplasmic pMAL-c2 vector provided with a protein fusion and purification system[18] from New England Biolabs (Beverly, MA). We also modified the pMAL-c2 vector to insert a tobacco

[10] E. R. LaVallie, E. A. DiBlasio, S. Kovacic, K. L. Grant, P. F. Schendel, and J. M. McCoy, *Bio/Technology* **11,** 187 (1993).

[11] J. M. Clement, A. Charbit, C. Leclerc, P. Martineau, S. Muir, D. O'Callaghan, O. Popescu, S. Szmelcman, and M. Hofnung, *Antonie van Leeuwenhoek* **61,** 143 (1992).

[12] S. Y. Chun, S. Strobel, P. Bassford, Jr., and L. L. Randall, *J. Biol. Chem.* **268,** 20855 (1993).

[13] G. Liu, T. B. Topping, and L. L. Randall, *Proc. Natal. Acad. Sci. U.S.A.* **86,** 9213 (1989).

[14] J. M. Betton and M. Hofnung, *EMBO J.* **13,** 1226 (1994).

[15] D. R. Davies, *Annu. Rev. Biophys. Chem.* **19,** 189 (1990).

[16] G. E. Conner and J. A. Udey, *DNA Cell Biol.* **9,** 1 (1990).

[17] X. L. Lin, R. N. Wong, and J. Tang, *J. Biol. Chem.* **264,** 4482 (1989).

[18] C. V. Maina, P. D. Riggs, A. G. d. Grandea, B. E. Slatko, L. S. Moran, J. A. Tagliamonte, L. A. McReynolds, and C. D. Guan, *Gene* **74,** 365 (1988).

etch virus (TEV) protease site[19] downstream of the factor Xa site. The TEV site is inserted using a double-stranded oligonucleotide, resulting in the obliteration of the *Xmn*I site in the polylinker of pMAL-c2 and adding a unique *Stu*I site. The vector is designated pMAL$_c$TEV. It retains the other features of the original pMAL-c2 vector.[18]

The cDNA encoding procathepsin D is amplified by polymerase chain reaction (PCR) using pSKCMVPCDMf1[20] as template, a phosphorylated primer to create a ligatable 5' blunt end, and a lower primer that adds a UAA stop codon at the 3' end plus an *Xba*I site. PCR uses Vent polymerase (New England Biolabs), which produces blunt-ended products. The DNA is gel purified and digested with *Xba*I. pMAL-c2 is digested with *Stu*I and *Xba*I and gel purified. The 1.3-kb blunt to the *Xba*I fragment of procathepsin D is then ligated between the *Stu*I and the *Xba*I sites of the vector. The resulting expression plasmid is called pMAL$_c$TEV-proCpD and is shown in Fig. 1. It expresses the fusion protein MBP-procathepsin D. The cDNA encoding pepsinogen is similarly amplified using a human pepsinogen cDNA plasmid pSKCMVPGMf1 as template. The 1.3-kb fragment is gel purified and ligated to the vector as described previously.

Transformation and Screening

Plasmids are transformed into *E. coli* XL1B (Stratagene, La Jolla, CA) and plated on Luria-Bertani (LB) plates with 0.2% glucose and 100 μg/ml ampicillin (LBD$^{0.2}$Ap^{100}). Plates are incubated at 37° overnight Colonies are picked and grown in 2 ml LBD$^{0.2}$Ap^{100} for 20 hr with shaking at 225 rpm and 37°. They are screened by miniscale plasmid DNA preparations followed by restriction enzyme mapping and by induction of protein expression on a small scale (2 ml culture) followed by analysis of uninduced and induced samples on SDS–PAGE; the bands are visualized with Coomassie blue staining.

Expression and Purification

LBD$^{0.2}$Ap^{100} (225 ml) is inoculated with 2 ml of an overnight culture of cells containing fusion plasmid. The cells are grown at 37° and 225 rpm until an OD$_{600}$ of 0.5 and are then induced with the addition of 0.3 mM isopropyl-β-thiogalactopyranoside (IPTG). The temperature is reduced to 30°, and cells are grown for another 2–3 hr. The cells are harvested by centrifugation at 6800g for 5 min, the supernatant discarded, and the pellet

[19] T. D. Parks, K. K. Leuther, E. D. Howard, S. A. Johnston, and W. G. Dougherty, *Anal. Biochem.* **216,** 413 (1994).
[20] J. S. Schorey, S. C. Fortenberry, and J. M. Chirgwin, *J. Cell Sci.* **108,** 2007 (1995).

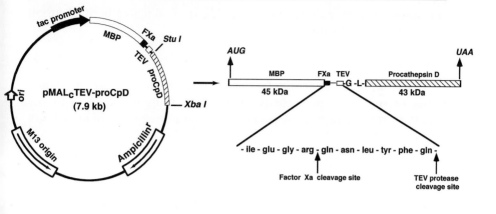

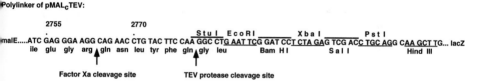

FIG. 1. pMAL$_c$TEV-proCpD expression vector and MBP-procathepsin D fusion protein. pMAL$_c$TEV is a vector that expresses a cloned gene as a fusion protein to MBP. The pMAL$_c$TEV-proCpD vector shown on the left expresses the fusion protein MBP-procathepsin D. The vector contains the inducible ptac promoter. The lac repressor encoded by the lacI gene suppresses transcription until IPTG is added. Arrows indicate direction of transcription. Restriction sites shown are unique. A gene of interest can be cloned at the end of the MBP gene between the blunt StuI site and any other site in the polylinker. The polylinker of pMAL$_c$TEV[24] is the same as that of pMAL-c2 vector[18] except that the XmnI site is replaced by StuI. The fusion protein MBP-procathepsin D is shown on the right and consists of bacterial MBP followed by eukaryotic procathepsin D. Cleavage sites for factor Xa protease and TEV protease are present between the two fusion partners.

suspended in 10 ml column buffer (20 mM Tris, pH 7.4, 200 mM NaCl, 1 mM EDTA, 0.02% Tween 80), and frozen at $-20°$ overnight. The sample is thawed in cold water, diluted with 10 ml of column buffer, and sonicated on ice with a microtip for 4 min at 75% pulse and maximum power. The sonicate is centrifuged at 20,000g for 15 min at 4°. The supernatant (crude extract) is diluted with another 10 ml of column buffer. Amylose resin (New England Biolabs, Beverly, MA) is poured into Econocolumns (Bio-Rad, Richmond, CA) to give a bed volume of about 10 ml, and the extract is applied to the resin at 0.5 ml/min. The resin is washed with several column volumes of buffer, and the protein is eluted with column buffer containing 10 mM maltose. Typical yields are 6–8 mg of purified fusion protein per 225-ml culture.

Cleavage of Fusion Partners

Proteins are incubated with factor Xa (New England Biolabs) in column buffer or with TEV protease (GIBCO-BRL) in buffer supplied by the manufacturer at a ratio of 50 μg protein : 1 μg protease for 18 hr at room temperature in the absence or presence of 1% SDS.

Refolding and Enzymatic Activity Assays

MBP fusions are denatured by the addition of solid urea to a concentration of 8 M urea and refolded following the protocol of Kuhelj et al.[21] Enzymatic activity[22] is determined by monitoring the proteolysis of ^{14}C-labeled hemoglobin as described.[23]

Results and Discussion

Expression Vector pMAL$_c$TEV

Figure 1 shows the pMAL$_c$TEV-proCpD plasmid used for the expression of MBP-procathepsin D. The vector pMAL$_c$TEV allows any DNA of interest to be inserted downstream of the MBP sequence between the StuI and any other restriction site 3′ in the polylinker.[24] Cleavage at the StuI site leaves a blunt end after the second base in the glycine codon of the TEV recognition sequence. Blunt-ended, phosphorylated PCR fragments can be ligated at this site, provided that one base is provided by the primer to recreate the glycine codon for TEV recognition and to preserve the reading frame. The fusion gene is transcribed from a tac promoter, and protein expression is induced with IPTG. No specialized bacterial host strain is required.

Soluble Expression of Fusions of MBP to the N Terminus of Eukaryotic Proteins

The aspartic proteinases procathepsin D and pepsinogen, when expressed in bacteria with MBP fused to their N termini, accumulated in the

[21] R. Kuhelj, M. Dolinar, J. Pungercar, and V. Turk, Eur. J. Biochem. **229**, 533 (1995).
[22] A. J. Barrett, in "Proteinases in Mammalian Cells and Tissues" (A. J. Barrett, ed.), p. 209. Elsevier/North Holland Biomedical Press, Amsterdam, 1977.
[23] D. Sachdev and J. M. Chirgwin, Protein Expr. Purif. **12**, 122 (1998).
[24] D. Sachdev and J. M. Chirgwin, Biochem. Biophys. Res. Commun. **244**, 933 (1998).

soluble fraction in the bacterial cytosol.[24] No insoluble fusion protein was detected. We have reproducibly expressed three human aspartic proteinases, with or without their propeptides, as MBP fusions in the soluble fraction of induced cells with yields ranging from 6 to 8 mg of pure protein from a 225-ml culture. After a one-step purification by amylose affinity chromatography, the fusions are about 90% pure as judged by SDS–PAGE.[24] The fusions remain soluble even on extended refrigeration at protein concentrations as high as 10 mg/ml. We have found it advisable to use freshly transformed *E. coli* strain XLIB for expression of the fusion proteins, as expression levels drop considerably with older transformed cells.

Cleavage of the Protein of Interest from the Fusion Partner

Initially, we used the original pMAL-c2 vector to express the aspartic proteinases in *E. coli.* However, factor Xa was unable to cleave the fusion protein. Steric hindrance may have impeded access of factor Xa to its cleavage site. Therefore, we constructed the modified vector pMAL$_c$TEV, which includes a TEV cleavage site downstream of the factor Xa site. This introduced an extra eight amino acid spacer between the factor Xa site and the amino terminus of the eukaryotic partner.[24] Fusions with cleavage sites for both endoproteinases were resistant to cleavage with TEV but were cleaved (inefficiently) with factor Xa. We were unable to optimize cleavage of procathepsin D or pepsinogen from MBP in the fusions because of apparent proteolysis of the eukaryotic partner. Others have also reported inefficient cleavage with factor Xa and have had better success with thrombin.[25] Inefficient and nonspecific digestion has been reported when using factor Xa.[26] We, therefore, used the fusion proteins without cleavage of the MBP partner to study the folding and properties of procathepsin D and pepsinogen.[27] Further biophysical characterization of the fusions revealed that, although the aspartic proteinase moiety is soluble, it is not in its native conformation.[27]

Although the ability to cleave MBP fusions depends on the eukaryotic partner, the attachment of MBP to the N terminus does not interfere with the function of some proteins, and several groups have used fusion proteins without cleavage of the MBP for structure–function studies. Two proteins essential for nucleotide excision repair in humans were expressed solubly

[25] C. Wang, A. F. Castro, D. M. Wilkes, and G. A. Altenberg, *Biochem. J.* **338,** 77 (1999).

[26] Y. H. Ko, P. J. Thomas, M. R. Delannoy, and P. L. Pedersen, *J. Biol. Chem.* **268,** 24330 (1993).

[27] D. Sachdev and J. M. Chirgwin, *J. Protein Chem.* **18,** 127 (1999).

in *E. coli* as fusions to MBP and complemented excision repair activity in mutant cell-free extracts.[28] An MBP-DnaJ fusion protein expressed in *E. coli* and purified to homogeneity was able to carry out the functions of DnaJ, such as stimulating ATPase activity of DnaK and preventing aggregation of denatured protein substrate,[29] indicating that fusion to MBP did not affect the functions of DnaJ. Hence, while cleavage of the protein of interest from MBP may not always be successful, the advantages of using MBP as a fusion partner may outweigh the problem. In a small percentage of proteins expressed as fusions to MBP, cleavage of MBP from the protein results in precipitation of the protein of interest, and thus, in those cases it is desirable to use the fusion protein itself for structure–function studies.[30] From a practical point of view, the cleavage of fusion proteins *in vitro* and their repurification to remove the bacterial partner are rate-limiting steps in terms of both time and expense. Because the structure and folding of MBP are well characterized, many experiments can be carried out with purified but uncleaved MBP fusions.

Role of MBP Fused to N Terminus of Eukaryotic Proteins

We found that when MBP was fused to the N terminus of aspartic proteinases, the fusions were all solubly expressed in bacteria.[24] In contrast, when aspartic proteinases were expressed unfused in *E. coli,* they accumulated in inclusion bodies. Inclusion bodies formed even when the growth temperature was lowered. The accumulation of aspartic proteinases into inclusion bodies was also seen by several laboratories using different bacterial expression vectors and strains of *E. coli.*[16,31,32] Thus, fusions of maltose-binding protein to the N terminus of aspartic proteinases efficiently block the aggregation into inclusion bodies seen when unfused aspartic proteinases are expressed in *E. coli.* As described in the next section and reported previously by us,[23] when MBP was fused to the C terminus of aspartic proteinases, the aspartic proteinase-MBP fusions still accumulated in inclusion bodies. Data suggest that the solubility conferred by N-terminal MBP fusions is a kinetic rather than an equilibrium effect.

The position of MBP in the fusion influences its efficiency in controlling the solubility and folding of the eukaryotic partner. This suggests that the

[28] C. H. Park and A. Sancar, *Nucleic Acids Res.* **21,** 5110 (1993).
[29] Y. Ishii, S. Sonezaki, Y. Iwasaki, E. Tauchi, Y. Shingu, K. Okita, H. Ogawa, Y. Kato, and A. Kondo, *J. Biochem.* **124,** 842 (1998).
[30] S. Sonezaki, Y. Ishii, K. Okita, T. Sugino, A. Kondo, and Y. Kato, *Appl. Microbiol. Biotechnol.* **43,** 304 (1995).
[31] P. E. Scarborough, G. R. Richo, J. Kay, G. E. Conner, and B. M. Dunn, *Adv. Exp. Med. Biol.* **306,** 343 (1991).
[32] T. J. Cottrell, L. J. Harris, T. Tanaka, and R. Y. Yada, *J. Biol. Chem.* **270,** 19974 (1995).

normal folding of the first sequence to emerge from the ribosome determines the folding pathway in *E. coli*. The order of the fusions does not alter the intrinsic folding properties of the partners, as we have shown that MBP-pepsinogen[24] and pepsinogen-MBP[23] are both proteolytically active and able to bind amylose following refolding from urea. Thus, the fusions appear to fold cotranslationally in bacteria.

During the biosynthesis of the MBP-aspartic proteinases in bacteria, the MBP moiety probably folds before the nascent aspartic proteinase emerges from the ribosome. The folded MBP could then function as a *cis*-acting chaperone for the aspartic proteinase and protect the latter from those interactions that would otherwise lead to inclusion body formation. The efficient and robust folding of the MBP moiety could drive the folding of the aspartic proteinase down a pathway to a soluble but nonnative state. This pathway is obviously different from the one followed when unfused aspartic proteinases are expressed in *E. coli*.

In addition to its role in controlling the solubility and folding of eukaryotic partners expressed in *E. coli*, MBP has been suggested to have other functions. Human T-cell leukemia virus type I transmembrane protein gp21 could not be crystallized as an MBP-gp21 fusion protein when the fusion partners were separated by a linker; however, the fusion protein was crystallized readily when the MBP C-terminal α helix was connected directly to the predicted N-terminal α helix of gp21.[33] Thus, MBP may have a general application for the crystallization of proteins containing N-terminal α-helical sequences. Because overexpression of an eukaryotic protein in bacteria is often undertaken to obtain the large quantities of the protein required for crystallization studies, the ability of MBP to facilitate crystal growth may make it a useful fusion partner, although the size of the data set to be solved will be increased substantially by the addition of MBP.

MBP fusions have also been used in characterizing phage-displayed peptides[34] and as immunogens for raising antibodies against eukaryotic proteins, as it is easy to obtain large quantities of pure protein. Antibodies directed against MBP can be removed readily from polyclonal antisera by passage over amylose resin bearing unfused MBP. It has been reported that in some cases an MBP-peptide fusion may be more antigenic and a better mimic of an epitope in a viral particle than a conjugated synthetic peptide or β-galactosidase-peptide fusion.[35]

Another advantage of using MBP as a fusion partner is that the fusions

[33] R. J. Center, B. Kobe, K. A. Wilson, T. The, G. J. Howlett, B. E. Kemp, and P. Poumbourios, *Protein Sci.* **7,** 1612 (1998).

[34] M. B. Zwick, L. L. Bonnycastle, K. A. Noren, S. Venturing, E. Leong, C. F. R. Barbas, C. J. Noren, and J. K. Scott, *Anal. Biochem.* **264,** 87 (1998).

[35] A. Benito and M. H. van Regenmortel, *FEMS Immunol. Med. Microbiol.* **21,** 101 (1998).

TABLE I

COMPARISON OF THE SOLUBILITY STATE OF VARIOUS CONSTRUCTS EXPRESSED IN *E. coli*

Protein	Expression state in *E. coli*	State after *in Vitro* refolding
Procathepsin D	Inclusion bodies	Mostly aggregated; 1–2% native yield
MBP-procathepsin D	Soluble in bacterial cytosol	Soluble; only 1–2% native yield
Procathepsin D-MBP	Inclusion bodies	Soluble and mostly nonnative
Pepsinogen	Inclusion bodies	Soluble and native
MBP-pepsinogen	Soluble in bacterial cytosol	Soluble and native
Pepsinogen-MBP	Inclusion bodies	Soluble and native

can also be targeted to the periplasm, where MBP folds efficiently in its normal environment. Expression of eukaryotic proteins in the periplasm often results in the soluble expression of heterologous protein compared to the bacterial cytosol, as the periplasm is not as reducing an environment as the cytosol and contains a unique complement of protein-folding catalysts. Yields are reduced substantially, however, by the relatively small volume of the periplasmic compartment compared to the bacterial cytoplasm. Only MBP was able to give soluble expression of some members of the ABC transporter gene family. Fusions to either thioredoxin or GST did not increase their soluble expression.[25] The choice of fusion partner used for expressing eukaryotic proteins does influence the solubility of the protein. A GST-interleukin 6 fusion accumulated in *E. coli* inclusion bodies, whereas a similar thioredoxin-interleukin 6 fusion was found in the *E. coli* periplasm in biologically active form.[36] MBP may be more efficient than GST or thioredoxin in assisting and controlling the solubility of its fusion partners, perhaps because the folding of MBP is more efficient and robust. Because it folds efficiently to its native state even on overexpression, it may in general be able to lead its fusion partners along a soluble folding pathway.

Role of MBP Fused to C Terminus of Eukaryotic Proteins

Addition of MBP as a C-terminal extension to the aspartic proteinases did not change their accumulation into inclusion bodies (Table I).[23] However, fusions to either *E. coli* maltose-binding protein or thioredoxin resulted in the quantitative recovery of soluble procathepsin D and pepsinogen on refolding from 8 M urea.[23] Refolding *in vitro* of these proteins in unfused form can result in insoluble aggregation, particularly in the case of procathepsin D.[16]

[36] B. Frorath, C. C. Abney, H. Berthold, M. Scanarini, and W. Northemann, *Biotechniques* **12,** 558 (1992).

The participation of MBP in the folding of fusion proteins suppresses intermolecular aggregation.[14] MBP renatures efficiently and may fold first when aspartic proteinase-MBP fusions are refolded *in vitro*. The presence of a highly soluble partner within the fusion may prevent aggregation of intermediates, which otherwise tend to associate during the slower folding of the aspartic proteinase. However, when the aspartic proteinase moieties are released first from the bacterial ribosomes during translation, their rate of aggregation apparently outruns the protective effect of MBP, resulting in fusions accumulating in inclusion bodies. High concentrations of nascent polypeptide chains favor aggregation, and aggregated polypeptides then collect in inclusion bodies.[37]

Fusions of MBP to the C termini of eukaryotic proteins serve an entirely different role from that described for fusions to the N termini of eukaryotic proteins and may be of general value in the biophysical characterization of nonnative folding states of partner proteins, without interference from aggregation and insolubility. Furthermore, for proteins that cannot tolerate N-terminal fusions, MBP at the C terminus provides a suitable affinity tag for purification.[38] C-terminal MBP also provides a convenient means of separating full-length protein from products of premature translational termination, as only full-length protein will bind to amylose.

Concluding Remarks

Results from the bacterial expression of fusions of MBP to eukaryotic proteins suggest that rapid folding of the first sequence to emerge from the ribosome determines posttranslational fate in *E. coli*. Data suggest that the solubility of fusion proteins is controlled by whether the protein domains emerging first from the ribosome fold into soluble or insoluble states in the bacterial cytosol.

Fusions to maltose-binding protein efficiently block the aggregation into inclusion bodies seen when unfused aspartic proteinases are expressed in *E. coli*. MBP appears to be generally more efficient than thioredoxin or GST in improving the solubility of the eukaryotic partner.

Acknowledgment

This work was supported by Merit and Associate Career Research Scientist Awards from the Veterans Administration Research Service to JMC.

[37] R. Rudolph and H. Lilie, *FASEB J.* **10,** 49 (1996).
[38] L. Henning and E. Schafer, *Protein Expr. Purif.* **14,** 367 (1998).

[21] Thioredoxin as a Fusion Partner for Production of Soluble Recombinant Proteins in *Escherichia coli*

By Edward R. LaVallie, Zhijian Lu, Elizabeth A. DiBlasio-Smith, Lisa A. Collins-Racie, and John M. McCoy

Introduction

Escherichia coli is often the first choice for investigators seeking a host organism in which to produce recombinant proteins for research or commercial purposes. The popularity of *E. coli* stems from its ease of culture, its well-characterized physiology and genetics, and the many excellent recombinant gene expression systems that have been developed for the organism since the 1980s.[1] Experimentalists choosing to express their particular genes of interest in *E. coli* enjoy a number of advantages. Recombinant gene expression in *E. coli* is fast; it is not unusual to progress from an isolated cDNA fragment to a purified recombinant protein in just a few days. Expression yields are high; recombinant protein typically may represent anywhere from 1 to 20% of the total *E. coli* cell protein, and success at the bench usually is predictive of success at a larger scale. Finally, the ease and convenience of *E. coli* expression systems allow for great operational flexibility, such that multiple experimental approaches can often be carried out simultaneously.

Escherichia coli has been used widely and successfully to produce recombinant proteins for a variety of purposes, including tertiary structure determinations, enzymological studies, structure–function experiments, and for pharmaceutical manufacturing. Nevertheless, *E. coli* expression systems possess several inherent disadvantages. Translation initiation of eukaryotic mRNAs on bacterial ribosomes can often be inefficient,[2] *E. coli* can sometimes fail to remove effectively the N-terminal initiator methionine residue from recombinant proteins,[3] and each native recombinant protein produced in *E. coli* usually requires the development of a custom purification scheme. Perhaps the biggest disadvantage of all, however, is the tendency of *E. coli* for producing recombinant proteins as misfolded cytoplasmic aggregates known as "inclusion bodies."[4] This tendency is particularly strong for com-

---

[1] P. O. Olins and S. C. Lee, *Curr. Opin. Biotechnol.* **4** (1993).

[2] G. D. Stormo, T. D. Schneider, and L. Gold, *Nucleic Acids Res.* **10**, 2971 (1982).

[3] P. H. Hirel, M. J. Schmitter, P. Dessen, G. Fayat, and S. Blanquet, *Proc. Natl. Acad. Sci. U.S.A.* **86**, 8247 (1989).

[4] A. Mitraki and J. King, *Bio/Technology* **7**, 690 (1989).

plex multidomain proteins, for proteins requiring specialized posttranslational modifications or prosthetic groups, and for eukaryotic-secreted glycoproteins.

A popular strategy to address many of these disadvantages of recombinant protein production in *E. coli* has been to utilize gene fusions.[5] In these approaches the protein of interest is fused to a "partner" protein, with the "partner" protein usually placed at the amino terminus. Successful fusion partners are typically well translated, they carry determinants that allow for convenient purifications, they can be cleaved easily and completely from the desired protein product, and ideally they reduce the incidence of inclusion body formation.

Previously we described an *E. coli* gene fusion expression system utilizing *E. coli* thioredoxin (TrxA) as the fusion partner.[6] Thioredoxin fusions have proved to be especially useful in avoiding inclusion body formation, particularly for the production of small, normally secreted, mammalian cytokines in an active form in the *E. coli* cytoplasm (see Fig. 1). *Escherichia coli* thioredoxin is a compact, highly soluble, thermally stable protein with robust folding characteristics.[7] These properties perhaps allow the molecule, when fused to a protein of interest, to serve as a covalently joined molecular chaperon. Thioredoxin may thus act to prevent the aggregation and precipitation of fused nascent proteins, giving them an extended opportunity to adopt their correct tertiary folds. Thioredoxin also possesses a number of additional characteristics that suit it to a role as a fusion partner. It is small, highly translated, and its tertiary structure reveals that both its amino and carboxyl termini are accessible for potential fusions to other molecules.[8] Moreover, its active-site comprises a surface-accessible loop that can be utilized for internal peptide insertions. Purifications of thioredoxin fusion proteins can be facilitated by making use of the remarkable ability of the molecule to be released from the bacterial cytoplasm by simple osmotic shock,[6] by taking advantage of the inherent thermal stability of the molecule,[6] by using avidin or streptavidin matrices to capture thioredoxin variants modified to allow for *in vivo* biotinylation,[9] or by using engineered forms of thioredoxin with affinity to metal chelate column matrices.[10] De-

[5] E. R. LaVallie and J. M. McCoy, *Curr. Opin. Biotechnol.* **6,** 501 (1995).
[6] E. R. LaVallie, E. A. DiBlasio, S. Kovacic, K. L. Grant, P. F. Schendel, and J. M. McCoy, *Bio/Technology* **11,** 187 (1993).
[7] A. Holmgren, *Annu. Rev. Biochem.* **54,** 237 (1985).
[8] S. K. Katti, D. M. Le Master, and H. Eklund, *J. Mol. Biol.* **257,** 167 (1990).
[9] P. A. Smith, B. C. Tripp, E. A. DiBlasio-Smith, Z. Lu, E. R. LaVallie, and J. M. McCoy, *Nucleic Acids Res.* **26,** 1414 (1998).
[10] Z. Lu, E. A. DiBlasio-Smith, K. L. Grant, N. W. Warne, E. R. LaVallie, L. A. Collins-Racie, M. T. Follettie, M. J. Williamson, and J. M. McCoy, *J. Biol. Chem.* **271,** 5059 (1996).

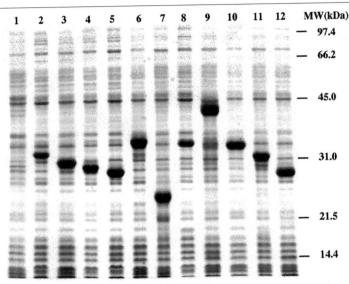

Fig. 1. Proteins found in *soluble* fractions derived from *E. coli* cells expressing 11 different thioredoxin gene fusions. Numbers in parentheses refer to the growth temperature chosen for expressing each particular fusion. Lane 1, host *E. coli* strain GI724 (negative control, 37°); lane 2, murine IL-2 (15°); lane 3, human IL-3 (15°); lane 4, murine IL-4 (15°); lane 5, murine IL-5 (15°); lane 6, human IL-6 (25°); lane 7, human MIP-1a (37°); lane 8, human IL-11 (37°); lane 9, human M-CSF (37°); lane 10, murine LIF (25°); lane 11, murine SF (37°); and lane 12, human BMP-2 (25°). A 10% SDS–PAGE gel, stained with Coomassie blue, is shown. Reproduced from LaVallie *et al., Bio/Technology* **11,** 187 (1993).

tailed protocols are provided for the use of the thioredoxin gene fusion expression system. We describe a variety of suitable *E. coli* expression strains and a number of thioredoxin expression vectors. We furnish an expression protocol, offer a number of procedures for thioredoxin fusion protein purification, and present a method for specific cleavage of thioredoxin fusions by enterokinase.

E. coli Expression Strains

The plasmid vectors described in the following section all utilize the major leftward promoter of bacteriophage λ, pL, for driving transcription of thioredoxin fusion genes. Several *E. coli* host strains have been specifically constructed to allow for the use of pL at any culture growth

TABLE I
E. coli STRAINS FOR PRODUCTION OF THIOREDOXIN FUSION PROTEINS AT
VARYING TEMPERATURES

Strain	Desired production temperature (°C)	Preinduction growth temperature (°C)	Induction period (hr)
GI698	15	25	20
GI698	20	25	18
GI698	25	25	10
GI724	30	30	6
GI724	37	30	4
GI723	37	37	5

temperature (Table I). All are based on *E. coli* K12 strain RB791 (= W3110 *lacI*q*lacPL8*).[11] In a manner similar to the gene control scheme described by Mieschendahl *et al.*,[12] each strain carries the bacteriophage λ repressor gene (cI) stably integrated into the chromosomal *ampC* locus under the transcriptional control of a synthetic *Salmonella typhimurium* *trp* promoter. This promoter is integrated upstream of cI in *ampC*. Levels of λ repressor protein in these strains are dependent on two factors: (1) the strength of the ribosome-binding site (RBS) positioned upstream of the integrated cI gene (where RBS strength decreases in the order GI723 > GI724 > GI698) and (2) the level of exogenous tryptophan present in the culture medium. Growth in minimal media or in rich media devoid of tryptophan (such as those containing casamino acids as the rich component) results in cI synthesis and repression of pL. Addition of tryptophan to the growth culture represses cI synthesis, leading to a gradual depletion of cI protein and a concomitant induction of pL-driven transcription. As shown in Table I, the choice of strain to use depends on the desired temperature of induction. Lower induction temperatures are sometimes preferable so as to maximize the amount of soluble protein produced.[13]

Thioredoxin Gene Fusion Expression Vectors

Several expression vectors were created to facilitate the use of thioredoxin, thioredoxin derivatives, and thioredoxin homologues as fusion part-

[11] R. Brent and M. Ptashne, *Proc. Natl. Acad. Sci. U.S.A.* **78,** 4204 (1981).
[12] M. Mieschendahl, T. Petri, and U. Hanggi, *Bio/Technology* **4,** 802 (1986).
[13] C. H. Schein and M. H. M. Noteborn, *Bio/Technology* **6,** 291 (1988).

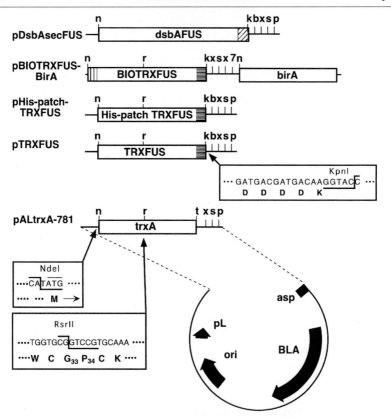

FIG. 2. Expression vectors for fusions to thioredoxin and thioredoxin variants. asp, *E. coli* aspA transcription terminator; BLA, β-lactamase gene; ori, colE1 origin of replication; pL, bacteriophage λ major leftward promoter; 7, bacteriophage T7 gene 10 ribosome-binding site; t, thioredoxin translation termination codon. A number of restriction endonuclease sites are marked with lowercase letters: b, *Bam*HI; k, *Kpn*I; p, *Pst*I; n, *Nde*I; r, *Rsr*II; s, *Sal*I; x, *Xba*I. The linker peptide "-GSGSGDDDDK-," containing an enterokinase recognition sequence, is shown by horizontal hatch marks. The linker peptide "-GSGSGHHHHHHDDDDK-," containing an enterokinase recognition sequence and a hexahistidine purification tag, is shownby diagonal hatch marks. The biotinylation peptide "MASSLRQILDSQKMEWRS-NAGGS-" is shown by vertical hatch marks.

ners. Plasmid maps for five vectors are shown in Fig. 2; each is discussed in more detail later. All the vectors share certain common features; all are based on pUC-18[14] and contain a colE1 origin of replication, a β-lactamase gene as a selectable marker, and the bacteriophage λ pL promoter[15] located

[14] J. Norrander, T. Kempe, and J. Messing, *Gene* **26,** 101 (1983).
[15] H. Shimatake and M. Rosenberg, *Nature* **292,** 128 (1981).

upstream of the thioredoxin gene or thioredoxin-related sequence. To ensure good translation initiation, the *trx*A-related gene is typically positioned 5' of the gene of interest. A short DNA sequence encoding a linker peptide of sequence "-GSGSGDDDDK-" is positioned immediately 3' of the *trx*A-related gene in place of its normal translational termination codon (but not in pALtrxA-781, which produces native thioredoxin). This short linker not only serves as a flexible peptide spacer between the TrxA-related protein and any downstream protein sequence, but also provides a specific enteropeptidase (also called enterokinase or EK[16]) site to allow for subsequent fusion protein cleavage. EK is particularly well suited for the cleavage of caboxyl-terminal fusions, as its recognition site, the sequence "-DDDDK-", lies entirely on the amino-terminal side of the scissile bond. Thus carboxyl-terminal-fused proteins liberated by EK cleavage bear their native amino termini. Immediately following the sequence encoding the linker peptide lies a polylinker DNA sequence containing convenient restriction sites, which allow for precise translational fusions of coding sequences to the upstream *trx*A-related gene. The entire thioredoxin fusion transcription unit culminates with a transcription terminator sequence based on that found in the *E. coli asp*A gene.

pALtrxA-781

This vector carries the wild-type thioredoxin gene and is useful for constructing internal peptide fusions into the thioredoxin active-site loop.[6] A unique *Rsr*II restriction site lying within the sequence encoding this loop is perfectly positioned to allow for convenient insertion of sequences encoding short peptides. When expressed as thioredoxin "loop" fusions, these peptides are positioned in an accessible location on the surface of the protein.[6] Thioredoxin active-site loop fusions have proven to be useful in a number of binding studies and as antigens.[9,17,18,19]

5' fusions to wild-type thioredoxin can also be constructed using pALtrxA-781, taking advantage of the unique *Nde*I site at the extreme 5' end of the thioredoxin-coding sequence. Be aware that changes made at the 5' end of the thioredoxin gene may have deleterious effects on the efficiency of translation initiation. Also, because the thioredoxin gene found

---

[16] S. Maroux, J. Baratti, and P. Desnuelle, *J. Biol. Chem.* **246**, 5031 (1971).
[17] Z. Lu, K. S. Murray, V. van Cleave, E. R. LaVallie, M. L. Stahl, and J. M. McCoy, *Bio/Technology* **13**, 366 (1995).
[18] P. Colas, B. Cohen, T. Jessen, I. Grishina, J. McCoy, and R. Brent, *Nature* **380**, 548 (1996).
[19] S. J. Lombardi, A. Truong, P. Spence, K. J. Rhodes, and P. G. Jones, *J. Biol. Chem.* **273**, 30092 (1998).

in pALtrxA-781 carries its native translation termination codon, the vector is not readily amenable to carboxyl-terminal thioredoxin fusions.

pTRXFUS

The pTRXFUS plasmid allows for in-frame fusions of any coding sequence of interest to the 3′ end of the wild-type *E. coli* thioredoxin gene.[6] Carboxyl-terminal fusions to thioredoxin can be cleaved with enterokinase using the cleavage site positioned in the intervening linker peptide. pTRXFUS can also be used to construct 5′ fusions and active-site loop internal fusions.

The pTRXFUS plasmid produces thioredoxin fusion proteins at high levels, and very often these fusions are found in the soluble cellular fraction. However, the purification advantages afforded by wild-type thioredoxin are limited. Certain wild-type thioredoxin fusions can be partially purified by osmotic shock procedures (see later) or by protocols utilizing the high thermal stability of particular thioredoxin fusions relative to host protein contaminants (see later). However, to provide more generic, efficient, and high-throughput affinity purification schemes, thioredoxin variants were developed with improved purification properties (see later).

pHis-patch-TRXFUS

The pHis-patch-TRXFUS plasmid contains a modified *E. coli* *trx*A gene in which codons for two surface-exposed amino acids have been changed to histidine codons (E30H and Q62H). The resulting "histidine-patch" thioredoxin (hpTRX) can bind to nickel ions immobilized on iminodiacetic acid- and nitrilotriacetic acid-Sepharose resins, providing a very convenient purification step.[10] An additional mutation (D26A) introduced into hpTRX restores the thermal stability of the mutant protein to wild-type levels.[10] "Histidine-patch" thioredoxin retains all of the desirable attributes of wild-type thioredoxin as a fusion partner, but provides the added benefit of a specific purification handle. As with wild-type thioredoxin, fusions to hpTRX can be made at the amino terminus, the carboxyl terminus, and within the active-site loop. pHis-patch-TRXFUS also encodes an enterokinase cleavage site at the hpTRX carboxyl terminus, allowing for cleavage of carboxyl-terminal protein fusions.

Proteins fused to hpTRX often can be purified from *E. coli* lysates in a single step at high yield using immobilized metal affinity chromatography (IMAC).[10] In a purification scheme known as the "IMAC mirror," an initial IMAC column performed on a clarified cell lysate is followed by an enterokinase cleavage reaction and then by a second IMAC column identical to the first. All IMAC-binding contaminants remaining from the first column, all uncleaved hpTRX fusion protein, and all released hpTRX protein are retained by the second IMAC column, whereas the desired protein product is found in the flow through, highly purified and in high yield.

The affinity of histidine-patch thioredoxin or metal ions can also be exploited for binding studies utilizing immobilized hpTRX fusion proteins. In this regard it is worth noting that the IMAC-binding surface of hpTRX lies on the face of the molecule opposite to carboxyl terminus, ensuring that C-terminally fused proteins will be accessible even when bound to an IMAC matrix.

pBIOTRXFUS-BirA

This vector produces a form of recombinant thioredoxin (BIOTRX) that can be quantitatively biotinylated by *E. coli in vivo*.[9] The vector takes advantage of a novel strategy for *in vivo* biotinylation originally reported by Schatz,[20] who identified a series of small, 14–23 residue peptides that mimic the folded conformation of the biotin carboxyl-carrier protein subunit of *E. coli* acetyl-CoA carboxylase (BCCP), and which are themselves substrates for biotinylation *in vivo* at specific lysine residues by *E. coli* biotin holoenzyme synthetase (BirA). In pBIOTRXFUS-BirA, a sequence encoding the biotinylation peptide "MASSLRQILDSQ<u>K</u>MEWRSNAG-GS-" is fused to the 5' end of the wild-type *E. coli* thioredoxin gene. This peptide is *in vivo* biotinylated (on the lysine residue underlined in the sequence just given) to a limited extent by endogenous levels of *E. coli* BirA; however, the presence of the *bir*A gene in the vector (joined in an operon with the BIOTRX gene), along with the addition of exogenous biotin to the growth medium, results in quantitative biotinylation of BIOTRX or BIOTRX fusion proteins, which can then be captured from lysates using immobilized avidin or streptavidin. Similar to the wild-type and his-patch thioredoxin expression constructs described earlier, the presence of an enterokinase cleavage site at the BIOTRX carboxyl terminus allows for convenient cleavage of carboxyl-terminal protein fusions.

The high-affinity binding interaction of biotin to avidin or streptavidin has been used widely in biochemistry and molecular biology.[21] However, *in vitro* chemical techniques for protein biotinylation are not always successful, with some common problems being a lack of reaction specificity, inactivation of amino acid residues critical for protein function, and low levels of biotin incorporation. *In vivo* biotinylation avoids these issues. BIOTRX, like hpTRX, retains all of the desirable attributes of wild-type thioredoxin

---

[20] P. J. Schatz, *Bio/Technology* **11,** 1138 (1993).
[21] M. Wilchek and E. A. Bayer, *Methods Enzymol.* **184,** 5 (1990).

as a fusion partner, but provides the added benefit of a high specificity tag, which can be used either as a purification handle or for sensitive detection purposes. Biotinylated BIOTRX fusions are especially useful in binding studies involving antibodies or ligand–receptor pairs, where the high-affinity biotin–avidin/streptavidin interaction is a distinct advantage. Moreover, in BIOTRX the biotinylated peptide is linked to the amino terminus of thioredoxin, leaving both the carboxyl terminus and the active-site loop free for fusions. The thioredoxin tertiary structure[8] shows that it is unlikely that binding of avidin/streptavidin to the amino-terminal biotinylation tag will be hindered by peptides or proteins fused at either the active-site loop or the carboxyl terminus.

pDsbAsecFUS

Because thioredoxin is a cytoplasmic protein, fusions to thioredoxin are produced in the *E. coli* cytoplasm. There are circumstances, however, when secretion of a fusion protein to the *E. coli* periplasmic space is essential in order to provide the appropriate oxidizing environment for correct protein folding and/or activity, as was the case for recombinant enterokinase produced in *E. coli*.[22] pDsbAsecFUS is an expression vector that, in place of thioredoxin, utilizes the secreted thioredoxin homologue, DsbA,[23] as a partner protein for directing fusions to the periplasm. Fusions to DsbA are typically constructed to the carboxyl terminus of the protein, as amino-terminal fusions are complicated by the presence of the secretory signal peptide. DsbA active-site loop fusions may be possible but have not yet been attempted. Carboxyl-terminal DsbA fusions compare favorably to thioredoxin fusions both in levels of expression and in solubility.[24] pDsbAsecFUS encodes a carboxyl-terminal spacer/ EK site (-GSGSGHHHHHHDDDDK-), which contains an additional hexa-histidine sequence not found in pTRXFUS. This hexa-histidine sequence confers metal ion binding affinity to DsbA fusion proteins.

We have also fused various proteins to the carboxyl terminus of *mature* DsbA, using a vector (pDsbAmatFUS) in which the DNA sequence encoding the DsbA signal peptide has been replaced with a simple initiator methionine codon. These proteins are localized to the *E. coli* cytoplasm and behave very much like fusions of these same proteins to thioredoxin.[24]

Secreted DsbA fusions produced in *E. coli* using pDsbAsecFUS can be isolated by selectively releasing the periplasmic contents of the cell by

[22] L. A. Collins-Racie, J. M. McColgan, K. L. Grant, E. A. DiBlasio-Smith, J. M. McCoy, and E. R. LaVallie, *Bio/Technology* **13**, 982 (1995).

[23] J. C. Bardwell, K. McGovern, and J. Beckwith, *Cell* **67**, 581 (1991).

[24] Edward R. LaVallie, unpublished results (1995).

osmotic shock or, alternatively, the cells can be lysed by mechanical means and the fusion purified by IMAC chromatography (see later).

Table II summarizes the properties of the various thioredoxin fusion partners discussed earlier.

Recombinant Gene Expression Using Thioredoxin Gene Fusion Vectors

This section provides a protocol for construction of a gene fusion between thioredoxin (or a thioredoxin variant) and any gene encoding a particular protein or peptide of interest and a procedure for fusion protein production in *E. coli*. The method is described in terms of host strain GI724 grown at 30°, although it may also be applied to strains GI698 and GI723 for growth at other temperatures according to the parameters specified in Table I.

Materials

A DNA fragment encoding the desired protein

A thioredoxin expression vector: pALtrxA-781, pTRXFUS, pHis-patch-TRXFUS, pBIOTRXFUS-BirA or pDsbAsecFUS

E. coli strain GI724, grown in LB medium and made transformation competent

$10\times$ M9 salts: 60 g Na_2HPO_4 (0.42 M final), 30 g KH_2PO_4 (0.24 M final), 5 g NaCl (0.09 M final), 10 g NH_4Cl (0.19 M final), deionized H_2O to 1 liter. Adjust to pH 7.4 with 1 M NaOH, autoclave or filter sterilize through a 0.45 μm filter, store $\leq$6 months at room temperature.

LB medium: 10 g tryptone, 5 g yeast extract, 5 g NaCl (0.09 M final), 1 ml 1 M NaOH, deionized H_2O to 1 liter. Autoclave or filter sterilize through a 0.45-μm filter.

IMC medium containing 100 μg/ml ampicillin: 200 ml 12% (w/v) casamino acids (Difco-certified, sterile; 0.4% final), 100 ml $10\times$ M9 salts (sterile; $1\times$ final), 40 ml 20% (w/v) glucose (sterile; 0.5% final), 1 ml 1 M MgSO$_4$ (sterile; 1 mM final), 0.1 ml 1 M CaCl$_2$ (sterile; 0.1 mM final), 1 ml 2% (w/v) vitamin B$_1$ (sterile; 0.002% final), 658 ml deionized H_2O (sterile), 10 ml 10 mg/ml ampicillin (optional, sterile; 100 μg/ml final). Use fresh.

IMC plates containing 100 μg/ml ampicillin: 15 g agar [Difco; 1.5% (w/v) final], 4 g casamino acids [Difco-certified; 0.4% (w/v) final], 858 ml deionized H_2O (sterile). Autoclave for 30 min, cool in a 50° water bath, and then mix with 100 ml $10\times$ M9 salts (sterile; $1\times$

TABLE II
PROPERTIES OF THIOREDOXIN FUSION PARTNERS

Fusion partner	N-terminal fusion?	C-terminal fusion?	Active-site loop fusion?	EK-cleavable fusion?	Osmotic shock or heat purification?	IMAC purification?	Avidin purification?
TrxA	Yes	Yes	Yes	Yes	Sometimes	No	No
hpTRX	Yes	Yes	Yes	Yes	Sometimes	Yes	No
BIOTRX	No	Yes	Yes	Yes	Sometimes	Yes	Yes
DsbA	Yes	Yes	?	Yes	Yes	Yes	No

final), 40 ml 20% (w/v) glucose (sterile; 0.5% final), 1 ml 1 M MgSO$_4$
(sterile; 1 mM final), 0.1 ml 1 M CaCl$_2$ (sterile; 0.1 mM final), 1 ml
2% (w/v) vitamin B$_1$ (sterile; 0.002% final), 10 ml 10 mg/ml ampicillin
(sterile; optional; 100 μg/ml final). Mix well and pour into petri
plates, store wrapped ≤1 month at 4°.

CAA/glycerol/ampicillin 100 medium: 800 ml 2% (w/v) casamino acids
(sterile; 1.6% final), 100 ml 10× M9 salts (sterile; 1× final), 100 ml
10% (v/v) glycerol (sterile; 1% final), 1 ml 1 M MgSO$_4$ (sterile; 1
mM final), 0.1 ml 1 M CaCl$_2$ (sterile; 0.1 mM final), 1 ml 2% (w/v)
vitamin B$_1$ (sterile; 0.002% final), 10 ml 10 mg/ml ampicillin (sterile;
100 μg/ml final). Prepare fresh.

10 mg/ml tryptophan (sterile)

SDS–PAGE sample buffer: 15% (v/v) glycerol, 0.125 M Tris–Cl, pH
6.8, 5 mM Na$_2$EDTA, 2% (w/v) SDS, 0.1% (w/v) bromphenol blue,
1% (v/v) 2-mercaptoethanol (2-ME; add immediately before use).
Store indefinitely at room temperature.

Cell lysis buffer: 50 mM Tris–HCl, pH 7.5, containing 1 mM PABA
(p-aminobenzamidine) and 1 mM phenylmethylsulfonyl fluoride
(PMSF)

Osmotic shock buffer: 20 mM Tris–HCl, 2.5 mM EDTA, pH8

Constructing Thioredoxin Gene Fusion

Standard molecular biological techniques are used in the construction
of thioredoxin gene fusions; these are described in detail by Sambrook *et
al.*[25] Use the unique *Kpn*I site in the vector polylinker of plasmids
pTRXFUS, pHis-patch-TRXFUS, pBIOTRXFUS-BirA, or pDsbAsecFUS
to make a precise fusion of the 5′ end of the desired gene to the enterokinase
linker sequence. The cleaved *Kpn*I site should be trimmed to a blunt end
with the Klenow fragment of *E. coli* DNA polymerase in the presence of
all four deoxynucleotide triphosphates. Usually, the desired gene can then
be precisely adapted to this blunt end to form an in-frame fusion by utilizing
a synthetic oligonucleotide duplex ligated between it and any convenient
downstream restriction site close to the 5′ end of the gene of interest. Use
one of the other sites downstream of the *Kpn*I site in the thioredoxin vector
polylinker to ligate to the 3′ end of the desired gene fragment. In designing
the in-frame fusion junction, researchers should note that enterokinase is
able to cleave --DDDDK↓X--, where X can be any amino acid residue
except proline.

Fusions to the amino terminus of thioredoxin may be constructed by

[25] J. Sambrook, E. F. Fritsch, and T. Maniatis, *in* "Molecular Cloning: A Laboratory Manual,"
2nd Ed. Cold Spring Harbor Laboratory Press, Cold Spring Harbor, NY, 1989.

utilizing the unique *Nde*I site in vectors pALtrxA-781, pTRXFUS, or pHis-patch-TRXFUS. This *Nde*I site encompasses the thioredoxin initiator methionine codon. Researchers should note that since inserts into the *Nde*I site can occur in both orientations, the desired orientation should be verified by restriction enzyme digests or DNA sequencing.

Synthetic oligonucleotides encoding short peptides for insertion into the thioredoxin active-site loop can be ligated into the unique *Rsr*II site in vectors pALtrxA-781, pTRXFUS, pHis-patch-TRXFUS, or pBIOTRXFUS-BirA. Peptide insertions of up to 14 amino acids into the thioredoxin active-site loop are very successful in terms of generating soluble fusions (95%). This success rate decreases as the size of the peptide insertion increases, although soluble thioredoxin fusions with active-site insertions of over 40 amino acids have been constructed.[26] Note that oligonucleotide insertions into the *Rsr*II site will orient themselves in only *one* direction as a consequence of the asymmetrical three base *Rsr*II cohesive end.

The ligation mixture containing the new thioredoxin fusion plasmid should be used to transform competent GI724 cells, and transformants can be selected by plating onto IMC plates containing 100 μg/ml ampicillin. It is important to use IMC plates as they lack tryptophan, and thus pL will not be induced prematurely. Incubate the plates at 30° in a convection incubator until colonies appear. Individual clones should be picked and grown overnight at 30° in 5 ml CAA/glycerol/ampicillin 100 medium. Plasmid minipreps should then be performed and the correct construct identified and verified by restriction enzyme digests and DNA sequence analysis (if appropriate). The *E. coli* culture containing the correct expression construct can be stored indefinitely by mixing 1 ml with 70 μl of 100% dimethyl sulfoxide and freezing quickly at −80°. Viable bacteria can be recovered from this stock by scraping the frozen surface with an inoculating loop and streaking to single colonies on an IMC plate containing 100 μg/ml ampicillin. The frozen stock itself should not be thawed.

Induction of Thioredoxin Fusion Protein Synthesis

The level of transcription initiated from the pL promoter is determined by the level of cI in the host *E. coli* strain. In GI724, as described earlier, cI levels are controlled by cytoplasmic tryptophan levels. In the absence of tryptophan, cI synthesis is maximal and pL is off. In the presence of tryptophan, cI synthesis is repressed, and transcription from the pL promoter is induced. In the following protocol, host strain GI724, containing a pL-controlled thioredoxin fusion expression vector, is first

---

[26] John M. McCoy, unpublished results (1995).

grown to a moderate cell density in the absence of tryptophan and is then switched to tryptophan-rich conditions in order to induce fusion protein synthesis.

1. Streak out a fresh colony of GI724 containing the thioredoxin expression vector from the frozen stock culture on an IMC plate containing 100 μg/ml ampicillin. Grow for 20 hr at 30°. Pick a single colony and inoculate 5 ml of fresh IMC medium containing 100 μg/ml ampicillin in a 150 × 18-mm culture tube. Grow the culture overnight at 30° to saturation on a roller drum (New Brunswick Scientific).

2. Add 0.5 ml of this culture to 50 ml fresh IMC medium containing 100 μg/ml ampicillin in a 250-ml culture flask (1:100 dilution). Grow at 30° with vigorous aeration until the absorbance at 550 nm reaches 0.4–0.6 OD/ml (~3.5 hr). Larger scale cultures can be grown, but be sure to maintain a 5:1 culture flask to media volume ratio to ensure good aeration.

3. Remove a 2-ml aliquot of the culture (uninduced). Measure the absorbance at 550 nm and harvest 1 ml of cells by pelleting in a microcentrifuge for 1 min at maximum speed. Carefully remove all of the spent medium with a pipette and store the cell pellet at −80°.

4. Induce pL by adding 0.5 ml of 10 mg/ml tryptophan (100 μg/ml final) to the culture.

5. Incubate for 4 hr at 37°. At hourly intervals during this period, remove 2-ml aliquots of the culture, measure the absorbance at 550 nm, and harvest 1 ml of cells as described earlier.

6. Four hours following induction, harvest the cells from the remaining culture by centrifugation (10 min at 3000 rpm in a Beckman J6 rotor). Store the cell pellet at −80°.

7. Resuspend the cell pellets from the induction intervals in SDS–PAGE sample buffer (200 μl of sample buffer/OD$_{550}$ cells; e.g., the cell pellet from 1 ml of cells at 0.5 OD$_{550}$ should be resuspended in 100 μl sample buffer). Heat for 5 min at 70° to completely lyse the cells and denature the proteins. Check for production of the desired protein by running the equivalent of 0.15 OD$_{550}$ cells per lane (30 μl) on an SDS–polyacrylamide gel.

Most thioredoxin fusion proteins are produced at levels between 5 and 20% of the total cell protein. The desired fusion protein should exhibit the following characteristics: it should migrate on the gel with a mobility consistent with its calculated molecular weight, it should be absent prior to induction, and it should gradually accumulate during the induction period with a maximum accumulation occurring 3 hr postinduction at 37°. Other induction temperatures may be used; note that lower temperatures require longer induction periods for acceptable levels of the desired fusion protein to accumulate.

Releasing Target Proteins from *E. coli* Cells

Conventional mechanical lysis methods can be used for releasing thioredoxin fusion proteins from *E. coli* cells. Preferred methods use either a French press (SLM Instruments Inc.) or a microfluidizer (Microfluidics, Newton, MA). As an alternative to mechanical lysis, some thioredoxin fusion proteins can be released from *E. coli* by a simple osmotic shock procedure.[6]

Mechanical Lysis

1. Resuspend *E. coli* cells containing the thioredoxin fusion protein in ice-cold cell lysis buffer.
2. Passage the cell suspension once through a French press pressure cell at 20,000 lb/in² or twice through a microfluidizer at 10,000 lb/in².
3. Clarify the lysate by ultracentrifugation at 35,000 rpm for 30 min in a Beckman Ti-50 rotor (or equivalent) and carefully collect the supernatant for subsequent purification steps.
4. Run a small aliquot of the clarified supernatant on an SDS–polyacrylamide gel to check for fusion protein solubility.

Osmotic Shock Release

1. Resuspend *E.coli* cells containing the thioredoxin fusion protein in ice-cold osmotic shock buffer, containing 20% sucrose, to a density of 50 OD_{550}/ml.
2. Incubate the cell suspension on ice for 20 min.
3. Collect the cells by centrifugation at 3000g for 10 min at 4° and gently resuspend in osmotic shock buffer (*with no sucrose*). Keep the cell suspension on ice for 20 min and then centrifuge at 3000g for 30 min. Harvest the supernatant (shockate).

Purification of Thioredoxin Fusion Proteins

Three protocols are presented for the purification of thioredoxin fusion proteins: heat treatment (suitable for certain thermally stable thioredoxin and hpTRX fusions), immobilized metal ion affinity chromatography (suitable for hpTRX and DsbA fusions), and avidin bead capture (suitable for BIOTRX fusions). Depending on the purity level desired for particular applications, additional conventional column chromatography may additionally be required. Because thioredoxin fusions are typically well expressed, their purification can usually be conveniently monitored by SDS–PAGE.

Heat Treatment

The extraordinary heat stability of thioredoxin enables heat treatment to be used as an efficient initial purification step for certain thioredoxin fusion proteins. It should be emphasized that only a subset of thioredoxin fusions are amenable to purification by heat treatment, so researchers should examine the suitability of their particular fusion for this procedure. A pilot experiment will be necessary to determine empirically the optimum duration of heat treatment.

1. Resuspend *E. coli* cells containing the thioredoxin fusion protein in cell lysis buffer to a density of 100 OD_{550}/ml. This high cell density is important to ensure efficient precipitation of contaminants during the subsequent heating step.
2. Lyse the cells using a French press or microfluidizer (described earlier).
3. Transfer 2 ml of the crude, unclarified lysate into a thin-walled 10-ml glass test tube. Incubate for 10 min at 80°, remove 100-μl aliquots after 30 sec, 1 min, 2 min, and 5 min, place the aliquots into glass tubes, and plunge them immediately into ice water. At 10 min plunge the remaining heated lysate into ice water. Centrifuge the aliquots for 10 min at maximum speed in a microcentrifuge and analyze the pellets and supernatants by SDS–PAGE to determine the heat stability of the fusion protein and the minimum heat treatment time required to obtain a good purification.

For successful scale-up of this protocol, strict and efficient control of heating and cooling processes is essential. Use glass-walled vessels rather than plastic because of the better thermal conductivity of glass, and at larger scales ensure that lysates are well mixed continuously during both heating and cooling stages.

IMAC Protocol for Purification of His-patch Thioredoxin Fusions

Substantial purification of hpTRX and DsbA fusions can be achieved using immobilized metal affinity chromatography. Note that osmotic shockates contain EDTA, which must be removed completely by dialysis before attempting IMAC purifications.

1. Lyse the cells as described earlier (French pressure cell or microfluidizer).

2. Add NaCl and imidazole to the lysates to 500 and 4 mM, respectively, and keep the lysates on ice for 30 min. Remove cell debris by ultracentrifugation at 35,000 rpm in a Beckman Ti-50 rotor for 30 min.

3. Gently recover the clarified supernatant and load onto an IDA (iminodiacetic acid) column, charged previously with Ni^{2+} and equilibrated in buffer A (25 mM Tris–HCl, pH 7.5, 2 mM imidazole, 200 mM NaCl). Load the column at a flow rate of 20% of the resin volume per minute.

4. Wash the column thoroughly with buffer A and then elute the bound proteins with a linear gradient of buffer B (25 mM Tris–HCl, pH 7.5, 100 mM NaCl, and 100 mM imidazole). Fractions containing fusion proteins are usually those eluted between 30 and 60 mM imidazole. Those fractions containing the fusion protein at highest purity (as determined by SDS–PAGE) can be pooled and dialyzed against buffers suitable for enterokinase digestion (see later).

Capturing Biotinylated BIOTRX Fusions Using Avidin Beads

BIOTRX fusions, bearing a biotinylation sequence at their amino termini, can be quantitatively biotinylated *in vivo* under certain conditions.[9] These proteins can be efficiently recovered from crude lysates by capture on avidin agarose beads using the following protocol.

1. Either mechanically lyse the bacteria containing the BIOTRX fusion protein or release the fusion by osmotic shock (procedures described later).

2. Clarify the lysate or shockate by centrifugation at 15,000g for 10 min at 4°.

3. To the supernatant add avidin-agarose beads, preequilibrated with cell lysis buffer containing 200 mM NaCl and 0.1% Triton X-100. Allow binding to occur for 1 hr at 4° with gentle agitation.

4. Sediment the beads by brief centrifugation, washing three times with lysis buffer containing 200 mM NaCl and 0.1% Triton X-100.

BIOTRX fusion-loaded agarose beads can be used in binding studies or for solution capture of a desired binding protein. BIOTRX fusions are tightly bound to the beads, but may be released for analysis by SDS–PAGE by heat treatment at 70° in SDS–PAGE sample buffer.

Cleavage of Fusion Proteins by Enterokinase

Enterokinase (also called enteropeptidase or EK) is a mammalian trypsin-like serine protease that displays a high degree of specificity for the sequence "-DDDDK-", cleaving on the carboxy-terminal side of the lysine residue of the recognition sequence. Although in mammals the enzyme has evolved to recognize and cleave this sequence found close to the amino termini of trypsinogens,[16] EK can also be used to cleave fusion proteins, such as thioredoxin fusions, which contain this recognition sequence inserted between the "carrier" protein and the carboxy-terminal fusion partner. EK is capable of cleaving fusion proteins under a wide range of reaction conditions, with pH ranging from 4.5 to 9.5 and temperatures ranging from 4° to 45°. The enzyme is also extremely tolerant of the nature of the amino acid residue in the P1′ position, although the Lys-Pro bond is totally refractory to cleavage.[27] A cDNA encoding the catalytic subunit of bovine enterokinase has been cloned and expressed,[28] and the recombinant enzyme can be purchased from companies such as Invitrogen, Promega, New England Biolabs, Stratagene, and others. Nonrecombinant commercial preparations of enterokinase (bovine or porcine) are often extremely impure and tend to be contaminated with (among other things) trypsin and chymotrypsin, which can degrade the fusion protein. It is recommended that only commercial *recombinant* EK of the highest purity be used. The enzymatic reaction is very efficient provided that the substrate (fusion protein) is fairly concentrated and ionic strength is kept to a minimum. Complete cleavage of fusion proteins can often be accomplished in a 16-hr digestion at EK:fusion protein ratios of 1:5000 to 1:100,000 (w/w). The following protocol describes the use of recombinant bovine EK in this application. Because cleavage efficiency does vary, pilot reactions should first be performed to determine empirically the optimum ratio of EK to fusion protein.

1. Provide at least partially purified fusion protein at $\geq$1 mg/ml in 50 mM Tris–Cl, pH 8.0, 1 mM CaCl$_2$ (EK is inactive in crude *E. coli* lysates).
2. Mix 20 μl ($\approx$20 μg) aliquots of fusion protein with 0.1 unit, 0.5 unit, 1 unit, 2 units, 5 units, and 10 units of enterokinase in separate 1.5-ml microfuge tubes. Bring the total volume of all samples to 30 μl with 50 mM Tris–Cl, pH 8.0, 1 mM CaCl$_2$.
3. Mix another 20-μl aliquot of fusion protein with 10 μl of 50 mM Tris–Cl, pH 8.0, 1 mM CaCl$_2$ in a 1.5-ml microfuge tube (mock digestion).
4. Incubate samples at 37° for $\geq$16 hr.
5. Terminate the reactions by adding 30 μl of 2× SDS–PAGE sample buffer to each aliquot. Heat at 100° for 10 min. (For larger-scale reactions, EK can be irreversibly inactivated by 1 mM PMSF or competitively inhibited by 5 mM *p*-aminobenzamidine.)
6. Load 10 μl of each sample onto an SDS–PAGE gel to analyze the

---

[27] Edward R. LaVallie and Lisa A. Racie, unpublished results (1995).
[28] E. R. LaVallie, A. Rehemtulla, L. A. Racie, E. A. DiBlasio, C. Ferenz, K. L. Grant, A. Light, and J. M. McCoy, *J. Biol. Chem.* **268**, 23311 (1993).

progress of the cleavage reaction. Adjust enterokinase concentration and length of incubation accordingly to accomplish complete digestion.

7. Scale up the reaction components linearly to digest a larger amount of fusion protein.

Concluding Remarks

Although current systems for heterologous gene expression in *E. coli* are quite sophisticated,[1] there is no one system, even today, that guarantees success for every application. A prudent experimental approach toward achieving high-level production in *E. coli* of any particular protein of interest would be to try a variety of expression systems, with the prospect that one or more may succeed while others may fail. With their established capability for producing heterologous proteins in a soluble and bioactive form, their high expression levels, and the convenient purifications they afford, thioredoxin gene fusion systems should be an important component of any such approach.

[22] Affinity Purification of Recombinant Proteins Fused to Calmodulin or to Calmodulin-Binding Peptides

By Peter Vaillancourt, Chao-Feng Zheng, Danny Q. Hoang, and Lisa Breister

Introduction

The interaction between calmodulin (CaM) and various of its ligands has been exploited for affinity purification of recombinant proteins from *Escherichia coli* extracts in a number of systems. CaM is a calcium-binding protein that plays a central role in regulating a wide range of calcium-dependent intracellular processes. CaM binds to at least 30 proteins in eukaryotic cells and interacts with a wide range of peptides and hydrophobic ligands with high affinity in a calcium-dependent manner.[1,2] The use of CaM and its ligands for affinity purification is attractive for a number of reasons. CaM is small (17 kDa), stable, and highly soluble in *E. coli* cells

[1] A. R. Means, I. C. Bagchi, M. F. A. Vanberkum, and C. D. Rasmussen, *in* "Cellular Calcium, A Practical Approach." IRL Press, Oxford, 1991.

[2] J. F. Head, *Curr. Biol.* **2,** 609 (1992).

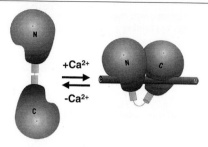

Fig. 1. Schematic of the interaction between CaM and its target peptide. The N- and C-terminal lobes of CaM are linked by a flexible tether. In the presence of threshold levels of Ca^{2+}, CaM undergoes a conformational shift in which hydrophobic surfaces within the lobes are exposed. These surfaces interact with the hydrophobic face of the amphipathic CaM-binding peptide. When Ca^{2+} is removed from the complex, CaM is returned to its original conformation and the peptide is released. Adapted from J. F. Head, *Curr. Biol.* **2,** 609 (1992), with permission from Current Biology Publications.

and lysates. The CaM-coding sequence for at least one species (chicken) is translated efficiently in *E. coli,* and expression of its gene from high-copy plasmids often results in the production of lysates in which ≥50% of the protein in the soluble fraction is CaM. This feature is important for the high-level expression of fusion proteins in systems that employ CaM as an N-terminal affinity tag, as well as for systems based on calmodulin-binding peptide fusions for which the cost-effective production of high-capacity CaM affinity resin is dependent on the ease with which large quantities of highly purified CaM can be produced. Perhaps the most impor-tant feature of CaM-based purification systems is that despite the strong interaction of CaM for its ligands ($K_d = 10^{-9}$), fusion proteins may be purified to near homogeneity in a single step using gentle buffer conditions. Binding requires only the presence of a low concentration of calcium (≥0.2 mM CaCl$_2$), and elution requires a calcium-chelating agent such as EGTA (2 mM) in an otherwise neutral buffer (Figs. 1 and 2).

The focus of this article is on the purification of fusion proteins con-taining the 26 amino acid calmodulin-binding peptide (CBP) derived from the C terminus of skeletal muscle myosin light-chain kinase[3] using a solid support resin containing covalently bound CaM. Several excellent alterna-tive systems in which CaM is employed as a fusion partner are described at the end of the article. The high degree of specificity of CBP for CaM (there are no endogenous *E. coli* proteins that interact with CaM), together with the stringent wash conditions allowed due to the strength of the CBP–CaM interaction, results in the consistent recovery of highly purified

[3] R. E. Stofko-Hahn, D. W. Carr, and J. D. Scott, *FEBS Letts.* **302,** 274 (1992).

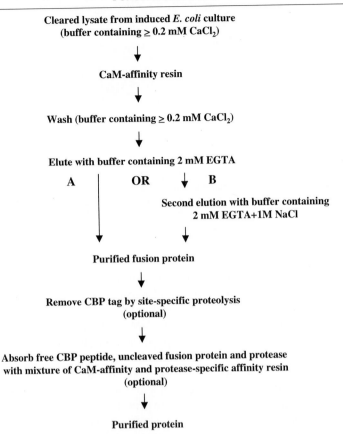

Fig. 2. Purification of CaM-binding peptide fusion proteins from *E. coli* lysates.

CBP fusions. The small size of the peptide (<4 kDa) allows recovery of soluble proteins of greater size than is attainable with the larger of the currently popular protein tags. In addition, the CBP is less likely to affect the biological function of the protein of interest compared with the larger tags. We have found that the CBP is translated efficiently in *E. coli*,[4] and thus a wide range of proteins are consistently expressed to high levels when the CBP is positioned at the N terminus of the fusion.

Preparation of CaM-Sepharose Resin

There are several commercially available CaM resins. We have found that resins that employ Sepharose 4B (Pharmacia) have the highest capacity

4 C. F. Zheng, T. Simcox, L. Xu, and P. Vaillancourt, *Gene* **186,** 55 (1997).

(approximately 2 mg CBP fusion protein purified/ml settled resin) and consistently give rise to eluted material of high purity. Sepharose 4B-based CaM affinity resins are available that contain native bovine CaM (Pharmacia) or recombinant chicken CaM[1] produced in *E. coli* (Stratagene); the quality of these resins is essentially the same, although we have found that the salt requirements in the elution buffer may vary for some proteins depending on the choice of resins (see "Expression and Purification of CBP Fusion Proteins"). The following protocol is based on expression and purification of recombinant CaM from a 1-liter culture of *E. coli* and typically yields 70–150 mg purified CaM per 4 g cell paste.

Expression of Recombinant CaM in E. coli

The coding sequence for the chicken CaM gene is inserted between *Nco*I and *Bam*HI restrictions sites of the the T7-based vector pET-11d[5] (for a general description of pET vectors and strains, see "Vectors and Strains") to generate the expression vector pETCaM11. CaM expression from this vector is driven from the T7*lac* promoter, which contains the recognition sequence for T7 RNA polymerase in conjunction with the binding site for the Lac repressor. The host strain BL21(DE3) contains a chromosomally integrated IPTG-inducible expression cassette for T7 RNA polymerase.

The day prior to induction, 20 ml LB containing 50 μg/ml carbenicillin or ampicillin is inoculated with a single colony of BL21(DE3) cells harboring the plasmid pETCaM11 and shaken overnight at 37°. The following morning, both 2 × 2.5-liter shake flasks containing 500 ml LB and 50 μg/ml carbenicillin each are inoculated with 10 ml/flask of the overnight culture. The flasks are shaken at 37° until an OD$_{600}$ of 0.6–1.0 is reached, at which time 10 μl of the culture is removed from the flask and added directly to a mixture of 20 μl Laemmli sample buffer and 10 μl water and stored at −20°. Immediately following removal of the culture sample the flasks are induced with 1 m*M* IPTG for 3–5 hr (for scale up of this procedure longer times may be used; we have found for fermentation of 200-liter cultures that a 10-hr induction gives optimal yields of CaM). Following induction, an additional 10-μl sample is removed and processed as described earlier. The cultures are harvested by centrifugation, and the pellets may be stored at −20° until the quality of the induction is determined or may be processed immediately. The 40-μl "uninduced" and "induced" culture samples are boiled for 2 min and analyzed by SDS–PAGE. A prominent band of

[5] F. W. Studier, A. H. Rosenberg, J. J. Dunn, and J. W. Dubendorff, *Methods Enzymol.* **185**, 60 (1991).

approximately 17 kDa should appear for IPTG-induced cultures, indicating a successful induction.

Purification of CaM

CaM is purified by two cycles of phenyl-Sepharose chromatography essentially as described[1] with some modifications.

Buffers

Buffer 1: 40 mM Tris–HCl, pH 7.5, 0.1 mM EDTA, 50 mM KCl, 1 mM dithiothreitol (DTT)

Buffer 2: 10 mM Tris–HCl, pH 7.5, 6 mM CaCl$_2$, 1 mM DTT

Buffer 3: 10 mM Tris–HCl, pH 7.5, 1 mM CaCl$_2$, 1 mM DTT

Buffer 4: 10 mM Tris–HCl, pH 7.5, 1 mM CaCl$_2$, 0.5 M KCl, 1 mM DTT

Buffer 5: 10 mM Tris–HCl, pH 7.5, 1 mM EDTA

Buffer 6: 10 mM Tris–HCl, pH 7.5, 1 mM EDTA, 100 mM KCl, 1 mM DTT

The cell pellet (3.5–5 g wet paste) is resuspended in 16 ml of buffer 1 containing 1 mg lysozyme per gram of cells, mechanically rotated 20 min at 10–15°, and cooled to 4°. The cells are sonicated to completion, and DTT is added to 1 mM. The sonicate is cleared by centrifugation at 25,000g for 1 hr, and CaCl$_2$ from a 1 M stock is slowly stirred in to a final concentration of 6 mM. The material is then loaded onto a phenyl-Sepharose column (Pharmacia) preequilibrated with buffer 2 at room temperature and is washed extensively with buffer 2 until a baseline at OD$_{280}$ is reached. The column is sequentially washed with buffer 3 and buffer 4 until baseline at OD$_{280}$ is achieved, and then the column is washed with several bed volumes of buffer 3 to lower the KCl concentration. The CaM is eluted by removal of the CaCl$_2$ with buffer 5, and peak fractions are identified by SDS–PAGE.

At this point the CaM may be further purified by preparative HPLC[1]; however, we have found that a second pass over phenyl-Sepharose will remove all detectable remaining impurities while allowing the purified CaM to pass through. The peak fraction are pooled and dialyzed against buffer 6 and then applied to a second phenyl-Sepharose column preequilibrated with buffer 6 at room temperature. The CaM is collected in the flow-through fraction, concentrated as needed, and stored frozen at −20°. CaM is stable for at least 2 years at this temperature.

Preparation of CaM-Sepharose

CaM is covalently attached to CNBr-activated Sepharose 4B resin (Pharmacia) essentially in accordance with the manufacturer's recommended protocols.

CaM is extensively dialyzed against coupling buffer (0.1 M NaHCO$_3$, 1 mM CaCl$_2$, 0.5 M NaCl). CNBr-Sepharose 4B resin is swollen in 200 ml of 1 mM HCl/g of resin for 15 min at room temperature in a sintered glass funnel. The resin capacity is 2 mg CaM/ml settled resin, and 1 g of resin will swell to 3.5 ml, thus 143 mg resin should be used per milligram CaM to be coupled. The resin is washed extensively in coupling buffer and resuspended in 2 volumes coupling buffer per volume resin. The CaM and resin slurry are mixed and mechanically rotated slowly for 16 hr at 4°. The resin slurry is then applied to a sintered glass funnel, and excess uncoupled CaM is removed by washing the resin with 5 bed volumes of coupling buffer. Any remaining active sites in the gel are blocked by transferring the resin to 2 volumes of 1 M ethanolamine, pH 8.0, and mechanically rotating slowly for 16 hr at 4°. The gel is washed sequentially with 5 bed volumes of a low pH buffer (0.1 M acetate buffer, pH 4.0, 0.5 M NaCl) followed by a high pH buffer (0.1 M Tris–HCl, pH 8.0, 0.5 M NaCl). This cycle is repeated three times. The gel is washed with several volumes of storage buffer (20% ethanol, 0.1 mM CaCl$_2$, 20 mM Tris, pH 7.5, 0.5 M NaCl) and stored at 4°.

Vectors and Strains

Similar to the original CBP fusion vector pET-MAP2(d1-154)-kfc described by Scott and co-workers,[3] the five "pCAL" vectors described here are derived from the T7-based vectors pET-11d[5] (pCAL-c and pCAL-kc) or pET-11a (pCAL-n, pCAL-n-EK, and pCAL-n-FLAG) (Fig. 3). The vectors contain the strong ϕ10 promoter from bacteriophage T7 used in conjunction with the binding site for the Lac repressor (*lac O*) to form the hybrid promoter T7*lac*. The plasmid also contains a copy of the *lacI*^{q} gene. Together these regulatory elements confer tight control and high level-induced expression in *E. coli* strains that contain a chromosomally integrated IPTG-inducible expression cassette for T7 RNA polymerase (described later). The vectors also include the T7 ϕ10 ribosome-binding site (RBS), which facilitates highly efficient translation of the downstream protein-coding sequence.

Fusion of CBP at C Terminus

The vectors pCAL-c and pCAL-kc (Fig. 3) contain the CBP positioned 3′ to the cloning sites for C-terminal fusion. The ATG codon within the restriction site *Nco*I is positioned relative to the RBS for optimal initiation of translation. The protein-coding sequence is typically inserted between *Nco*I and *Bam*HI sites in these vectors. Because the efficiency of translation

pCAL-n

Calmodulin-Binding Peptide

```
  M   K   R   R   W   K   K   N   F   I   A   V   S   A   A   N   R   F   K   K   I   S   S   S   G
CATATGAAGCGACGATGGAAAAAGAATTTCATAGCCGTCTCAGCAGCCAACCGCTTTAAGAAAATCTCATCCTCCGGG
Nde I
```
    Thrombin
```
  A   L   L   V   P   R   G   S   P   G   I   L   D   S   M   G   R   L   E   L   K   L   R   S   A
GCACTTCTGGTTCCGCGTGGATCCCCGGGAATTCTAGACTCCATGGGTCGACTCGAGCTCAAGCTTAGATCCGCC
           BamH I  Sma I  EcoR I        Nco I        Sal I   Xho I  Sac I  Hind III
```

pCAL-n-EK Calmodulin-Binding Peptide   Thrombin

```
  M   K   R   R   W   K   K   N   F   I   A   V   S   A   A   N   R   F   K   K   I   S   S   S   G   A   L   L   V   P   R   G
CATATGAAGCGACGATGGAAAAAGAATTTCATAGCCGTCTCAGCAGCCAACCGCTTTAAGAAAATCTCATCCTCCGGGGCACTTCTGGTTCCGCGTGGA
Nde I            EK
  S   G   S   G   D   D   D   D   K                         LIC overhang
TCTGGTTCTGGTGATGACGACGACAAGGGAAGAGGATCCGAATTCTCTTCCCGGGTCTTGTTCCATGGGTCGACTCGAGCTCAAGCTT
     LIC overhang   Eam 1104 I BamHIEcoRIEam1104 I Sma I        Nco I   Sal I  Xho I Sac I Hind III
```

pCAL-n-FLAG Calmodulin-Binding Peptide   Thrombin

```
  M   K   R   R   W   K   K   N   F   I   A   V   S   A   A   N   R   F   K   K   I   S   S   S   G   A   L   L   V   P   R   G
CATATGAAGCGACGATGGAAAAAGAATTTCATAGCCGTCTCAGCAGCCAACCGCTTTAAGAAAATCTCATCCTCCGGGGCACTTCTGGTTCCGCGTGGA
Nde I            EK
  S   D   Y   K   D   D   D   D   K                      LIC overhang
TCTGACTACAAGGATGACGACGACAAGGGAAGAGGATCCGAATTCTCTTCCCGGGTCTTGTTCCATGGGTCGACTCGAGCTCAAGCTT
     LIC overhang   Eam 1104 I BamHIEcoRIEam1104 I Sma I      Nco I   Sal I  Xho I Sac I Hind III
              FLAG
```

pCAL-c

            Thrombin
```
  M   A   S   M   T   G   G   Q   Q   M   G   G   S   M   Y   P   R   G   N   G   T   K   R   R
CCATGGCTAGCATGACTGGTGGACAGCAAATGGGCGGATCCATGTATCCACGTGGGAATGGTACCAAGCGACGA
Nco I  Nhe I                          BamH I                    Kpn I
              Calmodulin-Binding Peptide
  W   K   K   N   F   I   A   V   S   A   A   N   R   F   K   K   I   S   S   S   G   A   L   stop
TGGAAAAAGAATTTCATAGCCGTCTCAGCAGCCAACCGCTTTAAGAAAATCTCATCCTCCGGGGCACTTTGATCC
```

pCAL-kc

           Kemptide   Thrombin
```
  M   A   S   M   T   G   G   Q   Q   M   G   G   S   L   R   R   A   S   L   G   R   S   M   Y   P   R   G   N
CCATGGCTAGCATGACTGGTGGACAGCAAATGGGCGGATCCCTTAGACGCGCATCACTTGGTAGATCCATGTATCCACGTGGGAAT
Nco I  Nhe I                          BamH I
              Calmodulin-Binding Peptide
  G   T   K   R   R   W   K   K   N   F   I   A   V   S   A   A   N   R   F   K   K   I   S   S   S   G   A   L   stop
GGTACCAAGCGACGATGGAAAAAGAATTTCATAGCCGTCTCAGCAGCCAACCGCTTTAAGAAAATCTCATCCTCCGGGGCACTTTGATCC
Kpn I
```

initiation in *E. coli* is affected greatly by codon usage for the N-terminal coding sequence of an open reading frame, expression levels may vary according to the N-terminal sequence of the gene of interest. However, bidirectional insertion at the *Bam*HI site allows fusion of the efficiently translated ϕ10-s10 peptide at the N terminus, thereby increasing the likelihood of efficient translation of the "tribrid" fusion. (It should be noted that the ϕ10-s10 peptide and the CBP are in separate frames in these vectors, thus when cloning bidirectionally at the *Bam*HI site, care should be taken that the coding sequence of interest is fused in frame with both peptides. Likewise if inserting bidirectionally at the *Nco*I or *Nhe*I sites, the coding sequence should be in frame with the CBP and not with the ϕ10-s10 peptide.)

The vector pCAL-c contains the five amino acid recognition target for the site-specific protease thrombin positioned immediately downstream of the *Bam*HI cloning site, followed by the coding sequence for the CBP. Cleavage by thrombin occurs between arginine and glycine residues within the target sequence, thus thrombin-treated fusions from this vector contain the four amino acid extension YPRG fused at the C terminus, in addition to any other amino acids coded for by the cloning sites and intervening sequences.

Although we have found fortuitously that the CBP-thrombin fusion tag in the vector pCAL-n serves as an excellent target for isotopic labeling with $[\gamma\text{-}^{32}P]ATP$ using protein kinase A[4] (PKA), there are some applications for which the presence of the CBP tag on a ^{32}P-labeled protein probe may contribute to background due to its interaction with CaM (e.g., "pull-down"-type immunoprecipitation experiments from eukaryotic cell extracts[6] or direct screening of λ cDNA expression libraries for protein–protein interactions). The vector pCAL-kc contains the nine amino acid PKA "kemptide" substrate[3] between the cloning sites and the thrombin recognition target, thus high specific activity probes may be generated from purified fusions following removal of the CBP with thrombin.

---

[6] B. Derijard, M. Hibi, I-H. Wu, T. Barrett, B. Su, T. Deng, M. Karin, and R. J. Davis, *Cell* **76**, 1025 (1994).

Fig. 3. Maps of the CBP fusion vectors. The circular map depicts the general structure of the vector. The middle insert shows an expanded view of the pT7*lac* expression cassette, including the Lac I-repressible pT7*lac* promoter, the T7 ϕ10 ribosome-binding site (RBS), the calmodulin-binding peptide (CBP) fusion cassette, and the T7 transcriptional terminator. The top insert shows the nucleotide sequence, amino acid sequence, multiple cloning site, and other relevant features for each of the five pCAL vectors.

Fusion of CBP at N Terminus

The vectors pCAL-n, pCAL-n-EK, and pCAL-n-FLAG all contain the CBP sequence positioned for N-terminal fusion to the coding sequence of interest. We have found that the CBP is translated efficiently in *E. coli,* and as described earlier for the ϕ10-s10 peptide, the majority of proteins that we have tested express high levels when fused downstream of the CBP in these vectors. It is interesting that although there is no clear consensus PKA target sites within the CBP (although the sequence KKIS may be a potential substrate), as mentioned earlier the tag is efficiently phosphorylated with PKA for the generation of ^{32}P-labeled probes of high specific activity.

The vector pCAL-n contains a thrombin recognition site immediately downstream of the CBP. Cleavage of fusions produced from this vector contains a single glycine fused at the N terminus derived from the thrombin target sequence, in addition to any additional amino acids coded for by the cloning sites.

There are many applications for which the presence of any extraneous amino acids derived from the cloning site or the the protease target site are undesirable. This issue is of particular importance for the production of antibodies to be used for the recognition of native proteins. In addition, stability, intracellular targeting, and/or functional activity is often dependent on the presence of natural, appropriately processed amino termini. The vector pCAL-n-EK[7] was designed to allow the production of "seamless" fusion proteins in which any amino acid sequence may be fused directly downstream of the protease cleavage site, thus allowing the recovery of proteins with native amino acid sequence from purified CBP fusions (Fig. 4). This is achieved by combining the use of the site-specific protease enterokinase (EK) and the type IIs restriction enzyme *Eam*1104 I, which cleave outside their respective (amino acid and nucleotide) target sequences. The 5' *Eam*1104 I site is positioned such that the 3 base 5' single-stranded overhang, which may be composed of any sequence, coincides with the C-terminal lysine codon of the EK recognition sequence. Polymerase chain reaction (PCR) primers are designed so that the coding sequence for the desired N-terminal amino acids of the protein of interest are fused directly to the lysine codon for insertion by either direct ligation of *Eam*1104 I-restricted DNA[8] (Fig. 4B)(insertion is directional because the 3 base overhangs for the two *Eam*1104 I sites are different) or by ligation-indepen-

[7] D. L. Wyborski, J. C. Bauer, C.-F. Zheng, K. Felts, and P. Vaillancourt, *Prot. Express. Purif.* **16,** 1 (1999).

[8] K. A. Padgett and J. A. Sorge, *Gene* **168,** 31 (1996).

A.

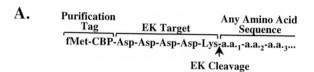

B. Seamless insertion by ligation

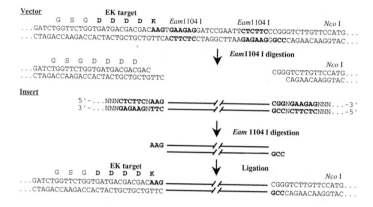

C. Seamless insertion by LIC

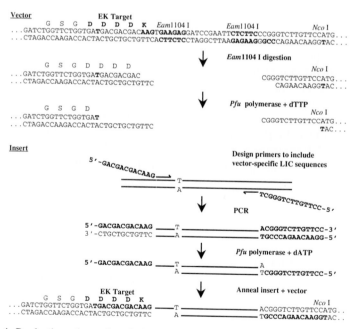

Fig. 4. Production of seamless fusions using the vector pCAL-n-EK. (A) N-terminal structure of fusion proteins produced from the vector pCAL-n-EK. Seamless fusions may be produced by ligation using the type IIs restriction enzyme *Eam*1104 I (B) or by LIC (C). Adapted from D. L. Wyborski *et al.*, *Prot. Express. Purif.* **16**, 1 (1999).

dent cloning (LIC)[9] (Fig. 4C). Because EK cleaves C-terminal to the lysine residue, there are no vector-encoded amino acids fused at the N terminus of the EK cleavage product.

The vector pCAL-n-FLAG is identical to pCAL-n-EK except that the three amino acids immediately N-terminal to the EK site have been changed to produce the coding sequence for the FLAG epitope. Because the FLAG epitope is positioned between the two protease targets in this vector, CBP fusions may be cleaved with EK to produce proteins with no vector-derived extraneous amino acids or may be cleaved with thrombin to produce fusions that have the FLAG epitope fused at the N terminus in the absence of CBP. Similar to the use of kemptide for pCAL-kc, this feature is particularly useful for applications involving the analysis of protein–protein interactions in which the purified protein of interest is used to probe eukaryotic cell extracts or screen bacteriophage λ-based expression libraries.[10]

Seamless Cloning of Inserts into pCAL-n-EK and pCAL-n-FLAG

For seamless insertion by ligation, PCR primers are designed as depicted in Fig. 4B. The directionality of the recognition sites for the type IIs restriction enzymes allows for the *Eam*1104 I sites within the primers to be oriented such that digestion of the PCR product results in the complete removal of the restriction site sequence. There are two oppositely oriented *Eam*1104 I sites within the multiple cloning site whose cleavage results in the production of vector ends with two different 5′ overhangs (the CTT lysine anticodon for the 5′ or "leftward" site, and CGG for the "rightward site"), thus care should be taken that the correct triplet overhang sequence is positioned immediately adjacent to the coding sequence of interest when designing the primers. Cleavage of *Eam*1104 I sites within the gene-specific sequence of the PCR product may be blocked by the addition of an excess of 5-methyl deoxycytosine (^{m5}dCTP) into the PCR reaction. Typically, PCR reactions containing 200 μM each of dATP, dTTP, dGTP, and dCTP are subjected to 10 cycles of amplification according to a standard protocol. The reaction is then spiked with a mixture containing an additional 200 μM each of dATP, dTTP, and dGTP plus 1 mM ^{m5}dCTP, and amplification is continued for an additional five cycles.

Seamless insertion by LIC is performed as outlined in Fig. 4C. LIC has the advantage that inserts can be cloned efficiently without the use of restriction enzymes or ligase, thus obviating concerns associated with ligation-dependent cloning such as contaminating endonucleases or the presence of undersirable restriction sites internal to the gene-specific sequence.

[9] C. Aslandis and P. J. de Jong, *Nucleic Acids Res.* **18**, 6069 (1990).
[10] K. Felts (Stratagene), unpublished data.

Insert-specific PCR primers are designed such that the 5' ends of the primers correspond to the 12 and 13 base vector-specific sequences shown, with the 5' primer engineered such that the desired N-terminal codon is positioned immediately downstream of the Lys_{AAG} codon of the EK target sequence. PCR reactions are performed according to standard protocol, and the PCR products are gel purified. The PCR product (10–100 fmol) is incubated in a 10-μl reaction with 1 U of cloned *Pfu* DNA polymerase (Stratagene) in 1× cloned *Pfu* buffer [10 mM KCl, 10 mM $(NH_4)_2SO_4$, 10 mM Tris–Cl, pH 8.8, 2 mM $MgSO_4$, 0.1% Triton X-100, and 100 μg/ml bovine serum albumin (BSA)] in the presence of 1 mM dATP at 72° for 10 min. Under these conditions, 3'-5' exonucleolytic digestion of the PCR product occurs until the first dA residue in the strand is encountered, revealing the 5' single-stranded overhangs shown in the diagram. Similarly, 20–50 ng of *Eam*1104 I-digested vector is treated with *Pfu* polymerase in the presence of dTTP to produce 5' single-stranded ends that are complementary to the vector-specific single-stranded ends in the PCR product. The *Pfu* reaction products are cooled briefly at room temperature, and vector (20 ng) and insert are mixed and annealed at room temperature for ≥1 hr and then used to directly transform *E. coli*-competent cells. Note that in most instances a single-stranded gap will occur in the annealed product whose length corresponds to the distance from the 5' end of the insert-specific sequence to the position of the first dT residue; these gaps are repaired efficiently *in vivo*.[7]

Expression Strains

Several *E. coli* expression strains have been described that contain the replication-defective "locked" λ derivative DE3,[5] which contains an IPTG-inducible T7 RNA polymerase expression cassette. The most commonly used strains are based on the *E. coli* B strain BL21 (F⁻ *ompT* r_B^- m_B^-), which is defective for the proteases *ompT* and *lon*. Uninduced background expression in the strain BL21(DE3) is extremely low, and induction with IPTG often results in the production of lysates in which the protein of interest constitutes a large fraction of the cellular protein. The background expression of toxic genes is reduced further in derivatives of BL21(DE3) in which low levels of T7 lysozyme are expressed. T7 RNA polymerase is specifically inhibited stoicheometrically by this enzyme; however, on induction the increased expression of polymerase overrides the effect of the lysozyme, and high expression levels are attainable. The pACYC184-based plasmids pLysS and pLysE express low to moderate levels of T7 lysozyme, respectively, and therefore inhibit uninduced (and induced) expression to varying degrees. For highly toxic plasmids whose uninduced

A.   **B.**

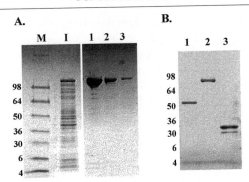

FIG. 5. Purification of CBP fusion proteins. (A) Coding sequence for the full-length *E. coli* β-galactosidase gene was inserted between *Nco*I and *Bam*HI sites in the vector pCAL-c. BL21(DE3)pLysS cells harboring the resulting construct were induced with IPTG, and the 120-kDa CBP fusion was purified. A Coomassie Brilliant Blue-stained SDS–PAGE gel is shown. I: total lysate from the induced culture. Lanes 1–3 represent the first three fractions in the buffer D elution step. M: molecular mass standards (kDa). (B) Examples of purified CBP fusion proteins. Lane 1: c-jun N-terminal kinase (50 kDa) produced from the vector pCAL-n-EK. Lane 2: MoMLV reverse transcriptase (80 kDa) produced from the vector pCAL-n. Lane 3: λ phosphatase produced from the vector pCAL-c. Numbers at the left represent molecular masses (kDa) of the standards.

expression is not tolerated even in pLysE-containing BL21 (DE3) cells, BL21 cells harboring the toxic plasmid may be infected with the T7 RNA polymerase-expressing λ phage CE6.[5]

The availability of improved BL21(DE3)-derived strains that allow increased plasmid yield [BLR (Novagen)], higher transformation efficiency and improved quality of DNA preparations (BL21-Gold[11]), disulfide bond formation in the cytoplasm (AD494[12]), and enhanced expression of genes containing codons that are rarely used in *E. coli* [BL21(DE3)-codon+[13]] have substantially increased the range of functional proteins that can be obtained from T7-based expression systems.

Expression and Purification of CBP Fusion Proteins

The following protocol is for expression and purification of CBP fusion protein from a 1-liter culture using the "batch-binding" method, for which we often obtain 10–20 mg of highly pure protein from cultures in which the gene of interest is efficiently expressed (Fig. 5). Due to the versatility

[11] A. Greener (Stratagene), unpublished data.
[12] A. I. Derman, W. A. Prinz, D. Belin, and J. Beckwith, *Science* **262,** 1744 (1993).
[13] C.-P. Carstens (Stratagene), unpublished data.

of the system, a number of variations of the protocol are tolerated with respect to buffer constituents, scale, and column method (i.e., batch binding versus the standard "top-loading" column method). Some of these parameters are explored in the "Discussion" section.

IPTG Induction of Gene Expression

The day prior to induction, 20 ml LB containing 50 μg/ml carbenicillin or ampicillin is inoculated with a single colony of BL21(DE3) or BL21(DE3) pLysS cells harboring the pCAL vector [25 μg/ml chloramphenicol is included if BL21(DE3)pLysS cells are used] and shaken overnight at 37°. The following morning, both 2 × 2.5-liter shake flasks containing 500 ml LB each are inoculated with 10 ml/flask of the overnight culture. The flasks are shaken at 37° until an OD_{600} of 0.6–1.0 is reached, at which time 10 μl of the culture is removed from the flask and added directly to a mixture of 20 μl Laemmli sample buffer and 10 μl water, and stored at −20°. Immediately following removal of the culture sample, the flasks are induced by the addition of 1 mM IPTG and shaken for an additional 3–5 hr. Following induction, a second 10-μl sample is removed and processed as described earlier. The cultures are harvested by centrifugation, and the pellets may be stored at −80° until the quality of the induction is determined or may be processed immediately. The 40-μl "uninduced" and "induced" culture samples are boiled for 2 min and analyzed by SDS–PAGE. A prominent band of the expected molecular weight should appear for IPTG-induced cultures, indicating a successful induction.

Purification of Fusion Protein

The general purification scheme is outlined in Fig. 2. As mentioned previously, the ionic strength of the EGTA elution buffer may vary for some proteins, depending on the source of the CaM used to prepare the CaM affinity resin. In our experience, when using resins containing CaM purified from native sources (Pharmacia), CBP fusion proteins elute consistently and efficiently with EGTA buffer containing physiological NaCl, whereas with resins containing recombinant CaM purified from *E. coli* (Stratagene), high NaCl concentrations in addition to EGTA are often required. Because we use the latter type of resin, two elution steps are included in the protocol, employing EGTA buffers containing 150 mM NaCl and 1 M NaCl, respectively. Soluble proteins elute efficiently in the first two or three fractions of either the first or the second of these elution steps (Fig. 5A).

Buffers

Buffer A: 50 m*M* Tris–HCl, pH 8.0, 150 m*M* NaCl, 10 m*M* 2-mercapto-
    ethanol, 1 m*M* magnesium acetate, 1 m*M* imidazole, 2 m*M* CaCl₂
Buffer B: 50 m*M* Tris–HCl, pH 8.0, 150 m*M* NaCl, 10 m*M* 2-mercapto-
    ethanol, 1 m*M* magnesium acetate, 1 m*M* imidazole, 0.1 m*M* CaCl₂
Buffer C: 50 m*M* Tris–HCl, pH 8.0, 150 m*M* NaCl, 10 m*M* 2-mercapto-
    ethanol, 1 m*M* magnesium acetate, 1 m*M* imidazole, 2 m*M* EGTA
Buffer D: 50 m*M* Tris–HCl, pH 8.0, 1 *M* NaCl, 10 m*M* 2-mercaptoetha-
    nol, 1 m*M* magnesium acetate, 1 m*M* imidazole, 2 m*M* EGTA

All steps are carried out at 4° wherever possible. The cell pellet is resuspended in 30 ml buffer A containing 0.2 mg/ml lysozyme (Sigma), rotated gently for 30 min, and sonicated to completion. The lysate is cleared by centrifugation at 25,000*g* for 15 min. At this point it is prudent to remove a small sample of the cleared lysate to determine the proportion of the induced CBP fusion that remains in the soluble fraction. The amount of CaM-Sepharose resin to be used may be determined by estimating the amount of fusion protein in the lysate as a percentage of the total by visual inspection of an SDS–PAGE gel containing the cleared lysate sample. The resin capacity of CaM-Sepharose is 2 mg CaM/ml settle resin; 1 ml resin typically binds 1.5–3.0 mg CBP fusion protein. However, in order to reduce the amount of time that the fusion protein is exposed to proteases in the lysate, as a general rule we use 10 ml resin for every liter of induced culture harvested. The resin is equilibrated ahead of time during the lysozyme treatment with several washes of buffer A.

The equilibrated CaM-Sepharose is mixed with the lysate in a 50-ml polystyrene tube (Falcon) and rotated gently for 1 hr. Unbound material is removed following a low-speed spin of the slurry, and the resin is washed once with 40 ml of buffer A, resuspended in 20–30 ml of buffer A, and loaded into an Econo column (Bio-Rad). A sample of the unbound fraction is removed for later analysis to assess the efficiency with which the CBP fusion is removed from the flow-through fraction. The resin is washed with 5 bed volumes of buffer A followed by 5–10 bed volumes of buffer B until a baseline at OD$_{280}$ is reached. Although we have detected no endogenous *E. coli* proteins that bind specifically to the resin, there are situations for which it is desirable to include a buffer A/1 *M* NaCl wash (prior to the buffer B wash) to remove *E. coli* contaminants that interact with the fusion protein. The resin may be washed with buffer containing 1 *M* NaCl in the presence of 2 m*M* CaCl₂ with minimal leaching of the fusion protein.

The resin is washed with 5 bed volumes of buffer C followed by 5 volumes of buffer D, and the fractions are analyzed by SDS–PAGE.

Removal of CBP Using Site-Specific Proteases

Although we have purified several CBP fusion proteins whose activity is unaffected by the presence of the tag[4] (i.e., when compared with native protein or recombinant protein from which the CBP has been removed), it is often desirable to remove the CBP following purification. The following protocols describe proteolytic removal of the fusion tag with the site-specific proteases EK or thrombin, followed by the removal of protease (and the free CBP tag) or inactivation of the protease (Figs. 2 and 6).

Cleavage with EK and Recovery of Purified Cleavage Product

The CBP fusion proteins produced from the vectors pCAL-n-EK and pCAL-n-FLAG can be cleaved with EK to removal the fusion tag. As described in "Vectors and Strains," these vectors were designed specifically for the production of fusions from which proteins with native amino may be recovered.

In many cases the material in the elution fractions is concentrated enough to allow dilution directly into cleavage buffer (50 mM Tris–HCl, pH 8.0, 50 mM NaCl, 2 mM $CaCl_2$, 0.1% Tween 20) by adding a compensatory amount of $CaCl_2$ to neutralize the EGTA (i.e., 4 mM $CaCl_2$ neutralizes 2 mM EGTA, thus the addition of 6 mM $CaCl_2$ will result in an effective concentration of 2 mM $CaCl_2$). However, for proteins whose elution requires the presence of 1 M NaCl, a dialysis step is usually necessary. The high specific activity of commercially available recombinant EK preparations[14] allows a relatively low enzyme-to-substrate ratio. It is wise to perform one or two pilot time course or EK dilution curve experiments to determine the minimum amount of EK required to cleave the target protein to completion in the desired amount of time. In general, a good starting point is to use a 1:1000 EK-to-substrate ratio and incubate at room temperature for 12–20 hr.

If desired, EK and the free CBP tag, as well as any uncleaved fusion protein remaining in the reaction, can be removed quantitatively by absorption with a mixed resin slurry containing CaM-Sepharose and soybean trypsin-inhibitor (STI)-agarose (Sigma, Stratagene), the latter of which specifically absorbs EK[15] (Fig. 6). When the cleavage reaction is complete, NaCl is added to the reaction to 200 mM to prevent nonspecific adsorption of the protein of interest to the resin. The required volumes of CaM-Sepharose (0.5 ml resin/mg fusion protein cleaved) and STI-agarose (resin

[14] L. A. Collins-Racie, J. M. McColgan, K. L. Grant, E. A. Diblasio-Smith, J. M. McCoy, and E. R. LaVallie, *Bio/Technology* **13**, 982 (1995).
[15] P. Vaillancourt, T. G. Sincox, and C. F. Zheng, *BioTechniques* **22**, 451 (1997).

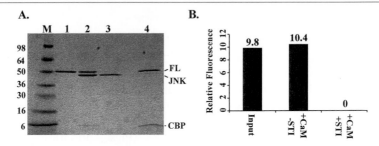

FIG. 6. Removal of EK, free CBP, and uncleaved CBP fusion protein from EK reactions. (A) Purified CBP-EK-JNK (50 μg) was digested with 100 ng of EK in 100 μl of EK reaction buffer for 3 min at 37° and cooled on ice. Following removal of a portion of the reaction for SDS–PAGE analysis, the sample was diluted to 500 μl with binding buffer (50 mM Tris–HCl, pH 8.0, 4 mM CaCl$_2$, 200 mM NaCl, and 0.1% Tween 20) and added to a mixture of settled resin containing 50 μl CaM affinity resin and 20 μl STI-agarose. The slurry was rotated mechanically at 4° for 1 hr. Following removal of unbound material, a portion of the resin was boiled in Laemmli sample buffer. Samples containing uncleaved CBP-EK-JNK (lane 1), EK-cleaved CBP-EK-JNK (lane 2), EK-cleaved CBP-EK-JNK absorbed with mixed resin (lane 3), and postabsorption resin (lane 4) were boiled in Laemmli sample buffer and electrophoresed on a 4–20% Tris–glycine SDS polyacrylamide gel (Novex). The gel was stained with Coomassie Brilliant Blue dye. M: molecular mass standards (kDa), JNK: JNK polypeptide EK cleavage product, CBP: calmodulin-binding peptide EK cleavage fragment. (B) Determination of enterokinase activity by fluorogenic peptide substrate assay. Purified CBP-EK-JNK (100 μg) was digested with 200 ng EK for 5 min at 37° in a 1.0-m reaction. The reaction was split into two 500-μl portions that were added to either 50 μl CaM-affinity resin (+CaM, $-$STI) or a mixture of 50 μl CaM-affinity resin plus 20 μl STI-agarose (+CaM, +STI). The resin slurries were rotated mechanically for 1 hr at 4°, and the resin was removed from the slurry by centrifugation. Unbound material was analyzed for enterokinase activity using a fluorogenic peptide assay. Volumes of both fractions were adjusted to 420 μl with binding buffer, and the samples were mixed with 1.6 ml of a solution containing 0.5 mM of the fluorogenic enterokinase substrate in 70 mM Tris–HCl, pH 8.0, and 10% dimethyl sulfoxide. Samples were mixed and analyzed immediately using a Shimadzu RF-1501 spectrofluorometer. Enzyme activity was determined by measuring increased fluorescence (excitation 337 nm, emission 420 nm) due to the release of β-naphthylamine over a 1-min interval. For a positive control, 30 ng of EK was assayed (Input), and the increased fluorescence value was normalized to the amount of enzyme used in the experimental reactions. Reprinted from P. Vaillancourt *et al., BioTechniques* **22,** 451 (1997), with permission from Eaton Publishing.

capacity: 2 mg EK bound/ml resin; for most applications, 10–20 μl is sufficient) are mixed and washed extensively with cleavage buffer containing 200 mM NaCl. The resin and cleavage reaction products are then mixed together with a 4:1 buffer-to-resin volume ratio and slowly rotated at 4° for 30 min. The resin is removed by low-speed centrifugation, and the reaction product is analyzed by SDS–PAGE.

Cleavage with Thrombin

As described earlier for EK, concentrated fractions containing purified protein may be diluted directly into thrombin cleavage buffer (20 mM Tris–HCl, pH 8.4, 150 mM NaCl, 2.5 mM CaCl$_2$), providing care is taken to neutralize the EGTA with a compensatory amount of CaCl$_2$. Because the efficiency of thrombin cleavage will vary from protein to protein, pilot experiments should be performed to determine the optimal enzyme-to-substrate ratio (starting at 1 : 500) and/or reaction time. Reactions are typically performed at room tmperature, although higher temperatures may be used.

Following digestion, thrombin may be inactivated by adding 0.5 mM phenylmethylsulfonyl fluoride to the reaction mixture or may be removed from the mixture by the addition of antithrombin III resin (Sigma) or α_2-macroglobulin resin (Boehringer Mannheim). The mixed resin strategy described earlier for EK may be employed to remove the free CBP tag and uncleaved fusion protein in the same step.

Detection of CBP Fusion Proteins and Interacting Proteins

There are many applications for which the specific detection of CBP fusion proteins is required. The ability to detect poorly expressed fusion proteins in crude lysates or throughout the purification protocol, to assess the stability of proteins, and to monitor the presence and location proteins throughout certain experimental protocols is highly desirable, particularly when no protein-specific antibody is available. In addition, the use of fusion tags for the detection and identification of interacting proteins has proven useful for mapping out signaling pathways and other protein networks.

Detection of CBP Fusions by Blot Overlay with Biotinylated CaM

The use of biotinylated-CaM (bio-CaM) for the detection of natural CaM-binding proteins in crude extracts has been described.[16] The following protocol describes the detection of CBP fusion proteins following electrophoretic transfer from SDS–PAGE gels to nitrocellulose filters. Although the following method is based on detection using colorimetric alkaline phosphatase substrates, with which we consistently detect ≤10 ng fusion protein, we have had success using chemiluminescent substrates as well.

A nitrocellulose or PVDF membrane containing electrophoretically fractionated protein is blocked with BLOTTO [20 mM Tris–HCl, pH 7.5, 150 mM NaCl, 0.01% thimerosal (Sigma), 5 g/100 ml nonfat dry milk] at

[16] R. L. Kincaid, M. L. Billingsley, and M. Vaughan. *Methods Enzymol,* **159,** 605 (1988).

room temperature for 30 min (all incubation times, with the exception of the color development reaction, may be extended with incubation at 4°). The filter is washed twice with TBSCT (20 mM Tris–HCl, pH 7.5, 150 mM NaCl, 1 mM CaCl$_2$, 0.05% Tween 20) and once with TBSC (20 mM Tris–HCl, pH 7.5, 150 mM NaCl, 1 mM CaCl$_2$). A 1.0-ml solution of TBSC containing 300 ng bio-CaM (CalBiochem, Stratagene) is spotted onto a smooth sheet of plastic wrap, and the membrane is overlayed "gel side" down, taking care to avoid bubbles. The membrane is wrapped in plastic wrap and incubated under slight pressure for 30 min. The membrane is washed as described earlier and is then incubated with appropriately diluted streptavidin alkaline phosphatase (strept-AP) for 1 hr. Prior to development, nitroblue tetrazolium (NBT) is added to 20 ml color development solution (100 mM Tris–HCl, pH 9.5, 100 mM NaCl, 5 mM MgCl$_2$, 1 mM CaCl$_2$) to 0.3 mg/ml (volumes are for a standard 8 × 6-cm membrane and may be scaled up). Immediately prior to development, 5-bromo-4-chloro-indolyl phosphate (BCIP) is added to a final concentration of 0.15 mg/ml. The membrane is washed and immersed in the color development solution with gentle agitation until the desired signal intensity is reached (appearing as dark bands on the membrane), and then the reaction is terminated by immersion in stop solution (20 mM Tris–HCl, pH 2.9, 1 mM CaCl$_2$). It should be noted that although there are no endogenous *E. coli* proteins that interact with CaM, there is a single biotinylated *E. coli* protein with a molecular mass of 22.5 kDa that may obscure the detection of CBP fusions of a similar size and may also contribute to background in dot blot experiments if crude lysates are used.

Fusion proteins produced from the vector pCAL-n-FLAG may also be detected with α-FLAG monoclonal antibodies using standard immunodetection methods.[10]

Detection of Interacting Proteins Using ^{32}P-Labeled Probes

Purified CBP fusions may be labeled isotopically with [γ-^{32}P]ATP using PKA. These probes are then used in blot overlay-type experiments to detect interacting proteins that are present in complex mixtures[17] or to screen bacteriophage expression libraries. In the protocol described here we use the his$_6$-tagged recombinant PKA catalytic subunit (Stratagene), which can be removed from kinase reactions with nickel resin.

Typically, 1 μg of CBP fusion protein is incubated in a 40-μl reaction containing 400 μM rATP, 1 μl [γ-^{32}P]ATP (10 μCi/μl), and 0.25 U PKA in 1× PKA buffer (25 mM HEPES, pH 7.4, 10 mM magnesium acetate) at 30° for 30 min. A parallel reaction containing 1 μg Histone IIs is often

---

[17] D. W. Carr, R. E. Stofko-Hahn, I. D. C. Fraser, S. M. Bishop, T. S. Acott, R. G. Brennan, and J. D. Scott, *J. Biol. Chem.* **266,** 14188 (1991).

included as a positive control for the labeling reaction. At this point the efficiency of the labeling can be determined by analyzing a portion of the reaction by SDS–PAGE followed by autoradiography. The specific activity of the probe is determined by excising the gel slice containing the labeled protein and measuring counts per minute in a scintillation counter.

If desired, the his-tagged PKA may be removed by absorption with a small amount of Ni-NTA-agarose (Qiagen) equilibrated with 1X PKA buffer. It should be noted that certain target proteins, particularly those with exposed histidine-rich sequences, may bind to the Ni-NTA resin during this step. This binding may be inhibited by including imidazole in the equilibration and absorption steps. For CBP fusions that are found to bind the Ni-NTA resin, it is recommended that a set of small-scale absorption reactions be performed to determine the optimal imidazole concentration that will inhibit the target protein–Ni-NTA interaction yet allow removal of the PKA.

Although we have not determined the extent to which the interaction between the CBP and the CaM contributes to background when CBP fusions are used to probe complex samples derived from eukaryotic cells, there are certain applications (e.g., probing for calcium-dependent interactions) for which removal of the CBP is required. Fusions derived from the vector pCAL-kc (Fig. 3) may be treated with thrombin to remove the CBP and then labeled to high specific activity due the the presence of the PKA "kemptide" substrate, which remains fused to the protein of interest.

Detection of Interacting Proteins by Immunoprecipitation

Detection and analysis of interacting proteins may also be acheived using immunoprecipitation "pull-down" experiments. In these experiments the purified target protein is spiked into complex mixtures of proteins such as crude lysates, and then the target protein and any interacting proteins are then selectively removed by immunoprecipitation using a target-specific antibody and then analyzed by SDS–PAGE. Pull-down experiments may be performed using CBP fusion proteins as targets. The mixture is incubated with CaM-Sepharose resin in the presence of 2 mM CaCl$_2$ and 200 mM NaCl for 30 min at 4°, centrifuged at low speed, and washed several times in neutral buffer containing 2 mM CaCl$_2$ and 200 mM NaCl. However, because the protein mixtures to be probed are often derived from eukaryotic extracts that contain CaM and CaM-binding proteins, background due to CBP–CaM interactions preclude the use of CaM-Sepharose for many applications. Fusions produced from the vector pCAL-n-FLAG retain the FLAG epitope at their N termini following removal of the CBP by thrombin treatment. Immunoprecipitions using FLAG fusions can now be carried out using α-FLAG monoclonal antibodies according to standard procedures.

Discussion

We have successfully purified a wide range of functional CBP fusion proteins to apparent homogeneity, ranging in size from 17 kDa (the SH2 domain from the mammalian signaling protein Grb2) to 120 kDa (*E. coli* β-galactosidase). These proteins include mammalian kinases, phosphatases, and other signaling proteins, a retroviral reverse transcriptase, prokaryotic proteins including a phosphatase, an endonuclease, and a site-specific recombinase, and several thermostable archae DNA polymerases.[18] The calmodulin-CBP purification system is relatively tolerant of a variety of buffer constituents, provided ≥0.2 m*M* CaCl$_2$ is present in the binding and wash steps and ≥0.2 m*M* EGTA is present in the elution. Reducing agents such as dithiothreitol and 2-mercaptoethanol (up to 5 and 10 m*M*, respectively), detergents including Nonidet P-40 and Triton X-100 (up to 0.1%), and all protease inhibitors that we have tested (with the exception of metal ion-chelating agents) are compatible with the system. Proteins from preparations containing high concentrations of potassium chloride and ammonium sulfate are efficiently bound to the resin. In addition, we have found that in general the inclusion of 1 m*M* imidazole in the binding and wash buffers reduces the required wash volumes, presumably by inhibiting nonspecific protein–protein interactions between histidine residues. We have yet to determine the tolerance of the system for purification of fusion proteins from preparations containing residual amounts of sarkosyl or chaotropic agents such as urea or guanidinium hydrochloride, which are often present following recovery of proteins from inclusion bodies.

Resin-dependent differences in the salt requirements for the elution buffer may be due to differences in posttranslational modifications between native CaM and recombinant CaM purified from *E. coli*. It is well established that calmodulin undergoes at least two posttranslational modifications in eukaryotic cells: acetylation at the N-terminal alanine residue and trimethylation at lysine 115.[19] For example, it is quite possible that the presence of a basic residue at lysine 115 may account for electrostatic interactions between *E. coli*-derived calmodulin and target proteins that do not occur with native calmodulin for which this residue is modified. However, it is clear that high NaCl in the absence of EGTA is not sufficient to elute proteins; in fact, prior to elution the resin may be washed with buffer containing 1 *M* NaCl in the presence of calcium with minimal loss of fusion protein from the resin.[3,20]

[18] L. B., D. H, C.-P. Carstens, J. Cline, C. Hansen, H. Hogrefe, D. McMullan, and M. Simcox (Stratagene), unpublished data.

[19] D. M. Roberts, P. M. Rowe, F. L. Siegel, T. J. Lukas, and D. M. Watterson, *J. Biol. Chem.* **261**, 1491 (1986).

[20] P. Vaillancourt, unpublished observations.

Studies are underway to determine the efficiency with which CBP fusion proteins may be purified from eukaryotic cell extracts. The purification is obviously complicated by the presence of naturally occurring CaM-binding proteins. For proteins produced from pCAL-n-EK, we routinely recover 80–90% of input fusion protein as mature EK-cleavage product following absorption of EK digests with calmodulin affinity resin to remove the cleaved CBP tag.[15] Such a strategy should also allow the removal of native calmodulin-binding proteins that coelute with the desired CBP fusion protein. The efficient two-step recovery of recombinant proteins in eukaryotic systems employing yeast, baculovirus, or adenoviral expression vectors should allow purification of proteins of native amino acid sequence for which improper folding or lack of requisite posttranslational processing events preclude the use of *E. coli* for expression and purification of functional protein.

Use of CaM as Fusion Partner for Affinity Purification

There are several systems in which CaM is used as a fusion partner for affinity purification using either solid supports that contain immobilized CaM ligand or that take advantage of the calcium-dependent interaction between CaM and phenyl-Sepharose resin. Ishizaka and co-workers fused the coding sequence for the T-cell receptor α chain (TCRα) at the C terminus of the rat CaM gene in a vector from which the fusion is expressed from the trp promoter.[21] Although previous attempts at purifying TCRα resulted in the production of insoluble protein, the CaM-TCRα fusion comprised 30% of the soluble lysate using this system. TCRα was purified by either a single pass through phenyl-Sepharose, cleavage with thrombin, and fractionation of the CaM and TCRα by gel filtration or was cleaved directly from the phenyl-Sepharose with thrombin. Using the latter scheme, 10 mg TCRα was recovered from 1 liter of culture.

In another system, Winter and colleagues used both the pelB secretory leader and CaM to produce "tribrid" fusions in which antibody fragments were secreted to the *E. coli* periplasm and purified via the CaM moiety.[22] The CaM fusions were efficiently purified using a resin containing the natural CaM ligand N-(6-aminohexyl)-5-chloro-1-naphthalenesulfonamide, to which the fusions bound and eluted with calcium dependence. The authors also found that the fusions could be efficiently purified by anion-exchange chromatography due to the highly acidic nature of CaM in the absence of calcium.

---

[21] Y. Ishii, T. Nakano, N. Honma, N. Yuyama, Y. Yamada, H. Watarai, T. Tomura, M. Sato, H. Tsumura, T. Ozawa, T. Mikayama, and K. Ishizaka, *J. Immunol. Methods* **186,** 27 (1995).
[22] D. Neri, C. de Lalla, H. Petrul, P. Neri, and G. Winter, *Bio/Technology* **13,** 373 (1995).

Daunert and co-workers purified CaM-protein A fusions using another natural CaM ligand, phenothiazine, coupled to a solid matrix.[23] Using this system, CaM fusions were also efficiently purified to high yield in a calcium-dependent manner using mild buffer conditions.

The choice of whether to use CaM or CaM-binding peptide fusions may depend on the size of the protein to be purified (larger proteins may be more soluble fused to the CBP, whereas small peptides may be consistently more soluble fused to CaM), the solvent accessibility of the termini (small tags such as polyhistidine and CBP are more likely to be inaccessible than larger tags such as CaM or GST), the requirement for the removal of the tag (protein function is less likely to be hindered using the smaller tag, an important point when purification of a poorly expressed protein results in low yields), and the desired use of the purified protein. Taken together, the versatility of the systems described here should allow for the relatively simple and inexpensive production of a wide range of proteins using *E. coli* as host.

Acknowledgments

We thank T. Simcox, M. Simcox, A. Stevens, J. D. Scott, and D. Carr for useful discussions and technical assistance; J. Bauer, J. Sorge, and A. Greener for support; and Diane Beery for excellent artwork.

[23] N. G. Hentz, V. Vukasinovic, and S. Daunert, *Anal. Chem.* **68**, 1550 (1996).

[23] The S·Tag Fusion System for Protein Purification

By RONALD T. RAINES, MARK McCORMICK, THOMAS R. VAN OOSBREE, and ROBERT C. MIERENDORF

Fusion Proteins

The detection, immobilization, and purification of proteins is idiosyncratic and can be problematic. Fortunately, these processes can be generalized by using recombinant DNA technology to produce fusion proteins in which target proteins are fused to carrier polypeptides. The affinity of the carrier for a specific ligand enables the facile detection, immobilization, and purification of a fusion protein.

Ribonuclease S

Bovine pancreatic ribonuclease A (RNase A; EC 3.1.27.5) catalyzes the cleavage of RNA.[1] The protease subtilisin prefers to cleave a single peptide bond in native RNase A. The product of this cleavage, RNase S (where "S" refers to subtilisin), consists of two tightly associated fragments: S-peptide (residues 1–20) and S-protein (residues 21–124).[2–4] Although neither fragment alone has any ribonucleolytic activity, RNase S is approximately as active as intact RNase A. The three-dimensional structure of RNase S is essentially identical to that of RNase A.[5,6]

The S-peptide fragment of RNase A has played an important role in the history of biochemistry. Before molecular biologists were able to use recombinant DNA technology to explore protein structure–function relationships, organic chemists synthesized analogs of S-peptide and studied their complexes with S-protein.[7,8] These studies provide much information on the role of individual residues in RNase S. Most significantly, only residues 1–15 of S-peptide were found to be necessary to form a fully functional complex with S-protein.[9] This shorter fragment is called "S15" or the "S·Tag" sequence.[10] (S·Tag, pBAC, and Perfect Protein are trademarks of Novagen, Inc., Madison, WI.)

In addition to structural information, extensive data have been acquired on the stability of RNase S. The value of K_d for RNase S is dependent on pH, temperature, and ionic strength.[1] Isothermal titration calorimetry has shown that the $K_d = 1.1 \times 10^{-7}$ M for the S-protein–S·Tag complex at 25° in 50 mM sodium acetate buffer, pH 6.0, containing NaCl (0.10 M).[11] Only a low yield of native S-protein (which contains four disulfide bonds) is isolable from the air oxidation of reduced S-protein.[12] The recovery of native S-protein is complete, however, if

[1] R. T. Raines, *Chem. Rev.* **98,** 1045 (1998).

[2] F. M. Richards, *C. R. Lab Carlsberg (Sér. Chim.)* **29,** 322 (1955).

[3] F. M. Richards and P. J. Vithayathil, *J. Biol. Chem.* **234,** 1459 (1959).

[4] F. M. Richards, *Protein Sci.* **1,** 1721 (1992).

[5] H. W. Wyckoff, K. D. Hardman, N. M. Allewell, T. Inagami, L. N. Johnson, and F. M. Richards, *J. Biol. Chem.* **242,** 3984 (1967).

[6] H. W. Wyckoff, K. D. Hardman, N. M. Allewell, T. Inagami, D. Tsernoglou, L. N. Johnson, and F. M. Richards, *J. Biol. Chem.* **242,** 3749 (1967).

[7] E. A. Barnard, *Annu. Rev. Biochem.* **38,** 677 (1969).

[8] F. M. Richards and H. W. Wyckoff, *Enzymes* **IV,** 647 (1971).

[9] J. T. Potts, Jr., D. M. Young, and C. B. Anfinsen, *J. Biol. Chem.* **238,** 2593 (1963).

[10] M. McCormick and R. Mierendorf, *Novations* **1,** 4 (1994).

[11] P. R. Connelly, R. Varadarajan, J. M. Sturtevant, and F. M. Richards, *Biochemistry* **29,** 6108 (1990).

[12] E. Haber and C. A. Anfinsen, *J. Biol. Chem.* **236,** 422 (1961).

A

```
LysGluThrAlaAlaAlaLysPheGluArgGlnHisMetAspSer
 +   -  0   0   0   0   +  0    -   +   0 0/+ 0   -   0
```

B

```
AAAGAAACCGCTGCTGCTAAATTCGAACGCCAGCACATGGACAGC
LysGluThrAlaAlaAlaLysPheGluArgGlnHisMetAspSer
```
pET

```
AAAGAAACGGCGGCGGCGAAATTTGAACGCCAACACATGGACAGC
LysGluThrAlaAlaAlaLysPheGluArgGlnHisMetAspSer
```
pBAC

Fig. 1. (A) Amino acid sequence of S·Tag, which corresponds to the first 15 amino acid residues of ribonuclease A. The net charge on each residue at neutral pH is indicated by +, 0, or −. (B) Nucleotide sequences of S·Tag coding regions used in expression vectors. Sequences are shown that optimize codon usage in *Escherichia coli* (pET vectors) and baculovirus (pBAC transfer vectors).

oxidation is performed in the presence of S-peptide, which serves as a template for proper folding.[13]

The S·Tag Carrier

The wealth of information that has been accumulated on RNase S has enabled the development of this noncovalent complex as the basis for a fusion protein system.[14–17] In this system, S·Tag is the carrier and S-protein is the ligand. The S·Tag carrier combines a small size (15 amino acid residues) with a high sensitivity of detection (20 fmol, which is 1 ng of a 50-kDa fusion protein, in solution or on Western blots). S·Tag has several additional properties that are desirable in a carrier. For example, S·Tag is composed of four cationic, three anionic, three uncharged polar, and five nonpolar residues (Fig. 1A). This composition makes S·Tag an excessively soluble peptide with little structure and net charge near neutral pH. The S·Tag carrier is therefore unlikely to interfere with the proper folding or function of a fused target protein. The S-peptide portion of RNase A is

[13] I. Kato and C. B. Anfinsen, *J. Biol. Chem.* **244**, 1004 (1969).

[14] J.-S. Kim and R. T. Raines, *Protein Sci.* **2**, 348 (1993).

[15] V. N. Senchenko, M. V. Dianova, V. Y. Kanevskii, A. L. Bocharova, and M. Y. Karpeiskii, *Mol. Biol.* **27**, 565 (1993).

[16] M. Y. Karpeisky, V. N. Senchenko, M. V. Dianova, and V. Y. Kanevsky, *FEBS Lett.* **339**, 209 (1994).

[17] J.-S. Kim and R. T. Raines, *Anal. Biochem.* **219**, 165 (1994).

not antigenic.[18] The exacting nature of the interaction between the S·Tag peptide and the S-protein minimizes the likelihood of interference from naturally occurring molecules. Also, the topology of RNase S is such that target proteins fused to either terminus of S·Tag allow for binding to S-protein. It appears that the peptide can also reside internally within fusion proteins and remain fully accessible for S-protein binding. For example, fusion proteins in which the S·Tag sequence is located between N-terminal glutathione *S*-transferase, thioredoxin, or cellulose-binding domains, and C-terminal β-galactosidase domains are detected on Western blots with the same sensitivity as fusion proteins that carry the S·Tag peptide at either terminus (data not shown). (Although the S·Tag peptide is hydrophilic and thus expected to also be solvent accessible under native conditions, it is possible that the folding of certain target proteins could mask the peptide and prevent efficient binding of S-protein, which would compromise the S·Tag rapid assay and affinity purification procedures described herein.) Finally, the affinity between S·Tag and S-protein can be fine-tuned by rational mutagenesis.[14] Together, these properties make S·Tag an extremely useful and versatile carrier in fusion protein systems.

Expression Vector Design

A variety of vectors has been used successfully for the expression of S·Tag fusion proteins in prokaryotic, eukaryotic, and *in vitro* systems. Expression levels appear to be unaffected by the presence of the S·Tag sequence in all constructs we have used. Examples of codon-optimized nucleotide sequences for expressing the S·Tag peptide in bacterial and baculovirus vectors are shown in Fig. 1B. Transcripts of both sequences are also efficiently translated in rabbit reticulocyte lysates. If the potential for translation initiation at Met13 is a concern, it can be replaced by the isoleucine codon ATC with no effect on the interaction with S-protein. Expression vectors having various configurations of S·Tag fusion sequences, as well as S·Tag detection and purification reagents, are available commercially from Novagen.

S·Tag Rapid Assay

A unique feature of this fusion system is that the S·Tag sequence confers the ability to quantify target proteins using a simple enzymatic assay, which is based on the reconstitution of ribonucleolytic (RNase S) activity. A crude or purified sample containing the fusion protein is added to

---

[18] G. W. Welling and G. Groen, *Biochim. Biophys. Acta* **446,** 331 (1976).

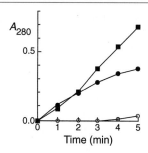

FIG. 2. Time course of an S·Tag rapid assay. Reactions with the indicated samples were performed as described in the text. (■) Rabbit reticulocyte lysate *in vitro* translation of S·Tag β-galactosidase. (●) Crude *E. coli* extract of a pET construct that directs the expression of S·Tag β-galactosidase. (○) Blank—*in vitro* translation reaction with no template.

a buffer containing excess purified S-protein and the ribonuclease substrate poly(C). After a brief incubation, the reaction is stopped with trichloroacetic acid (TCA) and the resulting precipitate is removed by centrifugation. Activity is measured by reading the absorbance of the supernatant at 280 nm, which increases as the poly(C) is broken down into acid-soluble nucleotides by the enzyme.[19] By comparing the results with a known S-peptide standard, the molar concentration of target protein in the sample can be determined. With this assay, as little as 20 fmol of target protein can be detected in a 5-min incubation.

Typical assay profiles are shown in Fig. 2. A linear signal was obtained during a 5-min incubation at 37°. In this experiment, 2 μl of crude translation mix and 2 μl of a 1/100 dilution of *Escherichia coli* lysed with 1% (w/v) SDS were used as samples. Appropriate blanks serve as controls for any endogenous ribonucleolytic activity, which appears to be minimal under the conditions of the assay.

Unlike amino acid incorporation assays (both radioactive and nonradioactive), the S·Tag rapid assay is independent of protein size, amino acid composition, and endogenous amino acid pool size. The method is also extremely versatile for measuring protein expression in cells, and it can be applied to both soluble and insoluble proteins. Multiple samples can be screened easily for expression levels by preparing crude extracts of whole cells in 1% (w/v) SDS. Because the assay detects as little as 20 fmol target protein in a 5-min incubation, even poorly expressed proteins can be measured with a high degree of accuracy.

The S·Tag rapid assay relies on ribonucleolytic activity. One might expect some level of background activity due to the presence of endogenous ribonucleases, especially in crude extracts. In practice, however, we have

[19] S. B. Zimmerman and G. Sandeen, *Anal. Biochem.* **10**, 444 (1965).

observed negligible background signals using crude *E. coli* extracts, insect cell extracts, and rabbit reticulocyte lysates. In many applications the expression levels are such that crude protein samples are diluted 10- to 500-fold prior to use in the assay, which further mitigates any background activity. Although we have less experience with mammalian cell extracts, we would expect that with few exceptions (e.g., pancreatic or liver tissue) most sources would not pose a significant background problem. The use of appropriate control samples, prepared in a manner identical to test samples but lacking expressed target protein, allow the direct measurement of any background signal. Recommended control samples for the rapid assay are uninduced crude extracts (prokaryotic), nontransfected crude extracts (eukaryotic), or *in vitro* translation reactions lacking template.

S·Tag Rapid Assay Protocol

Materials needed to perform 100 S·Tag rapid assays are as follows.
1 ml S·Tag grade S-protein 50 ng/μl (Novagen)
4 ml 10× S·Tag assay buffer [0.20 M Tris–HCl buffer, pH 7.5, containing NaCl (1 M) and poly(C) (1 mg/ml)]
50 μl S·Tag standard (20 amino acid residues; 0.05 pmol/μl) (Novagen)
40 ml sterile deionized water
10 ml 25% (w/v) TCA at 4°
Microcentrifuge at 4°
UV spectrophotometer set at 280 nm
Samples can be total, soluble, or insoluble protein fractions dissolved in denaturing buffers (see later). In general, crude cellular samples need to be diluted from 1:10 to 1:500 in water to be in the linear range of the assay. Up to 10 μl of a 1/10 dilution of 1% (w/v) SDS or 10 μl of a 1/100 dilution of 6 M urea or 6 M guanidine–HCl can be added to the assay with little effect. For each set of samples, a blank without added target protein and the S·Tag standard are run in parallel.

1. Assemble the following components in a set of sterile 1.5-ml microcentrifuge tubes. This example uses one unknown (in tube 3).

Component	Tube 1	Tube 2	Tube 3
Sterile water	348 μl	346 μl	348 μl
S·Tag standard	—	2 μl	—
Sample extract	—	—	2 μl
Blank extract	2 μl	2 μl	—
10× S·Tag assay buffer	40 μl	40 μl	40 μl
S·Tag grade S-protein	10 μl	10 μl	10 μl

2. Start the reactions by adding the S-protein at timed intervals (e.g., every 20 sec).

3. Incubate the tubes at 37° for exactly 5 min.

4. Stop the reactions by adding 100 μl ice-cold 25% (w/v) TCA, vortex, and place on ice for 5 min.

5. Centrifuge the tubes at 14,000g for 10 min.

6. Read the absorbance of the supernatants at 280 nm. Zero the spectrophotometer with sample 1. If the absorbance of the sample is greater than 1.5, the assay should be repeated with a more dilute sample to stay within the linear range.

7. To calculate the concentration (in pmol/μl) of S·Tag protein in the sample, use the following equation:

$$[S \cdot Tag] = (A_{280} \text{ of tube } 3/2 \ \mu l) \times (0.1 \text{ pmol S} \cdot \text{Tag standard}/A_{280} \text{ of tube 2})$$

For example, if the A_{280} for the S·Tag standard (No. 2) is 0.5 and the A_{280} of a 1:200 dilution of the sample (No. 3) is 1.0, then [S·Tag] = (1.0 × 200/2 μl) × (0.1 pmol/0.5) = 20 pmol/μl. If the target protein has a molecular mass of 50 kDa, then its concentration is 20 pmol/μl × 0.05 μg/pmol = 1 μg/μl.

Notes

a. The blank extract for tube 1 should be prepared under conditions identical to the test sample but without expression of the S·Tag fusion protein (e.g., a blank *in vitro* translation reaction lacking target mRNA, extracts of cells lacking expression vector constructs). This preparation provides a control for the enzymatic and nonenzymatic contribution to the absorbance reading from endogenous sample components. In addition, the buffer composition of the blank should be as similar to that of the unknown as possible.

b. Best results are obtained using a special grade of S-protein ("S·Tag grade") that has been purified to remove residual RNase A and RNase S, which commonly contaminate commercial preparations of S-protein.

c. When assaying purified proteins, use a buffer containing 10 mM Tris–HCl buffer, pH 7.5, containing Triton X-100 (0.1%, v/v) to dilute the sample to avoid loss of material on tube surfaces.

d. The assays just described are designed for using standard 1-ml cuvettes; the assay volume can be scaled down proportionately to accommodate smaller cuvettes.

e. Keep in mind that RNase is being reconstituted in this assay and could contaminate cuvettes used for reading the results. Clean the cuvettes with 0.5 N NaOH or a strong quaternary amine detergent to prevent carryover.

S·Tag Western Blots

S·Tag fusion proteins produced by *in vitro* translation, eukaryotic, or prokaryotic expression are detected quickly and with high sensitivity by Western blot analysis. Blots are prepared by conventional SDS–PAGE and Western transfer protocols, blocked briefly with nonfat dry milk, and then incubated with S-protein horseradish peroxidase (HRP) or alkaline phosphatase (AP) conjugate. Target bands are visualized with colorimetric or chemiluminescent substrates. An optimized protocol allows fully developed blots to be produced in 45 min with detection of nanogram amounts of target proteins.

A time course of induction of β-galactosidase expressed from the bacterial vector pET-30b(+) is shown in Fig. 3. The S·Tag Western blot (Fig. 3, right) shows high specificity for target proteins (and their amino-terminal-containing breakdown products) with very low background staining of other *E. coli* proteins. This gel also contains a marker lane containing a set of seven S·Tag fusion proteins having defined sizes at convenient intervals, which serve as precise internal standards for S·Tag Western blots.

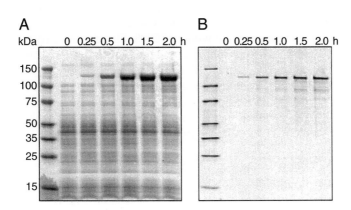

Fɪɢ. 3. SDS–PAGE and S · Tag Western blot analysis of an S · Tag fusion protein expressed in *E. coli*. An *E. coli* host carrying a pET construct encoding S · Tag β-galactosidase (118 kDa) was grown in culture. Whole cell extracts were prepared at the indicated times following the addition of IPTG to induce expression of the fusion protein. The marker lane on the left side of each panel contains a mixture of S · Tag fusion proteins having defined sizes (Perfect Protein Markers, Novagen). (A) SDS–PAGE gel stained with Coomassie blue. (B) S · Tag Western blot. A gel was loaded with a 1/100 dilution of the samples used in A. Proteins were transferred to nitrocellulose and incubated with S-protein alkaline phosphatase conjugate followed by color development with NBT/BCIP substrates.

S · Tag Western Blot Protocol

Materials needed for the development of 25 medium size (10 × 10 cm) blots are as follows.

50 μl S-protein conjugate with HRP or AP (Novagen)

25 g nonfat dry milk

200 ml 10× TBST (0.10 M Tris–HCl buffer, pH 8.0, containing NaCl (1.5 M) and Tween 20 (1% v/v)

HRP or AP buffer and development substrates (colorimetric or chemi-luminescent)

The blot development procedure begins with a membrane (preferably nitro-cellulose) containing proteins transferred from an SDS–polyacrylamide gel. Standard procedures for SDS–PAGE and transfer are suitable.

S · Tag Western Blot Development

1. Prepare blocking solution by dissolving 1.25 g nonfat dry milk in 25 ml of 1× TBST with stirring. This solution will block one blot.

2. Remove the nitrocellulose from the blotting apparatus and incubate in blocking solution at room temperature for 15–30 min to block excess protein-binding sites.

3. Prepare 250 ml of 1× TBST per blot by diluting the 10× stock with deionized water. Rinse the membrane for 1 min in 25 ml TBST at room temperature to remove excess blocking reagent.

4. Incubate the membrane with a 1 : 5000 dilution of S-protein HRP or AP conjugate in TBST for 30 min at room temperature. Use enough reagent to cover the membrane, usually about 10 ml.

5. Wash the membrane five times in 25–50 ml TBST at room temperature. This washing can be done in 1–2 min by adding the wash solution, briefly shaking, and decanting. It is important to wash thoroughly the membrane at this point to achieve maximum signal : noise ratios.

6. Remove excess liquid and cover the membrane in blot development solution containing the appropriate buffer and substrates. Suitable development systems for alkaline phosphatase include a solution of NBT and BCIP for a colorimetric end point [10 mM Tris–HCl buffer, pH 9.4, containing NBT (0.33 mg/ml), BCIP (0.17 mg/ml), NaCl (10 mM), and MgCl$_2$ (1 mM)] and CDP-Star (Tropix) for a chemiluminescent end point. SuperSignal (Pierce) substrate is suitable for sensitive HRP-based chemiluminescent detection. Make sure that the entire surface of the membrane has been wetted with the substrate. For colorimetric substrates, incubate the blot until satisfactory signals are obtained and stop the reaction by rinsing in water. For chemiluminescent systems, incubate the blot in the substrate at room temperature for 1 min and then proceed to step 7.

7. Remove the membrane from the substrate and cover with plastic wrap. Remove any bubbles between the plastic and the membrane. Gently remove any liquid from the exterior of the plastic.

8. Place blot in a film cassette with autoradiographic film and expose for 1–10 min. Be careful not to move the film or blot after initial placement or multiple images can result. An initial exposure time of 1 min is recommended. Longer exposures can be performed, although the highest light output occurs in the first 5 min. Light output continues over several hours.

S·Tag Affinity Purification

S·Tag fusion proteins are purified rapidly by affinity chromatography using immobilized S-protein, such as S-protein covalently coupled to agarose.[1] Several purification strategies are possible, depending on the application. When fusion proteins are expressed from vectors that also encode a site-specific protease cleavage site (e.g., thrombin, LeuValProArg ↓ GlySer; or enterokinase, AspAspAspAspLys ↓) between the S·Tag sequence and the cloning region, the target protein can be released from an S-protein matrix under native conditions simply by protease digestion. The efficiency of digestion is somewhat dependent on the target protein, as well as on the sequence context surrounding the protease cleavage site. Optimal contexts are present in relevant pET bacterial and pBAC baculovirus expression vectors available from Novagen. An example of the purification achieved under these conditions is shown in Fig. 4. Methods have been developed to allow specific, quantitative removal of the protease in the eluted fraction. For example, biotinylated thrombin is fully active and removed easily after digestion with streptavidin agarose, leaving the purified protein in solution. Under these conditions, homogeneous target proteins lacking the S·Tag peptide are recovered, and the S-protein matrix is not reused.

As an alternative, fusion proteins can be eluted from S-protein agarose under conditions that disrupt the S·Tag:S-protein interaction [e.g., 3 M guanidinium thiocyanate; 0.2 M potassium citrate buffer, pH 2; or 3 M MgCl$_2$]. By this method, the S·Tag peptide remains attached to the fusion protein, and the S-protein agarose can be recycled. If the target protein accumulates in cells as inclusion bodies, the insoluble fraction can be prepared and dissolved in 6 M urea. The sample is then diluted three fold (to 2 M urea) and applied to the S-protein agarose equilibrated in the same buffer. Fusion proteins are bound to the S-protein agarose under these conditions and can be eluted either with biotinylated thrombin digestion in the presence of urea or with the partially denaturing conditions listed earlier.

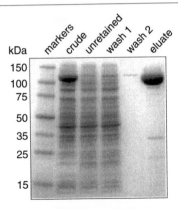

FIG. 4. S · Tag affinity purification. S · Tag β-galactosidase expressed from a pET construct was purified from a crude soluble fraction using S-protein agarose under native conditions. Elution of the target protein from the agarose was performed by digestion with thrombin, which is visible as a minor band (~33 kDa) in the eluted fraction. Use of biotinylated thrombin enables its removal with streptavidin agarose following digestion (see text). The identity of each fraction is indicated.

Binding can be performed in a column or batch mode. The capacity of S-protein agarose varies somewhat based on the size and folding characteristics of a given target protein. Under native conditions, commercial preparations of S-protein agarose exhibit a minimum binding capacity of 500 μg/ml for S · Tag β-galactosidase.

S · Tag Affinity Purification Protocol

Materials needed for the purification of 1 mg of target protein are as follows.

2 ml S-protein agarose [50% (v/v) slurry in 50 mM Tris–HCl buffer, pH 7.5, containing NaCl (0.15 M), EDTA (1 mM), and sodium azide (0.02%, w/v); Novagen]. Note that stated volumes of resins are settled bed volumes.

150 ml bind/wash buffer [20 mM Tris–HCl buffer, pH 7.5, containing NaCl (0.15 M) and Triton X-100 (0.1%, v/v)]

3 ml 10× thrombin cleavage buffer [for thrombin elution; 0.20 M Tris–HCl buffer, pH 8.4, containing NaCl (1.5 M) and CaCl$_2$ (25 mM)]

50 U biotinylated thrombin (for thrombin elution; Novagen)

0.8 ml streptavidin agarose [50% (v/v) slurry in sodium phosphate buffer, pH 7.5, containing sodium azide (0.02%, w/v); Novagen]

Spin filters, 5-ml capacity (optional)

Binding to S-Protein Agarose

The purification procedure begins with a cell extract (bacterial, insect, or mammalian) or an *in vitro* translation reaction containing the S·Tag fusion protein. Note that elution with thrombin or enterokinase requires the presence of the appropriate protease cleavage site between the S·Tag peptide and the target protein sequence.

1. Gently suspend the S-protein agarose by inversion and add 2 ml of the slurry (equivalent to 1 ml settled resin) to the desired amount of soluble protein extract. If binding from partially denatured protein in 2 *M* urea, add an equal volume of 4 *M* urea in bind/wash buffer to 2 ml of resin prior to adding it to the protein extract. The resin is transferred most conveniently with a 1-ml wide-mouth pipette tip. Mix thoroughly and incubate at room temperature on an orbital shaker for 30 min. Do not shake vigorously as this will tend to denature protein.

2. Centrifuge the entire volume at 500*g* for 10 min and decant supernatant carefully.

3. Resuspend the S-protein agarose, which now contains bound S·Tag fusion protein, in 5 ml bind/wash buffer (include 2 *M* urea if using partially denatured protein). Mix by gently vortexing or by repeated inversion.

4. Repeat steps 2 and 3 twice more to wash away unbound proteins. Remove the final supernatant and elute the target protein either with biotinylated thrombin or using guanidinium thiocyanate, pH, or MgCl$_2$ as described later.

Elution with Biotinylated Thrombin

1. Prepare 15 ml thrombin cleavage buffer by diluting 1.5 ml of the 10× stock in 13.5 ml deionized water (remember to add 2 *M* urea if purifying proteins from solubilized inclusion bodies).

2. Resuspend the washed S-protein agarose containing the bound target protein in 5 ml of 1× thrombin cleavage buffer from step 1. Centrifuge at 500*g* for 10 min and remove the supernatant carefully. Resuspend and centrifuge again to fully equilibrate the resin in 1× thrombin cleavage buffer. Remove as much supernatant as possible.

3. Resuspend the washed, equilibrated agarose pellet in a final volume of 2 ml of 1× thrombin cleavage buffer (plus 2 *M* urea if necessary). Add 25 U biotinylated thrombin and incubate for up to 2 hr at room temperature on an orbital shaker. The target protein released from the agarose no longer contains the S·Tag peptide. The biotinylated thrombin is removed with streptavidin agarose. Note that biotinylated thrombin is fully active in the presence of 2 *M* urea. The recommended cleavage conditions (25 U, 2 hr)

are optimal for a variety of proteins; however, if secondary cleavage is observed with a particular protein, less enzyme can be used or the incubation duration decreased.

4. Thoroughly resuspend the streptavidin agarose by inversion. Add 800 μl of the slurry to the cleavage reaction (see note below if using urea in the buffers). Mix thoroughly and incubate for 10 min at room temperature on an orbital shaker. Note that the streptavidin agarose, in its listed buffer, can be added directly to the biotinylated thrombin/target protein/S-protein agarose mixture without the need for preequilibration. If urea has been included in the procedure, first bring the urea concentration in the streptavidin agarose slurry to 2 M by adding an appropriate volume from a concentrated stock solution. If desired, the streptavidin agarose can be preequilibrated in 1× thrombin cleavage buffer to avoid the addition of other components (such as inorganic phosphate or azide) in the supplied storage buffer. Proceed to step 7 if not using a spin filter to remove the agarose.

5. Transfer the entire reaction to a spin filter that has been placed in a collection tube. Centrifuge at 500g for 5 min.

6. Without removing the filtrate in the lower chamber, add 1.25 ml of 1× thrombin cleavage buffer (plus urea, if necessary) to the "cake" of resin in the upper chamber and centrifuge at 500g for 5 min. The clear filtrate contains the purified target protein, which can be used directly in many applications. See "Processing the Sample after Elution" for suggested procedures for concentration and changing the buffer.

7. Optional (if a spin filter is not used): Centrifuge at 500g for 5 min and transfer the supernatant, which contains the target protein, to a fresh tube. Wash the agarose pellet with an additional 1–2 ml of 1× thrombin cleavage buffer, centrifuge, and pool the second supernatant with the previous supernatant.

Elution with Guanidinium Thiocyanate, pH, or MgCl$_2$

1. Resuspend the washed resin containing the bound target protein in 1.5× settled resin volumes of one of the following solutions: bind/wash buffer containing guanidinium thiocyanate (3 M); 0.2 M potassium citrate buffer, pH 2; or 3 M MgCl$_2$. Incubate for 10 min at room temperature; mix gently every few minutes to keep the resin suspended. To make the sodium citrate buffer, prepare a 2 M stock of citric acid, adjust the pH to 2.0 with 10 M KOH, and dilute to 0.2 M.

2. Transfer the entire reaction to a spin filter that has been placed in a collection tube. Centrifuge at 500g for 5 min.

3. Without removing the filtrate, add 1.25 ml elution buffer to the

"cake" of resin in the upper chamber and centrifuge at 500g for 5 min. The clear filtrate contains the purified target protein.

4. Optional (if a spin filter is not used): Centrifuge at 500g for 5 min and transfer the supernatant, which contains the target protein, to a fresh tube. Wash the agarose pellet with an additional 1–2 ml of elution buffer, centrifuge, and pool the second supernatant with the previous supernatant.

5. Change the buffer in the eluted sample by one of the methods described in the next section.

6. The S-protein agarose may be recycled by washing three more times with elution buffer and then three times with bind/wash buffer. Store at 4° in bind/wash buffer containing 0.02% (w/v) sodium azide or other preservative.

Processing the Sample after Elution

The buffer of the purified sample may be changed or the sample concentrated by one of several methods. Note that, depending on the solubility characteristics of target protein, changing the buffer may result in precipitation. Three alternative procedures are as follows.

1. Dialyze into the buffer of choice. After dialysis, the sample may be concentrated by sprinkling solid polyethylene glycol (15–20 kDa) or Sephadex G-50 (Pharmacia) on the dialysis tubing. Use dialysis tubing with an exclusion limit of 6 kDa or less and leave the solid in contact with the tubing until the desired volume is reached, replacing it with fresh solid as necessary.

2. Use plastic disposable microconcentrator units (e.g., Centricon; Amicon) as directed by the manufacturer to both desalt and concentrate the sample by ultrafiltration.

3. Desalt the sample by gel filtration on Sephadex (G-10, G-25, G-50; Pharmacia) or Bio-Gel (P6DG, P-10, P-30; Bio-Rad).

Like most high-affinity chromatography methods, purification of S·Tag fusion proteins using immobilized S-protein has some disadvantages. For example, the elution of S·Tag fusion proteins from S-protein agarose requires the use of proteases or mildly chaotropic conditions to release the target protein. The use of chaotropes carries with it the risk of partial or total denaturation or inactivation of the target protein, which can be problematic when downstream applications require the isolation of a properly folded, active protein. Proteolytic elution avoids denaturing conditions, but results in the removal of the S·Tag peptide, making it unavailable for further detection or purification procedures. Proteolytic elution also carries the risk of secondary, nonspecific cleavage of the target protein. Secondary cleavage can be reduced or eliminated in many instances with the use of lower amounts of protease, but the molecular weight and integrity of cleavage products should be confirmed by SDS–PAGE.

Prospectus

The S·Tag fusion system uniquely combines small tag size, antibody-like ligand-binding specificity, and the ability to confer an easily measured enzymatic activity to fusion proteins. The S-protein ligand is also small, relatively inexpensive, and can be used in a variety of formats to enable many applications with a single tagging system. Virtually every technique used with antibodies and their epitope tags can be applied to the S-protein:S·Tag interaction, including the blotting and purification methods described here, fusion protein immobilization, affinity capture of interacting molecules, and use of fluorophore-labeled S-protein for *in situ* affinity localization and cell sorting. The reconstitution of enzymatic activity by the simple addition of S-protein to any S·Tag fusion protein provides the platform for the development of novel assays. In particular, the discovery of a hypersensitive fluorogenic substrate for RNase A makes assays of S·Tag fusion proteins amenable to automation and thus useful in a variety of high-throughput screening applications.[20]

Acknowledgment

The initial development of the S·Tag fusion system at the University of Wisconsin–Madison was supported by Grant GM44783 (NIH).

[20] B. R. Kelemen, T. A. Klink, M. A. Behlke, S. R. Eubanks, P. A. Leland, and R. T. Raines, *Nucleic Acids Res.* **27,** 3696 (1999).

[24] Fusions to Self-Splicing Inteins for Protein Purification

By MING-QUN XU, HENRY PAULUS, and SHAORONG CHONG

Introduction

Protein splicing involves the self-catalyzed excision of an intervening polypeptide segment, the intein, from a precursor protein, with the concomitant ligation of the flanking polypeptide sequences, the exteins, to yield a functional protein. The catalysis of protein splicing, which was reviewed recently,[1] is entirely mediated by the intein and involves three distinct reaction steps (Fig. 1). Elucidation of the sequence of steps that underlie protein splicing and studies on the effect of amino acid substitutions in the intein and adjacent residues on these steps led to the realization that catalysis of each of the steps in the protein splicing pathway is relatively independent and opened the way for modulating the protein splicing process as a protein engineering tool.[2–4] This article describes how inteins can be used to effect the self-catalyzed cleavage of fusion proteins at highly specific sites. The various strategies by which this objective was accomplished can best be appreciated by a brief review of the mechanism of protein splicing in the context of the amino acid residues surrounding the splice junctions (Fig. 2).[5]

Mechanism of Protein Splicing

Step 1 in protein splicing is an N-S or N-O acyl rearrangement at the upstream splice junction, involving the side chain of the conserved Cys or Ser residue in position 1 of the intein, to yield an intermediate in which one of the peptide bonds has been replaced by an ester. Ordinarily, the amide-ester equilibrium favors amide bond formation at neutral pH. However, crystallographic evidence suggests that the peptide bond at the upstream splice junction is in the *cis* conformation[6]; the relief of strain when an energetically unfavorable *cis* peptide bond is replaced by an ester would shift the amide-ester equilibrium toward ester formation. According to this scenario, the C-terminal amino acid of the N-extein, residue -1, which participates in the *cis* peptide bond, should also play a critical role in the first step of protein splicing. Indeed, replacement of the amino acids in positions 1 and -1 affects protein splicing profoundly. No other amino acid can substitute for the conserved Cys or Ser residues in position 1,

[1] H. Paulus, *Chem. Soc. Rev.* **27**, 375 (1998).

[2] M.-Q. Xu and F. B. Perler, *EMBO J.* **15**, 5146 (1996).

[3] S. Chong and M. Q. Xu, *J. Biol. Chem.* **272**, 15587 (1997).

[4] S. Chong, K. S. Williams, C. Wotkovicz, and M.-Q. Xu, *J. Biol. Chem.* **273**, 10567 (1998).

[5] Nomenclature: Amino acid residue numbering—The N-extein residue adjacent to the first amino acid of the intein is labeled -1 and numbering proceeds toward the N terminus of the extein (i.e., . . . $N_{-2}P_{-1}$-intein). Intein residues are numbered sequentially starting with the N-terminal amino acid (C_1). C-extein amino acids are numbered beginning with the residue immediately following the intein (i.e., intein-$C_{+1}G_{+2}$ . . .). Amino acid substitutions—Amino acid replacements are indicated using the residue numbering system just described. For example, replacement of Asn_{454} by Ala is Asn454Ala and replacement of Pro_{-1} by Gly is Pro(-1)Gly. Intein nomenclature—The nomenclature used is that proposed by Perler *et al., Nucleic Acids Res.* **22**, 1125 (1994).

[6] T. Klabunde, S. Sharma, A. Telenti, W. R. Jacobs, and J. C. Sacchettini, *Nature Struct. Biol.* **5**, 31 (1998).

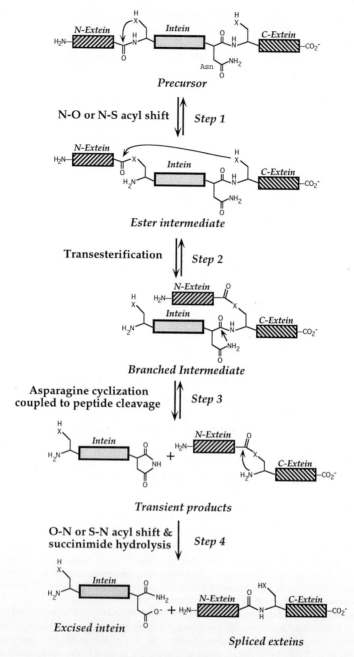

FIG. 1. Mechanism of protein splicing. The amino acid residues that participate directly in the chemical transformations are shown (X = O or S). The remainder of the intein and the exteins are shown by boxes, which are not to scale. The first three steps occur at the catalytic center of the intein, whereas the fourth step proceeds spontaneously.

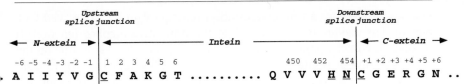

FIG. 2. Amino acid residues flanking the splice junctions of the *Sce* VMA intein illustrating the residue numbering system used in this article.[5] Splice junctions are indicated by vertical lines and highly conserved amino acid residues are underlined.

whereas the amino acids in position −1 that can support protein splicing depend to a large extent on the intein. The N-S acyl rearrangement involving inteins with a Cys residue at position 1 leads to the formation of a peptide thioester, which can be induced to undergo cleavage by nucleophilic displacement with a thiol or water, as illustrated in Fig. 3. It occurs even when all subsequent steps in the protein splicing pathway are blocked by amino acid replacements at the downstream splice junction. This forms the basis for the self-cleavage of fusion proteins in which the target protein is cloned at the N terminus of an intein, as described in the first part of this article.

In step 2 of protein splicing, the linear ester intermediate formed in the

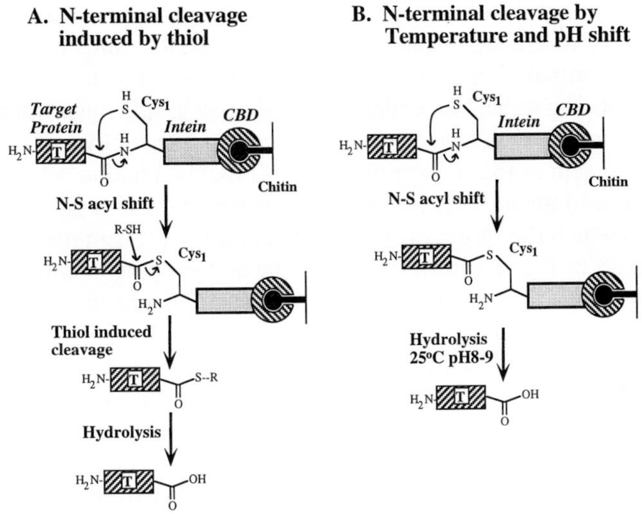

A. N-terminal cleavage induced by thiol

B. N-terminal cleavage by Temperature and pH shift

FIG. 3. Reactions that underlie the use of inteins as components of fusion proteins that undergo self-cleavage at the upstream splice junction. The amino acid residues that participate directly in the reactions are shown, whereas the remainder of the intein and the exteins are shown by boxes or disks with the same shading as in Fig. 1. (A) N-terminal cleavage at the upstream splice junction induced by a thiol, R-SH. (B) N-terminal cleavage at the upstream splice junction induced by a temperature or pH shift.

first step undergoes a transesterification reaction with the conserved Cys, Ser, or Thr residue in position +1 at the downstream splice junction. It involves the nucleophilic attack by the side chain of the +1 residue on the ester carbonyl of residue −1 at the upstream splice junction. As a result, N-extein is transferred to the side chain of the +1 residue to yield a branched ester intermediate. If the +1 amino acid is replaced by any amino acid other than Cys, Ser, or Thr, transesterification cannot occur and protein splicing is arrested at step 1.

Step 3 in the protein splicing process involves cyclization of the conserved Asn residue at the C terminus of the intein (Asn_{454} in the case of the *Sce* VMA intein shown in Fig. 2) with the concomitant cleavage of the peptide bond at the downstream splice junction. Replacement of Asn_{454} by any other amino acid completely prevents the C-terminal cleavage reaction but allows the rearrangements that lead to ester intermediates. The penultimate amino acid at the C terminus of the intein is usually a His residue, whose function seems to be related to step 3 of protein splicing, as its replacement by other amino acids attenuates the C-terminal cleavage reaction but not the other steps in protein splicing. An interesting feature of step 3 of protein splicing is its relative independence of preceding steps. In the *Sce* VMA intein, occurrence of the N-S rearrangement, followed by either chemical or intein-catalyzed transesterification, enhances the rate of C-terminal cleavage significantly,[4] but in *Mth* RIR1[7] and the *Ssp* DnaB[8] inteins, C-terminal cleavage can proceed rapidly even in the absence of the other splicing steps. Moreover, step 2 of protein splicing is not a prerequisite for C-terminal cleavage, which can therefore occur even if the nucleophilic amino acid at the N terminus of the C-extein, residue +1, is replaced by other amino acids. This forms the basis for the self-cleavage of fusion proteins in which the target protein is cloned at the C terminus of an intein, as illustrated in Fig. 4 and described in the second part of this article.

The products of these three catalyzed reactions are the excised intein with a C-terminal aminosuccinate residue and the exteins linked by an ester bond. Both products are intrinsically unstable, with the aminosuccinimide undergoing slow hydrolysis to yield an asparagine or isoasparagine residue and the ester bond isomerizing by an S-N or O-N acyl rearrangement to yield a more stable amide bond. The latter rearrangement, which is the final step in the protein splicing process, proceeds extremely rapidly and essentially irreversibly in the absence of catalysis[9] and yields the spliced exteins linked by a stable, natural peptide bond.

[7] T. C. Evans, J. Benner, and M.-Q. Xu, *J. Biol. Chem.* **274,** 3923 (1999).

[8] S. Mathys, T. C. Evans, I. C. Chute, H. Wu, S. Chong, J. Benner, X.-Q. Liu, and M.-Q. Xu, *Gene* **231,** 1 (1999).

[9] Y. Shao and H. Paulus, *J. Peptide Res.* **50,** 193 (1997).

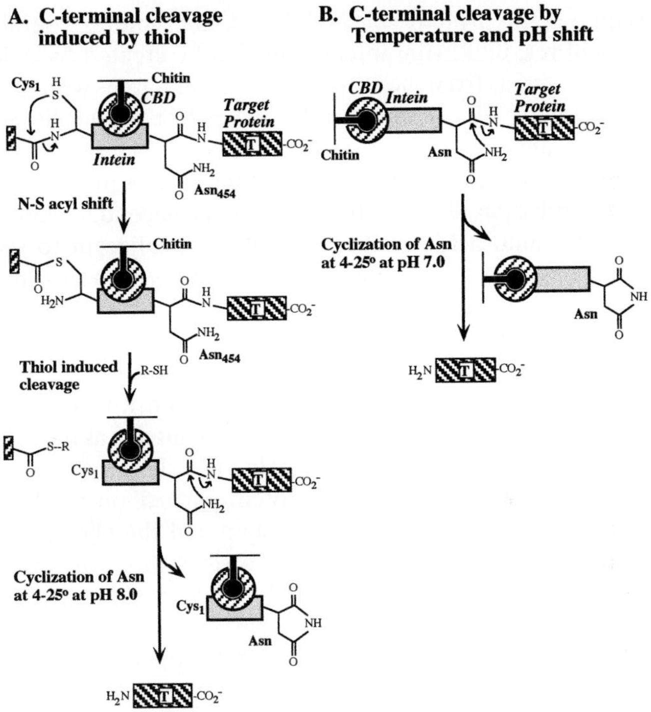

FIG. 4. Reactions that underlie the use of inteins as components of fusion proteins that undergo self-cleavage at the downstream splice junction to allow the one-step purification of a target protein. The amino acid residues that participate directly in the reactions are shown, whereas the remainder of the intein and the exteins are shown by boxes or disks with the same shading as in Fig. 1. (A) C-terminal cleavage at the downstream splice junction triggered by thiol-induced cleavage at the upstream splice junction. (B) C-terminal cleavage at the downstream splice junction induced by a termperature or pH shift.

Comparison of Different Inteins as Tools for Protein Purification

There are a significant number of common sequence motifs in the approximately 100 known inteins[10] that suggest a common evolutionary origin of inteins. Nevertheless, every intein has evolved for some time in the context of its host protein to optimize its splicing and therefore has distinct preferences for particular adjacent amino acids. These differences, which may be relatively subtle, must be kept in mind when using inteins as protein engineering tools. Of special interest is the handful of inteins that lack an endonuclease domain and consist of just the two terminal

[10] F. B. Perler, Nucleic Acids Res. 27, 346 (1999).

regions required for protein splicing. These mini-inteins contain 134 to 198 amino acid residues. Mini-inteins can also be created by deleting the endonuclease domains from the more typical larger inteins without affecting self-splicing activity. Mini-inteins, either natural or artificially generated, offer certain advantages over the 454-residue *Sce* VMA intein as tools for protein purification. First, the use of a mini-intein as a fusion partner may increase the expression levels of fusion proteins. Second, different inteins favor different amino acid residues or sequences adjacent to the scissile peptide bond for cleavage and may cleave at different rates in response to various expression and cleavage conditions, such as thiols, pH, and temperature. Some mini-inteins have been engineered to undergo cleavage in the absence of thiols simply by pH and temperature shifts, thus avoiding exposure to thiols during the purification of thiol-sensitive proteins. Third, the extent of *in vivo* cleavage differs with different inteins as fusion partners, thus allowing optimization of fusion protein yield. It is therefore advantageous to select an intein as a fusion protein partner on the basis of the sequence and properties of the target protein and the choice of reagents to achieve maximum yield. This article describes three or four different modified inteins for each type of self-catalyzed cleavage reaction. Because the various expression vectors described have similar multiple cloning sites, it is possible and recommended to insert a single target protein into several different vectors so as to compare their performance and choose the expression system that gives the most satisfactory results.

In order to make use of the self-cleaving activity of inteins as tools for protein purification, it is necessary to prevent protein splicing by strategic amino acid replacements at either the upstream or the downstream splice junction as described in the two main sections of this article. Depending on the type of amino acid substitution used, protein cleavage can be induced at either the upstream or the downstream splice junction. Different inteins vary in the efficiencies with which N- or C-terminal cleavage can be induced and in the compatibility of these cleavage reactions with the extein residues adjacent to the cleavage site. The decision of whether to fuse the C or the N terminus of the target protein to the intein also has important implications on the efficiency of protein expression and purification yield. The expression of heterologous proteins in *Escherichia coli* sometimes results in the formation of inclusion bodies or is inefficient due to the incompatibility of codon usage in the foreign gene with *E. coli* coding preferences. Both of these problems are most pronounced when the foreign polypeptide constitutes the N-terminal segment of the fusion protein and can often be reduced by fusing the target protein to the C terminus of the intein.

Although this article focuses on the use of intein self-cleavage as a tool in protein purification, it is appropriate to mention that the methods described in this article can also yield precursors for *in vitro* protein ligation.[7,8,11-13] Intein-mediated protein ligation takes advantage of the transesterification of a reactive polypeptide thioester with the N-terminal cysteine residue of another polypeptide. The product of this uncatalyzed reaction is a peptide ester analogous to that formed in the course of protein splicing, which rapidly undergoes an S-N acyl rearrangement to form a normal polypeptide, analogous to the final step of protein splicing (see Fig. 1). Reactive peptide thioesters suitable for subsequent transesterification can be prepared by inducing cleavage of a target protein fused to an intein N terminus with thiophenol or 2-mercaptoethanesulfonic acid. Such thioesters react efficiently with an N-terminal cysteine residue of another peptide, which can be a synthetic or natural polypeptide or a protein produced by intein-mediated self-cleavage of a target protein fused to an intein C terminus.[7,8]

The advantages and disadvantages of various inteins and the two modes of target protein fusion are presented in some detail in the remainder of this article and are summarized in Table I.

The two sections that follow describe the experimental procedures for the expression systems based on the *Sce* VMA intein in considerable detail. Procedures for the other expression systems are similar and are detailed only where they differ significantly from the *Sce* VMA intein systems. It should also be pointed out that, although the experimental systems discussed in this article focus on the expression of self-cleaving fusion proteins in *E. coli,* the methods described are applicable to protein expression in any host, ranging from yeast to animal cells, provided that the appropriate expression vectors are adapted for cloning the coding sequences of the target protein–intein–chitin-binding domain fusion proteins described in this article.

Fusion of Proteins at Their C Terminus to a Modified Intein and Release by Cleavage at the Intein N Terminus

Use of *Sce* VMA Intein (pTYB Vectors)

This method uses an *Sce* VMA intein modified to undergo cleavage only at its N terminus. A mutation, Asn454Ala, at the C-terminal splice junction of the *Sce* VMA intein blocks splicing and C-terminal cleavage,

---

[11] T. C. Evans, J. Benner, and M.-Q. Xu, *Protein Sci.* **7,** 2256 (1998).

[12] T. W. Muir, D. Sondhi, and P. A. Cole, *Proc. Natl. Acad. Sci. U.S.A.* **95,** 6705 (1998).

[13] K. Severinov and T. W. Muir, *J. Biol. Chem.* **273,** 16205 (1998).

TABLE I
Comparison of Intein-Mediated Protein Purification Systems

Cleavage site	Intein	Intein length (amino acids)	Intein mutations	Plasmid	Site of target protein fusion	Compatible target protein termini	Induction of cleavage	Application to in vitro protein ligation
Upstream splice junction	Sce VMA	454	N454A[a]	pTYB1,2,3,4	C terminus	All except P C N D R[a]	Thiol[b,c]	Yes[c,d]
Upstream splice junction	Mxe GyrA	198	N198A	pTXB1,3	C terminus	All except S P E D	Thiol[b]	Yes[d]
Upstream splice junction	Mth RIR1	134	P(−1)G N134A	pTRB5	C terminus	G or A but not P, others not tested	Thiol[b]	Yes[d]
Upstream splice junction	Ssp DnaB	154	N154A	pTSB1,3	C terminus	G, others not tested	Thiol[b] or pH and temperature	Yes[d]
Downstream splice junction	Sce VMA/CBD	510	H509Q	pIMC1,2 pTYB11,12	N terminus	All except P S C	Thiol	No
Downstream splice junction	Mth RIR1	134	P(−1)G C1A	PBRC1,3	N terminus	C, others not tested	pH and temperature[b]	Yes[e]
Downstream splice junction	Ssp DnaB	154	C1A	pBSC1,3	N terminus	All except Q N L I R K P	pH and temperature[b]	Yes[e]

[a] Amino acids are abbreviated using the single letter notation.
[b] Other nucleophiles may be substituted for the thiol reagent. Cleavage *in vivo* and during purification may occur to various extents depending on the residue(s) flanking the cleavage site.
[c] Cleavage with MESNA or thiophenol is not as efficient as with the other inteins described in this table.
[d] Can generate thioester-tagged protein for ligation with a protein or peptide possessing an N-terminal cysteine.
[e] Intein cleavage yields a protein possessing an N-terminal cysteine for ligation with a thioester-tagged protein or peptide when cysteine is present at the +1 position.

but allows N-terminal cleavage mediated by an N-S acyl shift at Cys_1. Nucleophiles that react with thioesters, such as thiols (1,4-dithiothreitol, 2-mercaptoethanol, or cysteine) or hydroxylamine, can effectively shift the N-S equilibrium by attacking the thioester and thereby induce the N-terminal cleavage of the intein. In the case of cysteine-induced cleavage, a spontaneous S-N shift occurs after the nucleophilic attack by the sulfhydryl group of cysteine, resulting in formation of a native peptide bond between the cysteine and the C terminus of the N-extein, analogous to the final step in protein splicing (Fig. 1).[14]

To express and purify a target protein, the C terminus of the target protein is fused to the N terminus of the modified Sce VMA intein, which is in turn linked to a small (5 kDa) affinity tag, the chitin-binding domain

[14] S. Chong, Y. Shao, H. Paulus, J. Benner, F. B. Perler, and M. Q. Xu, *J. Biol. Chem.* **271,** 22159 (1996).

(CBD) from *Bacillus circulans*.[15] The CBD allows tight binding of the fusion protein from crude cell extracts to chitin beads. Stringent wash conditions (e.g., high salt concentration or use of nonionic detergents) can be used to remove nonspecifically bound material. The intein of the immobilized fusion protein is then induced to undergo self-cleavage by overnight incubation at 4° in the presence of dithiothreitol (DTT) or 2-mercaptoethanol.[16] The target protein is thereby specifically released from chitin beads and is recovered as a pure protein.

pTYB Vectors

pTYB1-4 vectors are general cloning vectors for expression and purification of target proteins in *E. coli*. The target gene is cloned into the multiple cloning sites (MCS) upstream of the intein-CBD fusion. pTYB vectors contain either an *Nde*I or an *Nco*I cloning site in their MCS for translation initiation. pTYB1 (7280 bp) and pTYB2 (7277 bp) contain an *Nde*I site for cloning the 5' end of a target gene, and pTYB3 (7280 bp) and pTYB4 (7277 bp) contain an *Nco*I for cloning the 5' end of a target gene. pTYB1 and pTYB3 contain a *Sap*I site for cloning the 3' end of a target gene, which allows target gene insertion directly adjacent to the intein; this results in the target protein being cleaved without leaving any additional vector-derived residues at its C terminus. pTYB2 and pTYB4 contain a *Sma*I cloning site that yields a target protein with a single glycine residue added to its C terminus after cleavage. The arrangement of the control and cloning sites of the pTYB vectors and the corresponding nucleotide sequences are shown in Fig. 5.

These vectors use an inducible bacteriophage T7 promoter to provide stringent control of the fusion gene expression.[17] The vectors carry their own copy of the *lacI* gene encoding the *lac* repressor. Binding of the *lac* repressor to a *lac* operator sequence immediately downstream of the T7 promoter suppresses basal expression of the fusion gene in the absence of isopropyl-β-D-thiogalactopyranoside (IPTG) induction. Four tandem transcription terminators (*rrnB* T1) placed upstream of the promoter minimize background transcription. A T7 transcription terminator downstream of the CBD prevents continued transcription. The vectors also contain the origin of DNA replication from bactriophage M13, which allows for the

[15] T. Watanabe, Y. Ito, T. Yamada, M. Hashimoto, S. Sekine, and H. Tanaka, *J. Bacteriol.* **176,** 4465 (1994).

[16] S. Chong, F. B. Mersha, D. G. Comb, M. E. Scott, D. Landry, L. M. Vence, F. B. Perler, J. Benner, R. Kucera, C. A. Hirvonen, J. J. Pelletier, H. Paulus, and M. Q. Xu, *Gene* **192,** 271 (1997).

[17] J. W. Dubendorff and F. W. Studier, *J. Mol. Biol.* **219,** 45 (1991).

pTYB vectors

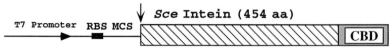

pTYB1

```
5′  GGT TTA AAC CGG GGA TCT CGA TCC CGC GAA ATT AAT ACG ACT CAC TAT AGG GGA ATT
        PmeI                                      T7 Promoter

GTG AGC GGA TAA CAA TTC CCC TCT AGA AAT AAT TTT GTT TAA CTT TAA GAA GGA GAT ATA
   lac operator              XbaI                                  Shine Dalgarno

    Met Ala Ser Ser Arg Val Asp Gly Gly Arg Glu Phe Leu Glu Gly Ser Ser Cys1
CAT ATG GCT AGC TCG CGA GTC GAC GGC GGC CGC GAA TTC CTC GAG GGC TCT TCC TGC
 NdeI    NheI    NruI    SalI        NotI    EcoRI   XhoI    SapI    Intein.

    ... Sce VMA Intein (454 aa)...
TTT GCC AAG GGT ACC AAT GTT TTA ATG GCG GAT GGG TCT ATT GAA TGT ATT ...3′
            KpnI
```

pTYB2

```
    Met Ala Ser Ser Arg Val Asp Gly Gly Arg Glu Phe Leu Glu Pro Gly Cys1
CAT ATG GCT AGC TCG CGA GTC GAC GGC GGC CGC GAA TTC CTC GAG CCC GGG TGC
 NdeI    NheI    NruI    SalI        NotI    EcoRI   XhoI    SmaI
```

pTYB3

```
    Met Ala Ser Ser Arg Val Asp Gly Gly Arg Glu Phe Leu Glu Gly Ser Ser Cys1
ACC ATG GCT AGC TCG CGA GTC GAC GGC GGC CGC GAA TTC CTC GAG GGC TCT TCC TGC
 NcoI    NheI    NruI    SalI        NotI    EcoRI   XhoI    SapI
```

pTYB4

```
    Met Ala Ser Ser Arg Val Asp Gly Gly Arg Glu Phe Leu Glu Pro Gly Cys1
ACC ATG GCT AGC TCG CGA GTC GAC GGC GGC CGC GAA TTC CTC GAG CCC GGG TGC
 NcoI    NheI    NruI    SalI        NotI    EcoRI   XhoI    SmaI
```

FIG. 5. Schematic representation of pTYB vectors and the nucleotide sequence of control elements and multiple cloning sites. The diagram is not to scale, and arrows indicate the sites of thiol-induced cleavage. Vectors pTYB2-4 differ from pTYB1 only in the MCS, which is shown.

production of single-stranded DNA by helper phage, M13KO7, superinfection of cells bearing the plasmid. pTYB vectors carry an Amp^r marker (the *bla* gene), which conveys ampicillin resistance to the host strain. A map of pTYB1 showing the unique restriction sites and the coding regions, as well as major control elements, is given in Fig. 6.

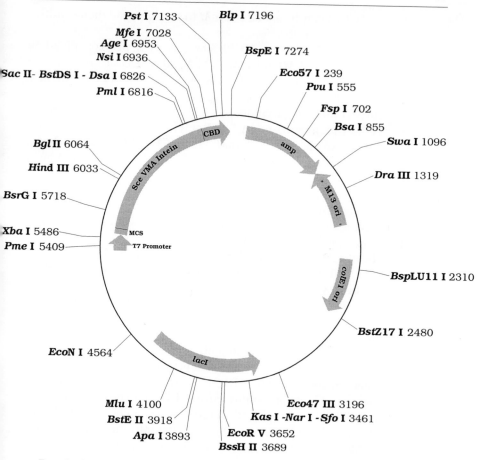

Pst I 7133
Mfe I 7028
Age I 6953
Nsi I 6936
Sac II - BstDS I - Dsa I 6826
Pml I 6816
Blp I 7196
BspE I 7274
Eco57 I 239
Pvu I 555
Fsp I 702
Bsa I 855
Swa I 1096
Dra III 1319
Bgl II 6064
Hind III 6033
BsrG I 5718
Xba I 5486
Pme I 5409
BspLU11 I 2310
BstZ17 I 2480
EcoN I 4564
Mlu I 4100
BstE II 3918
Apa I 3893
Eco47 III 3196
Kas I -Nar I -Sfo I 3461
EcoR V 3652
BssH II 3689

CBD
Sce VMA Intein
amp
+ M13 ori -
colE1 ori
lacI
MCS
T7 Promoter

FIG. 6. pTYB1 vector. All unique restriction endonuclease sites are shown, with the coordinates referring to the 5'-nucleotide in each recognition sequence.

Reagents

Lysis/column buffer: 20 mM Na–HEPES or Tris–HCl, pH 8.0; 500 mM NaCl; 0.1 mM EDTA. In addition, the use of 0.1% Triton X-100 or Tween 20 is recommended to prevent nonspecific adsorption to the chitin beads, unless the target protein is known to be inactivated by these nonionic detergents. Furthermore, protease inhibitors such as phenylmethylsulfonyl fluoride do not interfere with subsequent purification steps and may also be used. The reducing agents tris-(2-carboxyethyl) phosphine (TCEP) or tris-(2-cyanoethyl) phosphine (TCCP) may be added at 1 mM to stabilize oxidation-sensitive proteins during purification.

Cleavage buffer: 20 mM Na–HEPES or Tris–HCl, pH 8.0; at least 50 mM NaCl (up to 2 M); 0.1 mM EDTA; 30 mM DTT

Stripping buffer: 20 mM Na–HEPES or Tris–HCl; pH 8.0; 500 mM NaCl; 1% SDS (kept at room temperature)

LB broth: Per liter, 10 g tryptone, 5 g yeast extract, 5 g NaCl, adjusted to pH 7.2 with about 1 ml of 1 N NaOH

SDS–PAGE sample buffer: 70 mM Tris–HCl, pH 6.8; 33 mM NaCl; 1 mM Na$_2$EDTA; 2% SDS; 40 mM DTT; 0.01% bromphenol blue; 40% glycerol

1 M DTT: Dissolve 3.09 g of DTT in 20 ml of 0.01 M sodium acetate, pH 5.2. Sterilize by filtration and store in 1- to 5-ml aliquots at $-20°$.

pTYB vectors: Constructed as described earlier, available from New England Biolabs (Beverly, MA)

Chitin beads: Available from New England Biolabs (Beverly, MA). Prepare a 10- to 20-ml column for a 1-liter culture and equilibrate at 4° with 10 volumes of the lysis/column buffer.

Anti-CBD antibody: Rabbit serum raised against the *B. circulans* chitin-binding domain is available from New England Biolabs for Western blot analysis.

E. coli hosts: The clone can be initially established in a nonexpression host, especially if the gene of interest is toxic and the clone may not be stable. *E. coli* ER2566 or BL21(DE3) (or other derivatives) can be used as expression strains. Strain ER2566 (provided by New England Biolabs with pTYB vectors) carries a chromosomal copy of the T7 RNA polymerase gene inserted into the *lacZ* gene and thus under the control of the IPTG-inducible *lac* promoter.

Cloning the Target Gene into pTYB Vectors

Primer Design. Normally, a target gene is amplified by the polymerase chain reaction (PCR) before it is inserted in-frame in the MCS of one of the pTYB vectors. Appropriate restriction sites, absent in the target gene, are incorporated in the forward and reverse primers. The choice of restriction sites in the primers determines the extra amino acid residues that may be attached to the N terminus of the target protein after cleavage of the intein. For instance, to obtain a target protein with no extra vector-derived residues, one can clone a target gene between *Nde*I and *Sap*I sites in pTYB1 (Fig. 7). Cloning the 3′ end of a target gene into the *Sma*I site of pTYB2 adds an extra glycine residue to the target protein. Table II gives some examples of forward and reverse primers with different restriction sites for cloning into pTYB vectors.

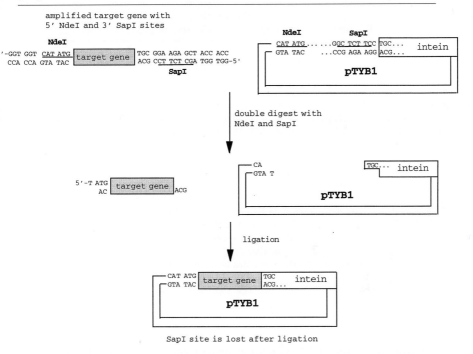

Fig. 7. Diagram illustrating the cloning of a target gene by insertion into the *Nde*I and *Sap*I sites of pTYB1.

Cloning Procedure. The protocol described here clones an amplified target gene fragment using restriction enzymes that create different cohesive ends. For blunt-end or single-site cloning, the vector may need to be treated with an alkaline phosphatase and larger amounts of DNA ligase should be added to the ligation mixture.

The reaction mixture containing the amplified target gene fragment (about 100 μl) is loaded directly onto a 1% low-melting agarose gel and subjected to electrophoresis. A gel slice (100–150 μl, the lower the volume, the better) containing the correct fragment is cut out with a razor blade and melted at 65° for 7 min. After the agarose has cooled at 42° for 7 min, β-agarase (2 μl per 100 μl agarose) is added and incubation is continued at 42° for 1 hr. The digested agarose containing the target gene is directly used for restriction digestion with no prior DNA precipitation. A sample of the gel-purified gene fragment (80 μl) is digested with an appropriate pair of the restriction endonucleases in a 100-μl volume. The pTYB vector (0.5 μg) is digested in parallel with the same enzymes in a 50-μl reaction mixture. After 2–4 hr digestion, the two reaction mixtures are subjected to

TABLE II

EXAMPLES OF PRIMER DESIGN FOR CLONING IN pTYB VECTORS

Cloning site		Primer sequence[a]	Cloning vector
Forward primer	NdeI	5'-GGT GGT <u>CAT ATG</u> NNN NNN . . . -3'	pTYB1 and 2
	NcoI	5'-GGT GGT A<u>CC ATG G</u>NN NNN . . . -3'	pTYB3 and 4
Reverse primer	SapI[b]	5'-GGT GGT T<u>GC TCT TCC</u> GCA NNN NNN . . . -3'	pTYB1 and 3
	SapI[b,c]	5'-GGT GGT T<u>GC TCT TCC</u> GCA ACC NNN NNN . . . -3'	pTYB1 and 3
	SmaI[d]	5'-NNN NNN . . . -3'	pTYB2 and 4
	XhoI[c]	5'-GGT GGT <u>CTC GAG</u> NNN NNN . . . -3'	pTYB2 and 4

[a] The target gene starts at "5'-N(or G)NN NNN. . . ." Restriction sites are underlined. Use of a "GGT GGT" sequence at the 5' end of the primer is to ensure efficient DNA cleavage by the restriction enzyme when the restriction site is close to the 5' end.
[b] SapI site is not regenerated after the cloning.
[c] SapI and XhoI sites allow insertion of extra codon(s) (e.g., glycine) between the C terminus of the target protein and the intein.
[d] Blunt-end cloning of the PCR product. The SmaI site is not regenerated. Any PCR primer can be used in conjunction with a DNA polymerase that yields a product with blunt ends rather than 3' A overhangs.

electrophoresis on a 1% low-melting agarose gel. The gel slices (150–200 μl) containing the digested gene fragment and digested pTYB vector are combined in the same microcentrifuge tube and melted at 65° for 10 min. After cooling at 42° for 7 min, 2 μl per 100 μl gel slice of β-agarase is added to digest the gel for 1 hr at 42°. Ligation is conducted at 16° overnight in the same mixture after adding T4 DNA ligase (2 μl) and an appropriate volume of 10× T4 DNA ligase buffer.

The ligation mixture (15 μl) is used to transform 150 μl competent cells of a nonexpressing E. coli strain such as DH5α. To reduce the background caused by self-ligation of the vector, the ligation sample can be treated prior to transformation with an enzyme whose site is deleted from the polylinker during cloning and is also absent in the insert, thereby linearizing any residual parental vectors without insert. Plasmids are isolated from each transformant colony and those containing the correct target gene insert are identified by digesting the plasmid DNA with the same restriction enzymes used for cloning the target gene fragment, except when SapI is used, because the SapI site is not regenerated after the insert is ligated to the vector. Alternatively, one can use colony PCR or colony hybridization to screen a large number of transformants for the presence of target gene inserts. Immunodetection can also be used with anti-CBD to detect the full-length fusion precursor in total cell lysates. The correct clones should be further confirmed by DNA sequencing before proceeding to cell culture and protein expression.

Expression and Purification Procedures

For expression, the plasmid is transferred to an *E. coli* strain that carries a chromosomal copy of the T7 RNA polymerase gene under the control of the *lac* promoter, such as *E. coli* ER2566 (New England Biolabs). For the expression of genes with a codon usage significantly different from that of wild-type strains of *E. coli,* strain BL21-CodonPlus(DE3)-RIL (Stratagene) may offer an advantage. LB medium (1 liter) containing 100 μg/ml ampicillin is inoculated with a freshly grown colony or 10 ml of a fresh starter culture and incubated on a rotary shaker at 37° to a culture density ($A_{600\ nm}$) of 0.5–0.8. IPTG is then added to a final concentration of 0.3–0.5 mM and the culture is transferred to a rotary shaker at 15° and incubated overnight. The conditions for induction and expression may need to be optimized for every protein, such as induction at 30° for 3–6 hr or at 20–25° overnight. Cells are collected by centrifugation at 5000g for 10 min at 4° and the cell pellet can be used directly or stored at −20°.

Cells from a 1-liter culture are resuspended in 50 ml ice-cold lysis/column buffer and disrupted either by sonication or with a French press. The cell extract is clarified by centrifugation at 12,000g for 30 min (supernatant). Samples (40 μl) of the total cell extract (before centrifugation) and the supernatant (after centrifugation) are mixed with 20 μl of 3× SDS–PAGE sample buffer and analyzed by SDS–PAGE and/or Western blotting. If the fusion precursor is detected in the total cell extract but not in the supernatant, the fusion protein is probably expressed in inclusion bodies rather than in soluble form. If the fusion protein is soluble, the clarified extract is loaded onto a chitin column (15–20 ml per liter of culture) at a flow rate not exceeding 0.5–1.0 ml/min. An undiluted sample (40 μl) from the flow through is mixed with 20 μl of 3× SDS–PAGE sample buffer and is analyzed by SDS–PAGE. Comparison of the samples from the flow through and the original supernatant fraction provides an indication of the binding efficiency of the fusion protein to the chitin column. The high affinity of the CBD for chitin beads allows subsequent washing of the column to be carried out at higher flow rates (about 2 ml/min) and more stringent wash conditions. Loading and washing at high salt (e.g., 1 M NaCl) or in the presence of nonionic detergents may reduce nonspecific binding of other *E. coli* proteins. Normally, at least 10 bed volumes of lysis/column buffer are required for washing the column thoroughly. Care should be taken that all traces of the crude cell extract have been washed off the sides of the column.

The target protein is released from the chitin column by inducing the chitin-bound intein tag to undergo self-cleavage in the presence of DTT 2-mercaptoethanol, or cysteine. Induction of on-column cleavage is conducted by flushing the column quickly with 3 bed volumes of cleavage buffer containing 30–50 mM DTT 2-mercaptoethanol, or cysteine and then stopping the column flow. This step evenly distributes thiols throughout the column in the absence of significant cleavage, provided the column flushing period does not exceed 30 min. Cysteine and DTT result in similar cleavage efficiencies. Cysteine is a weaker reducing agent than DTT, and, if the target protein is sensitive to high DTT concentrations, cysteine (50 mM) may be used but will lead to the addition of a Cys residue to the C terminus of the target protein. Composition of the cleavage buffer can be adapted to that of the final storage buffer for the target protein. For example, if the target protein is stable only under certain buffer conditions, then that buffer, supplemented with 30–50 mM DTT or cysteine, may be used to induce the cleavage reaction and for subsequent elution. After induction of the cleavage reaction, the column is left at 4–23° for 16–40 hr to release the target protein from the intein tag. The target protein is then eluted from the column using the lysis/column buffer or a specific storage buffer. The target protein is normally eluted within 2–3 bed volumes. However, any uncleaved fusion protein and the intein tag remain bound to the chitin resin during the target protein elution and can be stripped from the resin by 1% SDS. To examine residual proteins, 300 μl of resin slurry is gently removed from the column and mixed with 100 μl of 3× SDS–PAGE sample buffer and boiled for 2 min. The resin is removed by centrifugation, and a sample (3–10 μl) of the supernatant is used directly for SDS–PAGE analysis. Alternatively, bound proteins can be eluted from the column with 3 bed volumes of stripping buffer. The elution should be conducted at room temperature to prevent precipitation of SDS.

Limitations and Critical Parameters

Successful expression and purification of a target protein in a pTYB vector depend on the following factors: (a) *E. coli* strain; (b) culture conditions (e.g., temperature, aeration, cell density); (c) induction and protein expression conditions (temperature, time, IPTG concentration); (d) expression level of the fusion protein; (e) solubility of the fusion protein; (f) efficiency of thiol-induced cleavage; and (g) solubility of the target protein after the cleavage.

An appropriate *E. coli* host strain is important for achieving optimal expression of a given target gene. pTYB vectors require the use of *E. coli* host strains that carry the T7 RNA polymerase gene [e.g., BL21(DE3) or

ER2566]. The use of hosts carrying a pLysS plasmid encoding T7 lysozyme, a natural inhibitor of T7 RNA polymerase, can reduce basal transcription in uninduced cells, which may be critical when a toxic target protein is expressed. Expression of a toxic target protein may also call for a lower level of induction by using less IPTG or a lower culture temperature to prevent loss of the vector. The expression level of the fusion protein is highly dependent on the target protein and its coding sequence. Problems of codon usage, mRNA degradation, or proteolysis due to protein misfolding may all contribute to poor expression. Different protease-deficient hosts should be tested to minimize proteolysis. The fact that the target protein constitutes the N-terminal segment of the fusion protein may contribute to the formation of inclusion bodies, especially if the target protein is of eukaryotic origin. Normally, protein expression should be induced at 15–30°. Low induction temperatures may reduce the formation of inclusion bodies, improve the folding and solubility of the fusion protein, and increase the cleavage efficiency of the intein. If all attempts to prevent inclusion body formation fail, the procedures described in the second part of this article, where the target protein represents the C-terminal segment of the fusion protein and is released by self-catalyzed cleavage at the C terminus of the intein, provide a good alternative.

The yield of a target protein is determined primarily by the recovery of the fusion protein in cell extracts and its cleavage efficiency after binding to chitin beads. Cleavage efficiency is significantly affected by the amino acid residue of a target protein that is immediately adjacent to the intein cleavage site (Cys1 at the N terminus of the Sce VMA intein). Studies with maltose-binding protein as the target protein showed that the majority of 20 natural amino acids, when located at the C terminus of the target protein, allow both purification of fusion proteins and efficient cleavage with DTT (Table III). However, Asp, Glu, Arg, His, and Thr as the C-terminal residue of MBP lead to significant *in vivo* cleavage with a concomitant decrease in the recovery of fusion protein. With Leu, Ile, and Val at the MBP C terminus, a higher temperature or longer incubation time is needed to achieve quantitative cleavage of the fusion protein. When Pro, Cys, or Asn is adjacent to the intein, no *in vitro* cleavage with DTT can be observed. If the target protein has any of these unfavorable residues at its C terminus, the insertion of an extra Gly residue adjacent to the cleavage site (e.g., by using the *Sma*I site in pTYB2) can often reduce *in vivo* cleavage or increase *in vitro* cleavage efficiency.

In some cases, the target protein is not eluted after on-column cleavage but is subsequently found in SDS elution fractions, suggesting that the target protein becomes insoluble after induced on-column cleavage. Such a problem can be minimized by using higher salt concentrations (0.5–2 *M*

TABLE III
EFFECT OF C-TERMINAL AMINO ACID RESIDUE OF
TARGET PROTEIN ON N-TERMINAL CLEAVAGE OF
FUSION PROTEINS BASED ON pTYB1 VECTOR[a]

| Residue −1 | % in vivo cleavage | % in vitro cleavage induced with DTT | |
		4°	16°
Gly	<10	92	95
Ala	<10	94	94
Met	<10	78	95
Phe	<10	78	95
Tyr	<10	80	95
Trp	<10	79	95
Gln	<10	82	93
Lys	<10	79	95
Ser	<10	62	90
Leu	<10	31	91
Val	<10	33	62
Ile	<10	31	41
Thr	25	50	91
Glu	50	66	84
His	50	66	72
Arg	75	n.d.[b]	n.d.
Asp	>95	n.d.	n.d.
Asn	<10	<5	<5
Cys	<10	<5	<5
Pro	<10	<1	<1

[a] E. coli MBP was used as the target protein (MYB) with amino acid substitutions at the position (−1) immediately upstream of the cleavage site in the sequence ($L_{-3}E_{-2}X_{-1}/C_1$). The approximate percentage of in vivo cleavage at the N terminus was estimated from SDS–PAGE as the ratio of MBP versus the MYB precursor. Cleavage reactions were conducted in 30 mM HEPES, pH 8.0, 0.5 M NaCl, and 40 mM DTT for 16 hr under the conditions indicated.
[b] Not determined.

NaCl) or adding a nonionic detergent to the cleavage buffer to improve the solubility of the target protein. A number of nonionic detergents examined (0.1–0.5% Triton X-100 or 0.1–0.2% Tween 20) were compatible with binding to chitin beads and cleavage. The use of urea to prevent protein precipitation is more problematical because some of the intein tag may coelute

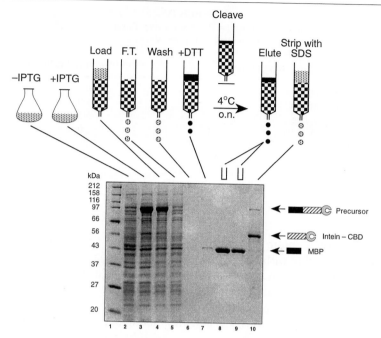

FIG. 8. Expression and purification of the *E. coli* maltose-binding protein (MBP) by cloning into pTYB1. Lane 1, protein markers; lane 2, crude extract from uninduced cells; lane 3, crude extract from cells induced with IPTG for 15 hr at 15°; lane 4, clarified crude extract from induced cells; lane 5, chitin column flow through; lane 6, chitin column wash; lane 7, quick flush with 30 m*M* DTT; lanes 8 and 9, fractions of eluted MBP after overnight incubation with 30 m*M* DTT at 4°; and lane 10, SDS stripping of remaining proteins bound to chitin column.

with the target protein, necessitating repurification and refolding of the target protein.

Representative Results

An example of the purification of *E. coli* MBP using the pTYB1 vector is shown in Fig. 8. The range of prokaryotic and eukaryotic proteins that can be expressed and purified using pTYB vectors and a single chitin column illustrated by the examples summarized in Table IV.

Use of *Mxe* GyrA Mini-intein (pTXB Vectors)

The 198 residue intein from the *gyrA* gene of *Mycobacterium xenopi* (*Mxe* GyrA intein) was the first mini-intein to be engineered for thiol-inducible cleavage of a protein of interest from a C-terminal affinity tag.[11]

TABLE IV
RECOMBINANT PROTEINS EXPRESSED AND PURIFIED BY pTYB
C-TERMINAL FUSION VECTORS

Target protein	Yield (mg/liter of culture)
E. coli maltose-binding protein	20.0
E. coli McrBC	6.0
Bacteriophage T4 DNA ligase	5.0
Bst DNA polymerase Fragment	2.0
BamHI restriction endonuclease	1.0
BglII restriction endonuclease	4.0
Human cyclin-dependent protein kinase	1.0
Calmodulin-dependent protein kinase II	0.8

The *Mxe* GyrA intein was modified by replacing the intein C-terminal asparagine residue with an alanine (Asn 198 Ala) to block splicing and C-terminal cleavage. Advantages of the pTXB vectors using such engineered *Mxe* GyrA inteins include higher expression levels, less *in vivo* cleavage, and higher sensitivity to cleavage by thiols such as sodium salt of 2-mercaptoethanesulfonic acid (MESNA) and thiophenol that can be used for protein ligation and labeling.

pTXB Vectors

pTXB vectors are 6509-bp plasmids, which are essentially identical to pTYB plasmids except for the replacement of the 454 residue *Sce* VMA intein with the 198 residue *Mxe* GyrA intein. The MCS of pTXB vectors are designed for the in-frame insertion of target genes just upstream of the *Mxe* GyrA intein/chitin-binding domain-coding region. The MCS of pTXB1 and pTXB3 are identical to those of TYB1 and pTYB3, respectively. Note, however, that the *Sal*I sites in the MCS of pTXB1 and pTXB3 are not unique. pTXB1 allows cloning a gene of interest using NdeI and *Sap*I, as in pTYB1; pTXB3 allows cloning a gene of interest using NcoI and *Sap*I, as in pTYB3. Thus both vectors allow the coding sequences of target proteins to be fused to the N terminus of the intein for subsequent cleavage from all vector-derived sequences. However, the *Sap*I site also permits insertion of additional amino acids between the C terminus of the protein and the intein when needed to improve expression or cleavage. The compatible MCS allows cloning the same insert into both pTXB and pTYB vectors and comparison of these constructs for the efficiency of expression and purification. Expression of the fusion gene in pTXB is also under control of an IPTG-inducible bacteriophage T7 promoter. A diagram of the control

pXB1,3

pXB1

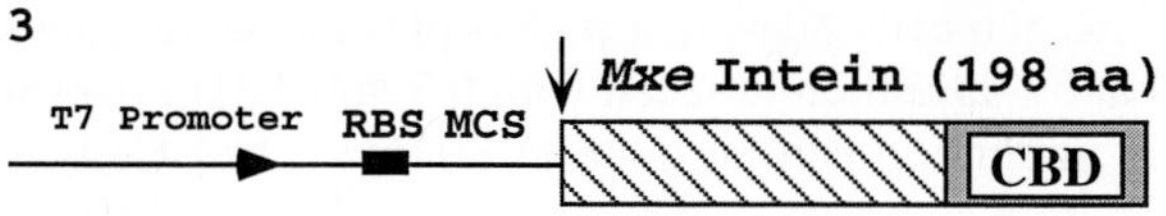

Met Ala Ser Ser Arg Val Asp Gly Gly Arg Glu Phe Leu Glu Gly Ser Ser Cys1
<u>AT ATG</u> <u>GCT AGC</u> <u>TCG CGA</u> <u>GTC GAC</u> GGC GGC CGC <u>GAA TTC</u> <u>CTC GAG</u> <u>GGC TCT TCC</u> TGC
 NdeI   NheI   NruI   SalI   NotI   EcoRI   XhoI   SapI   Intein...

.. Mxe GyrA Intein (198 aa)...
ATC ACG GGA GAT GC<u>A CTA GT</u>T GCC CTA CCC ... 3'
                 SpeI

pXB3

Met Ala Ser Ser Arg Val Asp Gly Gly Arg Glu Phe Leu Glu Gly Ser Ser Cys1
ACC ATG GCT AGC TCG CGA <u>GTC GAC</u> GGC GGC CGC <u>GAA TTC</u> <u>CTC GAG</u> <u>GGC TCT TCC</u> TGC
 NcoI   NheI   NruI   SalI   NotI   EcoRI   XhoI   SapI

Fig. 9. Schematic representation of pTXB vectors and the nucleotide sequence of the multiple cloning sites. Unique sites are in bold. The diagram is not to scale, and arrows indicate the sites of thiol-induced cleavage.

and cloning sites of the pTXB vectors and the corresponding nucleotide sequences are shown in Fig. 9.

Reagents

Lysis/column buffer: 20 mM Tris–HCl or Na–HEPES, pH 8.0, and 0.5 M NaCl. (The permissible ranges are as follows: buffer concentration, 10–50 mM; pH, 7.0–8.5; NaCl concentration, 0.1–1.0 M.)

Cleavage buffer: 20 mM Tris–HCl or Na–HEPES, pH 8.0, 0.5 M NaCl, and 30 mM DTT or MESNA. (The permissible ranges are as follows: buffer concentration, 10–50 mM; pH, 8.0–8.5; NaCl concentration, 0.1–1.0 M; MESNA concentration, 30–100 mM.)

pTXB1 or pTXB3 vectors: Available from New England Biolabs (Beverly, MA)

Chitin beads: Available from New England Biolabs. Prepare a 10- to 20-ml column for a 1-liter culture and equilibrate at 4° with 10 volumes of the lysis/column buffer.

Procedures

Cell growth, expression, and purification procedures are essentially the same as the protocols described in the preceding section for the *Sce* VMA intein (pTYB vectors). An example follows.

Escherichia coli ER2566 cells bearing a pTXB plasmid were grown to an $A_{600 \, nm}$ of 0.5–0.6 at 37° and then induced with 0.5 mM IPTG overnight at 15°. (Alternatively, induction can be done at 30° for 3 hr.) Cells were harvested by centrifugation and disrupted by sonication. This and all subsequent steps were carried out at 4°. The fusion protein was bound to chitin beads (10 ml bed volume) equilibrated with 50 mM Tris, pH 7.4, and 500 mM NaCl, and washed with 10 column volumes of the same buffer to remove unbound material. Cleavage was initiated using a buffer of 50 mM MESNA, 50 mM Tris, pH 8.0, and 100 mM NaCl. After overnight incubation at 4°, the cleaved target protein was eluted from the column.

Critical Parameters and Limitations

Like the *Sce* VMA intein derived from pTYB vectors, the *Mxe* GyrA intein cleaves efficiently with thiol reagents such as DTT and 2-mercaptoethanol. However, the *Mxe* GyrA intein cleaves more efficiently with MESNA or thiophenol to allow subsequent labeling with cysteine or ligation to peptides with an N-terminal cysteine. The *Sce* VMA intein cleaves much less efficiently at 4° with these compounds, resulting in lower yields of target protein. The C-terminal thioester tag generated with MESNA is relatively stable at pH 8.0–8.5 at 4°, and protein products isolated after 16- to 24-hr cleavage reactions can be ligated at efficiencies greater than 90% to peptides possessing an N-terminal cysteine.[11]

Thiol-induced cleavage at the N terminus of the *Mxe* GyrA intein can be affected dramatically by the C-terminal residue of the target protein. Cleavage at the N terminus of the wild-type *Mxe* GyrA intein (without the Asn198Ala mutation) was examined with the 20 naturally occurring amino acids at the −1 position.[18] Most amino acids allowed precursor purification and significant DTT-induced cleavage, except for Ser, Pro, or Glu, which blocked thiol-induced cleavage, and Asp, which resulted in >90% *in vivo* cleavage. The specificity for cleavage of the *Mxe* GyrA intein by various thiols is also influenced by the C-terminal residue of a target protein. For example, when Tyr, the native residue, is present at the −1 position, efficient cleavage occurs at 4° with DTT, 2-mercaptoethanol, and MESNA. In contrast, with glycine at the −1 position, cleavage with MESNA is much less efficient than with DTT and 2-mercaptoethanol. As in the case of the *Sce* VMA intein, amino acids adjacent to the −1 position may also affect cleavage efficiency. If the amino acids at C terminus of a target protein do not permit efficient cleavage, additional residues, such as Tyr, Met, or Gly, may be inserted at the −1 position to improve thiol-induced cleavage.

[18] M. W. Southworth, K. Amaya, T. C. Evans, M.-Q. Xu, and F. B. Perler, *Biotechniques* **27,** 110 (1999).

Representative Results

The engineered *Mxe* GyrA intein (Asn198Ala) was used successfully to produce two potentially cytotoxic proteins, bovine pancreatic ribonuclease A (RNase A) and the type II restriction endonuclease *Hpa*I, by first expressing inactive N-terminal fragments fused to the *Mxe* intein in *E. coli*.[11] DNA sequences encoding the inactive truncated forms of these proteins, the first 109 or 223 amino acids of RNase A or *Hpa*I, respectively, were cloned into pTXB vectors using *Nde*I–*Sap*I sites or *Nco*I–*Sap*I sites. Fusion proteins were expressed in *E. coli* ER2566 and isolated by passage of the clarified cell extracts to columns containing chitin beads. Cleavage of fusion proteins on chitin columns was induced by MESNA to release the target proteins with a C-terminal thioester moiety. MESNA-tagged fragments were then ligated with synthetic peptides corresponding to the missing C-terminal sequence to generate enzymatically active mature products. The turnover number and K_m for ligated and renatured RNase A were 8.2 sec^{-1} and 1.5 mM, in good agreement with the reported values of 8.3 sec^{-1} and 1.2 mM.[19] Ligated *Hpa*I had a specific activity of 0.5–1.5 $\times$ 10^6 units/mg, which compared favorably with the expected value of 1–2 $\times$ 10^6 units/mg.[11] This procedure resulted in the isolation of 3–6 mg/liter of culture of *Hpa*I and 0.3 mg/liter of culture of RNase A. A major cause of the large differences in the yields of the two proteins was the cleavage efficiency by MESNA adjacent to the C-terminal residues of the target proteins, which, in the case of the restriction endonuclease *Hpa*I fragment, was methionine (Met$_{223}$) and, in the case of the RNase A fragment, alanine (Ala$_{109}$).

Use of *Mth* RIR1 Mini-intein (pTRB5 Vector)

The smallest intein known is the 134 amino acid intein found in the ribonucleoside diphosphate reductase gene of *Methanobacterium thermoautotrophicum* (*Mth* RIR1 intein).[20] Interestingly, this intein possesses a proline as the last N-extein residue (Pro^{-1}), which has been shown to block splicing of the *Sce* VMA intein and the mini-intein derived from the *Ssp* DnaB intein.[4,8] The coding region for the *Mth* RIR1 intein along with five native N- and C-extein residues was assembled from synthetic oligonucleotides, using 61 silent mutations to adapt 49 of the 134 codons to *E. coli*

[19] R. S. Hodges and R. Merrifield. *J. Biol. Chem.* **250,** 1231 (1975).
[20] D. Smith, L. Doucette-Stamm, C. Deloughery, H. Lee, J. Dubois, T. Aldredge, R. Bashirzadeh, D. Blakely, R. Cook. K. Gilbert, D. Harrison, L. Hoang, P. Keagle, W. Lumm, B. Pothier, D. Qiu, R. Spadafora, R. Vicaire, Y. Wang, J. Wierzbowski, R. Gibson, N. Jiwani, A. Caruso, D. Bush, and J. Reeve, *J. Bacteriol.* **179,** 7135 (1997).

pTRB5

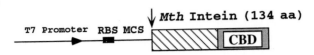

```
        Met Ala Ser Ser Arg Val Asp Gly Gly Arg Glu Phe Leu Glu Gly Cys1...
      ACC ATG GCT AGC TCG CGA GTC GAC GGC GGC CGC GAA TTC CTC GAG GGC TGC GTA TCC
         NcoI    NheI    NruI    SalI    NotI    EcoRI   XhoI       Intein...

      ... Mth RIR1 Intein (134 aa)...
      GGT GAC ACC ATT GTA ATG ACT AGT GGC GGT...3'
                             SpeI
```

FIG. 10. Schematic representation of the pTRB5 vector and the nucleotide sequence of the multiple cloning sites. The diagram is not to scale, and arrows indicate the sites of thiol-induced cleavage.

coding preferences.[7] The splicing activity of the *Mth* RIR1 intein was investigated by expressing it as an in frame fusion between *E. coli* MBP and the *B. circulans* CBD. This fusion protein was found to splice poorly in *E. coli* with the naturally occurring Pro$_{-1}$ adjacent to the N-terminal cysteine of the intein, but splicing proficiency increased significantly when this residue was replaced by alanine or glycine. Pro(−1)Ala or Pro(−1)Gly mutants also exhibited some cleavage at the N- and C-terminal junctions of the intein. Introduction of an Asn134Ala substitution, along with the Pro(−1)-Gly mutation, blocked C-terminal cleavage and splicing and yielded a mutant that underwent efficient cleavage at the N-terminal splice junction at 4° in the presence of thiol reagents such as DTT, 2-mercaptoethanol, or MESNA. As described in a later section, the *Mth* RIR1 intein could also be altered by a Cys1Ala mutation, suppressing N-terminal cleavage and yielding a construct capable of undergoing efficient C-terminal cleavage.

pTRB5 Vector

The pTRB5 vector employs the *Mth* RIR1 intein with Pro(−1)Gly and Asn134Ala substitutions, but otherwise has the same backbone as pTYB1, with an IPTG-inducible bacteriophage T7 promoter followed by an MCS and the *Mth* RIR1 intein/CBD translational fusion, as illustrated in Fig. 10. The NcoI site downstream of the ribosome-binding site contains the translation-initiating ATG codon. A gene of interest can be cloned into pTRB5 in-frame with the *Mth* intein/chitin-binding domain-coding region. The use of NcoI and XhoI site leaves Leu-Glu-Gly at the C terminus of a target protein after cleavage of the fusion protein.

Reagents

Lysis/column buffer: 20 mM Tris–HCl or Na–HEPES, pH 8.0, and 0.5 M NaCl. (The permissible ranges are as follows: buffer concentration, 10–50 mM; pH, 7.0–8.5; NaCl concentration, 0.1–1.0 M.)

Cleavage buffer: 20 mM Tris–HCl or Na–HEPES, pH 8.0, 0.5 M NaCl, and 30 mM DTT or MESNA. (The permissible ranges are as follows: buffer concentration, 10–50 mM; pH, 8.0–8.5; NaCl concentration, 0.1–1.0M; MESNA concentration, 30–100 mM.)

pTRB5 vector: Available from New England Biolabs (Beverly, MA) on request

Chitin beads: Available from New England Biolabs. Prepare a 10- to 20-ml column for a 1-liter culture and equilibrate at 4° with 10 volumes of the lysis/column buffer.

Procedures

Cell growth, expression, and purification procedures are essentially the same as the protocol described in the preceding sections for the *Sce* VMA intein (pTYB vectors) and the *Mxe* GyrA intein (pTXB vectors). An example follows.

Escherichia coli ER2566 cells bearing pMRB encoding the MBP–*Mth* intein–CBD fusion protein are grown at 37° in LB broth containing 100 μg/ml ampicillin to an $A_{600 \text{ nm}}$ of 0.5–0.8, followed by induction with IPTG (0.3–0.5 mM), either overnight at 15° or for 3 hr at 30°. The cells are disrupted in 20 mM Tris–HCl, pH 7.5, containing 500 mM NaCl. The clarified cell extract is applied to a column packed with chitin resin (10-ml bed volume per liter of culture) equilibrated with the same buffer. Unbound protein is removed by washing with 10 column volumes of buffer. Thiol-induced cleavage is initiated by rapidly equilibrating the chitin resin in column buffer containing 20 mM Tris–HCl, pH 8.0, 500 mM NaCl, and 30 mM DTT. For protein labeling and ligation, 50–100 mM MESNA is used in place of DTT.

Critical Parameters and Limitations

Some N-terminal cleavage occurred *in vivo* when protein synthesis was induced at either 15° or 37°. The yield of fusion protein was greater at higher induction temperatures (30–37°C) due to less *in vivo* cleavage, N-terminal cleavage *in vivo,* and *in vitro* being less affected by induction temperature than C-terminal cleavage. The fusion proteins underwent thiol-mediated cleavage at the N terminus of the intein when incubated at 4° overnight with 30–100 mM DTT, 2-mercaptoethanol, or MESNA and could be used to isolate thioester-tagged proteins for protein ligation. However, the fusion protein was stable in the absence of thiol on overnight incubation at pH 8.0 and 25°.

The small size (15 kDa), optimized codon usage, and bacterial origin of the synthetic *Mth* RIR1 intein make it an effective fusion partner for protein expression in *E. coli*. For example, use of the pTRB5 vector, based on the *Mth* RIR1 intein, to isolate the *Bacillus stearothermophilus* DNA polymerase large fragment (67 kDa) resulted in twice the yield of purified protein than when a pTYB vector, based on the *Sce* VMA intein, was used. However, the sensitivity of thiol-mediated cleavage activity to different residues or protein sequences adjacent to the N terminus of the intein has yet to be investigated. In the absence of this information, it is advisable to include a Gly residue or Leu-Glu-Gly between the protein of interest and the intein to achieve controllable cleavage. For comparison, parallel cloning experiments can be done in pTYB and pTXB vectors by using compatible polylinker regions.

Representative Results

pMRB vector was constructed by cloning the coding region of the *E. coli malE* gene using the *Nde*I and *Xho*I sites of pTRB5. Typically, pMRB yielded 15–20 mg of MBP per liter of culture on cleavage of the bound fusion protein with 30 m*M* DTT.

Use of a *Ssp* DnaB-Derived Mini-intein (pTSB Vectors)

The *dnaB* gene encoding the DNA helicase of the cyanobacterium *Synechocystis* sp. strain PCC6803 contains an intein of 429 amino acid residues (*Ssp* DnaB intein).[21] Constructs containing the *Ssp* DnaB intein with 5 native extein residues flanking its splice junctions are capable of splicing. A splicing-proficient minimal intein was created by deleting the central 275 amino acid residues containing the endonuclease domain, leaving 106 N-terminal and 48 C-terminal residues of the original intein.[22] The 154 residue *Ssp* DnaB mini-intein was modified by mutagenesis to allow protein purification by cleavage of the peptide bond at either the N-terminal or the C-terminal splice junction.[8] Proteins fused to either the N terminus or the C terminus of an engineered *Ssp* DnaB mini-intein could be purified in a single chromatographic step in conjunction with the use of the *B. circulans* chitin-binding domain. A major advantage of the *Ssp* DnaB intein system is that both N-terminal and C-terminal cleavage reactions can be

[21] S. Pietrokovski, *Trends Genet.* **12**, 287 (1996).
[22] H. Wu, M.-Q. Xu, and X.-Q. Liu, *Biochim. Biophys. Acta* **1387**, 422 (1998).

induced by incubation at an optimal pH and temperature (4–25°) without the use of a thiol reagent. Alternatively, *Ssp* DnaB mini-intein vectors can be cleaved in the presence of MESNA or thiophenol to generate proteins with a C-terminal thioester for protein ligation or labeling.

An expression vector (pTSB) for the purification of a fusion protein and subsequent self-catalyzed cleavage of the fusion protein was generated from the *Ssp* DnaB mini-intein by an Asn154Ala mutation, which prevented C-terminal cleavage and protein splicing, and by translational fusion to the N terminus of a chitin-binding domain. After binding to chitin beads, the target protein can be isolated either by thiol-induced cleavage of the fusion protein at 4° or by incubation at 25° at pH 8.0 in the absence of thiols.[8] Unlike the other vectors described in this article, this system allows purification of thiol-sensitive proteins. In addition, the *Ssp* DnaB mini-intein exhibited efficient cleavage induced by MESNA, generating a thioester at the C terminus of a target protein for ligation with proteins carrying an N-terminal cysteine.

pMSB17 and pTSB Vectors

The pMSB17 vector utilizes the 154 residue *Ssp* DnaB mini-intein carrying an Asn154Ala substitution to block the C-terminal cleavage reaction.[8] Expression of the fusion gene is under control of the IPTG-inducible T7 promoter as in pTYB1. The coding region of *E. coli* MBP was cloned in frame with the *Ssp* intein–CBD-coding region using *Nde*I and *Xho*I sites. The vector expresses an MBP–intein–CBD fusion protein. After cleavage of the fusion protein, the C terminus of MBP contains the additional C-terminal sequence Leu-Glu-Gly. A gene of interest can be inserted into the pMSB17 vector using the *Nde*I and *Xho*I sites, replacing the MBP-coding region. Other pMSB17 derivatives contain different amino acid substitutions at Asn_{154} (see "Critical Parameters"). More general cloning vectors were constructed by replacing the MBP-coding region with multiple cloning sites to yield pTSB1 and pTSB3, as illustrated in Fig. 11. *Nde*I and *Nco*I sites, in pTSB1 and pTSB3, respectively, contain a translation-initiating methionine codon (ATG). If the *Sap*I site is used, the C terminus of a target protein will carry no extra amino acids. It is also possible to add favorable residues, such as glycine, to the C terminus of the target protein to improve cleavage efficiency.

Reagents

Lysis/column buffer: 20 mM Tris–HCl or Na–HEPES, pH 6.0, and 0.5M NaCl. (The permissible ranges are as follows: buffer concentration, 10–50 mM; pH 6.0–7.0.)

pTSB1,3

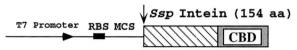

pTSB1

```
      Met Val Asp Gly Gly Arg Glu Phe Leu Glu Gly Ser Ser Cys1...
     CAT ATG GTC GAC GGC GGC CGC GAA TTC CTC GAG GGC TCT TCC TGC ATC AGT GGA GAT
     NdeI    SalI      NotI    EcoRI   XhoI    SapI      Intein...

     ... Ssp Mini-Intein (154 aa)...
     AGT TTG ATC AGC TTG GCT AGC ACA ...3'
                         NheI
```

pTSB3

```
      Met Val Asp Gly Gly Arg Glu Phe Leu Glu Gly Ser Ser Cys1...
     ACC ATG GTC GAC GGC GGC CGC GAA TTC CTC GAG GGC TCT TCC TGC ATC AGT GGA GAT
     NcoI    SalI      NotI    EcoRI   XhoI    SapI

     ... Ssp Mini-Intein (154 aa)...
     AGT TTG ATC AGC TTG GCT AGC ACA ...3'
                         NheI
```

FIG. 11. Schematic representation of pTSB vectors and the nucleotide sequence of multiple cloning sites. The diagram is not to scale, and arrows indicate the sites of thiol- or pH-induced cleavage.

Cleavage buffer A (for thiol-induced cleavage): 20 mM Tris–HCl or Na-HEPES, pH 8.5, 0.5M NaCl, and 30 mM DTT or MESNA. (The permissible ranges are as follows: buffer concentration, 10–50 mM; pH, 8.0–8.5; NaCl concentration, 0.1–1.0 M; MESNA concentration, 30–100 mM.)

Cleavage buffer B (for cleavage in the absence of thiols): 20 mM Tris–HCl or Na–HEPES, pH 8.5, and 0.5M NaCl. (The permissible ranges are as follows: buffer concentration, 10–50 mM; pH, 8.0–8.5.)

pTSB vector: Available from New England Biolabs on request

Chitin beads: Available from New England Biolabs. Prepare a 10- to 20-ml column for a 1-liter culture and equilibrate at 4° with 10 volumes of the lysis/column buffer.

Procedures

Cell growth, expression, and purification procedures are essentially the same as protocols described in earlier sections on other inteins. However, because *Ssp* DnaB mini-intein fusion proteins may be very sensitive to hydrolysis, it is important that the cell extract and the column buffers be kept at 4° and the purification procedure be accomplished within 4–6 hr. The use of lysis/column buffer at pH 6.0 may reduce cleavage during purification. An example follows.

A 1-liter culture of *E. coli* ER2566 cells bearing pMSB17 is induced with 0.3 m*M* IPTG at 30° for 3 hr (or at 12–15° overnight). SDS–PAGE analysis shows that some *in vivo* cleavage of the fusion protein has occurred, producing MBP (43 kDa) and intein–CBD conjugate (28 kDa). The clarified cell extract is loaded onto a column (10 ml) of chitin beads. After washing with 20 column volumes of 50 m*M* Tris, pH 6.0, 0.5 *M* NaCl at 4°, the column is flushed with 50 m*M* Tris, pH 8.5, 0.5 *M* NaCl at 4°, followed by overnight incubation at 25°. Alternatively, the column can be flushed with buffer containing 50 m*M* DTT or MESNA and cleavage achieved by over-night incubation at 4°. The target protein is recovered by elution with the corresponding buffer.

Critical Parameters and Limitations

The *Ssp* DnaB mini-intein carrying an Asn154Ala substitution has unique properties for inducible cleavage at the N-terminal splice junction. Efficient N-terminal cleavage can be affected by treatment with a thiol compound such as DTT or MESNA at 4°, but also in the absence of thiols by incubation at pH 8–9 at higher temperatures (16–25°). Cleavage without the use of a thiol reagent is beneficial in the purification of thiol-sensitive proteins.

Mutagenesis studies indicated that the natural glycine residue in the −1 position is essential for efficient splicing and N-terminal cleavage activity of the *Ssp* DnaB mini-intein.[8] Replacement of Gly$_{-1}$ with Ala, Leu, or Pro affected splicing and cleavage drastically. The presence of a glycine residue favored thioester formation at Cys$_1$ and resulted in the accumulation of substantial amounts of the linear thioester intermediate. Boiling of cell extracts prior to SDS–PAGE analysis, with or without DTT, can cause a loss of fusion protein and lead to an overestimation of *in vivo* cleavage. The N-extein residues in positions −2 to −5 had more subtle effects on the splicing and cleavage reactions, perhaps by influencing the folding of the intein and the conformation of the scissile peptide bond.

Ssp DnaB mini-intein (Asn154Ala) fusion proteins display significant *in vivo* cleavage during induction of expression, resulting in free target protein and intein–CBD conjugate. It is therefore important to determine the optimal expression conditions. Because the *in vitro* cleavage of the fusion protein is pH and temperature dependent, cell disruption and purifi-cation of the fusion protein should be carried out rapidly at 4° and pH 6.0.

Mutation of the intein or its flanking residues may not only modulate protein splicing by affecting catalytic residues but may alter the architecture of the protein-splicing active center by affecting the alignment of reacting groups as well as solvent accessibility. This is illustrated by the observation that the N-terminal cleavage reaction may be modulated by replacing the intein C-terminal asparagine residue (Asn_{154}) with different amino acids.[8] *Ssp* DnaB mini-intein mutants in which Asn_{154} is replaced by Leu, Lys, or Pro rather than Ala exhibit significantly reduced *in vivo* cleavage of MBP–*Ssp* intein–CBD fusion proteins, resulting in a 2- to 10-fold increase in the final yield of MBP after purification and cleavage in the presence of DTT at 4°. However, Asn154Leu and Asn154Lys mutants could not be induced to cleave efficiently without a thiol reagent. In contrast, the Asn154Pro mutant could undergo efficient thiol-independent cleavage at pH 8–9 at 25°. Replacement of Asn_{154} with Met, Val, Glu, Thr, Cys, Ser, or Gln resulted in extensive *in vivo* cleavage (>90%) of the MSB fusion proteins.

Representative Results

Escherichia coli MBP was purified from extracts of *E. coli* transformed with pMSB17, which encodes a tripartite fusion protein consisting of MBP–*Ssp* mini-intein(Asn154Ala)–CBD. After isolation of the fusion protein and cleavage in the absence of thiol at pH 8.0 with 50 mM DTT, 4–6 mg of protein was typically obtained from 1 liter of culture. Several other proteins have been purified with or without a thiol reagent by the same procedure.

Fusion of Proteins at Their N Terminus to a Modified Intein and Release by Cleavage at the Intein C Terminus

Use of a Modified *Sce* VMA Intein (pIMC Vectors and pTYB11/pTYB12)

This method uses an *Sce* VMA intein modified to undergo cleavage reaction at both termini.[23] The intein mutation, His453Gln, attenuates the cleavage reaction at the adjacent intein C terminus, Asn_{454}. In conjunction with mutation of the first C-extein residue [Cys(+1)Ala], this allows the isolation of the fusion precursor in which both splicing and cleavage are blocked *in vivo*. Cleavage at both splice junctions can be induced *in vitro*

[23] S. Chong, G. E. Montello, A. Zhang, E. J. Cantor, W. Liao, M.-Q. Xu, and J. Benner, *Nucleic Acids Res.* **26**, 5109 (1998).

by incubating the fusion protein with DTT, 2-mercaptoethanol, or free cysteine. To allow affinity purification of the fusion precursor, a CBD is inserted in the homing endonuclease domain of the intein without affecting the structure of the protein splicing domains and the cleavage reaction. A small N-extein sequence consisting of the first 10 residues of MBP is fused to the N terminus of the modified intein to provide a favorable translational start for protein expression.

To express and purify a target protein, its N terminus is fused to the C terminus of the modified *Sce* VMA intein. The intein is then induced to undergo on-column self-cleavage by incubation with DTT, 2-mercaptoethanol, or cysteine. The target protein is specifically released from the column and eluted along with the short N-extein peptide. Normally, a final dialysis step is performed to separate the purified target protein from the N-extein peptide and the excess of thiols.

pIMC Vectors and pTYB11/pTYB12

General cloning vectors, called pIMC, were constructed for expression and purification of target proteins in *E. coli*.[23] pIMC vectors use a *tac* promoter and the *lacI* gene to provide control of fusion gene expression, but were subsequently modified by transferring the fusion protein-coding region to pTYB vectors so as to be under the control of the T7 promoter and the *lacI* gene (Fig. 12). pIMC1 (6876 bp) and its T7 promoter equivalent, pTYB11 (7413 bp), contain an MCS, similar to that of pTYB1, with a *Sap*I cloning site, which allows the target gene to be cloned adjacent to the C-terminal cleavage site of the intein (Asn454); this results in the purification of a target protein without any extraneous residues attached to its N terminus. pIMC2 (6873 bp) and its T7 promoter equivalent, pTYB12 (7416 bp), have an MCS with similar restriction sites as the MCS of pTYB2. Use of pIMC2 yields a target protein with one or more extra residues added to its N terminus. For instance, cloning the 5′ end of a target gene using the *Nde*I site in pIMC2 adds three extra residues (Ala-Gly-His) to the N terminus the target protein.

Reagents

Lysis/column buffer: 20 mM Na–HEPES or Tris–HCl, pH 8.0; 500 mM NaCl; 0.1 mM EDTA. In addition, the use of 0.1% Triton X-100 or Tween 20 is recommended to prevent nonspecific adsorption to the chitin beads, unless the target protein is known to be inactivated by these nonionic detergents. Protease inhibitors such as phenylmethylsulfonyl fluoride do not interfere with subsequent purification steps and may also be used. The reducing agents TCEP or

pIMC1,2

Sce **Intein/CBD (510 aa)**

pIMC1 and pTYB11

```
   ...Intein ↓
...Val Gln Asn Arg Arg Ala Thr Ser Ser Arg Val Asp Gly Gly Arg Glu Phe
...GTA CAG AAC AGA AGA GCT ACT AGT TCG CGA GTC GAC GGC GGC CGC GAA TCC
              SapI         SpeI    NruI    SalI      NotI      EcoRI

Leu Glu Pro Gly ***
CTC GAG CCC GGG TGA CTG CAG
 XhoI   SmaI        PstI
```

pIMC2 and pTYB12

```
   ...Intein ↓
...Val Gln Asn Ala Gly His Met Thr Ser Ser Arg Val Asp Gly Gly Arg Glu Phe
...GTA CAG AAT GCT GGT CAT ATG ACT AGT TCG CGA GTC GAC GGC GGC CGC GAA TCC
        BsmI        NdeI    SpeI    NruI    SalI      NotI      EcoRI

Leu Glu Pro Gly ***
CTC GAG CCC GGG TGA CTG CAG
 XhoI   SmaI        PstI
```

FIG. 12. Schematic representation of pIMC vectors and the nucleotide sequence of multiple cloning sites. The diagram is not to scale, and arrows indicate the sites of thiol-induced cleavage.

TCCP may be added at 1 mM to stabilize oxidation-sensitive proteins during purification.

Cleavage buffer: 20 mM Na–HEPES or Tris–HCl, pH 8.0; at least 50 mM NaCl (up to 2 M); 0.1 mM EDTA; 30 mM DTT

Stripping buffer: 20 mM Na–HEPES or Tris–HCl, pH 8.0; 500 mM NaCl; 1% SDS

LB broth: Per liter, 10 g tryptone, 5 g yeast extract, 5 g NaCl, adjusted to pH 7.2 with about 1 ml of 1 N NaOH

SDS–PAGE sample buffer: 70 mM Tris–HCl, pH 6.8; 33 mM NaCl; 1 mM Na$_2$EDTA; 2% SDS; 40 mM DTT; 0.01% bromphenol blue; 40% glycerol

1 M DTT: Dissolve 3.09 g of DTT in 20 ml of 0.01M sodium acetate, pH 5.2. Sterilize by filtration and store in 1- to 5-ml aliquots at −20°.

pTYB11/pTYB12 vectors: Available from New England Biolabs

Chitin beads: Available from New England Biolabs. Prepare a 10- to

20-ml column for a 1-liter culture and equilibrate at 4° with 10 volumes of the lysis/column buffer.

Anti-CBD antibody: Rabbit serum raised against the *Bacillus circulans* chitin-binding domain is available from New England Biolabs for Western blot analysis

E. coli hosts: Clones using pTYB11 or pTYB12 can be established initially in a nonexpression host, especially if the gene of interest is toxic and the clone may not be stable. *E. coli* ER2566 or BL21(DE3) (or other derivatives) can be used as expression strains. Strain ER2566 (provided by New England Biolabs with pTYB vectors) carries a chromosomal copy of the T7 RNA polymerase gene inserted into the *lacZ* gene and is thus under the control of the IPTG-inducible *lac* promoter.

Cloning the Target Gene into pIMC Vectors

To obtain a target protein with no extra vector-derived residues, one can clone a target gene between the *Sap*I and the *Pst*I sites of pTYB11 (Fig. 13). Cloning the 5′ end of a target gene into *Bsm*I of pIMC2 adds an

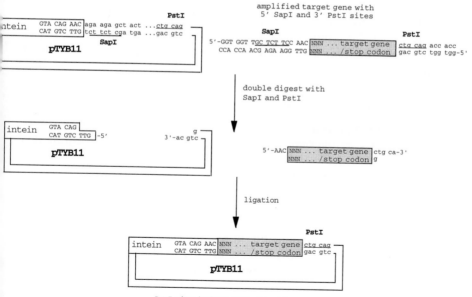

SapI site is lost after ligation

FIG. 13. Diagram illustrating the cloning of a target gene by insertion into the *Sap*I and *Pst*I sites of pIMC1.

TABLE V
EXAMPLES OF PRIMER DESIGN FOR CLONING IN pTYB11 AND pTYB12

	Cloning site	Primer sequence[a]	Cloning vector
Forward primer	*Sap*I[b]	5'-GGT GGT TGC TCT TCC AAC NNN NNN . . . −3'	pIMC1 or pTYB11
	*Bsm*I[c]	5'-GGT GGT GGG AAT GCG NNN NNN . . . −3'	
			pIMC2 or pTYB12
	*Nde*I	5'-GGT GGT CAT ATG NNN NNN . . . −3'	
Reverse primer	*Pst*I[d]	5'-GGT GGT CTG CAG TCA NNN NNN . . . −3'	pIMC1 or 2 pTYB11 or

[a] The target gene starts at "NNN NNN. . . ." Restriction sites are underlined. Use of a "GC GGT" sequence at the 5' end of the primer is to ensure efficient DNA cleavage by the restrictie enzyme when the restriction site is close to the 5' end.

[b] The *Sap*I site is not regenerated after cloning.

[c] *Bsm*I digestion is at 65°.

[d] An in-frame stop codon should be incorporated in the primer if the amplified gene lacks a translatic termination sequence.

extra Gly residue to the 5' end of the target protein. Table V gives some examples of forward and reverse PCR primers for the amplification of target genes. The amplified target genes are cloned into pIMC vectors by essentially the same protocol as that described for pTYB vectors in an earlier section.

Expression and Purification Procedures

Most of the procedures for expressing and purifying a target protein in pIMC vectors are the same as those for pTYB vectors. The following two steps are somewhat different.

Induction of On-Column Cleavage. Whereas in the case of pTYB vectors the on-column cleavage reaction is normally conducted at 4° overnight, use of pIMC vectors requires higher temperatures (16–23°) and longer cleavage times (40 hr). Data in Table VI can serve as guidelines for selecting an appropriate temperature and duration of the cleavage reaction. The cleavage reaction in pIMC vectors is more sensitive to pH than in pTYB vectors, with no cleavage occurring below pH 7.0. The on-column cleavage for pIMC vectors should therefore be conducted at a pH above 7.5.

Elution of the Target Protein. When using pIMC vectors, the short N-extein peptide (1.2 kDa) is also cleaved from the intein and coeluted with the target protein. Because of its small molecular weight, the cleaved pep-

TABLE VI
EFFECT OF N-TERMINAL AMINO ACID RESIDUE OF TARGET PROTEIN ON
C-TERMINAL CLEAVAGE OF FUSION PROTEINS BASED ON pIMC1 VECTOR[a]

Residue +1	Cleavage half-time (hr) at			% cleavage after 40 hr at		
	4°	16°	23°	4°	16°	23°
Met Ala Gln	14–24	5–10	2–6	60–90	>90	>95
Gly Leu Asn Trp Phe Tyr	30–60	10–20	<10	40–60	>90	>90
Val Ile Asp Glu Lys Arg His	>60	20–30	10–20	10–20	70–90	70–95
Pro	n.d.[b]	n.d.	n.d.	<10	<10	<10
Thr	100	20	10	20	80	>90
Ser	n.d.	n.d.	n.d.	n.d.	n.d.	n.d.
Cys	n.d.	n.d.	n.d.	n.d.	n.d.	n.d.

[a] Cleavage was conducted with 40 mM DTT in 30 mM HEPES, pH 8.0, 0.5 M NaCl under
the conditions indicated. Cleavage rates with Ser and Cys at the N terminus of the target
protein were not measured because these amino acids (and to a lesser extent Thr) may also
allow protein splicing to occur. n.d., Not determined.

tide cannot be detected on a regular SDS–PAGE gel but can normally be
separated from the target protein by dialysis.

Limitations and Critical Parameters

Cloning a target protein in pIMC1 using the *Sap*I site in the MCS places
the N-terminal residue of a target protein immediately adjacent to the
intein C-terminal cleavage site. The effect of the N-terminal residue of a
target protein on the intein C-terminal cleavage has been studied (Table
VI).[23] Most of the 20 amino acid residues allowed >50% cleavage at 16°
after 40 hr or at 23° after 16 hr. At 4°, cleavage occurred most efficiently
with Met, Ala, and Gln and somewhat less efficiently with other residues,
especially Val, Ile, Asp, Glu, Lys, Arg, and His. Pro inhibited intein C-
terminal cleavage, whereas N-terminal Ser and Cys led to *in vivo* splicing,

TABLE VII
RECOMBINANT PROTEINS EXPRESSED AND PURIFIED BY pIMC1,2 OR
pTYB11/12 N-TERMINAL FUSION VECTORS

Target protein	Yield (mg/liter of culture)
Bacteriophage T4 DNA ligase	8.4
Bacteriophage T4 gene 32 product	6.0
Bacteriophage T4 endonuclease VII	4.6
FseI restriction endonuclease	2.0
Green fluorescent protein	1.9
Calmodulin-dependent protein kinase II	2.2
Yeast invertase	1.7

with only low yields of the fusion protein. Some splicing was also observed in the case of Thr. In problematic cases, adding a few extra N-terminal amino acid residues by cloning into the BsmI or NdeI sites of pIMC2 may increase the cleavage efficiency and thus the yield of purified target protein.

Both cysteine and DTT result in similar cleavage efficiencies in pIMC vectors. Being a weaker reducing agent than DTT, cysteine may be preferred in some cases. Whereas cleavage of pTYB vectors with cysteine leads to the covalent attachment of the cysteine to the C terminus of the target protein, cysteine is not incorporated into the target protein when used for cleaving pIMC vectors but instead is attached to the N-extein peptide (Fig. 4A).

Representative Results

To illustrate the expression and purification of the pIMC vectors, a number of proteins from both prokaryotes and eukaryotes were cloned into these vectors and purified by a single chitin column step, as summarized in Table VII.

Use of *Mth* RIR1 Mini-intein (pBRC Vectors)

A C-terminal cleavage system based on the *Mth* RIR1 intein was generated by introducing the Pro(−1)Gly/Cys1Ala double mutation.[7] The Pro(−1)Gly mutation removes the unfavorable Pro$_{-1}$ residue and makes the intein competent for protein splicing, whereas the Cys1Ala mutation blocks protein splicing and N-terminal cleavage activities. The coding region for the protein of interest is fused in frame with the C-terminal intein codon, and the N terminus of the intein is fused to CBD with a C-terminal glycine residue. The full-length fusion protein, CBD–intein–target protein, can thus be isolated from cell extracts by binding to chitin resin. Fission of the peptide bond at C-terminal splice junctions leads to the release of the target protein from the chitin column. Unlike cleavage of fusion proteins involving the *Sce* VMA intein, C-terminal cleavage involving the *Mth* RIR1 intein does not need to be induced with a thiol. The C-terminal cleavage activity of the mutated *Mth* RIR1 intein potentially provides a means to isolate proteins possessing a wide range of N-terminal residues, although the specificity for the +1 residue has not yet been tested. For instance, proteins with an N-terminal cysteine were isolated as substrates for the *in vitro* fusion of large, bacterially expressed proteins.

pBRC Vectors

Mth RIR1 intein C-terminal cleavage vectors, termed pBRC, have the same plasmid backbone as pTYB1. The *B. circulans* chitin-binding domain is fused to the N terminus of the modified *Mth* intein and expression of the fusion gene is controlled by an IPTG-inducible bacteriophage T7 promoter. pBRC1 or pBRC3 contains a polylinker region just downstream of the intein, as shown in Fig. 14. pBRL contains the coding region of

FIG. 14. Schematic representation of pBRC vectors and the nucleotide sequence of multiple cloning sites. The diagram is not to scale, and arrows indicate the sites of pH- and temperature-induced cleavage.

bacteriophage T4 DNA ligase cloned in frame with the C terminus of the modified *Mth* intein.[7] It expresses a CBD–intein–T4 DNA ligase fusion protein, and cleavage of the C terminus of the intein yields T4 DNA ligase with a N-terminal cysteine residue that can be used for protein ligation. Other genes of interest can be inserted into the pBRL vector using *Age*I and *Pst*I sites to replace the T4 DNA ligase sequence.

Reagents

Lysis/column buffer: 20 mM Tris–HCl or Na–HEPES, pH 8.0, and 0.5M NaCl. (The permissible ranges are as follows: buffer concentration, 10–50 mM; pH, 6.0–8.5.)

Cleavage buffer: 20 mM Tris–HCl or Na–HEPES, pH 7.0, and 0.5 M NaCl

pBRC vectors: Available from New England Biolabs on request

Chitin beads: Available from New England Biolabs. Prepare a 10- to 20-ml column for a 1-liter culture and equilibrate at 4° with 10 volumes of the lysis/column buffer.

Expression and Purification Procedures

Procedures for expressing and purifying a target protein in pBRL vectors are essentially the same as those described in an earlier section for pTYB vectors. An example follows.

Escherichia coli ER2566 cells bearing the appropriate plasmid are grown at 37° in LB broth containing 100 μg/ml ampicillin to an $A_{600\,\text{nm}}$ of 0.5–0.8, followed by induction with IPTG (0.3–0.5 mM) overnight at 15°. The cells are harvested and resuspended in 20 mM Tris–HCl, pH 8.5, containing 0.5 M NaCl. The clarified cell extract is loaded onto a chitin column. Following flushing of the column with 20 mM Tris–HCl, pH 7.0, containing 500 mM NaCl, the cleavage reaction proceeds overnight at room temperature prior to elution of the protein product.

Critical Parameters and Limitations

Some *in vivo* cleavage occurs when fusion protein expression in *E. coli* transformed with pBRC1 vectors is induced at 15°. Although *in vivo* cleavage is diminished by induction at higher temperatures (30–37°), fusion proteins isolated at the higher temperatures have a much reduced ability to undergo C-terminal cleavage, perhaps due to misfolding of the *Mth* RIR1 intein, and induction at 15° is therefore recommended.[7] It is interesting that the double mutant, Pro(−1)Gly/Cys1Ser, showed more rapid *in vivo* cleavage at 15° than the Pro(−1)Gly/Cys1Ala double mutant, whereas neither underwent *in vivo* cleavage at 37°, suggesting that *in vivo* and *in vitro* cleavage activities can be modulated by induction temperature as well as by replacing Cys_1 with different amino acids. The sensitivity of C-terminal cleavage activity to different residues or protein sequences adjacent to the C terminus of the intein has not yet been investigated.

Representative Results

pBRL and analogous plasmids based on the modified *Mth* RIR1 intein were used to purify bacteriophage T4 DNA ligase and thioredoxin, with yields of 4–6 mg/liter cell culture and 5–10 mg/liter cell culture, respectively.

Use of a *Ssp* DnaB-Derived Mini-intein (pBSC Vectors)

The C-terminal cleavage system based on the 154 residue *Ssp* DnaB mini-intein is similar to that involving the *Mth* RIR1 intein described in the preceding section in that Cys_1 is replaced by Ala to block overall protein splicing and N-terminal cleavage so that only the C-terminal cleavage reaction can occur.[8] The modified *Ssp* DnaB mini-intein carrying a Cys1Ala mutation undergoes cleavage of the peptide bond between the intein and a C-terminal target protein in a pH-dependent manner, with an optimum between pH 6.0–7.5. In pBSC vectors, the CBD is cloned in frame with the N terminus of the modified *Ssp* DnaB mini-intein, controlled by an inducible bacteriophage T7 promoter. The target gene is cloned in frame into an MCS adjacent to the C terminus of the *Ssp* DnaB mini-intein. The fusion protein, CBD–intein–target protein, is isolated from induced *E. coli* cell extracts by binding to chitin resin and is induced to undergo cleavage by incubation at 25° and pH 7.0 for 16–24 hr, thus releasing the target protein. In contrast to the *Sce* VMA intein-based cleavage system, the *Ssp* DnaB mini-intein C-terminal cleavage system allows rapid purification of recombinant proteins without use of a thiol reagent and has been utilized to generate proteins with an N-terminal cysteine for ligation of large bacterially expressed proteins. Considerable information is available on the effect of N-terminal amino acids of the target protein on the ability of fusion proteins to undergo C-terminal cleavage.[8]

pBSC Vectors

These vectors have the same plasmid backbone as pTYB1 except for intein and MCS segments. pBSC vectors allow cloning a target gene in frame with the C terminus of the modified *Ssp* DnaB mini-intein, which in turn is fused at its N terminus to CBD and is controlled by an inducible bacteriophage T7 promoter. Diagrams of the pBSC plasmids and the MCS sequence of pBSC1 and pBSC3 are shown in Fig. 15. In addition to the *Sap*I site, which allows purification and isolation of a protein without addi-

pBSC vectors

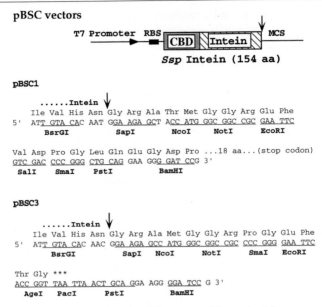

FIG. 15. Schematic representation of pBSC vectors and the nucleotide sequence of multiple cloning sites. The diagram is not to scale, and arrows indicate the sites of pH- and temperature-induced cleavage.

tional N-terminal amino acids, *Nco*I and *Not*I sites can be used for cloning, as well as the unique *Nru*I and *Bsr*GI sites, which are 20 and 4 bp, respectively, upstream of the 3' end of the intein. The pBSL-C155 plasmid expresses a CBD–intein–T4 DNA ligase fusion protein and can be modified by inserting a gene of interest using the *Age*I and *Pst*I sites, thereby replacing the coding sequence of T4 DNA ligase.

Reagents

Lysis/column buffer: 20 m*M* Tris–HCl or Na–HEPES, pH 8.5, and 0.5 *M* NaCl
Cleavage buffer: 20 m*M* Tris–HCl or Na–HEPES, pH 7.0, and 0.5 *M* NaCl
pBSC vectors: Available from New England Biolabs on request
Chitin beads: Available from New England Biolabs. Prepare a 10- to 20-ml column for a 1-liter culture and equilibrate at 4° with 10 volumes of the lysis/column buffer.

Procedures

Cell growth, expression, and purification procedures are essentially the same as protocols described for the modified *Sce* VMA intein. The cell extract is prepared in lysis/column buffer at pH 8.5, which inhibits cleavage during purification. The CBD–*Ssp* mini-intein (Cys1Ala)–target protein expressed from a *Ssp* DnaB mini-intein C-terminal cleavage vector can be isolated from an induced *E. coli* cell extract by binding to chitin resin (10–20 ml resin per liter of induced cell culture). A 20-ml chitin column should be washed with 500 ml of lysis/column buffer followed by flushing with 100 ml of cleavage buffer at pH 7.0. At this pH, overnight incubation at 25° will cleave the peptide bond between the intein and the target protein. Except for cleavage, all steps should be carried out at 4° with precooled buffers as rapidly as possible (4–6 hr) to prevent premature cleavage. An example follows.

Escherichia coli ER2566 transformed with pBSL is induced with 0.3 m*M* IPTG at 30° for 3 hr. Cells are harvested and disrupted in 10 m*M* Tris, pH 8.5, 0.5 *M* NaCl. The soluble fraction from 1 liter of culture is loaded at 4° onto a chitin column (10 ml). The column is washed at 4° with 20 column volumes of 10 m*M* Tris, pH 8.5, containing 0.5 *M* NaCl, flushed with 10 m*M* Tris, pH 7.0, 0.5 *M* NaCl, and incubated in this buffer at 25° for 16 hr. The cleaved protein is then eluted with the same buffer.

Critical Parameters and Limitations

The N-terminal sequence of a target protein affects both *in vivo* and *in vitro* cleavage of a fusion protein. These effects are to some extent at odds, as *in vivo* cleavage reduces the yield of fusion protein whereas *in vitro* cleavage is essential for isolating the pure target protein. The effect of the N-terminal residue of a target protein has been investigated by examining all 20 naturally occurring amino acid in the +1 position.[8] The presence of Ser, Cys, Ala, or His in the +1 position leads to the most rapid C-terminal cleavage, but target proteins with N-terminal Gly, Met, Glu, Asp, Trp, Phe, Tyr, Val, and Thr can also be cleaved from the fusion protein at reasonable rates. However, fusion proteins with Gln, Asn, Leu, Ile, Arg, Lys, and Pro in the +1 position do not cleave efficiently. Induction of expression conducted overnight at 15° or for 3 hr at 30° resulted in similar patterns of C-terminal cleavage, indicating that *in vivo* cleavage is not strikingly affected by temperature and that *Ssp* DnaB intein fusion proteins expressed at either low or high temperature are competent to undergo cleavage, in contrast to fusion proteins based on the *Mth* RIR1 intein described in the preceding section. C-extein residues at positions +2 to +5 can also influence protein splicing and cleavage. Replacement of the +2 residue (Ile) in the native DnaB C-extein with Pro inhibited protein splic-ing.[8] C-exteins with Cys-Gly at the +1 and +2 positions displayed greater than 90% of *in vivo* cleavage, whereas a Cys-Arg sequence in the same position reduced *in vivo* cleavage to 10–50%.[24] The C-terminal cleavage reaction is most favored at pH 6.0–7.5, being inhibited substantially below pH 5.5 or above pH 8.0.[8] The cleavage of purified fusion proteins with Ile, Thr, Gly, Trp, Tyr, or Phe at the +1 position showed essentially the same pH profile. Advantage can be taken of this narrow pH response by purifying fusion proteins at pH 8.5 and 4°, where little cleavage occurs, and carrying out the cleavage on the chitin column by overnight incubation at pH 7.0 and 25°. Studies with CBD–*Ssp* mini-intein–T4 DNA ligase fusion proteins showed that the cleavage reaction can be carried out at 4° for 3–5 days with comparable yields, which may be an advantage if the target protein is unstable.

Representative Results

The modified *Ssp* DnaB mini-intein has been used to purify T4 DNA ligase by the C-terminal cleavage reaction, using plasmid pBSL-C155, which encodes a tripartite fusion protein consisting of CBD, a modified *Ssp* mini-intein, and T4 DNA ligase.[8] Two liters of induced cultured *E. coli* ER2566 transformed with pBSL-C155 yielded 10 mg of cleaved T4 DNA ligase, which migrated as a single band on 12% SDS–PAGE stained with Coomassie blue.

Acknowledgment

The experimental work discussed in this article was carried out by the scientists at New England Biolabs, Inc., with the support and encouragement of Donald G. Comb.

[24] T. C. Evans and M.-Q. Xu, unpublished observations.

[25] Fusion Proteins Containing Cellulose-Binding Domains

By Jae-Seon Park, Hae-Sun Shin, and Roy H. Doi

Introduction

This article illustrates the construction of fusion proteins containing a cellulose-binding domain (CBD) and describes several potential uses of such fusion proteins for basic research and biotechnology.

CBDs are found as functional domains of many cellulolytic enzymes.[1] The primary function of the CBD is to bind the enzyme or enzyme complex to cellulose and facilitate the interaction of the catalytic site with the substrate. CBDs are found in a wide variety of bacterial and fungal cellulases, and to date have been classified into about 13 different families based on their amino acid sequence homologies.[2] Additional ones are being identified.

Most of the CBDs are part of active cellulolytic enzymes, but there are some CBDs that are part of the scaffolding protein of cellulosomes.[3,4] The common evolution of many of these CBDs is suggested by their conserved sequence homologies. However, significant sequence differences of CBDs also indicate that many types of CBDs have evolved independently and that their common feature is their ability to bind crystalline and/or amorphous forms of cellulose.

Properties of CBDs

An excellent review on the properties of CBDs and their applications for biotechnology has been presented by the foremost group working in this area.[2] Thus only a few salient features of CBDs will be presented here.

The CBDs that have been studied to date exist in two general forms: one illustrated by the CBDs of fungal cellulases and the other by the CBDs of bacterial cellulases. The fungal CBD domains are generally smaller (33–36 amino acids) than the bacterial CBD domains, which usually contain about 90–180 amino acids.[1] A number of their sequences have been determined, which can be retrieved from the data banks.

The affinity of CBD for crystalline forms of cellulose is very high. The dissociation constants (K_d) of CBDs for cellulose are in the range of 1.0×10^{-5} M to 2.9×10^{-7} M and depend on the actual cellulosic substrate that is used.[2,5] For *Clostridium cellulovorans* CBD the affinity was greater for crystalline-enriched forms of cellulose and, interestingly, the CBD had an equally high binding affinity for chitin.[5] The tight binding of CBD to a cellulose matrix can be an advantage or disadvantage; this problem is discussed in the next section.

[1] P. Tomme, R. A. J. Warren, and N. R. Gilkes, *Adv. Microb. Physiol.* **37,** 1 (1995).
[2] P. Tomme, A. Borason, B. McLean, J. Kormos, A. L. Creagh, K. Sturch, N. R. Gilkes, C. A. Haynes, R. A. J. Warren, and D. G. Kilburn, *J. Chromatogr. B* **715,** 283 (1998).
[3] O. Shoseyov, M. Takagi, M. Goldstein, and R. H. Doi, *Proc. Nat. Acad. Sci. U.S.A.* **89,** 3483 (1992).
[4] U. T. Gerngross, M. P. M. Romaniec, T. Kobayashi, N. S. Huskisson, and A. L. Demain, *Mol. Microbiol.* **8,** 325 (1993).
[5] M. A. Goldstein, M. Takagi, S. Hashida, O. Shoseyov, R. H. Doi, and I. H. Segel, *J. Bacteriol.* **175,** 5762 (1993).

CBD binds to cellulose at a moderately wide range of pHs. The pH range of *C. cellulovorans* CBD binding was reported to be pH 3.5 to pH 9.5, although pH 7 is used for most binding studies.[5]

Potential Uses for CBD Fusion Proteins

The versatility of CBD for the construction of fusion proteins is that it can be attached to the N or C terminus of the target protein. The fusion protein so expressed in *Escherichia coli* can then be purified by affinity to a cellulose matrix column. One problem is the high affinity of the fusion protein for cellulose, which has both advantages and disadvantages.

If the fusion protein is to be used affixed to a cellulose column, then the high affinity to cellulose is an advantage and not a problem. The affixed fusion protein-cellulose column can be used as a bioreactor, e.g., a CBD-glucose isomerase column can be used to convert glucose to fructose.

However, if the fusion protein is to be released from the cellulose column, buffers containing urea or guanidium hydrochloride or some other denaturing reagent must be used to release the fusion protein. The successful renaturation of the target protein fused to CBD will depend on the property of the target protein.

In another option to either release or use the fusion protein as a purification procedure for a target protein, a factor Xa site can be incorporated between the CBD and the target protein, and after affixing the fusion protein to a cellulose matrix, treatment with factor Xa will release the target protein. This is an ideal way to purify small peptides that may be difficult to purify from crude extracts or to purify enzymes that are sensitive to chemical procedures used in other purification methods.

The following procedure describes a model system in which a CBD from the scaffolding protein of *C. cellulovorans*[3] is used to make a fusion protein with an endoglucanase B (EngB) lacking a CBD.[6] The method describes the construction of a gene with CBD at the N terminus (CBD-EngB) or the C terminus (EngB-CBD) (Fig. 1), the expression of the fusion proteins in *E. coli,* and the subsequent use of the fusion protein as a purification procedure for EngB and as a bioreactor column.

Plasmids for Construction of Fusion Proteins

Construction of Plasmids pLTCM and pMC for Cloning and Expression of CBD-Containing Fusion Proteins

Two plasmids, pLTCM and pMC, are constructed for the synthesis of fusion proteins containing CBD at the N or C terminus of the fusion proteins (Figs. 2 and 3). Plasmid pLTCM contains the *Eco*RI–*Nco*I fragment carry-

[6] F. Foong, T. Hamamoto, O. Shoseyov, and R. H., *J. Gen. Microbiol.* **137,** 1729 (1991).

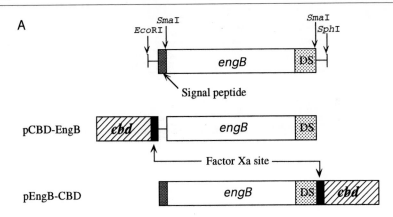

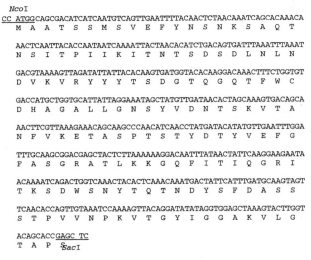

FIG. 1. Construction of CBD fusion proteins with endoglucanase B (EngB). (A) Plasmids pCBD-EngB and pEngB-CBD are constructed for the expression of CBD-Xa-EngB and EngB-Xa-CBD fusion proteins. For the subcloning of the *engB* gene into pLTCM or pMC, *Sma*I sites were constructed at the N terminus and C terminus of EngB using site-directed mutagenesis. The EngB contains a duplicated segment (DS) at its C terminus. The black box indicates the portion having the factor Xa recognition site. (B) The sequence of the CBD from the cellulose-binding protein A (CbpA) of *C. cellulovorans* [O. Shoseyov, M. Takagi, M. Goldstein, and R. H. Doi, *Proc. Nat. Acad. Sci. U.S.A.* **89,** 3483 (1992)].

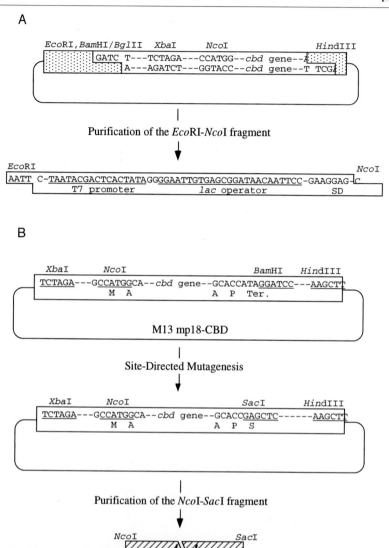

FIG. 2. Construction of plasmid pLTCM for production of CBD-proteins. (A) The *Eco*RI–*Nco*I fragment carrying the T7 promoter is prepared from pET-CBD. (B) The *Nco*I–*Sac*I fragment encoding the CBD is prepared using site-directed mutagenesis. (C) The linker carrying the Xa factor recognition site is inserted into pUC18 to prepare ligation with the *Eco*RI–*Nco*I fragment and the *Nco*I–*Sac*I fragment. (D) Plasmid pLTCM is constructed for the synthesis of fusion proteins containing CBD at the N terminus of the fusion proteins (CBD-proteins).

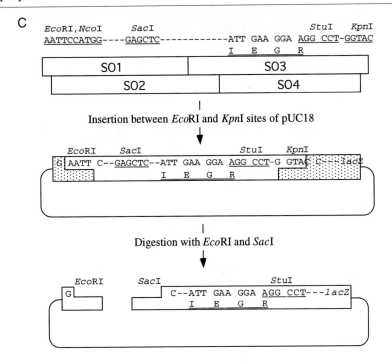

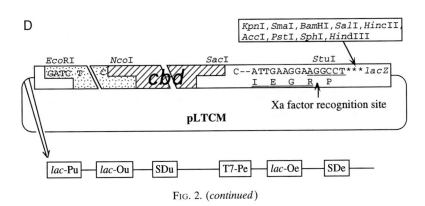

FIG. 2. (*continued*)

ing the T7 promoter of pET-8c, the *Nco*I–*Sac*I fragment encoding the CBD from the *C. cellulovorans* CbpA (Fig. 1B), the linker encoding the Xa factor recognition site (Ile-Glu-Gly-Arg), and restriction enzyme sites (*Eco*RI and *Kpn*I sites at the ends, and internal *Nco*I, *Sac*I, and *Stu*I sites). The *Eco*RI–*Nco*I fragment carrying the T7 promoter of pET-8c was

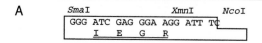

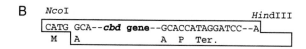

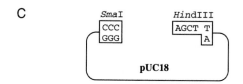

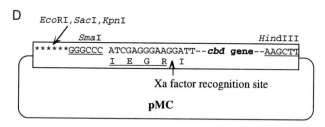

Fig. 3. Construction of plasmid pMC for production of protein-CBDs. (A) The linker carrying the factor Xa recognition site is prepared. (B) The NcoI–HindIII fragment encoding the CBD protein is prepared from pET-CBD. (C) pUC18 is digested with SmaI and HindIII for ligation with the linker and the NcoI–HindIII fragment. (D) Plasmid pMC is constructed for the synthesis of fusion proteins containing CBD at the C terminus of the fusion proteins (protein-CBDs).

prepared by insertion of the BglII–HindIII fragment of pET-CBD[5] between the BamHI and the HindIII sites of pMal-c2 and purification of the EcoRI–NcoI fragment after digestion of the constructed plasmid with EcoRI and NcoI (Fig. 2A). The NcoI–SacI fragment was prepared by creation of a SacI site at the translation termination codon of the cbd gene. For elimination of the termination codon of the cbd gene and creation of the SacI site, the XbaI–HindIII fragment of pET-CBD encoding the CBD protein is subcloned into M13mp18 and site-directed mutagenesis is conducted (Fig. 2B).

The linker carrying the Xa factor recognition site is constructed by the synthesis of four kinds of oligonuclotides and annealing of the synthesized oligonucleotides. As the first step for the construction of pLTCM, the linker containing the Xa factor is inserted between the *Eco*RI and the *Kpn*I sites of pUC18. For the construction of plasmid pLTCM, the *Eco*RI–*Nco*I fragment carrying the T7 promoter and the *Nco*I–*Sac*I fragment encoding CBD are ligated with the linker inserted pUC18 that had been digested with *Eco*RI and *Sac*I (Figs. 2C and 2D). In this construction, pLTCM has a T7 promoter (T7-Pe), *lac* operator (lac-Oe), and SD sequence (SDe) from pET-8c in addition to the *lac* promoter (*lac*-Pu), *lac* operator (*lac*-Ou), and SD sequence (SDu) from pUC18 (Fig. 2D). For the construction of CBD-fusion proteins, the *Stu*I site is used for cloning the gene encoding the N terminus of the target protein and several restriction sites (*Kpn*I, *Sma*I, BamHI, *Sal*I *Hinc*II, *Acc*I, PstI, SphI, and *Hind*III sites) can be used as multicloning sites for cloning genes encoding the C terminus of the target protein (Fig. 2D). The constructed pLTCM is used for the synthesis of fusion proteins containing CBD at the N terminus of the fusion proteins.

For plasmid pMC, the linker encoding the Xa factor recognition site (Ile-Glu-Gly-Arg) and carrying restriction enzyme sites (*Sma*I and *Nco*I sites at the ends and an internal *Xmn*I site) and the *Nco*I–*Hind*III fragment encoding the CBD protein from pET-CBD are prepared. The linker and the *Nco*I–*Hind*III fragment are ligated between *Sma*I and *Hind*III sites of pUC18 for the construction of pMC. For the construction of fusion protein, *Eco*RI, *Sac*I, and *Kpn*I sites can be used for cloning the gene encoding the N terminus of the target protein and the *Sma*I site can be used for cloning the gene encoding the C terminus of the fused protein (Fig. 3D). The constructed pMC will be used for the synthesis of fusion proteins containing CBD at the C terminus of the fusion protein.

Construction of CBD Fusion Proteins with Endoglucanase B (CBD-EngB): A Model System

Plasmids pCBD-EngB and pEngB-CBD are constructed for the expression of CBD fusion proteins such as CBD-Xa-EngB and EngB-Xa-CBD (Fig. 1A). Restriction enzyme sites and the termination codon within the *engB* gene are created by site-directed mutagenesis for subcloning into pLTCM and pMC. The *Sma*I site is created in the gene at the 23rd and 24th amino acid residues to construct N-terminal CBD fusion proteins. The *Sma*I site is created in the gene at the 441 st amino acid to construct C-terminal CBD fusion proteins. The *Sma*I–*Sph*I fragment encoding EngB (from the 24th amino acid to the 441 st amino acid) is subcloned between *Stu*I and *Sph*I sites of pLTCM to construct pCBD-EngB. The *Eco*RI–*Sma*I fragment encoding EngB (from the 1st amino acid to the 441st amino acid) is subcloned into the *Eco*RI and *Sma*I sites of pMC to construct pEngB-CBD.

Properties of pLTCM

Plasmid pLTCM, constructed for the N-terminal CBD fusion proteins (CBD-Xa-target), has the fusion gene that codes for CBD-Xa-LacZ α-peptide fusion protein. *Escherchia coli* TG1 is deficient in β-galactosidase activity due to a deletion in the genomic copy of the *lacZ* gene. If the CBD does not interfere with the function of the α-peptide (LacZ), the deletion in the *lacZ* gene can be complemented by addition of a functional α-peptide encoded by pLTCM. Blue *E. coli* TG1 colonies were detected after transformation with pLTCM. This allowed for blue–white color selection of recombinant colonies supplemented with X-Gal and IPTG.

Expression and Purification of Fusion Proteins in *E. coli*

Preparation of Fusion Proteins

Plasmids pCBD-EngB and pEngB-CBD are used to transform *E. coli* TG1 or *E. coli* JM109(DE3). Plasmid-containing cells are incubated overnight with shaking in LB medium containing 100 μg/ml ampicillin at 37°. The overnight cultures are diluted 100 times in 40 ml of LB medium containing 100 μg/ml ampicillin in 250-ml flasks. The cultures are grown at 37° with shaking to a Klett reading of 160 (green filter). At this point, isopropyl-β-D-thiogalactopyranoside (IPTG) is added to a final concentration of 0.5 mM. After 4 hr, the cells are harvested by centrifugation. For lysis, *E. coli* cells are resuspended in 10 ml of TSE buffer (20 mM Tris, 200 mM NaCl, 1 mM EDTA, pH 8). Lysozyme is added to a final concentration of 1 mg/ml. The solution is kept on ice water for 30 min with occasional mixing. Triton X-100 is added to a final concentration of 0.1%. The solution is kept on ice water for 10 min with occasional mixing. DNase (10 units) is added to reduce the viscosity of the solution. The solution is kept on ice water for 10 min with occasional mixing. Cellular debris is removed by centrifugation (15 min at 10,000g and 4°). The supernatant is used for the test of purification and for the immobilization system.

Purification of Fusion Protein and Target Protein

Cellulose (fibrous, medium) is suspended in TSE buffer. The 1-ml volume of cellulose is packed in a column. The solution containing fusion protein (CBD-Xa-EngB or EngB-Xa-CBD) is applied to the cellulose column. The column is washed with 20 ml of TSE buffer. To check its binding specificity, proteins bound to the cellulose are eluted from a column by use of SDS–PAGE sample loading buffer. The eluted protein showed the molecular weight of CBD-Xa-EngB or EngB-Xa-CBD. Although CBD-fusion proteins have a binding specificity to the cellulose, a method has not been developed to get free CBD-fusion proteins from the cellulose column without loss of some activity.

Because CBD-fusion proteins contain an Xa factor recognition site between CBD and the target protein, the target protein can be collected from the cellulose column containing bound CBD-fusion protein by using Xa factor to cleave the recognition site. For purification of the target protein EngB, fusion proteins (CBD-Xa-EngB or EngB-Xa-CBD) bound to the cellulose columns are incubated with 1 ml of 500 times diluted factor Xa solution. The columns are kept overnight at room temperature. The flow-through fractions are collected, and the cleaved protein (EngB) is eluted from the column by three stepwise washes with 1 ml of TEN buffer. Analysis of the flow-through fractions with SDS–PAGE confirm that EngB protein is eluted from the column after overnight incubation.

Good endoglucanase activity is recovered after factor Xa treatment, indicating that the CBD-Xa-target or the target-Xa-CBD attached to a cellulose matrix can be used to purify target proteins (Table I). This is an ideal method to purify small peptides, as it is usually difficult to purify them from cell extracts. By making a fusion protein with CBD, one can bind the CBD-Xa-target peptide or the target peptide-Xa-CBD to a cellulose column, wash the column with buffer to get rid of contaminating proteins, and then, after treatment with factor Xa, elute off the desired target peptide.

TABLE I
ACTIVITY OF CBD FUSED TO ENGB

Fusion protein	CMCase activity[a]
CBD-Xa-EngB	+
EngB-Xa-CBD	+
CBD-Xa-EngB bound to cellulose	+
EngB-Xa-CBD bound to cellulose	+
EngB released from CBD-Xa-EngB[b]	+
EngB released from EngB-Xa-CBD[b]	+

[a] The plus sign indicates that activity was recovered.
[b] After treatment with factor Xa.

Immobilization and Recovery of Target Protein

Activity of Immobilized Fusion Protein Bound to Cellulose

Endoglucanase activity of fusion protein CBD-Eng was lost when 8 *M* urea solution was used for the elution of purified protein from cellulose columns. Although some activity was confirmed after elution with 1% triethylamine and neutralization with acetic acid, a significant amount of activity was lost. These results suggest that elution of CBD-fusion protein with various reagents may not always be suitable for the recovery of active target proteins.

To overcome this problem, the use of immobilized CBD-fusion protein is an excellent alternative. For immobilization of a fusion protein, 10 mg of cellulose (Sigmacell, type 100) is suspended in 1 ml of TSE buffer. Cellulose is collected by centrifugation and resuspended with the cell extract containing CBD-Xa-EngB or EngB-Xa-CBD fusion protein. After a 1-hr incubation with slow shaking at room temperature, the supernatant is removed by centrifugation. The cellulose pellet is washed five times with 1 ml of TSE buffer and resuspended with 1 ml of TSE buffer. One hundred microliters of the resuspended mixture is distributed into Eppendorf tubes and the supernatant is removed by centrifugation. The CBD-Xa-EngB bound pellet is resuspended with 1% CMC solution. After incubation at 37° for 10, 30, 60, and 90 min, the supernatant is obtained by centrifugation and checked for CMCase activity (Table I). The reducing sugar is monitered by the DNS method. Each CBD-Xa-EngB-bound cellulose pellet is washed with TE buffer and resuspended with SDS–PAGE sample loading buffer to confirm that CBD-Xa-EngB protein was bound to the cellulose pellet. These results indicate that the target protein bound to the cellulose retains its activity and that the CBD fusion protein is bound stably to a cellulose matrix. Also, the same results are obtained in the experiment with EngB-Xa-CBD fusion protein.

The Immobilized Fusion Protein as a Bioreactor Column

The immobilized fusion protein bound to cellulose can be made into a column and used as a bioreactor. The tight binding of the fusion protein to cellulose is stable and allows the column to be used for at least 48 hr without loss of activity. The stability of the column would depend on the fusion target protein constructed and would have to be studied for each target protein or enzyme. In our case the column consisted of a cellulose matrix with 3 μg of CBD-EngB bound/mg of cellulose.

TABLE II
EXAMPLES OF SOME CBD FUSION PROTEINS

CBD fusion proteins	References[a]
CBD-human hsp60 epitope	(1)
CBD-alkaline phosphatase	(2)
CBD-human interleukin 2	(3)
CBD-CTAP-III (heparanase)	(4)
Streptavidin-CBD	(5)
Factor Xa-CBD	(6)
β-Glucosidase-CBD	(7, 8)

[a] Key to references (1) E. Shpigel, D. Elias, I. R. Cohen, and O. Shoseyov, *Prot. Expr. Purif.* **14,** 185 (1998). (2) J. M. Greenwood, N. R. Gilkes, R. C. Miller, D. G. Kilburn, and R. A. J. Warren, *Biotech. Bioeng.* **44,** 1295 (1994). (3) E. Ong, J. Alimonti, J. M. Greenwood, R. C. Miller, Jr., R. A. J. Warren, and D. G. Kilburn, *Bioseparation* **5,** 95 (1995). (4) M. Rechter, O. Lider, L. Cahalon, E. Baharav, M. Dekel, D. Seigel, I. Vlodavsky, H. Aingorn, I. R. Cohen, and O. Shoseyov, *Biochem. Biophys. Res. Commun.* **255,** 657 (1999). (5) K. D. Le, N. R. Gilkes, D. G. Kilburn, R. C. Miller, J. N. Saddler, and R. A. J. Warren, *Enz. Microb Technol.* **16,** 496 (1994). (6) Z. Assouline, H. Shen, D. G. Kilburn, and R. A. J. Warren, *Prot. Engin.* **6,** 787 (1993). (7) E. Ong, N. R. Gilkes, R. C. Miller, R. A. J. Warren, and D. G. Kilburn, *Enz. Microb. Tech.* **13,** 59 (1991). (8) D. H. Ahn, H. Kim, and M. Y. Pack, *Biotechnol. Lett.* **19,** 483 (1997).

Fusion Proteins That Have Been Constructed

The construction of several different CBD-fusion proteins have been proposed[7] and reported[2,8] (Table II). CBD fusion proteins can be used for a variety of purposes, e.g., to purify proteins especially small peptides, to be used as a bioreactor column with an inexpensive cellulose matrix, and to be used for diagnostic strips for medical, industrial, bioremedial, and other diagnostic purposes.

Acknowledgment

The research was supported in part by Department of Energy Grant DE-FG03-92ER20069.

[7] R. H. Doi, J.-S. Park, C.-C. Liu, L. M. Malburg, Jr., Y. Tamaru, A, Ichiishi, and A. Ibrahim, *Extremophiles* **2,** 53 (1998).
[8] E. Shpigel, D. Elias, I. R. Cohen, and O. Shoseyov, *Prot. Expr. Purif.* **14,** 185 (1998).

[26] Biotinylation of Proteins *in Vivo* and *in Vitro* Using Small Peptide Tags

By MILLARD G. CULL and PETER J. SCHATZ

Introduction

Biotinylation is one of the most commonly used means of labeling proteins for easy detection, immobilization, and purification. The utility of biotin as a tag is a result of the interaction of biotin with the proteins avidin and streptavidin that bind biotin with very high affinity and specificity.[1] Historically, biotinylation has been carried out using chemical reagents that lack site specificity, which can lead to inactivation of some biological molecules.[2] This problem was initially addressed by the Cronan group, who used fusion of domains of naturally biotinylated proteins to achieve a generic method of labeling chimeric proteins with biotin at a single site.[3] This specificity was possible because of the extreme selectivity of biotin holoenzyme synthetases, also known as biotin ligases, enzymes that catalyze the covalent attachment of biotin to the ε amino group of particular lysine residues in target proteins.[4]

The clear advantage of protein domain-mediated site-specific biotinylation is the production of uniform reagents with predictable behavior. The disadvantage is the need to link a domain of at least 66 amino acids to the fusion partner,[5] which could possibly cause steric interference with the activity of the linked protein. We addressed this disadvantage through the use of peptide libraries to find smaller peptides that were substrates for the *Escherichia coli* biotin holoenzyme synthetase BirA.[6] These BirA substrate peptides have been used by a number of investigators to produce biotinylated fusion proteins for a variety of purposes.[7-14] Another publication[15]

[1] M. Wilchek and E. A. Bayer, *Methods Enzymol.* **184** (1990).
[2] E. A. Bayer and M. Wilchek, *Methods Enzymol.* **184,** 138 (1990).
[3] J. E. Cronan, Jr., *J. Biol. Chem.* **265,** 10327 (1990).
[4] D. Samols, C. G. Thornton, V. L. Murtif, G. K. Kumar, F. C. Haase, and H. G. Wood, *J. Biol. Chem.* **263,** 6461 (1998).
[5] J. Stolz, A. Ludwig, and N. Sauer, *FEBS Lett.* **440,** 213 (1998).
[6] P. J. Schatz, *Bio/Technology* **11,** 1138 (1993).
[7] J. D. Altman, P. A. H. Moss, P. J. R. Goulder, D. H. Barouch, M. G. McHeyzer-Williams, J. I. Bell, A. J. McMichael, and M. M. Davis, *Science* **274,** 94 (1996).
[8] F. Crawford, H. Kozono, J. White, P. Marrack, and J. Kappler, *Immunity* **8,** 675 (1998).
[9] S. Duffy, K.-L. Tsao, and D. S. Waugh, *Anal. Biochem.* **262,** 122 (1998).

describes a more complete characterization of these peptides, including identification of a specific sequence that is biotinylated more efficiently than two others examined. A truncation series of chemically synthesized peptides allowed identification of a 14 residue peptide as the minimal substrate for BirA, which, interestingly, shows biotinylation kinetics similar to those of the natural protein substrate. Thus, these peptides combine the attributes of small size with efficient biotinylation to yield an extremely useful method for labeling proteins.

This article gives practical advice to the investigator who wants to biotinylate a particular protein through the use of small peptide tags, which can be done either *in vivo* or *in vitro*. We describe reaction conditions for *in vitro* biotinylation, vectors and strains useful for *in vivo* biotinylation, and methods for measuring biotinylation efficiency.

Selection of a Sequence and Site of Fusion

Examination of three different sequences[6] revealed that fusion proteins containing one of these sequences are biotinylated with substantially faster kinetics.[15] This is true whether the peptide is fused to the N terminus or the C terminus of MBP, demonstrating that this is an intrinsically better sequence. A chemically synthesized truncation series of this sequence revealed that a 14-mer "minimal" substrate (GLNDIFEAQKIEWH) is biotinylated at a rate within twofold that of the natural protein substrate. Although this 14-mer works quite well, we recommend using a slightly extended 15-mer, termed the AviTag (GLNDIFEAQKIEWHE), that is consistently biotinylated at a rate slightly better than that of the natural substrate.[15]

Fusion works well at either the N terminus or the C terminus of a target protein. At the N terminus, an ATG initiation condon is obviously necessary, which, for expression in *E. coli*, we generally follow with an Ala or Ser codon to confer proteolytic stability if the N-terminal Met is removed by methionine aminopeptidase.[16,17] For fusion at the C terminus of a target

[10] D. R. Kim and C. S. McHenry, *J. Biol. Chem.* **271**, 20690 (1996).

[11] P. Saviranta, T. Haavisto, P. Rappu, M. Karp, and T. Lovgren, *Bioconj. Chem.* **9**, 725 (1998).

[12] P. A. Smith, C. Tripp, E. A. DiBlasio-Smith, Z. Lu, E. R. LaVallie, and J. M. McCoy, *Nucleic Acids Res.* **26**, 1414 (1998).

[13] H. Tatsumi, S. Fukuda, M. Kikuchi, and Y. Koyama, *Anal. Biochem.* **243**, 176 (1996).

[14] K.-L. Tsao, B. DeBarbieri, H. Michel, and D. S. Waugh, *Gene* **169**, 59 (1996).

[15] D. Beckett, E. Kovaleva, and P. J. Schatz, *Protein Sci.* **8**, 921 (1999).

[16] F. Sherman, J. W. Stewart, and S. Tsunasawa, *Bioessays* **3**, 27 (1985).

[17] J. W. Tobias, T. E. Shrader, G. Rocap, and A. Varshavsky, *Science* **254**, 1374 (1991).

protein, the 15-mer is generally connected to the protein through a short Gly-Gly linker, although it is not clear that the linker is necessary. When the AviTag is fused to a location on the protein other than the termini, the results will likely be extremely variable, depending on whether the 15-mer is folded in such a way that BirA can recognize it as a substrate.

In Vitro Biotinylation

Proteins containing the AviTag can be efficiently biotinylated *in vitro* using the BirA enzyme from *E. coli* (available commercially from Avidity, L.L.C., Denver, CO). This option is desirable when expression of soluble target protein in *E. coli* is inefficient, as in the case of proteins that form inclusion bodies when expressed in bacteria, or when using an expression system not yet developed for efficient biotinylation, e.g., in yeast or in mammalian cells. Whereas the *in vitro* reaction can be performed in a complex mixture of proteins, a more desirable option is to use *in vitro* biotinylation on purified proteins to minimize protease contamination and concentrate the substrate to achieve a more efficient reaction. Because the protein may be biotinylated poorly or not at all during expression, a protein purification scheme that does not depend on biotinylated protein is required. For instance, AviTags fused to MHC molecules, a common candidate for biotinylation, are expressed as inclusion bodies in *E. coli* for easy purification, followed by biotinylation *in vitro*.[7]

When a convenient purification is not possible, the AviTag can be used in conjunction with a purification tag such as Hexa-His[10] or Flag (Sigma, St. Louis, MO).

The most common problem in the *in vitro* biotinylation of purified proteins is the presence of proteases that can degrade the AviTag during purification. Protease inhibitors can be used, and we have found that a combination of 0.1 mM phenylmethylsulfonyl fluoride (PMSF), 1 μg/ml leupeptin, and 0.7 μg/ml pepstatin can be effective. Other protease inhibitors have not been tested. If protease contamination is a problem, a further purification step is also recommended. If a sizing column is used to remove protease contamination, the column should be equilibrated and run in a low salt, low buffer solution, such as 10 mM Tris, pH 7.5, or 10 mM Bicine at pH 8.3 if possible. Regardless of the purification scheme used, the substrate protein should be in a low salt, mildly buffered solution for best results.

The *in vitro* reaction has been optimized using the peptide LCDIFESQ-KIEWHSA.[6] *In vitro* biotinylation reactions show that a number of reagents commonly present in biological buffer can inhibit the activity of BirA enzyme. These include NaCl, glycerol, and ammonium sulfate (Fig. 1). Other reagents that are commonly found in biological buffers, such as

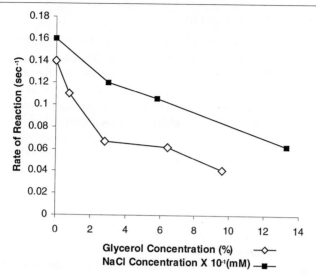

FIG. 1. Biotin holoenzyme synthetase reaction conditions. The rate of BirA activity was determined at varying glycerol and NaCl concentrations.

0.1% Tween 20, 5 mM mercaptoethanol, or 100 μM PMSF, do not affect BirA enzyme.

The amount of BirA enzyme added to the reaction may need to be varied to achieve biotinylation of the substrate within a reasonable time frame. More enzyme may be required if the peptide sequence is not optimal. Typically, 2.5 μg BirA enzyme will completely (>95%) biotinylate 10 nmol of substrate (at 40 μM) in 30–40 min at 30°. Lower substrate concentrations will lengthen the amount of time the reaction takes to reach completion. For example, whereas a substrate at 40 μM will be biotinylated in ~30 min, at 4 μM it will take ~5 hr using the same amount of BirA enzyme. To perform the biotinylation in 30 min (10 times faster), it is necessary to add 10 times more enzyme to the reaction mix.

The following conditions have been found to be optimal in reaction volumes from 20 μl to 10 ml both for the peptide substrate and for the tested AviTag fusion proteins: 38 μM protein substrate (stored in 10 mM Tris–HCl, pH 7.5), 50 mM Bicine buffer (pH 8.3), 10 mM magnesium acetate, 10 mM ATP, 40 μM biotin, 25 μg/ml BirA enzyme. The protein substrate is fully biotinylated (>95%) in 30 min at 30°. Less enzyme can be used for longer incubation times, and the reaction can be performed at room temperature overnight as is often done with MHC-AviTag molecules.[8]

Biotinylation of Proteins *in Vivo*

Most, but not all, AviTag fusion proteins can be efficiently biotinylated *in vivo,* saving time and eliminating the need for BirA enzyme. Biotinylated proteins can be purified efficiently by exploiting the high degree of specificity between avidin and biotin through the use of monomeric avidin columns, available from Pierce Chemical Company or Promega. Regardless of the expression vector and host used, efficient *in vivo* biotinylation of protein-AviTag fusions requires overexpression of BirA, as described later. In order to overproduce fusion proteins that would ordinarily be biotinylated inefficiently because they form insoluble inclusion bodies in *E. coli,* two alternatives are possible. Baculovirus vectors have been constructed that coexpress the fusion protein and BirA in *Spodoptera frugiperda* insect cells, avoiding *E. coli* expression altogether.[9] Alternatively, solubility in *E. coli* can be increased by using vectors that fuse the AviTag and the protein in question to thioredoxin.[12,18]

In *E. coli* with endogenous levels of the BirA enzyme, AviTag proteins expressed in low or moderate amounts are typically biotinylated at about 10–30% efficiency. Highly overexpressed proteins are biotinylated at only 2–6% in strain ARI 814, a robust strain derived from *E. coli* B that expresses endogenous levels of the BirA enzyme.

Biotinylation efficiency can be increased in *E. coli* by coexpressing the BirA enzyme with the overproduced protein fusion as a dicistronic message or expressing the BirA from its own promoter on the same plasmid as the overproduced fusion protein.[12,14] Still another strategy to boost intracellular BirA levels in *E. coli* employs a ColEl-compatible, low-copy number plasmid (Avidity's $_p$BirAcm, also called $_p$JS169) engineered to overexpress the BirA enzyme from a strong Tac promoter. the $_p$BirAcm plasmid is compatible with ColEl-derived plasmids such as $_p$BR322 and $_p$UC vectors constructed to express the biotinylation peptide fusions. Examples of plasmids useful for making biotinylated fusion proteins are the $_p$AN and $_p$AC vectors shown in Fig. 2, which are useful, respectively, for making N-terminal and C-terminal fusions to proteins of interest.

The $_p$BirAcm plasmid can be maintained in the overnight culture with low levels of chloramphenicol (10 μg/ml) added along with the antibiotic used to maintain the substrate expression vector. Induction conditions may need to be optimized for individual proteins and expression vectors. The following induction conditions work very well for the $_p$MAL-c vector (New England Biolabs, Beverly, MA) with a C-terminal AviTag fusion to the

---

[18] J. Tucker and R. Grisshammer, *Biochem. J.* **317,** 891 (1996).

maltose-binding protein (MBP) in strain AVB101 (Avidity, Denver CO), an *E. coli* B-derived strain that contains $_p$BirAcm.

TYH Media, 1 Liter

20 g tryptone
10 g yeast extract
11 g HEPES
5 g NaCl
1 g MgSO$_4$
pH 7.2–7.4 with KOH

1. Grow a 10-ml overnight culture from a single colony or glycerol stock in TYH media supplemented with 10 μg/ml chloramphenicol and the appropriate antibiotic (e.g., ampicillin) needed to maintain the expression vector with shaking at 37°.

2. Place 5 ml of the overnight culture into 1 liter of TYH media in a baffled Fernbach flask with 100 μg/ml ampicillin. Note: chloramphenicol is not included.

3. Add 20 ml of a 20% sterile glucose solution (0.5% final concentration) and shake vigorously at 37°.

4. When the OD$_{600}$ of the mixture reaches 0.7, remove 1.5 ml as a preinduction sample.

5. Add 10 ml of 5 mM biotin solution. The biotin solution is made by adding 12 mg of d-biotin to warm (microwaved) 10 mM Bicine buffer (pH 8.3) and sterilizing the solution with a 0.2-μm filter.

6. Add 15 ml of 100 mM IPTG (1.5 mM IPTG final) to induce for 3 hr.

7. Pellet cells in 4 × 250-ml centrifuge bottles at 5900g for 10 min.

8. Pour off media from cell pellets and resuspend each pellet in 10 ml B-PER (Pierce Chemical Company) (40 ml total volume).

9. Shake on a rotary shaker for 10 min at room temperature.

10. Combine suspensions into one bottle and centrifuge at 16,300g for 15 min.

11. Save supernatant. Resuspend pellet in 25 ml B-PER. Shake on a rotary shaker for 10 min at room temperature.

12. Centrifuge at 16,300g for 15 min.

13. Add supernatant to that saved previously. Discard pellet.

14. Purify the MBP on an amylose column according to the manufacturer's instructions (New England Biolabs).

15. This purification yields greater than 50 mg of MBP that is 70–80% biotinylated, with greater than 60% of the total cellular protein as MBP.

C-terminal Biotin Avitag Vector
pAC-4

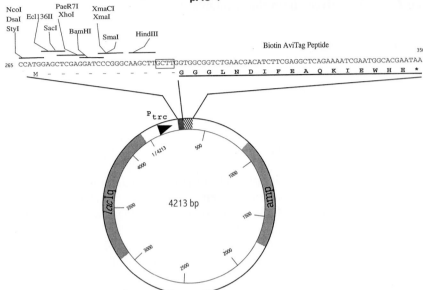

N-terminal Biotin AviTag Vector
pAN-4

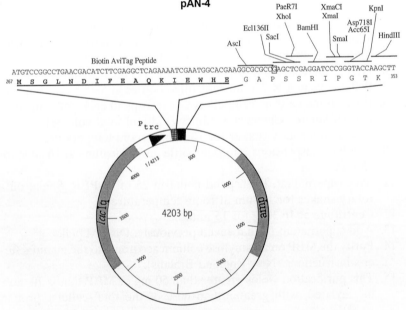

Determining the Extent of Biotinylation

A number of methods have been developed for determining the extent of biotinylation of a protein. These methods include a nondenaturing PAGE gel-shift assay, the 2-(4'-hydroxyazobenzene) benzoic acid (HABA) assay, the chromogenic biotin reaction timed assay (BRTA), and mass spectroscopy when the substrate is a peptide or small protein.[19]

The PAGE gel-shift assay is a quick, semiquantitative way of determining the extent of biotinylation that relies on the tight interaction of biotinylated proteins with avidin or streptavidin to form a complex of higher molecular weight than the unbiotinylated protein. The advantage of the gel-shift method is that it is relatively easy to perform, requiring only the equipment and materials needed to run a PAGE gel. The main drawback to this method is the imprecise visual quantitation of the amount of material that remains after biotinylated protein is shifted to a higher molecular weight. Use of a densitometer can improve quantitation.

The spectrophotometric HABA reaction is another method for quantitating biotinylated protein. The HABA assay shows a linear response from 0 to 7 nmol of biotin. There are drawbacks to using HABA in an assay to determine the extent of biotinylation of proteins biotinylated at a single site. The method requires copious amounts of sample protein, up to 300 μg of MBP-AviTag protein, to achieve a reasonable signal. All free biotin must be removed by desalting on a column or by using extensive dialysis.

Mass spectometry, such as MALDI-TOF, works well for peptides and small proteins, but is less useful for large proteins.

The chromogenic BRTA assay has a number of advantages. It is accurate with an error of less than 20%, requires only a small amount of test protein (pmoles), and tolerates free biotin in the sample.

BRTA Determination of the Extent of Biotinylation

This procedure follows a typical ELISA protocol. The extent of biotinylation of the test protein is measured by comparison with known quan-

[19] N. M. Green, *Biochem. J.* **94**, 23c (1965).

FIG. 2. Vectors for creating Biotin AviTag peptide fusions. C-terminal and N-terminal fusion vectors contain a multiple cloning site for insertion of the gene of interest. Three forms of each vector (pAN-4,5,6 or pAC-4,5,6) are available to accommodate the proper frame for fusion. The site of the reading frame difference between the different vectors is indicated by the sequence in the box. The promoter (P_trc) and Shine–Dalgarno are upstream from the *Nco*I site. The sequences of these vectors are available on the Avidity website (www.avidity.com).

tities of fully biotinylated MBP-AviTag protein. The standard protein and unknown proteins are adsorbed to the wells of a 96-well plate at known protein quantities. Biotin associated with the AviTag is detected by its interaction with streptavidin-conjugated alkaline phosphatase. Extraneous biotin is removed when the samples are washed after adsorption, eliminating exhaustive dialysis steps. The disadvantage of the BRTA is that some test proteins may not bind well to the plastic of the 96-well plate.

Reagents

Fully biotinylated MBP-AviTag fusion protein, 5 mg/ml (Avidity, L.L.C.)
Unbiotinylated MBP-AviTag fusion, 5 mg/ml (Avidity, L.L.C.)
Streptavidin-alkaline phosphatase, 2 mg/ml (Molecular Probes, Eugene, OR)

Solutions

Phosphate-buffered saline (PBS): 138 mM NaCl, 2.6 mM KCl in 10 mM potassium phosphate (pH 7.4)
PBST: PBS plus 0.05% Tween 20
Blocking solution: PBS plus 40 μg/ml bovine serum albumin (BSA)
Dilution buffer: PBS plus 0.15 μg/ml BSA
40 μg/ml streptavidin-alkaline phosphatase in PBS
TBS: 10 mM Tris, pH 7.5, 150 mM NaCl
Developing solution for alkaline phosphatase: p-nitrophenyl phosphate in diethanolamine buffer (Bio-Rad, Hercules, CA)

Equipment

Corning Costar Corp. (Corning, NY), 96-well polystyrene, high-binding, flat-bottom microtiter plates
Biokinetics Reader EL312e (Bio-Tek Instruments)

Procedure

1. For the standard curve, coat the wells of a 96-well polystyrene microtiter plate with 1 to 10 ng (1-ng increments) and 20 to 45 ng (5-ng increments) of fully biotinylated MBP-AviTag fusion protein using dilution buffer to bring the volume to 50 μl.
2. Coat wells of a 96-well polystyrene microtiter plate with 1.5 to 45 ng of the protein of interest. The proportion of contaminants in the

test protein samples must be determined to assess the degree of biotinylation accurately.

3. Forty-five nanograms of unbiotinylated MBP-AviTag should be added to one well as a negative control. Wells containing dilution buffer alone are used to blank the plate reader.

4. Allow the proteins to adsorb to the plate for at least 1 hr at room temperature with gentle rotational shaking.

5. Wash the plate four times with PBST.

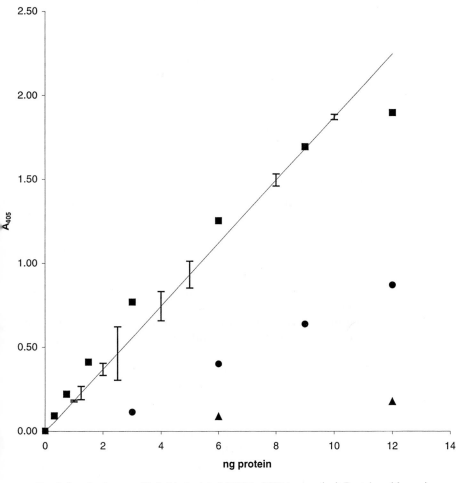

FIG. 3. Standard curve of fully biotinylated-MBP in BRTA assay (—). Proteins with varying degrees of biotinylation were evaluated in the BRTA assay. MBP-AviTag grown in AVB101 strain (■), MBP-AviTag that was 50% biotinylated (●), and MBP-AviTag grown in ARI814 (▲).

6. Incubate plate with 300 μl of blocking solution for 1 hr at room temperature to block nonspecific binding.
7. Rinse four times with PBST.
8. Pipette 50 μl of streptavidin-alkaline phosphatase solution into each well and incubate with gentle shaking for at least 1 hr at room temperature.
9. Wash four times with PBST.
10. Wash the plate twice with TBS.
11. Add 200 μl of the developing solution.
12. Develop the place with 200 μl of developing solution. Monitor absorption at 405 nm every minute to assure the linearity of the alkaline phosphatase reaction. Terminate the reaction by adding 50 μl of 2 M KOH.
13. A typical standard curve is shown in Fig. 3.

[27] Biotinylation of Proteins in Vivo: A Useful Posttranslational Modification for Protein Analysis

By John E. Cronan, Jr. and Kelynne E. Reed

The extensive use of biotin as a reagent in biotechnology, biochemistry, immunology, and cell biology has obscured the fact that biotin is a vitamin (vitamin H) required by all forms of life. Biotin is biologically active only when protein bound and is covalently attached to a class of important metabolic enzymes, the biotin carboxylases and decarboxylases.[1,2] These enzymes transfer CO_2 to and between metabolites using the biotin cofactor as a mobile carboxyl carrier and are catalyzed key steps of gluconeogenesis, lipogenesis, amino acid degradation, and energy transduction.[1] Biotin protein ligase (BPL), also known as holocarboxylase synthetase (EC 6.3.4.10), is the enzyme responsible for attaching biotin to these proteins. BPL catalyzes the posttranslational formation of an amide linkage between the carboxyl group of biotin and the ε amino group of a specific lysine residue

[1] D. Samols, C. G. Thornton, V. L. Murtif, G. K. Kumar, F. C. Haase, and H. G. Wood, *J. Biol. Chem.* **263**, 6461 (1988).
[2] J. R. Knowles, *Annu. Rev. Biochem.* **58**, 195 (1989).

of a carboxylase protein in a two-step reaction (Fig. 1). Genetic studies in microorganisms[3–7] and humans[8–10] indicate that each organism has a single BPL-encoding gene, a conclusion consistent with the complete genome sequences available to date.

Although the occurrence of biotin-dependent enzymes is ubiquitous in nature, biotinylation is a relatively rare modification, with between one and five biotinylated protein species found in different organisms.[11] Thus, BPLs catalyze a reaction of extraordinarily stringent specificity. This specificity can be exploited to tag recombinant proteins, which are readily detected, quantitated, and purified by use of the many techniques based on the high affinity binding of biotin by avidin and streptavidin.

Biotin Ligases

Escherichia coli BPL (BirA protein), a monomeric protein of 35.5 kDa, is the only BPL of known structure.[12] Unfortunately, several of the more interesting parts of the molecule are disordered in the crystal structure. It should be noted that BirA is multifunctional in that it also acts as the repressor of biotin biosynthetic (*bio*) operon,[13–15] thus BirA and the *Bacillus subtilis* homologue[5] are both DNA-binding repressor proteins and BPLs.

The mechanism of the biotin ligase reaction is straightforward (Fig. 1) and is similar to that of aminoacyl-tRNA synthetases (BirA shares structural features with type II enzymes).[16] In the first half reaction, BPL catalyzes the attack of an oxygen atom of the biotin carboxylate on P α of ATP to form biotinoyl-AMP (also called biotinoyl-adenylate) plus pyrophosphate.

---

[3] D. F. Barker and A. M. Campbell, *J. Mol. Biol.* **146**, 451 (1981).

[4] D. F. Barker and A. M. Campbell, *J. Mol. Biol.* **146**, 468 (1981).

[5] Bower, J. Perkins, R. R. Yocum, P. Serror, A. Sorokin, P. Rahaim, C. L. Howitt, N. Prasad, S. D. Ehrlich, and J. Pero, *J. Bacteriol.* **177**, 2572 (1995).

[6] J. E. Cronan, Jr., and J. C. Wallace, *FEMS Microbiol. Lett.* **130**, 221 (1995).

[7] U. Hoja, C. Wellein, E. Greiner, and E. Schweizer, *Eur. J. Biochem.* **254**, 520 (1998).

[8] Y. Suzuki, Y. Aoki, Y. Ishida, Y. Chiba, A. Iwamatsu, T. Kishino, N. Nikawa, Y. Matsubara, and K. Narisawa, *Nature Genet.* **8**, 122 (1994).

[9] Leon-Del-Rio, D, Leclerc, B. Akerman, N. Wakamatsu, and R. A. Gravel, *Proc. Natl. Acad. Sci. U.S.A.* **92**, 4626 (1995).

[10] L. Dupuis, A. Leon-Del-Rio, D. Leclerc, E. Campeau, L. Sweetman, J.-M. Saudubray, G. Herman, K. M. Gibson, and R. A. Gravel, *Hum. Mol. Genet.* **5**, 1011 (1996).

[11] J. E. Cronan, Jr., *J. Biol. Chem.* **265**, 10327 (1990).

[12] K. Wilson, L. M. Shewchuk, R. G. Brennan, A. J. Otsuka, and B. W. Matthews, *Proc. Natl. Acad. Sci. U.S.A.* **89**, 9257 (1992).

[13] D. Beckett and B. W. Matthews, *Methods Enzymol.* **279**, 362 (1997).

[14] M. A. Eisenberg, O. Prakash, and S.-C. Hsiung, *J. Biol. Chem.* **257**, 15167 (1982).

[15] J. E. Cronan, Jr., *Cell* **58**, 427 (1989).

[16] P. J. Artymiuk, D. W. Rice, A. R. Poirrette, and P. Willet, *Struct. Biol.* **1**, 758 (1994).

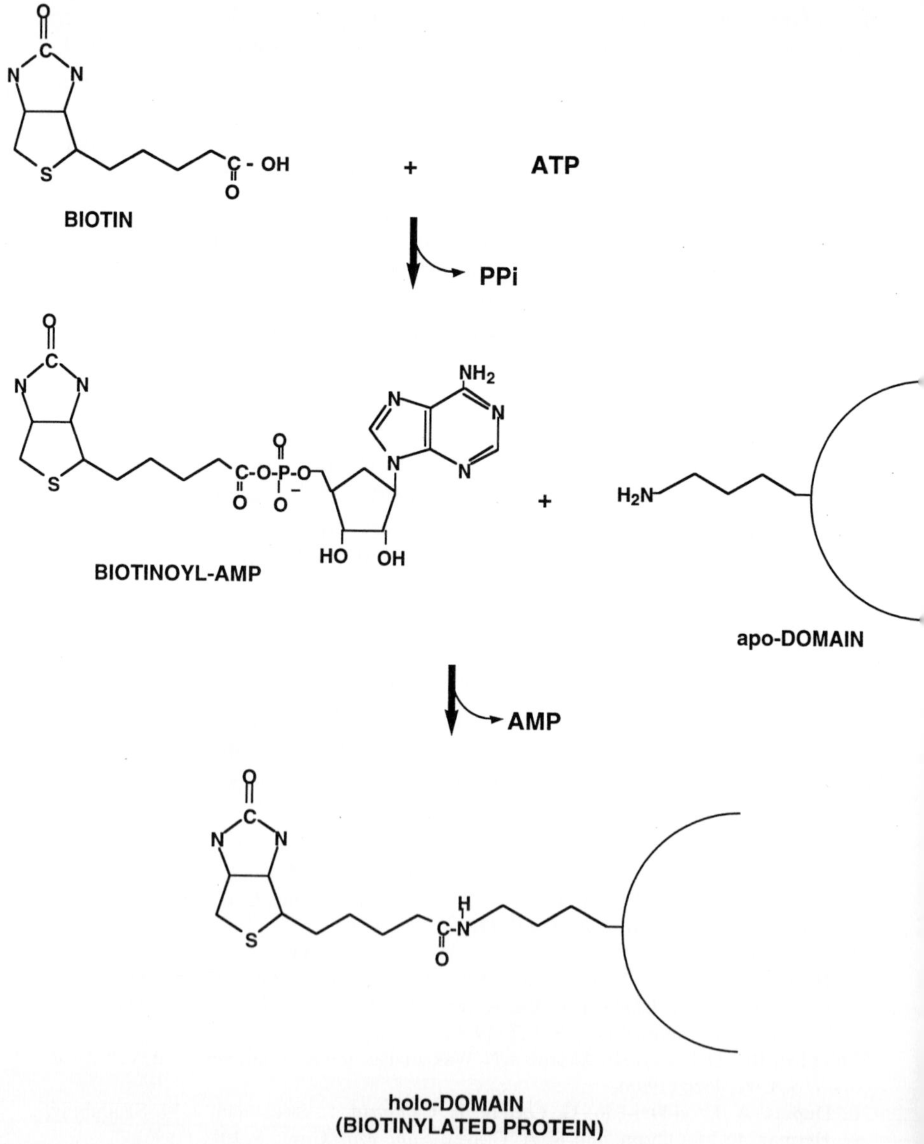

FIG. 1. The biotin protein ligase (BPL) reaction.

Biotinoyl-AMP remains bound in the active site. Biotinoyl-AMP is quite stable when bound (despite being a mixed anhydride), indicating that it is protected from solvent. In the presence of the apo form of the biotin-accepting domain of a biotin-requiring enzyme, the nucleophilic ε-amino group of the lysine to be modified attacks the mixed anhydride carbon atom, thus forming the amide bond between biotin and the lysine side chain with AMP as the other product. Once the amide bond is formed, the biotin moiety remains attached for the lifetime of the protein molecule.

The deduced amino acid sequences of many other BPLs have become available and several of these have been demonstrated to have biotinylation activity. The amino acid sequences are strongly related to BirA (Fig. 2) and include representatives of bacteria, archea, and eukaryota. Eukaryotic BPLs are about twice the size of bacterial and archael proteins and the similarity to BirA is restricted to the C-terminal halves.[6,8,9] The N-terminal halves of the eukaryotic BPLs share only modest sequence similarity and are of unknown function. An exception to the large size of the eukaryotic BPLs is that of the plant *Arabidopsis thalia,* which has a BPL only slightly larger than prokaryotic enzymes.[17] However, this BPL carries an N-terminal organelle targeting sequence (chloroplast or mitochondrial) and hence may be of prokaryotic origin consistent with its small size. The sequence conservation among BPLs (Fig. 2) has functional significance in that residues in contact with biotin in the crystal structure of *E. coli*[12] are invariant in all the proteins and those associated with mutations in BirA causing increased K_m values for biotin are highly conserved. Furthermore, the GRGRRG motif associated with ATP binding occurs in close proximity to the biotin-binding site in all cases. In addition, sequences showing homology to the biotin-binding protein, avidin, were identified in the C-terminal regions of both human and plant proteins, supporting the identification of the biotin-binding site.

Approaches used to isolate these coding sequences highlight the very strong conservation of both the ligase structure and the functional interaction between the enzyme and its protein substrate throughout evolution. Genome sequencing suggests that this conservation can be extended to the third kingdom, the archea. The genomes of *Methanococcus jannaschii, M. thermoautotrophicum,* and *Archaeoglobus fulgidus* contain sequences that encode proteins very similar to the BPLs and to biotinylated proteins found elsewhere in nature. In the case of *M. thermoautotrophicum,* a biotinylated protein has been shown to be a subunit of pyruvate carboxylase.[18]

None of the eukaryotic proteins contains sequences that suggest DNA-

[17] G. Tissot, R. Pepin, D. Job, R. Douce, and C. Alban, *Eur. J. Biochem.* **258,** 586 (1998).
[18] B. Mukhopadhyay, S. F. Stoddard, and R. S. Wolfe, *J. Biol. Chem.* **273,** 5155 (1998).

```
E. coli     - - - G D A C I A E Y Q Q A G R G R R G R K W F S P F G A N   130
H. sapiens  - - V I A A R Q T E G - - K G R G G - - N V W L S P V G C A   520
A. thalia   L P V G S V C V T D I Q F K G R G R T K N V W E S P K G C -   163
M. jann.    - - - - - - - - - D K Q N N G K G R W G R L V W Y S D E G -   52
S. cere.    S V V F V - - - - - - Q Y L S M L A Y C K A I L S Y A P G F   480

            L Y L - - - S M F W R L E Q G - - - P A A A T G L S L - V I   153
            L S T L L I S T P L R S Q L G Q R I P F V Q H L M S V A V V   550
            - - - L M Y S F T L E M E D G R V V P L I Q Y V V S L A V T   189
            L Y F - - - S M V L D S K L Y - - - N P K V T N L - L - V P   73
            - S D I P V R T K W P N D L Y A L S P T Y Y K R K N L K - -   508

            G I V M A E V L R K L G A D K V R V - K W P N D L Y L Q D R   182
            E A V R S I P E Y Q D - - - I N L R V K W P N D L Y L Y S D L   577
            E A V K D V C D K K G L P Y T D V K I K W P N D L Y V N G L L   219
            - I C I I E V L - K N Y V D K E L G L K F L P N D I M V K V N   101
            - L V N T G F E H T K L P L G - - - - - - - - D I E P A Y L   528

            - - - - K L A G I L V E L T G K T G D A A Q I V I G L A G I N   208
            M - - - K I G G V L V N S T L M - G E T F Y L I L G C G F N   603
            - - - - K V G G I L C T S T Y R - S K K F N V S V G V G L N   244
            D N Y K K L G G I L T E L T - - - D D Y M T - T G L G I N   126
            - - - - K I S G L L L V N T H F I - N N K Y C L L L G C G I N   553

            M A M R R V E E S V V N Q G W T T L Q E A - G I N L D R N T   237
            V T N S N P T L C I N - - - D L I T E Y N K Q H K A E L K P   630
            V D N G Q P T T C L N - - - A V L K G M A P E S N L - L K -   270
            V N N Q I R N E I R E I A - - I S L K E I T G K E L D K V E   154
            L T S D G P T T S L Q T W I D I L N E E R Q Q L H L D L L P   583

            L - - A A M L I R E L R A A L E L F E Q E G L A - - P Y - -   262
            L R - A D Y L I A R V V T V L E K L I K E F Q D K G P N S V   659
            - R - E E I L G A F F H K F - E K F F D L E M D Q G F K S L   296
            I - - L S N F L K T F E S Y L E K L K N K E I D - - D Y E I   180
            A I K A E K L Q A L Y M N N L E V I L K Q F I N Y G A A E I   613

            L S R W E K L D N F I N R P V K L I I - - G D K E T F - G -   288
            L P L Y Y R Y W V H S G Q Q V H L G S A E G P K V S - - - -   686
            E E L Y Y R T W L H S E Q R V I V E D K V E D Q V Q N V V   326
            L K K Y K K Y S I T I G K Q V K L L L - S N N E T I T G -   208
            L P S L Y Y E L W L H S N Q I V T L P D H G N T Q A M T T G I   643

            - I L S R G L I D K Q G A L L L - - - E Q D G L I I K P - - - W M   31C
            - I V - G L D D S G F L Q V H - - - Q E G G E V V T V H P D   71C
            T T Q - G L T S L G Y L L A V - - G D D N Q M Y E L H P D   352
            - K V Y L D I D F D G I V L G - - T E K G I E R I - - P S   23C
            T E D Y G L L I A K E L V S G S S T Q F T G N V Y N L Q P D   67:

            G G E I S - - - - - - L R S A E K   321
            G N S F D M L R N L I L P K R R .   72:
            G N S F D F F K G L V - - R R K I   367
            G L I C I H - - - - - - V R   23"
            G N T F D I F K S L I A K K V Q S   69C
```

FIG. 2. Alignment of BPLs. Residues identical to E. coli BirA are boxed. Only C termini of the proteins are aligned. S. cere., Saccharomyces cerevisiae; M. jann., Methanococcus jannaschii; K. pneu, Klebsiella pneumoniae.

binding activity. This is consistent with biotin metabolism in the different organisms. Unlike bacteria, both yeast and humans require an exogenous source of biotin and therefore would not be expected to possess repressor function. Plants, on the other hand, do synthesize biotin; however, the intracellular free biotin concentration is ca. 2000-fold greater than in bacteria, suggesting the lack of a strong regulatory mechanism.[19] Both human and yeast BPL proteins contain a large additional N-terminal domain having some sequence similarity to one another.[6] While there is as yet little indication of the function of this region, several mutations in the human BPL gene, which result in a biotin-responsive BPL deficiency, are located in this domain.[8,10]

Biotin Carrier Domains

The biotin carrier domain of biotin-containing enzymes is generally, but not invariably, located at the C-terminal end of the carboxylase, with the biotinyl-lysine located about 35 residues from the C terminus.[1,11] While several biotin enzymes consist of a single multifunctional polypeptide that is assembled into a tetrameric enzyme, others have the biotin carrier domain on a separate subunit of a multiprotein complex.[1] In *Propionibacterium shermanii*, transcarboxylase, an example of the latter class, biotinylation can occur before or after assembly of the complex.[20] A high degree of similarity is apparent in the primary structure of biotin attachment domains of the many biotin enzymes for which sequence data are now available. In particular, the specific biotinylated lysine residue occurs in a highly conserved AMKM tetrapeptide[1] (Fig. 3). Mutational analysis of the role of the Met residues flanking the biotin-lysine suggests that these residues are conserved because they are required for carboxylase function.[23]

Attachment of the biotin moiety by the ligase occurs posttranslationally within the context of a folded protein domain.[25] Studies of the efficiency with which truncated forms of biotin carrier proteins expressed as protein fusions are biotinylated have shown that a minimum of 70–80 residues (about 35–40 residues on either side of the biotin attachment site) is neces-

[19] G. Tissot, R. Douce, and C. Alban, *Biochem. J.* **323**, 179 (1997).
[20] N. H. Goss and H. G. Wood, *Methods Enzymol.* **107**, 261 (1984).
[21] C. Shenoy, Y. Xie, V. L. Park, G. K. Kumar, H. Beegen, H. G. Wood, and D. Samols, *J. Biol. Chem.* **267**, 18407 (1992).
[22] P. Reche and R. N. Perham, *EMBO J.* **18**, 2673 (1999).
[23] P. Reche and R. N. Perham, *Biochem. J.* **329**, 589 (1998).
[24] K. E. Reed and J. E. Cronan, Jr., *J. Biol. Chem.* **266**, 11425 (1991).
[25] S.-J. Li and J. E. Cronan, Jr., *J. Biol. Chem.* **267**, 855 (1992).

```
                                              *
E. coli   AccB   T P S P D A K A F I E V - - G Q K V N V G D T L C I V E A M K M M N Q I E A D K S G T V K A I
H. sapiens PCC   - P M P G V V V A V S V K P G D A V A E G Q E I C V I E A M K M Q N S M T A G K T G T V K S V
A. thalia AccB   S P A P G E P P F I K V - - G D K V Q K G Q V L C I V E A M K L M N E I E S L D H T G T V V D I
M. jann.         S P F R G M V T K L K V K E G D L K V K K G D V I V V L E A M K M E H P L E S P V E G T V L R I
S. cere.  Pyc2   A P M A G V I I E V K L H H K G L S L V K K G E S I A V L S A M K M E M V V S S P A D G L Q V K D V
K. pneu.  OLDC   A L P L A G T I W K V L A S E L Q T L A A L E V L L L T L L E A M K M E T E L R L A Q A L G T V R G L

                 V E S G Q L V E F L D E L V V I E
                 C Q A G D T V G E G D L L V E L E
                 A E D G K L V S L D T P L F L V Q P V E S A P
                 I D E G D A V N V G D V I M I T - - - - - - - - - K
                 L K D G E S V D A S D L L V V L E E E T L P P S Q K K
                 V K A G D A V A V G D T L M T L - - - - - - - - - A
```

FIG. 3. Alignment of biotin domains. Residues identical to *E. coli* AccB are boxed and the biotinylated lysine is marked with an asterisk. Only C termini of the proteins are aligned. Abbreviations are as given in Fig. 1 and Table I.

sary to specify biotinylation.[11,25,26] Further truncation results in a protein not recognized by BPL. Overexpression, biotinylation, and purification of isolated biotin domains from the biotin carboxyl carrier protein (BCCP) from *E. coli* acetyl-CoA carboxylase[27] and yeast pyruvate carboxylase[28] confirmed that the protein substrate recognized by BPL is a stably folded domain.

The three-dimensional structures of the apo and biotinylated forms of the biotin domain of BCCP from *E. coli* have been determined by both nuclear magnetic resonance[29–31] and X-ray crystallography,[32] giving essentially identical structures (Fig. 3). The protein has a barrel structure consisting of two antiparallel β sheets, each containing four strands. N and C termini are juxtaposed and interact at one end while the biotinyl-lysine is exposed on a tight β turn at the opposite end of the molecule. This geometry suggests that insertion of a biotin domain into another protein should result in protrusion of the domain from the surface of the fusion partner with the lysine targeted for biotinylation extended into the solvent. The similarities of the biotin domain sequences from different biological sources adjacent to the site of biotin attachment are almost certainly reflected in structural conservation, as BPL will biotinylate acceptor proteins across a wide variety of species (see later). Thus, it was anticipated that other biotin carrier proteins would have a three-dimensional structure very similar to the do-

[26] Leon-Del-Rio and R. A. Gravel, *J. Biol. Chem.* **269**, 22964 (1994).

[27] Chapman-Smith, D. L. Turner, J. E. Cronan, Jr., T. W. Morris, and J. C. Wallace, *Biochem. J.* **302**, 881 (1994).

[28] D. L. Val, A. Chapman-Smith, M. Walker, J. E. Cronan, Jr., and J. C. Wallace, *Biochem. J.* **312**, 817 (1995).

[29] X. Yao, D. Wei, C. Soden, Jr., M. F. Summers, and D. Beckett, *Biochemistry* **36**, 15089 (1997).

[30] X. Yao, C. Soden, Jr., M. F. Summers, and D. Beckett, *Protein Sci.* **8**, 307 (1999).

[31] L. Roberts, N. Shu, M. J. Howard, R. W. Broadhurst, A. Chapman-Smith, J. C. Wallace, T. Morris, J. E. Cronan, Jr., and R. N. Perham, *Biochemistry* **38**, 5045 (1999).

[32] F. K. Athappily and W. A. Hendrickson, *Structure* **3**, 1407 (1995).

main of *E. coli* BCCP, which has been confirmed by the structure of the 1.3S subunit of *Propionibacterium shermanii* transcarboxylase.[33] However, this structure, together with alignment of the sequences of biotin carrier proteins based on conservation of the residues crucial to formation (the folded structure), shows that the *E. coli* protein contains an inserted sequence not found in other biotin domains (Fig. 3). This sequence forms a protruding thumb in the three-dimensional structure that interacts with the biotin moiety.[31,32] Although the function of this segment is unknown, it is clearly not essential for the biotinylation reaction, as *E. coli* BPL readily biotinylates apoproteins lacking the thumb segment. The N-terminal half of the full-length BCCP molecule is linked to the biotin carrier domain by a flexible Pro-Ala-rich linker region and is required for BCCP dimerization and assembly of the functional carboxylase.[23]

The structured minimal biotin domain extends from the His-Ileu-Val motif at the N terminus to the C-terminal Val- Ileu-Glu residues, with the Ile and Val residues contributing to the formation of the hydrophobic core[29–32] (Fig. 3) and thus it is clear that truncations that remove these residues[11,25,26] would significantly destabilize the structure accounting for the observed lack of biotinylation. This supports the conclusion that BPL recognizes the Ala-Met-Lys-Met motif within the context of a folded protein domain. Although there is a minimum size required to form a stably folded functional biotin domain, affinity purification of peptide libraries has produced a consensus sequence of 23 amino acid residues that is sufficient to specify biotinylation.[34] The primary structure of these peptides has little resemblance to the sequence around the biotinylated Lys residue in the biotin carboxylases. Indeed, the only strictly conserved residue is the lysine itself. Shorter peptides of 14 residues retain the capacity to be efficiently biotinylated[35] *in vitro*, but have not been tested *in vivo*. It is probable that these peptides adopt a fold that mimics the conformation of a region in the biotin domain crucial for recognition by BirA. Because these peptides appear unstructured in solution,[35] they may adopt a transient fold that mimics the conformation of a region in the biotin domain crucial for recognition by the BirA ligase. However, it should be noted that these peptides do not seem to be general BPL substrates, but are active only with BirA, as unlike the *E. coli* biotin domain (see later), they fail to function with either insect[36] or plant BPLs.[17] Therefore, these mimics seem unlikely to

[33] D. V. Reddy, S. Rothemund, B. C. Shenoy, P. R. Carey, and F. D. Sonnichsen, *Protein Sci.* **7**, 2156 (1998).

[34] P. J. Schatz, *Biotechnology* **11**, 1138 (1993).

[35] D. Beckett, E. Kovaleva, and P. J. Schatz, *Protein Sci.* **8**, 921 (1999).

[36] S. Duffy, S., K. L. Tsao, and D. S. Waugh, *Anal. Biochem.* **262**, 122 (1998).

be useful in obtaining a fundamental understanding of the interactions of BPLs and biotin domains. This unexpected specificity for BirA also restricts the use of the peptide mimics to *E. coli* or to systems in which BirA can be coexpressed with the fusion protein.[36]

Given the strong conservation of sequence in both BPLs and their protein substrates, it would seem likely that BPLs should biotinylate acceptor proteins from widely divergent species. Indeed, this was first demonstrated by *in vitro* studies reported many years ago.[37] More recently, the high degree of conservation in the functional interaction between BPL and its protein substrate has been exploited to isolate yeast, human, and plant BPL-encoding genes by genetic complementation of *E. coli birA* mutants.[6,9,17] This genetic selection requires that the foreign BPL biotinylates the *E. coli accB* protein, an essential component of the lipid synthesis pathway. In other approaches, heterologous biotin domains were expressed in *E. coli* and were found to be biotinylated by direct analysis. Several of these were found by accident as "false positives" while screening genomic libraries in *E. coli* using biotinylated antibodies or DNA probes plus a steptavidin (or avidin)-enzyme conjugate to stain plaques or colonies.[38-40] These false positive clones were found to encode a biotinylated protein by DNA sequencing or by omission of the biotinylated antibody or DNA probe. Hence the observed staining was due to direct binding of the steptavidin (or avidin)-enzyme conjugate by a heterologous fusion protein expressed and biotinylated in *E. coli*. The known interactions of ligase with heterologous domains are given in Table I. In some cases, biotinylation has been demonstrated both *in vivo* and *in vitro*. Potentially interesting interactions that have not yet been tested are those of archeal ligases and biotin domains with eukaryotic and bacterial proteins. If the heterologous combinations produce biotinylated proteins, it would seem reasonable to assume that nature invented biotinylation of enzymes only once.

Construction of Biotinylated Protein Fusions

In vivo biotinylation provides a useful means to tag proteins for a variety of reasons. These are (i) the rarity of biotinylated protein species in all known cells types, (ii) the stability and discrete folding of biotin domains, (iii) the conservation of both the domains and the modifying ligases, and

---

[37] H. C. McAllister and M. J. Coon, *J. Biol. Chem.* **241**, 2855 (1966).
[38] N. E. Hoffman, E. Pichersky, and A. R. Cashmore, *Nucleic Acids Res.* **15**, 3928 (1987).
[39] M. E. Collins, M. T. Moss, S. Wall, and J. W. Dale, *FEMS Microbiol. Lett.* **43**, 53 (1987).
[40] D. Wang D, M. M. Waye, M. Taricani, K. Buckingham, and H. J. Sandham, *Biotechniques* **14**, 209 (1993).

(iv) avidin/streptavidin technology. Although the C terminus is the usual location of biotin domains in nature, these domains function as modules and can be attached to the N terminus of a target protein or even spliced within a coding sequence to give an internal fusion.[41-43] In any of these locations the domains seem to fold independently of the target protein and seldom disturb the activity of the target protein. Standard methods are used to construct the recombinant genes encoding the fusions, and the only difficulty is that fusions that are highly expressed in *E. coli* can be toxic to cell growth.[11,44] This is particularly true if a highly efficient biotin acceptor domain is used. The toxicity is due to decreased biotinylation of the essential *E. coli* lipid synthetic protein, AccB (e.g., Fig. 6), that results from competition for the endogeneous ligase.[11,44] Therefore, the original *E. coli* transformations and the subsequent growth of transformants should be done under conditions where expression of the fusion protein is minimal.

Although various biotin domains have been shown to be biotinylated in *E. coli*, some domains are biotinylated more efficiently than others. The most efficiently biotinylated domain we have found is derived from the α subunit of the oxalacetate decarboxylase (OLDC) of *Klebsiella pneumoniae*, a relative of *E. coli*.[24] During anaerobic growth on citrate, OLDC becomes a major protein of this organism and thus we thought that it might be a better ligase substrate than less well-expressed proteins such as BCCP. Indeed, the OLDC biotin domain is a better acceptor in *E. coli* than the BCCP domain (Fig. 4). Our first inkling of this somewhat unexpected behavior came from the difficulties we encountered in making fusion proteins with this domain. Even low-level expression of the fusion was extremely toxic to *E. coli*. Our original constructions were C-terminal fusions of a 99 residue OAADC domain to *E. coli* β-galactosidase (Fig. 5), but subsequent work showed that a 71 residue fusion was equally active, whereas a 54 residue domain was weakly biotinylated when fused to a full-length *E. coli* β-galactosidase and no biotinylation was seen when fusions to truncated forms of β-galactosidase were tested (Figs. 5 and 6). Stolz and co-workers[45] reported that a sequence of 66 OLDC domain residues was required to give detectable biotinylation of fusion proteins in *E. coli*. Because these workers used a λ phage display system in which biotinylation must occur prior to phage-induced cell lysis, we expect that the weak

---

[41] T. G. Consler, B. L. Persson, H. Jung, K. H. Zen, K. Jung, G. G. Privé, G. E. Verner, and H. R. Kaback, *Proc. Natl. Acad. Sci. U.S.A.* **90**, 6934 (1993).

[42] K. H. Zen, T. G. Consler, and H. R. Kaback, *Biochemistry* **34**, 3430 (1995).

[43] Y. Pouny Y, C. Weitzman C, and H. R. Kaback, *Biochemistry* **37**, 15713 (1998).

[44] Chapman-Smith, T. W. Morris, J. C. Wallace, and J. E. Cronan, Jr., *J. Biol. Chem.* **274**, 1449 (1999).

[45] J. Stolz, B. Darnhofer-Demar, and N. Sauer, *FEBS Lett.* **377**, 167 (1995).

TABLE I
CROSS-SPECIES FUNCTION OF BPLs AND ACCEPTOR DOMAINS[a]

BPL source	Biotin domain source	*In vivo*	*In vitro*	Ref.[b]
E coli	*P. shermanii* TC	+	+	1–3
S. cerevisiae	*P. shermanii* TC	+	+	4, 5
E. coli	*S. cerevisiae* PYC	+	+	6, 7
S. cerevisiae	*E. coli* ACC	+	+	8
H. sapiens	*E. coli* ACC	+	+	9, 10
E. coli	*H. sapiens* PC			11
A. thalia	*E. coli* ACC	+	+	12
S. frugiperda	*E. coli* ACC	+		13
S. frugiperda	*P. shermanii* TC	+		14
E. coli	*K. pneumoniae* OLDC	+		15
E. coli	*G. max* MCCC	+		16
E. coli	*L. esculentum* MCCC	+		17
E. coli	*M. tuberculosis* ACC/PC	+		18
E. coli	*M. leprae* ACC/PC	+		18
E. coli	*B. subtilis* ACC	+		19
B. subtilis	*E. coli* ACC	+		20
E. coli	*P. aeruginosa* ACC	+		21
E. coli	*V. parvula* MMCDC	+		22
S. cerevisiae	*K. pneumoniae* OLDC	+		23
S. pombe	*K. pneumoniae* OLDC	+		24
E. coli	*S. mutans* ?	+		25
P. shermanii	*R. rattus* PCC		+	5
L. timidus	*P. shermanii* TC		+	5
S. cerevisiae	*C. terrigena* MCCC		+	5
S. cerevisiae	*R. rattus* PCC		+	5
P. shermanii	*C. terrigena* MCCC		+	5

[a] The organisms are bacteria: *Escherichia coli, Propionibacterium shermani, Klebesellia pneumoniae, Bacillus subtilis, Mycobacterium tuberculosis,* and *Steptococcus mutans.* Archaea, *Methanococcus jannaschii.* Fungi: *Saccharomyces cerevisiae* (budding yeast), *Schizosaccharomyces pombe* (fission yeast). Plants: *Arabidopsis thalia, Lycopersicon esculentum* (tomato), and *Glycine max* (soybean). Mammals: *Homo sapiens* (human), *Rattus rattus* (rat), and *Lepus timidus* (rabbit). Insects: *Spodoptera frugiperda.* The enzymes are TC, transcarboxylase; ACC, acetyl-CoA carboxylase; PCC, propionyl-CoA carboxylase; PYC, pyruvate carboxylase; MCCC, β-methylcrotonyl-CoA carboxylase; MMCDC, methylmalonyl-CoA decarboxylase; OLDC, oxaloacetate decarboxylase.

[b] Key to references: (1) J. E. Cronan, Jr., *J. Biol. Chem.* **265,** 10327 (1990); (2) V. L. Murtif, C. R. Bahler, and D. Samols, *Proc. Natl. Acad. Sci. U.S.A.* **82,** 5617 (1985); (3) R. Buoncristiani and A. J. Otsuka, *J. Biol. Chem.* **263,** 1013 (1980); (4) S. H. Ackerman, J. Martin, and A. M. Tzagoloff, *J. Biol. Chem.* **267,** 7386 (1992); (5) H. C. McAllister and M. J. Coon, *J. Biol. Chem.* **241,** 2855 (1966); (6) D. L. Val, A. Chapman-Smith, M. Walker, J. E. Cronan, Jr., and J. C. Wallace, *Biochem. J.* **312,** 817 (1995); (7) S. W. Polyak, A. Chapman-Smith, T. D. Mulhern, J. E. Cronan, Jr., and J. C. Wallace, manuscript submitted (2000); (8) J. E. Cronan, Jr., and J. C. Wallace, *FEMS Microbiol. Lett.*

biotinylation of the 54 residue domain we observed is due to the increased time available for biotinylation.

Expression of Biotinylated Protein Fusions

In many cases, only modest expression of the fusion protein is desired and under such conditions the levels of ligase present in the host cell have been sufficient to give efficient biotinylation of the fusion protein. However, in cases where very high levels of expression are desired, the normal ligase levels generally do not suffice. In *E. coli,* several groups have noted that on high-level expression of biotin domain proteins, an appreciable fraction of the protein is not biotinylated, even with the addition of biotin to the growth medium.[11,41-44,46] This can be remedied either by co-overexpressing the ligase or by biotinylating the protein *in vitro.* The high-level *E. coli* expression system of Studier *et al.*[47] seems particularly prone to underbiotinylation. The phage T7 RNA polymerase expression is so active that RNA synthesis by the host RNA polymerase is inhibited, presumably due to a deficiency in the nucleotide triphosphate supply.[48] Because ATP is required for the biotin ligase reaction, biotinylation could be similarly

---

[46] J. B. Huder and P. Dimroth, *J. Bacteriol.* **177,** 3623 (1995).
[47] F. W. Studier, A. H. Rosenberg, J. J. Dunn, and J. W. Dubendorff, *Methods Enzymol.* **185,** 60 (1990).
[48] F. W. Studier and B. A. Moffatt, *J. Mol. Biol.* **189,** 113 (1986).

130, 221 (1995); (9) A. Leon-Del-Rio, D. Leclerc, B. Akerman, N. Wakamatsu, and R. A. Gravel, *Proc. Natl. Acad. Sci. U.S.A.* **92,** 4626 (1995); (10) A. Leon-Del-Rio, and R. A. Gravel, *J. Biol. Chem.* **269,** 22964 (1994); (11) Y. Suzuki, Y. Aoki, O. Sakamoto, X. Li, S. Miyabayashi, Y. Kazuta, H. Kondo, and K. Narisawa, *Clin. Chim. Acta* **251,** 41 (1996); (12) G. Tissot, R. Pepin, D. Job, R. Douce, and C. Alban, *Eur. J. Biochem.* **258,** 586 (1998); (13) E. Berliner, H. K. Mahtani, S. Karki, L. F. Chu, J. E. Cronan, Jr., and J. Gelles, *J. Biol. Chem.* **269,** 8610 (1994); (14) C. G. Lerner and A. Y. C. *Anal. Biochem.* **240,** 185 (1996); (15) K. Reed and J. E. Cronan, Jr., *J. Biol. Chem.* **266,** 11425 (1991); (16) J. Song, E. S. Wurtele, and B. J. Nikolau, *Proc. Natl. Acad. Sci. U.S.A.* **91,** 5779 (1994); (17) N. E. Hoffman, E. Pichersky, and A. R. Cashmore, *Nucleic Acids Res.* **15,** 3928 (1987); (18) M. E. Collins, M. T. Moss, S. Wall, and J. W. Dale, *FEMS Microbiol. Lett.* **43,** 53 (1987); (19) Morbidoni, H. R., D. de Mendoza, and J. E. Cronan, Jr., *J. Bacteriol.* **178,** 4794 (1996); (20) S. Bower, J. Perkins, R. R. Yocum, P. Serror, A. Sorokin, P. Rahaim, C. L. Howitt, N. Prasad, S. D. Ehrlich, and J. Pero, *J. Bacteriol.* **177,** 2572 (1995); (21) E. A. Best and V. C. Knauf, *J. Bacteriol.* **175,** 16881 (1993); (22) J. B. Huder and P. Dimroth, *J. Bacteriol.* **177,** 3623 (1995); (23) J. Stolz, B. Darnhofer-Demar, and N. Sauer, *FEBS Lett.* **377,** 167 (1995); (24) T. Caspari, I. Robl, J. Stolz, and W. Tanner, *Plant J.* **10,** 1045 (1996); (25) D. Wang, M. M. Waye, M. Taricani, K. Buckingham, and H. J. Sandham, *Biotechniques* **14,** 209 (1993).

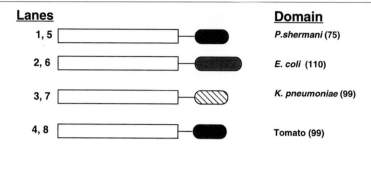

Lanes   **Domain**

1, 5   *P.shermani* (75)

2, 6   *E. coli* (110)

3, 7   *K. pneumoniae* (99)

4, 8   Tomato (99)

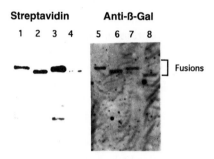

Streptavidin   Anti-β-Gal

1   2   3   4   5   6   7   8

Fusions

FIG. 4. Biotinylation efficiency in *E. coli* of β-galactosidase fused to various biotin domains. Fusions were made to a full-length enzymatically active β-galactosidase and biotinylated in a wild-type strain of *E. coli* as described previously.[11] (Bottom) Proteins were separated by SDS gel electrophoresis (run in duplicate), followed by transfer to PVDF membranes and probing with either streptavidin (left lanes 1–4) or anti-β-galactosidase (right lanes 5–8) with detection by enhanced chemiluminescence.[24] (Top) Lanes and domains tested.

limited. Indeed, on overexpression of the biotin domain of *E. coli* BCCP, incomplete biotinylation was observed even when the biotin ligase was also overexpressed.[27] However, lysis of overproducing cells in the ligase reaction buffer followed by the addition of biotin and ATP to the lysates and incubation resulted in quantitative biotinylation of the domain.[27] Given fusion proteins that are stable in cell extracts, this extract biotinylation approach seems likely to be generally useful. Extract biotinylation should proceed when a cell extract from a ligase overproducer is mixed with the cell extract containing the fusion protein, but this has not been tested. The *E. coli* biotin ligase is a robust protein and is purified readily in either its native state[12,13,22,23] or as a polyhistidine-tagged fusion protein.[49] As detailed

[49] A. O'Callaghan, M. F. Byford, J. R. Wyer, B. E. Wilcox, B. K. Jakobsen, A. J. McMichael, and J. I. Bell, *Anal. Biochem.* **266,** 9 (1999).

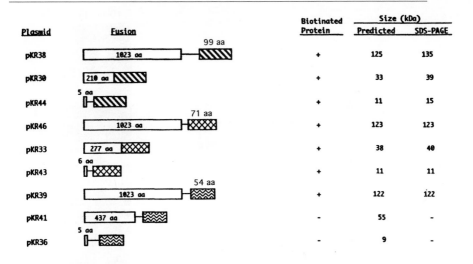

Fig. 5. Biotinylation efficiencies in *E. coli* of β-galactosidase fused to *K. pneumoniae* OLDC domains of various lengths. (A) The various fusions tested and (B) the sequence. Some supporting data are given in Fig. 6.

in Table I, biotinylated proteins can also be produced in eukaryotic cells. A concern is that the biotin ligase and fusion protein may reside in separate cellular compartments and be unable to interact. However, in mammalian cells, biotin ligase activity is present in both the cytosol and the mitochondria.[50,51]

Detection of Biotinylated Fusions

The most direct means to detect biotinylated proteins is the incorporation of radioactively labeled biotin *in vitro*.[11] In bacteria, use of a mutant strain requiring biotin for growth is most efficient, but wild-type strains

[50] H. I. Chang and H. D. Cohen, *Arch. Biochem. Biophys.* **225,** 237 (1983).
[51] F. Taroni and L. E. Rosenberg, *J. Biol. Chem.* **266,** 13267 (1991).

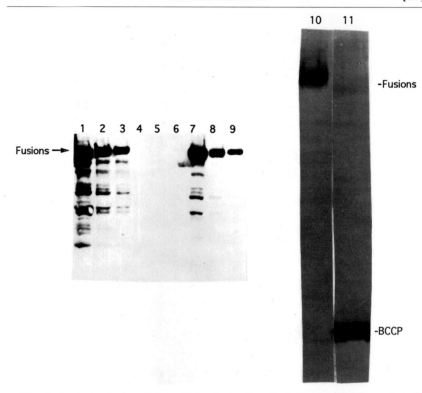

FIG. 6. Analysis of biotinylation efficiencies in *E. coli* of β-galactosidase fused to *K. pneumoniae* OLDC domains of various lengths. A fusion to a 99 residue *K. pneumoniae* OLDC domain were loaded in lanes 1–3. Lanes 2 and 3 are serial 1 : 5 dilutions of the extract in lane 1. Hence the concentration in lane 3 is one-twenty fifth of that in lane 1. A fusion to a 54 residue OLDC domain was loaded in lanes 4–6 with the same dilution series. Lanes 7–9 were loaded with a dilution series of a 71 residue OLDC domain fusion in the same manner. Detection by streptavidin binding and enhanced chemiluminescence.[24] Lanes 10 and 11 contain ³H-labeled biotin β-galactosidase fusion proteins to *K. pneumoniae* OLDC domains of 99 and 54 residues, respectively. Detection by fluorography.[11,24] Note the diminished labeling of BCCP (AccB) in lane 10 compared to lane 11, indicating competition for BPL.

can often be readily labeled due to repression of biotin biosynthesis by exogeneous biotin.[13] In the latter case the strain should be grown with excess unlabeled biotin to maximally repress biotin synthesis, washed free of the unlabeled biotin, and exposed to the radioactive species. Yeast and mammalian cells are natural biotin auxotrophs and thus labeling is straightforward. Tritiated biotin is the most useful isotopically labeled compound available commercially.¹⁴C-labeled biotin is generally of such low specific activity that awkwardly large volumes of cells must be labeled.[35]S-labeled biotin would be an ideal label, but unfortunately it is not available commercially. The usual means to detect the protein is SDS gel electrophoresis followed by fluorography or autoradiography of the gels.

Another method of even lower technology can be used to identify biotinylated proteins if the protein composition of a sample is simple or if the fusion protein is sufficiently highly expressed that it forms a distinct band on gel electrophoresis. Avidin (or streptavidin) is added to the protein sample prior to electrophoresis and this sample is run in parallel with an untreated sample (or better, a sample treated with the biotin-binding protein incubated previously with excess biotin). The greater molecular size (an increase of about 60 kDa) of the biotinylated fusion protein–avidin/ streptavidin complex results in retardation relative to the uncomplexed protein. If excess biotin-binding protein is present, the amount of protein running at the uncomplexed position in the avidin-treated sample compared to the control samples will give the biotinylation efficiency. If SDS gel electrophoresis is done, the samples must not be boiled in SDS. The biotin–avidin complex is stable to SDS, but not to boiling SDS.[52] This approach can also be used in nondenaturing or urea gels. In these cases the basic charge of avidin can prevent the complex from entering the gel and streptavidin may be the better choice.

The most popular approaches to detect biotinylated proteins involve the transfer of cellular proteins from a SDS gel to a membrane followed by detection of the biotinylated species by incubation of the membrane with avidin or streptavidin complexed to a reporter molecule.[53] The protocols are those used for immunoblotting with biotinylated primary antibodies except that the antibody is omitted. Because there are a large number of such procedures (many available in kit form) and because this technology is evolving rapidly, specific procedures will not be given. These methods can be extraordinarily sensitive such that the proteins from a single cell can be detected.[54]

Finally, in *E. coli* production of biotinylated proteins can be detected physiologically. As mentioned earlier, high-level expression of fusion proteins can block cell growth. However, a more elegant detection system results from the fact that *E. coli* biotin ligase (BirA) also represses transcription of the biotin biosynthetic operon detected.[15] BirA protein binds to the *bio* operator only when the protein is complexed to biotinoyl-AMP. This BirA–biotinoyl-AMP complex (Fig. 1) is formed in the first half reaction

[52] N. M. Green, *Methods Enzymol.* **184,** 51 (1990).
[53] See *Methods Enzymol.* **184,** (1990) (eds., M. Wilchek and E. A. Bayer) for several protocols.
[54] P. G. Gillespe and A. J. Hudspeth, *Proc. Natl. Acad. Sci. U.S.A.* **88,** 2563 (1991).

of the ligase and is a substrate for the second half reaction, modification of a biotin domain. Therefore, when there is an excess of unbiotinylated biotin domain, biotinoyl-AMP is consumed in protein biotinylation and BirA is unable to bind the *bio* operator. Therefore, production of a biotinylated fusion protein results in derepression of *bio* operon transcription that can be detected readily using a β-galactosidase gene driven by a *bio* operon promoter.[55]

Use of Biotinylation as a Localization Tag

In vivo biotinylation has been used as a sensitive reporter of protein sorting in both prokaryotes and eukaryotes (biotin domains seem readily able to transverse biological membranes).[24,36–58] Biotin domain fusions have been used to tag motor proteins in a defined manner such that the protein can be followed accurately in space. Good examples are the use of *in vivo*-biotinylated motor proteins, kinesin, and Cin8p to study the mechanisms of protein movement.[59,60]

Purification of Biotinylated Proteins

Avidin biotin technology also allows facile purification of biotinylated proteins. The proteins can be purified by harsh means to provide denatured proteins suitable for antibody production, protein chemistry, or refolding to the active state, as the biotin–avidin interaction is very robust. Conversely, biotinylated proteins can be purified by gentle and specific elution from columns of modified avidins.

Biotinylated proteins are purified in denatured form by dissolving whole cells in hot 1% SDS. This solution can then be passed over a column of native (tetrameric) avidin (or streptavidin) covalently coupled to a chromatographic matrix, such as Sepharose.[25,61] The column is then washed extensively with 1% SDS until no elution of unbound proteins can be detected. The column material is then extruded from the column and boiled in 1% SDS (to break the avidin–biotin interaction by denaturation of the avidin), reloaded into the column, and eluted with 1% SDS. This generally

[55] J. E. Cronan, Jr., *J. Biol. Chem.* **263**, 10332 (1988).
[56] S. H. Ackerman, J. Martin, and A. M. Tzagoloff, *J. Biol. Chem.* **267**, 7386 (1992).
[57] G. Jander, J. E. Cronan, Jr., and J. Beckwith, *J. Bacteriol.* **178**, 3049 (1996).
[58] E. Weiss, J. Chatellier, and G. Orfanoudakis, *Protein Expr. Purif.* **5**, 509 (1994).
[59] E. Berliner, H. K. Mahtani, S. Karki, L. F. Chu, J. E. Cronan, Jr., and J. Gelles, *J. Biol. Chem.* **269**, 8610 (1994).
[60] L. Ghebert, S. C. Kuo, and M. A. Hoyt, *J. Biol. Chem.* **274**, 9564 (1999).
[61] J. A. Swack, G. L. Zander, and M. F. Utter, *Anal. Biochem.* **87**, 114 (1978).

results in a mixture of endogeneous biotinylated proteins plus the fusion protein that is free of unbiotinylated protein species. However, it should be noted that in some commercial preparations of immobilized avidin, a portion of avidin monomers is not coupled directly to the column matrix and thus denaturation releases avidin monomers that contaminate the biotinylated protein fraction.

Biotinylated proteins cannot be eluted from columns of tetrameric avidin by the addition of biotin. The binding of biotin by avidin is too strong and the rate of dissociation is far to slow. However, if tetrameric avidin is coupled to a chromatographic matrix and washed with guanidine–HCl at low pH, the intersubunit interactions are broken, the uncoupled subunits are eluted, and a column of monomeric avidin results.[62] Monomeric avidin retains the ability to bind biotinylated proteins, but the binding is much weaker (a dissociation constant of about 10^{-7} M rather than 10^{-15} M for the tetrameric species) and can be competed readily by moderate concentrations of free biotin. The only difficulties encountered with such columns are residual tetrameric avidin (these sites are blocked readily by free avidin[62]) and ionic interactions with a highly basic avidin surface, which can be suppressed by the use of high salt buffers.[11] The technology of making monomer avidin columns was worked out many years ago, but columns of tetrameric avidin and streptavidin having binding properties similar to that of monomer avidin have been prepared by chemical modification.[63] Moreover, the crystal structures of avidin[64,65] and streptavidin[66,67] have demonstrated the mode of biotin binding by the proteins. Two of the major interactions in the biotin-binding site are (i) a hydrogen bond between the carbonyl group of the uriedo moiety of biotin and a tyrosine residue and (ii) tight packing of several tryptophan residues around the biotin, resulting in hydrophobic and van der Waals interactions. This tyrosine residue is the target of chemical modification,[63] whereas tryptophan residues have been targeted for site-directed mutagenesis.[68] In native avidin and streptavidin tetramers, one of these tryptophan residues is contributed from a neighboring monomer and it seems that loss of this interaction explains the loss of biotin-binding affinity on monomerization of avidin (this has not been

[62] R. A. Kohanski and M. D. Lane, *Methods Enzymol.* **184,** 194 (1990).
[63] E. Morag, M. Wilchek, and E. A. Bayer, *Anal. Biochem.* **243,** 257 (1996).
[64] O. Livnah, E. A. Bayer, M. Wilchek, and J. L. Sussman, *Proc. Natl. Acad. Sci. U.S.A.* **90,** 5076 (1989).
[65] L. Pugliese, A. Coda, M. Malcovati, and M. Bolognesi, *J. Mol. Biol.* **231,** 698 (1993).
[66] W. A. Henrickson, A. Pahler, J. L. Smith, Y. Satow, E. A. Merritt, and R. P. Phizackerley, *Proc. Natl. Acad. Sci. U.S.A.* **86,** 2190 (1989).
[67] P. C. Weber, D. H. Ohlendorf, J. J. Wendoloski, and F. R. Salemme, *Science* **243,** 85 (1989).
[68] Chilkoti, P. H. Tan, and P. S. Stayton, *Proc. Natl. Acad. Sci. U.S.A.* **92,** 1754 (1995).

tested in streptavidin, as monomeric streptavidin has not yet been prepared due to the great strength of the subunit interfaces). It seems likely that in the near future an ideal affinity matrix will be derived from avidin and/or streptavidin in which each tetramer will have four high affinity, but reversible (by free biotin), binding sites for biotinylated proteins. However, it should be noted that the monomer avidin columns currently available provide effective purification of a number of biotinylated fusion proteins.[11,41–43,45,69,70] Perhaps the most challenging application has been the purification of integral membrane proteins where other purification tags have failed or been problematic.[41–43,45]

[69] E. C. Young, E. Berliner, H. K. Mahtani, B. Perez-Ramirez, and J. Gelles, *J. Biol. Chem.* **270,** 3926 (1995).
[70] T. Caspari, I. Robl, J. Stolz, and W. Tanner, *Plant J.* **10,** 1045 (1996).

Section V

Hybrid Proteins for Detection and Production
of Antigens and Antibodies

[28] Methods for Generating Multivalent and Bispecific Antibody Fragments

By IAN TOMLINSON and PHILIPP HOLLIGER

Introduction: Isolation, Expression, and Purification of Antibody Fragments

Isolating Antibody Fragments by Phage Display

This article focuses on recent developments in antibody engineering that have facilitated the generation and tailoring of antibody-based reagents for specific purposes. One of the key advances has been the use of phage display to isolate antibody fragments directly from diverse repertoires of V genes.[1-4] In recent years the technology has become more robust and antibody fragments with nanomolar and even subnanomolar affinities for antigen can now be routinely obtained from phage–antibody libraries.[5-7] Protocols for the selection of antibodies from phage libraries have been published elsewhere[8] and several phage–antibody libraries are available to academic researchers (http://www.mrc-cpe.cam.ac.uk/phage/index.html).

The use of semisynthetic libraries[9,10] for the generation of antibody-based reagents is particularly attractive. These allow the inherent biases in natural repertoires to be controlled (e.g., by using equal ratios of the different germline V genes in library construction) and can lead to the creation

[1] G. Winter, A. D. Griffiths, R. E. Hawkins, and H. R. Hoogenboom, *Annu. Rev. Immunol.* **12**, 433 (1994).

[2] G. Winter, *FEBS Lett.* **430**, 92 (1998).

[3] A. D. Griffiths and A. R. Duncan, *Curr. Opin. Biotech.* **9**, 102 (1998).

[4] H. R. Hoogenboom, A. P. de Bruine, S. E. Hufton, R. M. Hoet, J. W. Arends, and R. C. Roovers, *Immunotechnology* **4**, 1 (1998).

[5] A. D. Griffiths, S. C. Williams, O. Hartley, I. M. Tomlinson, P. Waterhouse, W. L. Crosby, R. E. Kontermann, P. T. Jones, T. D. Prospero, H. R. Hoogenboom, A. Nissim, J. P. L. Cox, J. L. Harrison, M. Zaccolo, E. Gheradi, and G. Winter, *EMBO J.* **13**, 3245 (1994).

[6] T. J. Vaughan, A. J. Williams, K. Pritchard, J. K. Osbourn, A. R. Pope, J. C. Earnshaw, J. McCafferty, R. A. Hodits, J. Wilton, and K. S. Johnson, *Nature Biotech.* **14**, 309 (1996).

[7] M. D. Sheets, P. Amersdorfer, R. Finnern, P. Sargent, E. Lindqvist, R. Schier, G. Hemingsen, C. Wong, J. C. Gerhart, and J. D. Marks, *Proc. Nat. Acad. Sci. U.S.A.* **95**, 6157 (1998).

[8] J. L. Harrison, S. C. Williams, G. Winter, and A. Nissim, *Methods Enzymol.* **267**, 83 (1996).

[9] H. R. Hoogenboom and G. Winter, *J. Mol. Biol.* **227**, 381 (1992).

[10] A. Nissim, H. R. Hoogenboom, I. M. Tomlinson, G. Flynn, C. Midgley, D. Lane, and G. Winter, *EMBO J.* **13**, 692 (1994).

of large diverse libraries, which also yield antibodies with nanomolar affinities.[5] Disadvantages of such libraries are the high proportion of nonfunctional antibodies that arise due to the use of long oligonucleotides and polymerase chain reaction errors introduced during assembly as well as the incorporation of codons, which are poorly expressed in bacteria. A rationalized design of synthetic libraries can help overcome these problems and allows antibody scaffolds to be chosen that are particularly well folded, well expressed, and able to bind certain superantigens (e.g., Protein A and/ or Protein L), which can subsequently be used as affinity reagents for immunodetection in ELISA and Western blots, as well as for purification and as multimerization agents (see later). (I. Tomlinson *et al.,* unpublished data).

Expression of Antibody Fragments

Antibody scFv or Fab fragments isolated by phage display or cloned from hybridomas can be expressed conveniently in a range of different hosts, including bacteria[11] as well as yeast, plant, and insect cells (baculovirus) and both lymphoid and nonlymphoid mammalian cells.[12] This article focuses on antibody fragments produced by bacterial expression.

Antibody fragments can be expressed intracellularly as well as extracellularly by secretion to the bacterial periplasm. Intracellular expression of antibody fragments in *Escherichia coli* usually gives rise to nonfunctional aggregates (so-called inclusion bodies), presumably because the reducing environment of the cytoplasm precludes correct formation of the intradomain disulfide bridges. Inclusion bodies can be solubilized by strong denaturants, and both Fv and Fab fragments have been refolded successfully in good yields (20–80 mg/liter culture) from inclusion bodies. One drawback of refolding is the titration of optimal conditions, which are likely to differ from antibody to antibody.[13] Attempts have also been made either to engineer antibody fragments by removal of the intrachain disulfide bonds[14] or to select for functional cytoplasmic expression.[15] It remains to be seen if either approach is generally applicable.

We favor secretion of the antibody fragments to the periplasmic space of *E. coli* as a more convenient strategy. Although only a fraction of the secreted protein is soluble[16] while the rest forms inclusion bodies in the

[11] A. Skerra, *Curr. Opin. Immunol.* **5,** 256 (1993).
[12] R. Verma, E. Boleti, and A. J. George, *J. Immunol. Methods* **216,** 165 (1998).
[13] J. Buchner and R. Rudolph, *Bio/Technology* **9,** 157 (1991).
[14] K. Proba, A. Wörn, A. Honegger, and A. Plückthun, *J. Mol. Biol.* **275,** 245 (1998).
[15] P. Martineau, P. Jones, and G. Winter, *J. Mol. Biol.* **280,** 117 (1998).
[16] A. Skerra and A. Plückthun, *Protein Eng.* **4,** 971 (1991).

periplasm, sufficient amounts of functional antibody for most purposes can be obtained in shaker flask cultures. Attempts have been made to improve the yield of secreted antibodies through the use of different promotors,[17] signal sequences,[18] linker sequences,[19] coexpression of various chaperones,[20] framework mutations,[21] and CDR grafting.[22] Higher expression yields have also been selected using random mutagenesis in an *E. coli* mutator strain.[23] However, few general rules have emerged and for the best expression yields it remains necessary to optimize conditions for each antibody individually.

One of the few generalizations that can be made (in the authors experience) is that antibodies recloned directly from human or mouse hybridomas are poorly expressed in *E. coli* (typically 0.1–1 mg/liter in regular shaker flasks, although some can approach the 10–50 mg/liter level). In contrast, antibodies derived from phage display libraries tend to be better expressed with typical yields 1–10 mg/liter, with some approaching yields of up to 100 mg/liter. Presumably these have (to some extent) already been optimized for bacterial expression during phage selection. Even higher yields can be achieved, but this requires modification of the expression system (e.g., vector) and/or of the method of antibody production (e.g., using a fermentor in place of shaker flasks). Indeed, using fermentation technology expression, levels of up to 1 g/liter have been achieved in *E. coli*[24] for both a Fab fragment and a derivative bispecific diabody[25] (see later). Further improvements in bacterial secretion require changes to the antibody sequence and thus depend on the ease and speed with which mutations can be introduced, screened, and assayed for expression.

Purification of Antibody Fragments

Fragments can be purified using Protein G (binding to the CH1 domain of human Fab fragments)[24] or Protein A (binding to the V_H domain) or via a variety of C-terminal peptide tags, the most useful of which has proved

[17] A. Skerra, *Gene* **151**, 131 (1994).
[18] A. Skerra, I. Pfitzinger, and A. Plückthun, *Bio/Technology* **9**, 273 (1991).
[19] D. J. Turner, M. A. Ritter, and A. J. George, *J. Immunol. Methods* **205**, 43 (1997).
[20] A. Knappik, C. Krebber, and A. Plückthun, *Bio/Technology* **11**, 77 (1993).
[21] A. Knappik and A. Plückthun, *Protein Eng.* **8**, 81 (1995).
[22] S. Jung and A. Plückthun, *Protein Eng.* **10**, 959 (1997).
[23] G. Coia, A. Ayres, G. G. Lilley, P. J. Hudson, and R. A. Irving, *Gene* **201**, 203 (1997).
[24] P. Carter, R. F. Kelley, M. L. Rodrigues, B. Snedecor, M. Covarrubias, M. D. Velligan, W. L. Wong, A. M. Rowland, C. E. Kotts, M. E. Carver, *et al.*, *Bio/Technology* **10**, 163 (1992).
[25] Z. Zhu, G. Zapata, R. Shalaby, B. Snedecor, H. Chen, and P. Carter, *Bio/Technology* **14**, 192 (1996).

to be the hexahistidine tag that binds to immobilized metal chelates[26,27] and allows purification on Ni-chelate columns with particularly mild elution conditions (IMAC; see later). More recently, the production of recombinant Protein L (Affitech, Oslo, Norway) has enabled the direct detection of antibodies by binding to the V_κ domain. Because neither Protein A nor Protein L binding is thought to inhibit binding of the Fv to the target antigen, they provide effective reagents for the immobilization, detection, and/or purification of the smallest recombinant antibody fragments (Fabs, scFvs, Fvs, single V_H, and/or V_κ domains), provided the appropriate binding site is present (Protein A binds members of the human $V_H III$ family whereas Protein L binds members of the human $V_\kappa I$ and $V_\kappa III$ families[28]). Alternatively, instead of using Protein A and Protein L separately, a fusion protein, Protein LA (Affitech), can be used to bind antibodies that possess a binding activity for either Protein A or Protein L.[29]

Functionalizing Antibody Fragments

The technologies discussed earlier, phage selection for the isolation of antibody specificities, bacterial expression for antibody production, and IMAC, Protein A, and Protein L purification, are all now well established. Together they allow the generation of antibody-based laboratory reagents for the detection of a specific antigen in a matter of weeks.

Because it is often desirable to add further functionality to antibody fragments, we will focus on a list of simple and robust ways to functionalize antibody fragments that have proved useful in a range of diverse applications, such as immunostaining and tumor imaging.

Increasing Functional Affinity (Avidity) of Antibody Fragments

One crucial difference between recombinant antibody fragments and naturally occurring antibody fragments is the valency of antigen binding. All naturally occurring antibodies are multivalent, enabling them to bind to repetitive or solid-phase epitopes with greatly increased functional affinity, termed avidity.[30] Multivalency can have dramatic effects on the dissociation kinetics (k_{off}) of antigen binding as all interactions between antigen and antibody must be broken before dissociation takes place. This is of particu-

[26] E. Hochuli, W. Bannwarth, H. Döbeli, R. Gentz, and D. Stüber, *Bio/Technology* **6**, 1321 (1988).

[27] A. Skerra, I. Pfitzinger, and A. Plückthun, *Bio/Technology* **9**, 273 (1991).

[28] L. Björk, *J. Immunol.* **140**, 1194 (1988).

[29] H. G. Svensson, H. R. Hoogenboom, and U. Sjöbring, *Eur. J. Biochem.* **258**, 890 (1998).

[30] D. M. Crothers and H. Metzger, *Immunochemistry* **9**, 341 (1972).

lar importance when antibody–antigen interactions take place under non-equilibrium conditions (e.g., intravascularly) where slow dissociation kinetics may be of greater value than higher affinity.

In contrast, Fab and Fv antibody fragments are monovalent, whereas scFv fragments often occur as a mixture of mono- and multivalent species.[31] The tendency of scFvs to oligomerize can be enhanced, for example, by increasing the protein concentration, by elution of the scFv fragments from affinity-chromatography columns by high pH,[10] or by addition of organic solvents during a refolding step.[32] For many applications, it is desirable, however, to access the benefits of multivalent binding in a more controlled manner.

Avidity by Capture: Use of Multimeric Affinity Reagents. The detection of bound antibody in an ELISA or on a Western blot is usually performed using a highly specific affinity reagent that is linked covalently to an enzyme, such as alkaline phosphatase or horseradish peroxidase (HRP). Upon addition of a substrate, these enzymes are able to catalyze a colorimetric, fluorescent, or luminescent reaction. Affinity reagents for the detection of recombinant antibodies can themselves be antibodies (e.g., the anti-myc tag antibody 9E10) or superantigens that bind the framework regions of certain V_H and V_κ domains (such as Protein L that binds human $V_\kappa I$ and $V_\kappa III$ domains or Protein A that binds human $V_H III$ domains). Because the amount of affinity reagent bound after washing determines the ELISA signal or the strength of the banding pattern on a Western blot, it is essential that the affinity reagent has a high avidity for its target antibody. In practice, this avidity is achieved by a combination of high monovalent affinity and multimerization of the affinity reagent. Although natural IgG antibodies exist as dimers and therefore make ideal affinity reagents, higher order multimers, such as Protein L, which is a tetramer, or Protein LA, which is an octamer can give enhanced binding and thus greatly improved signals in Western blotting (see later and Fig. 1) and ELISA.

In addition to their use as enzyme conjugates for detection, affinity reagents can also be used to multimerize the antibody fragments themselves. Noncovalent multimerization of this type can produce a dramatic increase in the signal produced during ELISA and will profoundly influence any real time detection of binding, e.g., during BIAcore measurement.

Noncovalent multimerization removes the need for any vector manipulation and simply requires the antibody fragment to be mixed with the

[31] A. D. Griffiths, M. Malmqvist, J. D. Marks, J. M. Bye, M. J. Embleton, J. McCafferty, M. Baier, K. P. Holliger, B. D. Gorick, N. C. Hughes-Jones *et al.*, *EMBO J.* **12**, 725 (1993).

[32] M. Whitlow, D. Filpula, M. L. Rollence, S. L. Feng, and J. F. Wood, *Protein Eng.* **7**, 1017 (1994).

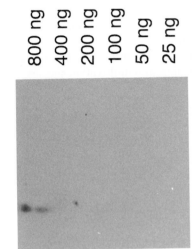

800 ng 400 ng 200 ng 100 ng 50 ng 25 ng

9E10-HRP

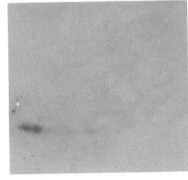

800 ng 400 ng 200 ng 100 ng 50 ng 25 ng

Protein A-HRP

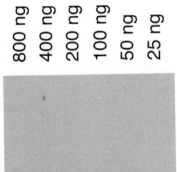

800 ng 400 ng 200 ng 100 ng 50 ng 25 ng

Protein L-HRP

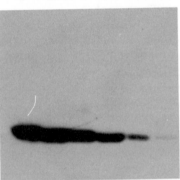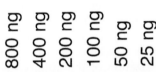

800 ng 400 ng 200 ng 100 ng 50 ng 25 ng

Protein LA-HRP

FIG. 1. Comparison of different detection reagents in Western blotting using decreasing amounts of antigen (given in ng). A fusion of Protein L and Protein A (Protein LA) is superior to all other detection reagents tested.

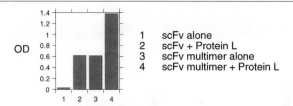

FIG. 2. Effect of increased avidity in ELISA of an anti-BSA scFv fragment. Multimerization by incubation with Protein L prior to binding to the ELISA plate increases the sensitivity of detection by a similar amount as multimerization by shortening of the interdomain linker (triabody formation). Combination of the two approaches provides a further increase in sensitivity.

multimeric affinity reagent in the right ratio, incubated for a short time, and then used to bind the immobilized antigen. Detection of the bound antibody is then performed using a second affinity reagent that does not interfere with the binding of the first (multimeric) affinity reagent. Thus, scFv fragments could be multimerized with Protein L and then detected using a HRP-labeled anti-tag antibody or Protein A-HRP (see later and Fig. 2). Alternatively, scFv fragments could be multimerized using an anti-tag antibody and then detected with Protein A-HRP or Protein L-HRP. An example of noncovalent multimerization of an anti-bovine serum albumin (BSA) scFv is shown in Fig. 2. Premixing of the scFv multimer with Protein L gives an even stronger signal. This method is therefore useful when the concentration of antigen is limiting (e.g., when detecting a protein in a cell lysate) or when the expression level of the recombinant antibody is low.

Avidity by Design: Diabodies and Triabodies. Bivalency can also be achieved by chemical cross-linking of Fab[24] or Fv fragments,[33] as well as by genetic fusions of multimerizing domains such as amphipathic helices[34] and antibody CH3 domains,[35] as well as streptavidin.[36] Another very simple approach is the diabody format in which a shortened peptide linker (five residues or less) between V_H and V_L domains precludes V_H–V_L association on the same polypeptide chain and drives dimerization to a divalent diabody molecule.[37] The homodimeric diabody structure[38] confirms that the two

[33] A. J. Cumber, E. S. Ward, G. Winter, G. D. Parnell, and E. J. Wawrzynczak, *J. Immunol.* **149,** 120 (1992).
[34] A. Plückthun and P. Pack, *Immunotechnology* **3,** 83 (1997).
[35] S. Hu, L. Shively, A. Raubitschek, M. Sherman, L. E. Williams, J. Y. Wong, J. E. Shively, and A. M. Wu, *Cancer Res.* **56,** 3055 (1996).
[36] S. M. Kipriyanov, M. Little, H. Kropshofer, F. Breitling, S. Gotter, and S. Dübel, *Protein Eng.* **9,** 203 (1996).
[37] P. Holliger, T. Prospero, and G. Winter, *Proc. Natl. Acad. Sci. U.S.A.* **90,** 6444 (1993).
[38] O. Peristic, P. A. Webb, P. Holliger, G. Winter, and R. L. Williams, *Structure* **2,** 1217 (1994).

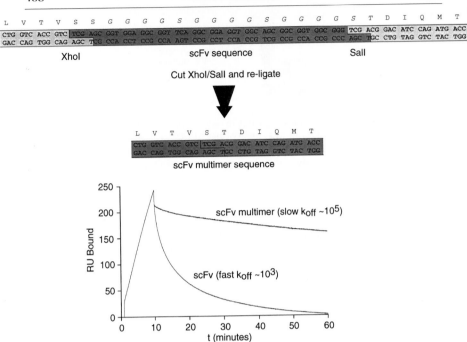

FIG. 3. One-step conversion of scFv fragments to scFv multimers (triabodies). Compatible restriction sites *Xho*I and *Sal*I flanking the 15 amino acid scFv linker allow linker deletion and multimer (triabody) formation in a single step. Multimerization has a dramatic effect on the dissociation rate (k_{off}) from solid-phase antigen as shown in the BIAcore trace of an anti-BSA scFv fragment and triabody.

antigen-binding sites of the diabody molecule are located at opposite ends of the molecule at a distance of about 70 Å apart. Despite the reduced span between binding sites in comparison to IgG, bivalent diabodies show dramatically reduced dissociation rates (k_{off}) as compared to the parental scFv molecules (see Fig. 3) and have been shown to be superior to scFv molecules for *in vivo* imaging of tumors.[39]

Recombinant antibodies derived from phage-display libraries can also be multimerized by shortening or removing the V_H–V_L linker, in which case the process of multimerization can be vastly simplified by the incorporation of restriction sites flanking the flexible linker, which are "in frame" and, once cut, can be immediately religated to excise the intervening linker sequence (because intramolecular ligation is favored, no gel purification of the cut fragments are required; see Fig. 3). A similar principle could be

[39] G. P. Adams, R. Schier, A. M. McCall, R. S. Crawford, E. J. Wolf, L. M. Weiner, and J. D. Marks, *Br. J. Cancer* **77**, 1405 (1998).

used to convert entire scFv libraries into multimer libraries without prior knowledge of the V_H and V_L sequences contained within. Such multimers have vastly improved off rates as observed during BIAcore measurement (see later and Fig. 3), even when the concentration of the coated antigen is very low.

In some cases, increased stability may also be desirable. Interchain disulfide bonds[40] have been found to greatly enhance diabody stability *in vitro* and tumor localization *in vivo*.[41] However, disulfide-stabilized diabodies were best produced using *Pichia pastoris* rather than *E. coli* as the expression host (see later). Further shortening of the linker to one to two residues or less promotes formation of a trimeric triabody molecule as shown by X-ray crystallography,[42] as well as gel filtration and ultracentrifugation.[43,44] Like diabodies, triabodies show drasically slower dissociation rates from solid-phase antigen. Their small size (<75 kDa), together with their increased valency, makes triabodies attractive candidate molecules for applications in imaging and immunotherapy.

Engineering of Bispecific Antibody Fragments: Bispecific Diabodies

Bispecific antibodies (BisAbs) can allow the recruitment of a whole range of effector functions, including cytotoxic T cells, natural killer cells, cytotoxic drugs, radionuclides, or even transducing retroviruses, and consequently bispecific antibodies are thought to have immense potential for both diagnostic and therapeutic applications.

Bispecific IgG antibodies have been made by hybridoma fusion[45] and chemical cross-linking,[46] but both methods give rise to complex product mixtures from which the desired bispecific antibody is often difficult to isolate. Protein engineering has been used to produce more defined bispecific reagents by chemical cross-linking of recombinant Fab fragments,[47,48] as well as interface engineering of the CH3 domain to drive bispecific

[40] Y. Reiter and I. Pastan, *Trends Biotech.* **16**, 513 (1998).
[41] K. FitzGerald, P. Holliger, and G. Winter, *Protein Eng.* **10**, 1221 (1997).
[42] X. Y. Pei, P. Holliger, A. G. Murzin, and R. L. Williams, *Proc. Natl. Acad. Sci. U.S.A.* **94**, 9637 (1997).
[43] A. A. Kortt, M. Lah, G. W. Oddie, C. L. Gruen, J. E. Burns, L. A. Pearce, J. L. Atwell, A. J. McCoy, G. J. Howlett, D. W. Metzger, R. G. Webster, and P. J. Hudson, *Protein Eng.* **10**, 423 (1997).
[44] P. Iliades, A. A. Kortt, and P. J. Hudson, *FEBS Lett.* **409**, 437 (1997).
[45] C. Milstein and A. C. Cuello, *Nature* **305**, 537 (1983).
[46] B. Karpovsky, J. A. Titus, D. A. Stephany, and D. M. Segal, *J. Exp. Med.* **160**, 1686 (1984).
[47] M. R. Shalaby, H. M. Shepard, L. Presta, M. L. Rodrigues, P. C. Beverley, M. Feldmann, and P. Carter, *J. Exp. Med.* **175**, 217 (1992).
[48] S. A. Kostelny, M. S. Cole, and J. Y. Tso, *J. Immunol.* **148**, 1547 (1992).

antibody formation in transfectomas.[49] BisAbs can also be constructed by fusing two scFv fragments in tandem.[50]

Another format of bispecific antibody is the bispecific diabody. Unlike other bispecific formats, diabodies can be produced by bacterial secretion with yields approaching 1g/liter[25] and can be displayed on phage and isolated directly from phage repertoires.[51] Bispecific diabodies form as heterodimers of two different scFv fragments in which the cognate V_H and V_L have been swapped. As with bivalent diabodies, heterodimerization is promoted by shortening of the V_H–V_L interdomain linker peptide.[37] While in principle both heterodimers and inactive homodimers can be formed, the correctly paired bispecific heterodimer is commonly the main product, presumably because the cognate V_H and V_L pairings are favored energetically. However, it as been shown that, when necessary, the fraction of active heterodimer can be increased further (from 75 to >95%) using either interchain disulfide bonds or interface engineering.[52] Bispecific diabodies have been shown to be effective in recruiting cytotoxic T cells,[25,53–55] complement,[56,57] and antibody-dependent effector functions such as ADCC, superoxide burst, and phagocytosis,[56] as well as enzymes for histology.[58]

Materials and Methods

We use the *E. coli* suppressor strain TG1 [K12, Δ(*lac-pro*), *sup*E, *thi*, *hsd*D5/F′ *tra*D36, *pro*A$^+$B$^+$, *lac*I^q, *lac*ZΔM15] for propagation of plasmids and antibody expression. Recipes for standard bacterial media, 2× TY for liquid culture, and TYE for plates, as well as standard buffers such as phosphate-buffered saline (PBS) and Tris-EDTA (TE), are as in Sambrook

[49] A. M. Merchant, Z. Zhu, J. Q. Yuan, A. Goddard, C. W. Adams, L. G. Presta, and P. Carter, *Nature Biotech* **16**, 677 (1998).

[50] A. J. George, J. A. Titus, C. R. Jost, I. Kurucz, P. Perez, S. M. Andrew, P. J. Nicholls, J. S. Huston, and D. M. Segal, *J. Immunol.* **152**, 1802 (1994).

[51] B. T. McGuinness, G. Walter, K. FitzGerald, P. Schuler, W. Mahoney, A. R. Duncan, and H. R. Hoogenboom, *Nature Biotech.* **14**, 1149 (1996).

[52] Z. Zhu, L. G. Presta, G. Zapata, and P. Carter, *Protein Sci.* **6**, 781 (1997).

[53] P. Holliger, J. Brissinck, R. L. Williams, K. Thielemans, and G. Winter, *Protein Eng.* **9**, 299 (1996).

[54] W. Helfrich, B. J. Kroesen, R. C. Roovers, L. Westers, G. Molema, H. R. Hoogenboom, and L. de Leij, *Int. J. Cancer* **76**, 232 (1998).

[55] S. M. Kipriyanov, G. Moldenhauer, G. Strauss, and M. Little, *Int. J. Cancer* **77**, 763 (1998).

[56] R. E. Kontermann, M. G. Wing, and G. Winter, *Nature Biotech.* **15**, 629 (1997).

[57] P. Holliger, M. Wing, J. D. Pound, H. Bohlen, and G. Winter, *Nature Biotech.* **15**, 632 (1997).

[58] R. E. Kontermann, P. Martineau, C. E. Cummings, A. Karpas, D. Allen, E. Derbyshire, and G. Winter, *Immunotechnology* **3**, 137 (1997).

et al.[59] For expression in *Pichia pastoris* we use the *Pichia* strain GS115 (Invitrogen). *Pichia pastoris* medium is based on YP (1% yeast extract, 2% peptone), supplemented with glucose (2%) for YPD and with sorbitol (1 M) for YPDS and 2% agar for plates.

Expression and Purification of Antibody Fragments

In order to analyze if an antibody fragment constructed from available V gene segments or isolated from a phage library has the desired binding activity, it is necessary to produce it in recombinant form. For a simple test of activity such as ELISA (see later), a small-scale culture will usually yield sufficient material, and both culture supernatant or periplasmic preparations (see later) are suitable. For more elaborate activity assays it is usually desirable to obtain purified material.

Basic E. coli Expression Protocol. The following protocol is suitable for expression of scFv fragments and diabodies using pUC-based expression plasmids in which expression is controlled by the lac promotor. The lac promotor is inhibited by glucose (Glu) and is induced with isopropyl β-D-thiogalactoside (IPTG).[60] Starting glucose concentrations of 1% or higher are sufficient to repress expression from the lac promotor even after the culture has reached saturation. Glucose concentrations of 0.1–0.2% will result in complete consumption of the glucose during the logarithmic phase of growth (log phase) and derepression of the lac promotor toward the end of the log phase, around about an OD_{600} of 0.8–1.0. The concentration of ampicillin (Amp) is 0.1 mg/ml unless stated otherwise. Because β-lactamase is secreted into the culture medium during growth, ampicillin selection will diminish as the culture reaches saturation. Nondegradable ampicillin derivatives such as carbenicillin (Sigma) offer an alternative, but in the authors experience this does not lead to improved yields of protein expression.

1. Starting from a single colony, grow an overnight culture in 2× TY, Amp (0.1 mg/ml), Glu (2%).
2. Dilute culture 1/100 into 2× TY, Amp, Glu (0.1%), and grow to OD 0.8–1.0 (this takes usually about 2 hr).
3. Induce expression by addition of IPTG to a final concentration of 1 mM and immediately transfer culture to a shaker with a temperature of 22°–33°. Depending on the antibody fragment expressed, yields of func-

---

[59] J. Sambrook, E. F. Fritsch, and T. Maniatis, "Molecular Cloning: A Laboratory Manual." Cold Spring Harbor Laboratory, Cold Spring Harbor, NY, 1990.
[60] D. De Bellis and I. Schwartz, *Nucleic Acids Res.* **18,** 1311 (1990).

tional protein can be dependent on temperature. Diabodies usually give higher functional yields at lower temperatures (e.g., 25°).

4. Incubate (a) for 12 hr (30°) to >24 hr (22°) to harvest antibody fragments secreted into the culture supernatant or (b) for 2 hr (30°) to up to 7 hr (22°) to harvest antibody fragments from the periplasm.

It is good practice (particularly for large-scale preparations) to investigate the build-up of expressed antibody fragment in periplasm and supernatant. Some antibody fragments build up quickly and to high levels in the periplasm prior to leaking into the supernatant. For such fragments a periplasmic prep is suitable. Other fragments build up more slowly. These are best harvested from the supernatant.

Harvesting Fragments from Culture Supernatant

1. Spin induced bacterial culture at 10,000g for 30 min at 4° and collect supernatant.

2. (a) For small-scale expression (<1 liter) the supernatant can be used directly for purification. (b) For large-scale preparations, supernatants are ultra filtered through a 0.16-μm tangential flow filter (Flowgen "Minisette" system) with the use of a peristaltic pump and then concentrated to a suitable final volume (0.5–1 liter) using the same system equipped with a tangential flow filter minisette with a 10-kDa cutoff (for scFv antibody fragments) or a 30-kDa cutoff (for Fab antibody fragments and diabodies).

Harvesting Fragments from Periplasm

1. Spin induced bacterial culture at 10,000g for 30 min at 4° and discard supernatant.

2. Resuspend bacterial pellet in 1/20 the original culture volume of 30 mM Tris, pH 8.0, 20% sucrose, 1 mM EDTA and let stand on ice for 20 min.

3. Spin again at 4000g for 15 min at 4° and collect supernatant (periplasmic fraction). Because of the low-speed spin the periplasmic fraction may look cloudy. However, higher speed centrifugation at this point often leads to substantial cell lysis and release of genomic DNA, making the periplasmic fraction very viscous and hard to handle.

4. Resuspend pellet in 5 mM MgSO$_4$ (1/20 original culture volume) and let stand on ice for 20 min.

5. Spin at 10,000g for 30 min at 4°. Collect supernatant (osmotic shock fraction) and pool with periplasmic fraction.

6. Clear periplasmic preparation with a high-speed spin: 20,000g for 10 min at 4°.

Expression in Yeast Pichia pastoris

The methylotrophic yeast *P. pastoris* is probably the most attractive alternative to *E. coli* as an expression host for antibody fragments. *Pichia pastoris* strains and expression vectors are available from Invitrogen, and *P. pastoris* protocols are available in great detail from their web site (http://www.invitrogen.com/manuals.html). Some antibody fragments, in particular those not derived by phage selection, are sometimes difficult to express in good yields in *E. coli,* but may be expressed in much higher yields from *P. pastoris.* We also found that engineered interdomain disulfide bonds in a diabody were formed more efficiently in *P. pastoris* than in *E. coli.*

P. pastoris Expression Protocol

1. Linearize miniprep DNA from *E. coli* recombinants with restriction enzymes *Avr*II (pGAPZa) or *Bst*X1 (pPICZ). Extract once with phenol/chloroform (1:1) and ethanol precipitate the DNA. Resuspend the precipitation pellet in 5 μl Tris–EDTA.

2. Add 2.5 μl DNA to electrocompetent *P. pastoris* GS115, electroporate (1.5 kV, 25 μFD, 200 Ω) (Bio-Rad Genepulser), resuspend in 1 ml of 1 M sorbitol, and incubate for 2 hr at 30°.

3. Plate on YPDS, 50 μg/ml Zeocin (Invitrogen), and incubate at 30°. Colonies will appear in 3–4 days.

4. Inoculate colonies into 1 ml YPD and grow overnight at 30°.

5. (a) For pGAPZ, dilute 1/100 into fresh YPD medium. (Zeocin selection during expression was found to be unnecessary and reduced the yield of expressed protein.) Protein expression takes place over 1–4 days at 30°. Maximum yields usually are obtained by harvesting on day 3. (b) For pPICZ, dilute 1/100 into fresh YP medium, grow for 24 hr at 30°, add methanol (MeOH) to 10% final concentration, and repeat MeOH addition every 24 hr. Protein expression takes place over 1–4 days at 30°. Maximum yields usually are obtained by harvesting on day 2 or 3.

6. Antibody fragments are harvested from culture supernatants as described for *E. coli* expression, filtered, and concentrated using tangential flow filtration (see earlier discussion).

Purification of Antibody Fragments

Antibody fragments expressed from *E. coli* or *P. pastoris* can be purified in a variety of ways, depending on the antibody fragment and the choice of C-terminal tag. This section focuses on two protocols that have proved to be useful in the laboratory.

Purification Using Immobilized Metal Affinity Chromatography (IMAC). IMAC is useful for the purification of antibody fragments with a polyhistidine tag appended to either the N or the C terminus. We generally use a hexahistidine tag appended to the C terminus. Antibody preparations from *E. coli* and *Pichia* media need to be dialyzed before use on IMAC columns as both commonly used *E. coli* media (2× TY) and *Pichia* media (YP and YPD) contain compounds that chelate metals and strip the metal off the IMAC column. The same is true for periplasmic preparations that contain EDTA.

1. Dialyze antibody preparation against two changes of PBS. For supernatants, dialysis can be performed using tangential flow filtration (by repeated addition of PBS during the concentration process). For smaller volumes, dialysis tubing with a 10-kDa cutoff is suitable.

2. Equilibrate Ni-NTA resin (Diagen) with 10 column volumes of loading buffer (50 m*M* phosphate buffer, pH 7.5, 0.5 *M* NaCl, 20 m*M* imidazole). One milliliter of Ni-NTA resin is usually sufficient to purify 2–3 mg of antibody fragment.

3. Load dialyzed antibody preparation.

4. Wash column with at least 10 column volumes of loading buffer. If the washing process can be observed using a UV flow cell, washing should continue until a stable baseline is reached.

5. Elute antibody fragments using loading buffer and an imidazole gradient from 35 to 200 m*M*. Most antibody fragments are eluted at a concentration of between 50 and 100 m*M* imidazole. Diabodies and triabodies, which have two or three hexahistidine tags, respectively, are usually eluted between 50 and 150 m*M* imidazole. Elution peak fractions should ideally be detected using a UV flow cell and confirmed by ELISA. Alternatively, they may be identified using a BCA protein assay (Pierce).

6. Antibody fractions should be dialyzed into the desired buffer (e.g., PBS) to remove imidazole and concentrated by ultrafiltration using a stirred cell (Amicon) with an appropriate cutoff (10 kDa for scFv fragments, 30 kDa for Fabs and diabodies).

7. Concentrated antibody preparations (>0.5 mg/ml) in PBS are suitable for freezing. As a rule, preparations should always be flash-frozen in dry ice or liquid nitrogen and never in a −20° freezer. Once frozen, a −20° freezer is suitable for short to medium term storage.

Purification Using Protein A or Protein L Columns

1. For a Protein A column, take 1.5 g protein A-Sepharose CL-4B (Amersham Pharmacia), add 10 ml distilled H$_2$O, and mix gently until a thick smooth paste is formed. For a Protein L column, use Protein L-agarose (Affitech).

2. (Steps 2–8 should be performed at 4°.) Pipette 0.5 ml of paste into

a 4-cm Poly-Prep chromatography (Bio-Rad). Let excess liquid drip through by gravity until it stops.

3. Apply 5 ml PBS to the column and let drip through by gravity until it stops.

4. Apply 2 × 10 ml (or more if required) of scFv supernatant (prefiltered through 0.45-μm filter), in each case, let drip through by gravity until it stops.

5. Apply 2 ml PBS and let drip through by gravity until it stops.

6. Apply 10 ml PBS and let drip through by gravity until it stops.

7. Position a 2-ml Eppendorf under the column containing 400 μl 0.1 M sodium phosphate buffer, pH 8.0 (make up according to Sambrook et al.[59]).

8. Apply 1.6 ml of the elution buffer (0.1 M Tris–glycine, pH 3.0) to the column and let drip through into a 2-ml Eppendorf by gravity until it stops.

9. Immediately mix gently and measure approximate concentration at OD 280 nm (1 OD unit is equivalent to between 0.7 and 0.8 mg/ml of protein). Purified scFvs are best stored at −20° and should not be freeze-thawed repeatedly.

Detection and Multimerization of Monovalent Antibody Fragments

Western Blotting: Detection Using Protein L-HRP

1. Perform gel electrophoresis with 4–12% NuPAGE Bis-Tris acrylamide gel using the Novex gel system (Invitrogen) and blot gel onto an Immobilon-P transfer membrane (Millipore) according to the Invitrogen instruction booklet.

2. Wet membrane by immersing in methanol for 5 min.

3. Block membrane in 5% Marvel (skimmed milk powder from a supermarket)–PBS (MPBS) for 1 hr.

4. Incubate with scFv or Fab in 5% MPBS for 1 hr (for supernatants, use 50:50 with blocking buffer, for purified antibody fragments, use at least 1 μg/ml final concentration, although if the antibody is very specific, up to 100 μg/ml can be used).

5. Wash 3 × 5 min with 0.1% Tween–PBS (TPBS).

6. Incubate with 1:5000 Protein L-HRP (Actigen) in 5% MPBS for 1 hr.

7. Wash 4 × 5 min with 0.1% TPBS.

8. Use ECL reagent (Amersham Pharmacia) to develop chemiluminescent signal according to the manufacturer's instructions.

9. Expose to X-ray film (Fuji) for 1 min or longer periods for weaker signals (note that the shortest exposure generally gives the sharpest and most specific signal).

ELISA: Multimerizing Recombinant Antibody Fragments Using Protein L

1. Coat antigen overnight in a 96-well Nunc maxisorp immunoplate (typical coating concentration is 1–100 μg/ml).

2. Wash plate three times with PBS.

3. Block plate by adding 200 μl of 2% Tween–PBS (TPBS) *or* 3% BSA–PBS to each well and incubate for 1 hr.

4. In a separate plate, mix 4 μg/ml of scFv or Fab fragment with 1 μg/ml of Protein L in blocking buffer (2% Tween–PBS *or* 3% BSA–PBS), 100 μl in each well. Incubate for 1 hr. If using supernatants, mix 50:50 with 1 μg/ml Protein L in blocking buffer.

5. Wash the antigen-coated plate with PBS.

6. Transfer the premixed antibody-Protein L solutions into the antigen-coated plate and incubate for 1 hr.

7. Wash plate three times with 0.1% TPBS.

8. To each well add 100 μl Protein A-HRP (Amersham Pharmacia) in blocking buffer (2% Tween–PBS *or* 3% BSA–PBS) and incubate for 1 hr.

9. Wash plate three times with 0.1% TPBS and once with PBS.

10. Develop by adding 100 μl of substrate solution to each well [100 μg/ml 3,3′,5,5′-tetramethylbenzidine (TMB) in 100 mM sodium acetate, pH 6.0, with 10 μl of 30% hydrogen peroxide added to 50 ml of this solution directly before use].

11. Incubate until a blue color develops (typically 1–15 min).

12. Stop the reaction by adding 50 μl of 1 M sulfuric acid. The blue color should change to yellow.

13. Read the OD at 650 nm and at 450 nm. Subtract the OD at 650 nm from the OD at 450 nm.

BIAcore Measurement

1. Take a research grade CM5, chip (BIAcore) and dock in a BIAcore machine according to the manufacturer's instructions.

2. Amine couple ~500 RU of Protein L. Briefly, the surface is first activated using EDC/NHS (typical injection is 30 μl at a 10-μl/min flow rate), Protein L (100 μg/ml in 100 mM sodium acetate, pH 4.0) is then coupled, and the remaining sites are then deactivated using 1.0 M ethanolamine. Alternatively, target antigen can be coupled at the appropriate pH (determined by performing a preconcentration step according to the manufacturer's instructions, do not rely on the pI value of the protein).

3. Pass filtered supernatants containing recombinant antibody (or puri-

fied antibody at known concentration) over the surface (typical injection times range from 1 to 10 min at flow rates of 5–50 μl/min).

4. Plot graph of resonance units bound vs time.

Diabodies/Triabodies

Testing Multivalency. When constructing bivalent diabodies or trivalent triabodies it is a good practice to verify the increased avidity for solid-phase antigen. Standard ELISA assays are usually not sufficient to provide this information, although in the case of a low-affinity monovalent K_{eq} ($<10^{-7} M^{-1}$), avidity is often indicated by a large increase in the ELISA signal. The tell-tale slower dissociation rate (k_{off}) can be measured directly using plasmon surface resonance and the BIAcore biosensor (see earlier discussion). Bivalent diabodies or trivalent triabodies directed against protein antigens typically will have 10- to 100-fold slower k_{off} rates than their monovalent equivalents.[41,61] (Fig. 3). For haptens, up to 10^4-fold reduced k_{off} can be observed (P. Holliger, unpublished results).

Agglutination. This is an inexpensive and simple assay to verify multivalency and has the advantage that the same modified red blood cells may be used in rosetting assays to verify bispecificity (see later). Both sheep and human erythrocytes (EC) are suitable. Human EC may be obtained from the local blood bank.

1. Dilute packed EC fourfold into PBS, spin at 1500 g for 10 min at 4°, discard supernatant. Repeat three times.

2. Prepare solution of desired protein in PBS at a concentration of 20 mg/ml and put on ice.

3. Prepare a solution of 100 mg/ml EDC [1-ethyl-3(3-dimethylamino-propyl)carbodiimide; Sigma] in PBS and put on ice.

4. In a polystyrene tube, carefully layer the EDC solution on top of the protein solution on top of packed EC in the ratio (1 volume EDC:4 volumes protein:1 volume packed EC).

5. Turn tube end over end for 1 hr at 4°.

6. Spin at 1500 g for 10 min at 4°, discard supernatant, and resuspend in an equal volume of PBS.

7. Repeat step 6 until no more EC lysis (red color) is apparent in the PBS supernatant; EC can now be stored for up to 2 weeks in PBS at 4°.

8. For agglutination, mix 50 μl of EC prepared as just described with 50 μl of diabody diluted to an appropriate concentration (e.g., 1 μg/ml) in PBS. Mix by gentle shaking and allow to stand for 1 hr at room temperature.

[61] G. P. Adams, R. Schier, K. Marshall, E. J. Wolf, A. M. McCall, J. D. Marks, and L. M. Weiner, *Cancer Res.* **58,** 485 (1998).

Nonagglutinated EC form a dense pellet at the bottom of the assay well, whereas agglutinated EC are visible as a spread-out coating of the well.

Testing Bispecificity. The bispecificity of a newly constructed diabody fragment can be assayed in a variety of ways. Usually, a bispecific diabody that binds both antigens separately, e.g., in ELISA, will turn out to be bispecific, i.e., able to bind both antigens simultaneously. If both antigens are available in recombinant form the following sandwich ELISA and BIAcore protocols can be used to confirm the simultaneous engagement of both antigen-binding sites and thus genuine bispecificity. In cases where recombinant antigens are not available, rosetting assays, as described in Holliger *et al.*,[56] can be used if both antigens are expressed at medium to high levels (>10,000 copies/cell) on cell lines.

Sandwich ELISA Assay

1. Coat antigen A overnight in a 96-well Nunc maxisorp immunoplate (typical coating concentration is 1–100 μg/ml).
2. Wash plate three times with PBS.
3. Block plate by adding 200 μl of 2% Marvel–PBS (MPBS) and incubate for 2 hr.
4. Add 100 μl diabody diluted in 2% MPBS to each well and incubate for 1 hr.
5. Wash plate three times with PBS.
6. To each well add 100 μl antigen B diluted in 2% MPBS and incubate for 1 hr.
7. Wash plate once with 2% Tween–PBS (TPBS) and twice with PBS.
8. To each well add 100 μl of an anti-antigen B antibody (IgGαB) diluted in 2% MPBS and incubate for 1 hr.
9. Wash plate three times with PBS.
10. To each well add 100 μl of an anti-IgGαB antibody–HRP conjugate diluted in 2% MPBS and incubate for 1 hr.
11. Wash plate three times with PBS and develop ELISA as described earlier.

Sandwich BIAcore Assay. This is useful when no anti-antigen B antibodies are available.

1. Amine couple ~500 RU of antigen A as described earlier to a CM5 BIAcore chip.
2. Pass filtered diabody preparation over the surface (typical injection times range from 1 to 10 min at flow rates of 5–50 μl/min).
3. Immediately after diabody injection has finished, pass the filtered antigen B preparation over the surface. In order to get clear capture of antigen B by the diabody, concentrations of antigen B should be higher than the diabody preparation (preferably >10 μg/ml)
4. Plot graph of resonance units bound vs time.

Future Directions

In nature, antibodies exist only as dimers and higher order multimers and are directed against a single antigen. As a result, monoclonal and polyclonal antibodies derived by animal immunization are not only highly specific, but can give extremely sensitive detection due to their prolonged retention on antigen. In contrast, most systems for bacterial expression of antibodies yield monomeric fragments that have off rates in the region of seconds or minutes and thus their utility as affinity reagents, for Western blotting, ELISA, or immunoprecipitation, is generally much poorer than their natural counterparts. Using simple tricks of recombinant DNA technology and protein engineering we can now create multivalent and multispecific antibody fragments to order. These have all the advantages of bacterial expression but are able to mimic (in the case of multivalency) or improve upon (in the case of multispecificity) naturally expressed antibodies.

Multivalent antibody fragments have a broad range of applications from enhanced signal detection as research reagents in immunochemistry to improved retention on cell surfaces when used as *in vivo* therapeutics or diagnostics. In many cases, the effect of multimerization is so astounding that a strong signal can be seen where previously there was none at all. Thus, the use of multimers is likely to be particularly important when the amount of antigen is limiting (e.g., when detecting very small amounts of a single protein in a cell lysate) or when the monomeric affinity of the parental clone is low. Because the conversion of monomers to multimers by vector manipulation can be laborious, we envisage the creation of phage-displayed multimer libraries that have "built-in" binding sites for superantigens such as Protein A and Protein L and are constructed from frameworks that are well expressed in bacteria. Such "single-pot" multimer libraries will provide a direct route for the production of multivalent antibody fragments to any given antigen, which can later be crossed with binders to other antigens to create multispecific antibody fragments, perhaps in the form of bispecific diabodies, triabodies, or even tetrabodies. Such multispecific fragments could be stabilized by incorporating disulfide bonds between V_H-V_L domains or by introducing additional mutations into the framework regions to enhance the folding characteristics of the expressed antibody fragments.[62] Indeed, these features could also be "built-in" to the initial multimer libraries. In summary, we believe that the use of multivalent and multispecific antibody fragments will continue to grow, particularly as phage display and other methods for *in vitro* selection become more widespread.

[62] P. Kristensen and G. Winter, *Folding Design* **3,** 321 (1998).

[29] Design and Use of Phage Display Libraries for the Selection of Antibodies and Enzymes

By Francesca Viti, Fredrik Nilsson, Salvatore Demartis, Adrian Huber, and Dario Neri

Introduction

Phage antibody technology, i.e., the display and use of antibody repertoires on the surface of bacteriophages, is a simple and inexpensive methodology that reliably yields specific monoclonal antibody fragments in 1–2 weeks of limited experimental work provided that a small amount of pure antigen is available.[1]

Phage display was originally described in 1985 by Smith,[2,3] who presented the use of the nonlytic filamentous bacteriophage fd for the display of specific binding peptides on the phage coat. The power of the methodology was further enhanced by the groups of Winter[4] and Wells,[5] who demonstrated the display of functional folded proteins on the phage surface (an antibody fragment and a hormone, respectively). The attraction of phage display relies on the fact that a polypeptide (capable of performing a function, e.g., an antibody fragment binding to its antigen) can be displayed as fusion protein on the phage surface by inserting the gene coding for the polypeptide in the phage genome (Fig. 1). On purification of a phage particle by virtue of the binding specificity displayed on its surface from a large repertoire of phage particles (e.g., a phage library of antibodies), one also isolates the genetic information coding for the binding protein and can amplify the corresponding phage by bacterial infection (Fig. 2). Phage remains infective when treated with acids, bases, denaturants, and even proteases, allowing a variety of selective elution protocols. Several rounds of selections can be performed. As a consequence, even very rare binding specificities present in large repertoires can be selected and amplified from a background of phages with irrelevant binding specificities.

Filamentous phage infect strains of *Escherichia coli* that harbor the F conjugative episome. On infection, phages translocate their genome (a

---

[1] G. Winter, A. D. Griffiths, R. E. Hawkins, and H. R. Hoogenboom, *Annu. Rev. Immunol.* **12**, 433 (1994).

[2] G. P. Smith, *Science* **228**, 1315 (1985).

[3] G. P. Smith, *Curr. Opin. Biotechnol.* **2**, 668 (1991).

[4] J. McCafferty, A. D. Griffiths, G. Winter, and D. J. Chiswell, *Nature* **348**, 552 (1990).

[5] H. B. Lowman, S. H. Bass, N. Simpson, and J. A. Wells, *Biochemistry* **30**, 10832 (1991).

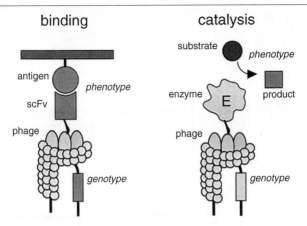

FIG. 1. Phage display of recombinant antibody fragments and enzymes. Proteins can be displayed on the surface of filamentous fd phage on fusion to the minor coat protein pIII. Phage displaying binding proteins, such as recombinant antibody fragments, can be physically isolated on antigen binding, together with the gene coding for the binding protein. Physical isolation of phage enzymes is more difficult, as reaction substrates and products are not anchored to the phage.

circular single-stranded DNA molecule) into the bacterial cytoplasm. The genome is replicated, involving both phage- and host-derived proteins, and packaged into elongated ("filamentous") viral particles of approximately 6 nm diameter and 900 nm length. For a review on phage biology, see Webster.[6]

Approximately 3000 copies of the major coat protein pVIII cover the majority of the phage surface. A minor coat protein, pIII, which is displayed in three to five copies and mediates the adsorption of the phage to the bacterial pilus, is the coat protein used most typically for the phage display of folded proteins.[2,7] Proteins can be displayed on phage by cloning the corresponding gene into phage vectors (essentially the phage genome with suitable cloning sites for pIII fusions and an antibiotic resistance gene). Phage vectors carry all the genetic information necessary for phage life. With pIII fusions in phage vectors, in the absence of proteolysis, each pIII coat protein displayed on phage is fused to the heterologous polypeptide.

Phagemids, a more popular vector for display, are plasmid vectors that carry gene III with appropriate cloning sites and a phage-packaging sig-

[6] R. Webster, in "Phage Display of Peptides and Proteins" (B. Kay, L. Winter, and J. McCafferty, eds.), p. 1. Academic Press, San Diego, 1996.
[7] S. F. Parmley and G. P. Smith, Gene 73, 305 (1988).

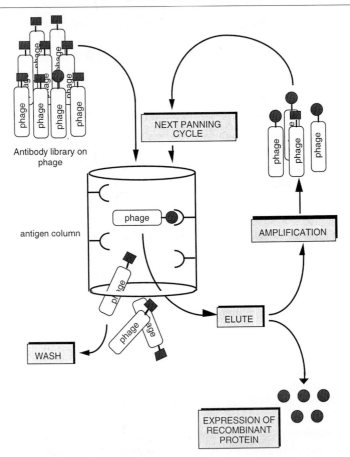

FIG. 2. Schematic representation of selection of phages from an antibody phage display library. Phage–antibodies with the desired binding specificity remain bound to the antigen immobilized on a support after several washing steps. They can be selected and amplified using successive rounds of panning. At the end of the process, monoclonal antibody fragments can be produced as recombinant proteins.

nal.[8–10] The phagemid encoding the polypeptide–pIII fusion is packaged preferentially into phage particles using a helper phage that contains a slightly defective origin of replication, such as M13K07 or VCS-M13, which supplies all the structural proteins. The resulting phage particles may incor-

[8] H. R. Hoogenboom, A. D. Griffiths, K. S. Johnson, D. J. Chiswell, P. Hudson, and G. Winter, *Nucleic Acids Res.* **19,** 4133 (1991).
[9] L. J. Garrard, M. Yang, M. P. O'Connell, R. F. Kelley, and D. J. Henner, *Biotechnology* **9,** 1373 (1991).
[10] S. Bass, R. Greene, and J. A. Wells, *Proteins* **8,** 309 (1990).

porate either pIII derived from the helper phage or the polypeptide–pIII fusion, encoded by the phagemid. Depending on the type of phagemid, growth conditions used, and the nature of the polypeptide fused to pIII, ratios of (polypeptide–pIII):pIII ranging between 1:9 and 1:1000 have been reported.[10–14] Proteolytic cleavage of antibody–pIII fusions may further contribute to elevated levels of full-length wild-type pIII.[4,15]

Antibody phage technology is particularly powerful when using "single-pot" libraries because they virtually contain all the possible binding specificities and are not biased for a particular antigen. These single-pot libraries are cloned once, with the aim to reach the largest possible diversity (ideally 10^9–10^{10} different clones). The corresponding phages are stored frozen in aliquots[16] and can be used directly in panning experiments against a variety of different antigens. Both library design and library size contribute to the performance of the library and to the quality of the antibodies isolated. Larger libraries have a higher probability of containing high-affinity antibodies.[17,18] However, the combinatorial diversity that can in practice be explored in panning experiments is limited by several factors, including the solubility of phage particles (typically $\leq 10^{13}$ transforming units/ml), the efficiency of antibody display on phage, and the phage recovery yields in biopanning experiments.[19]

It is convenient to classify single-pot libraries into "naive" repertoires[18,20–24] (in which V genes are isolated from unimmunized animals or

[11] H. Bothmann and A. Pluckthun, *Nature Biotechnol.* **16,** 376 (1998).

[12] S. Demartis, A. Huber, F. Viti, L. Lozzi, L. Giovannoni, P. Neri, G. Winter, and D. Neri, *J. Mol. Biol.* **286,** 617 (1999).

[13] P. Kristensen and G. Winter, *Fold Des.* **3,** 321 (1998).

[14] M. Tesar, C. Beckmann, P. Rottgen, B. Haase, U. Faude, and K. N. Timmis, *Immunotechnology* **1,** 53 (1995).

[15] J. McCafferty, in "Phage Display of Peptides and Proteins"(B. Kay, L. Winter, and J. McCafferty, eds.), p. 261. Academic Press, San Diego, 1996.

[16] D. Neri, A. Pini, and A. Nissim, *Methods Mol. Biol.* **80,** 475 (1998).

[17] A. D. Griffiths, S. C. Williams, O. Hartley, I. M. Tomlinson, P. Waterhouse, W. L. Crosby, R. E. Kontermann, P. T. Jones, N. M. Low, T. J. Allison *et al., EMBO J,* **13,** 3245 (1994).

[18] T. J. Vaughan, A. J. Williams, K. Pritchard, J. K. Osbourn, A. R. Pope, J. C. Earnshaw, J. McCafferty, R. A. Hodits, J. Wilton, and K. S. Johnson, *Nature Biotechnol.* **14,** 309 (1996).

[19] H. J. de Haard, B. Kazemier, M. J. Koolen, L. J. Nijholt, R. H. Meloen, B. van Gemen, H. R. Hoogenboom, and J. W. Arends, *Clin. Diagn. Lab. Immunol.* **5,** 636 (1998).

[20] J. D. Marks, H. R. Hoogenboom, T. P. Bonnert, J. McCafferty, A. D. Griffiths, and G. Winter, *J. Mol. Biol.* **222,** 581 (1991).

[21] M. D. Sheets, P. Amersdorfer, R. Finnern, P. Sargent, E. Lindqvist, R. Schier, G. Hemingsen, C. Wong, J. C. Gerhart, and J. D. Marks, *Proc. Natl. Acad. Sci. U.S.A.* **95,** 6157 (1998).

[22] R. Schier, J. D. Marks, E. J. Wolf, G. Apell, C. Wong, J. E. McCartney, M. A. Bookman, J. S. Huston, L. L. Houston, L. M. Weiner *et al., Immunotechnology* **1,** 73 (1995).

[23] A. M. Merchant, Z. Zhu, J. Q. Yuan, A. Goddard, C. W. Adams, L. G. Presta, and P. Carter, *Nature Biotechnol.* **16,** 677 (1998).

human donors, and are combinatorially assembled to create large arrays of antibodies) and "synthetic" repertoires [in which combinatorial diversity is introduced into a limited number of antibody scaffolds, e.g., by appending degenerate complementarity determining regions (CDRs) to a fixed antibody sequence].

While it is by now clear that high-affinity antibodies can be isolated rapidly from naive libraries if the corresponding pure antigen is available, potential disadvantages are (1) the lower affinity rescued when smaller repertoires are used; (2) the time and effort needed to construct these libraries; (3) the largely unknown and uncontrolled content of the library; and (4) the need to sequence the isolated antibodies and to design custom primers for affinity maturation strategies based on combinatorial mutagenesis of CDRs. Furthermore, it remains to be seen how well naive libraries perform against very conserved antigens for which the immune system is tolerant.

Synthetic repertoires[17,25–32] have the advantage that the content of the library (antibody structure, codon usage, definition of the antibody portions that are randomized and of those that are kept constant) is defined a priori. Furthermore, sequence diversity can be concentrated on those amino acid residues that are supposedly more important for antigen binding.[33] Because antibody genes have not undergone any immunological selection, the library is not biased against self-antigens. Indeed, synthetic libraries have already yielded good-quality antibodies against conserved antigens, such as calmodulin,[17] the ED-A[34] and ED-

[24] J. K. Osbourn, A. Field, J. Wilton, E. Derbyshire, J. C. Earnshaw, P. T. Jones, D. Allen, and J. McCafferty, *Immunotechnology* **2**, 181 (1996).

[25] A. Nissim, H. R. Hoogenboom, I. M. Tomlinson, G. Flynn, C. Midgley, D. Lane, and G. Winter, *EMBO J.* **13**, 692 (1994).

[26] C. F. D. Barbas, J. D. Bain, D. M. Hoekstra, and R. A. Lerner, *Proc. Natl. Acad. Sci. U.S.A.* **89**, 4457 (1992).

[27] J. de Kruif, E. Boel, and T. Logtenberg, *J. Mol. Biol.* **248**, 97 (1995).

[28] Y. Akamatsu, M. S. Cole, J. Y. Tso, and N. Tsurushita, *J. Immunol.* **151**, 4651 (1993).

[29] L. J. Garrard and D. J. Henner, *Gene* **128**, 103 (1993).

[30] A. Pini, F. Viti, A. Santucci, B. Carnemolla, L. Zardi, P. Neri, and D. Neri, *J. Biol. Chem.* **273**, 21769 (1998).

[31] M. Braunagel and M. Little, *Nucleic Acids Res.* **25**, 4690 (1997).

[32] E. Soderlind, M. Vergeles, and C. A. Borrebaeck, *Gene* **160**, 269 (1995).

[33] I. M. Tomlinson, G. Walter, P. T. Jones, P. H. Dear, E. L. Sonnhammer, and G. Winter, *J. Mol. Biol.* **256**, 813 (1996).

[34] L. Borsi, P. Castellani, G. Allemanni, D. Neri, and L. Zardi, *Exp. Cell Res.* **240**, 244 (1998).

B domain of fibronectin,[30,35,36] or against "difficult" antigens such as BiP.[37]

Using principles of protein design,[30,38] we produced a novel synthetic library containing $>5 \times 10^8$ individual clones (the "ETH-2 library"). The construction and use of this library are described in the first part of this article. Several dozens of different binding specificities have already been isolated in over 60 laboratories to which the library has been distributed.

Displaying enzymes on phage and selecting phage–enzymes on the basis of their catalytic properties is more difficult. In fact, reaction products (which represent the biochemical memory of the reaction catalyzed by the phage particle) diffuse away after the reaction is complete (Fig. 1). However, by anchoring the reaction products on the phage surface, it should be possible to physically isolate the catalytically active phages using specific antiproduct affinity reagents.[12,39–42]

We reported the display of enzyme–calmodulin fusions to gene III on the surface of filamentous phage as a means of providing the conditional anchoring of reaction substrates and products on phage.[12,40] The principle of the methodology is depicted in Fig. 3. The substrate of the reaction of interest is conjugated chemically to a recently isolated high-affinity calmodulin-binding peptide ($K_d = 2$ pM; Montigiani et al.[43]). The aim is to keep the substrate of reaction (and eventually the reaction product) noncovalently but tightly linked to the phage particles. We expect that the phage displaying a catalytically active enzyme will convert the substrate into the product efficiently with an intramolecular reaction. Antibodies (or other affinity reagents) against the reaction product are eventually used for the physical isolation of those clones that have catalyzed the reaction. As the

[35] D. Neri, B. Carnemolla, A. Nissim, A. Leprini, G. Querze, E. Balza, A. Pini, L. Tarli, C. Halin, P. Neri, L. Zardi, and G. Winter, *Nature Biotechnol,* **15,** 1271 (1997).

[36] B. Carnemolla, D. Neri, P. Castellani, A. Leprini, G. Neri, A. Pini, G. Winter, and L. Zardi, *Int. J. Cancer* **68,** 397 (1996).

[37] A. D. Griffiths, M. Malmqvist, J. D. Marks, J. M. Bye, M. J. Embleton, J. McCafferty, M. Baier, K. P. Holliger, B. D. Gorick, N. C. Hughes-Jones et al., *EMBO J.* **12,** 725 (1993).

[38] A. Pini, A. Spreafico, R. Botti, D. Neri, and P. Neri, *J. Immunol. Methods* **206,** 171 (1997).

[39] K. D. Janda, L. C. Lo, C. H. L. Lo, M. M. Sim, R. Wang, C. H. Wong, and R. A. Lerner, *Science* **275,** 945 (1997).

[40] D. Neri, D. Tawfik, G. Winter, S. Demartis, F. Viti, and A. Huber, International Patent Application No. PCT/GB97/01153, 1997.

[41] H. Pedersen, S. Holder, D. P. Sutherlin, U. Schwitter, D. S. King, and P. G. Schultz, *Proc. Natl. Acad. Sci. U.S.A.* **95,** 10523 (1998).

[42] J. L. Jestin, P. Kristensen, and G. Winter, *Angew. Chem. Int. Ed. Engl.* **38,** 1124 (1999).

[43] S. Montigiani, G. Neri, P. Neri, and D. Neri, *J. Mol. Biol.* **258,** 6 (1996).

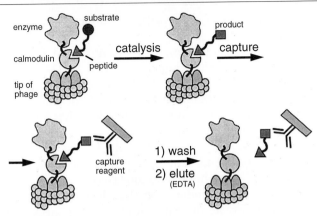

FIG. 3. Calmodulin-tagged phage enzyme for the selection of enzymatic acitvity. High-affinity calmodulin-binding peptides can be used for the noncovalent stable anchoring of reaction substrates and products on the surface of filamentous phage and for the selection of active phage enzymes with antiproduct affinity reagents. The calmodulin/ligand complex can be dissociated by the addition of calcium chelators.

complex between calmodulin and peptide derivative is calcium dependent, a mild and selective elution can be achieved by the addition of calcium chelators.

Protocols for the display of enzymes on calmodulin-tagged phage and for the selection of catalytic activities from mixtures of phage–enzymes are reported in the second part of this article.

Materials and Methods

Standard materials, culture media, bacterial strains, and helper phage for work with phage libraries have been described previously[44] and will not be reported here. Other relevant material and reagents are indicated in the following sections.

Unless otherwise stated in the text:
2× TY-Amp-Glu = 2× TY, 100 μg/ml ampicillin, 1% (w/v) glucose
2× TY-Amp-Kan = 2× TY, 100 μg/ml ampicillin, 33 μg/ml kanamicin
2× TY-Amp-Glu plates = 2× TY, 100 μg/ml ampicillin, 1% (w/v) glucose, 15 g/liter agar
Bacteria are grown in liquid media at 37° in an orbital shaker at 220 rpm
Agar plates are grown at 30° overnight
Bacterial infections with phages are carried out for 30–45 min in a 37° water bath without shaking

[44] J. L. Harrison, S. C. Williams, G. Winter, and A. Nissim, *Methods Enzymol.* **267,** 83 (1996).

Bacterial cultures OD is measured with a spectrophotometer at a wavelength of 600 nm

Bacterial cultures are defined here as exponentially growing at an OD of 0.4–0.5

Centrifugations are carried out at 4°

The ETH-2 Antibody Phage Library

The ETH-2 human antibody library is a synthetic library of scFv fragments displayed on the surface of filamentous phage as pIII fusion. In the scFv format chosen, the antibody fragment consists of a single polypeptide composed of a heavy chain variable domain, sequentially linked through a flexible polypeptide linker to a light chain variable domain. In the construction of this library, our aim has been

1. to choose a basic antibody structure, known to be properly expressed on phage and well expressed in solution, that allows an easy purification by chromatography on protein A resin
2. to concentrate combinatorial diversity in few, judiciously chosen amino acid positions in the antibody structure, while keeping the remaining parts of the antibody molecule constant
3. to clone the library in a phagemid vector that allows either the conditional antibody display on phage or the secretion of the antibody in the periplasmic space of *E.coli* bacteria in a form that can be detected easily by ELISA assay or labeled radioactively

The heavy chain variable domains of the ETH-2 library consist of a constant germline V segment (DP-47)[45] to which completely randomized sequences of four, five, or six amino acids are appended by polymerase chain reaction (PCR) with degenerate primers (Fig. 4). Using this approach, combinatorial diversity is introduced at the level of the complementarity determining region 3 (CDR-3), known to be the CDR, which makes most contacts with the antigen. The sequence continues further with a constant segment (FDYWGQGTLVTVSS) and is followed by a flexible peptide linker, connecting it to the light chains.

Two different types of light chain variable domains (one Vκ and one Vλ) can be paired to the VH domain in the ETH-2 library. These light chain variable domains are composed of a germline segment (DPK-22 or DPL-16, respectively)[20,46] to which a partially degenerate CDR3 is appended, followed by a constant segment (Fig. 4).

[45] I. M. Tomlinson, G. Walter, J. D. Marks, M. B. Llewelyn, and G. Winter, *J. Mol. Biol.* **227,** 776 (1992).
[46] J. P. Cox, I. M. Tomlinson, and G. Winter, *Eur. J. Immunol.* **24,** 827 (1994).

VH (based on DP-47)

```
 1      6        91      ╱ XXXX  ╲
EVQLLES.....YCAK  ╱─ XXXXX ─╲  FDYWGQGTLVTVSS-linker
  ──── DP-47 ────    ╲ XXXXXX ╱
```

Vk (based on DPK-22)

```
   1       6         90  ╱ XGXXPX ╲
linker-EIVLTQS.....YCQQ ╱           ╲ TFGQGTKVEIK
                         ╲ XXGXPX  ╱
          ──── DPK-22 ────
```

Vλ (based on DPL-16)

```
                              ╱ PXXXXX ╲
                             ╱  XPXXXX  ╲
   1       6         90  ╱   XXPXXX    ╲
linker-SSELTQD.....CNSS ╱    XXXXPX    ╲ VVFGGGTKLTVLG
                         ╲   XXXXXP    ╱
          ──── DPL-16 ────
```

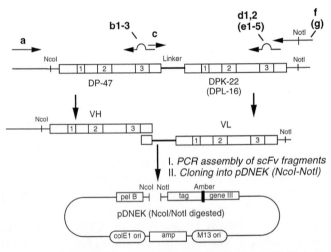

FIG. 4. Design of the ETH-2 antibody phage library. (Top) The sequence of relevant residues of the variable heavy and light chains, together with the human antibody germline segments from which they were derived. Amino acids are indicated with single letter nomenclature. Combinatorially mutated residues are indicated with the code "X". (Bottom) The cloning strategy used to introduce sequence diversity in scFv antibody genes. Oligonucleotides are indicated as arrows. The round portion of the arrow indicates the approximate locations in which degenerate codons are introduced. Antibody complementarity-determining regions are indicated as boxes with numbers 1, 2, and 3. The sequence of relevant oligonucleotides is presented in Table I. A more complete description of the vector pDNEK is presented in Fig 5.

Reasons for introducing combinatorial diversity at the positions indicated in Fig. 4 are described elsewhere.[30,38] The cloning strategy is depicted in Fig. 4; the sequences of relevant oligonucleotides are listed in Table I. The first part of this article describes how the ETH-2 library was constructed using a phagemid vector (Fig. 5), degenerate oligonucleotide primers, and PCR assembly methodologies.[30] Templates for the PCR reactions depicted in Fig. 4 were two previously described scFv fragments, CGS-2[35,36] and E1.[30] These clones contained a Vλ- and Vκ-based light chain, respectively.

The second part of this article covers the whole process of selecting phage antibodies, including antigen immobilization, panning procedures, ELISA screening of isolated antibody clones, and affinity purification of monoclonal human antibody fragments of interest.

Library Construction

In order to obtain a library of large dimension, a substantial amount of digested vector (either phagemid or phage vector) and of digested insert (coding for scFvs) needs to be prepared. To obtain the size of 5×10^8 clones of the ETH-2 library, more than 60 μg of digested vector was ligated with 40 μg of insert. The quality of digestion is essential, as a poorly digested vector leads to clones without insert and a poorly digested insert leads to low ligation efficiences. High amounts of pure double digested vector can be obtained if the phagemid contains a long "dummy" sequence inserted between the two restriction sites chosen for the cloning of the library. After double enzymatic digestion, it is easy to separate on agarose gel the double-digested vector from the long "dummy" insert and non- or single-digested vector. For the cloning of the ETH-2 library we used the phagemid pDNEK (Fig. 5), a pDN332 derivative[30] where we cloned a long "dummy" insert of 1600 bp between the restriction sites *Nco*I and *Not*I.

Good-quality PCR primers are also very important, both for the performance of the PCR reactions and for the quality of the insert's digestion. Ligation and bacterial transformation are crucial, as they directly influence the library's size.

This section describes the protocol used for the cloning of the ETH-2 library (see Fig. 4 for illustration of the cloning strategy). For basic molecular biology techniques, refer to Sanbrook *et al.*[47]

Phagemid Vector

1. Prepare 250 μg of purified vector pDNEK using the QIAfilter plasmid maxi kit (Qiagen, Basel, Switzerland).

[47] J. Sambrook, E. F. Fritsch, and T. Maniatis, "Molecular Cloning: A Laboratory Manual." Cold Spring Harbor Laboratory Press, Cold Spring Harbor, NY, 1989.

TABLE I
Sequence of Relevant Oligonucleotide Primers[a]

Primers for construction of ETH-2 library

A 5'-GCG GCC CAG CCG GCC ATG GCC GAG-3'

B1 5'-GT TCC CTG GCC CCA GTA GTC AAA MNN MNN MNN TTT CGC ACA GTA ATA TAC GGC C-3'

B2 5'-GT TCC CTG GCC CCA GTA GTC AAA MNN MNN MNN TTT CGC ACA GTA ATA TAC GGC C-3'

B3 5'-GT TCC CTG GCC CCA GTA GTC AAA MNN MNN MNN MNN TTT CGC ACA GTA ATA TAC GGC C-3'

C 5'-TTT GAC TAC TGG GGC CAG GGA ACC CTG GTC-3'

D1 5'-CAC CTT GGT CCC TTG GCC GAA CGT MNN CGG MNN ACC MNN CTG ACA GTA ATA CAC TGC-3'

D2 5'-CAC CTT GGT CCC TTG GCC GAA CGT MNN CGG MNN ACC MNN CTG ACA GTA ATA CAC TGC-3'

E1 5'-CTT GGT CCC TCC GCC GAA TAC CAC MNN MNN MNN GGG AGA GGA GTT ACA GTA ATA GTC-3'

E2 5'-CTT GGT CCC TCC GCC GAA TAC CAC MNN MNN MNN GGG MNN AGA GGA GTT ACA GTA ATA GTC-3'

E3 5'-CTT GGT CCC TCC GCC GAA TAC CAC MNN MNN MNN GGG MNN MNN AGA GGA GTT ACA GTA ATA GTC-3'

E4 5'-CTT GGT CCC TCC GCC GAA TAC CAC MNN GGG MNN MNN MNN AGA GGA GTT ACA GTA ATA GTC-3'

E5 5'-CTT GGT CCC TCC GCC GAA TAC CAC GGG MNN MNN MNN MNN AGA GGA GTT ACA GTA ATA GTC-3'

F 5'-TCA TTC TCG ACT TGC GGC CGC TTT GAT TTC CAC CTT GGT CCC TTG GCC GAA CG-3'

G 5'-GAG TCA TTC TCG ACT TGC GGC CGC TAG GAC GGT CAG CTT GGT CCC TCC GCC GAA-3'

Primers for sequencing of ETH-2 library

S1 5'-CAG GAA ACA GCT ATG ACC ATG ATT AC-3'

S2 5'-ACA TAC TAC GCA GAC TCC GTG AAG-3'

S3 5'-GAA TTT TCT GTA TGA GG-3'

[a] The use of primers for the construction of the ETH-2 library is depicted in Fig. 4. The use of primers for the DNA sequencing of recombinant antibody fragments selected from the ETH-2 library is indicated in Fig. 5.

| Not 1 | Phosphorylation site | FLAG-tag | His₆-tag |

R A A *D D D S D D D* <u>D Y K D D D D K</u> # H H H   H   H H amber
5'-<u>GCG GCC GCA</u> GAT GAC GAT TCC GAC GAT GAC TAC AAG GAC GAC GAC GAC AAG # CAC CAT CAC CAT CAC CAT TAG-3'

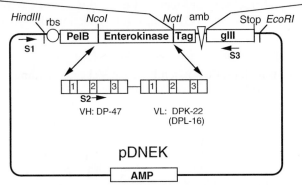

marks the cleavage site of the endopeptidase enterokinase.

FIG. 5. Schematic representation of the phagemid vector pDNEK, containing a dummy enterokinase insert. The sequence between the *Not*I cloning site and the amber codon, coding for the peptide tag of the antibody fragments, is indicated. The schematic structure of the scFv antibody gene to be cloned in the vector, as well as the position of oligonucleotide primers useful for antibody sequencing (short arrows), is indicated. PelB: leader peptide; rbs: ribosome binding site; gIII: gene III; S1–S4: sequencing primers; amb: amber codon.

2. Double digest pDNEK with *Nco*I/*Not*I [*Nco*I: 1 unit (U)/μg DNA; *Not*I: 2 U/μg DNA; DNA concentration in the digestion mix: lower than 0.1 μg/μl] for 4 hr at 37°.

3. Purify the double-digested vector by extraction from agarose gel using the QIAEX II gel extraction kit (Qiagen). In a 1.4% agarose gel, the double-digested vector can be separated easily from the other species (undigested, single digested vector and 1600-bp insert). The loading of undigested vector in another lane of the same gel will facilitate the recognition of all the DNA species.

4. If not used immediately, freeze the purified *Nco*I/*Not*I vector in liquid nitrogen and store at −80°.

VH, Vκ, and Vλ Genes

5. PCR amplify heavy and light chain V genes (Fig. 4) in parallel reactions as follows. In a total reaction volume of 50 μl, use 20 pmol nondegenerate primer, 40 pmol degenerate primer, <30 ng scFv template, 5 m*M* dNTPs, appropriate *Taq* polymerase buffer, and mineral oil (Sigma).

   a. On a thermal cycler, start the program: 94° (3 min) − [94° (1 min) − 60° (1 min) − 72° (1 min)]₂₅ cycles − 65° (2 min).

   b. Add 3 U *Taq* polymerase (Appligen Oncor, Basel, Switzerland) when the temperature of 94° of the first step has been reached.

   c. In different tubes, use the following 10 pairs of oligonucleotide primers (Fig. 4, Table I): a,b1; a,b2; a,b3; (for the VH), c,d1; c,d2 (for the Vκ), and c,e1; c,e2; c,e3; c,e4; c,e5 (for the Vλ).

6. Purify the correct size product of the 10 different PCR reactions by extraction from agarose gel (QIAquick gel extraction kit, Qiagen).

PCR Assembly of VH and VL Chains into VH-VL scFv

7. In parallel reactions, PCR assemble each VH chain-coding sequence with each VL chain-coding sequence (21 different reactions), using as reaction templates the PCR products purified in step 6 and the following primer pairs: a,f (for VH-Vκ assembly) and a,g (for VH-Vλ assembly). PCR assembly conditions and program are as in step 5, *but* use 10 pmol of both the primers and 10 ng of template for each reaction. Perform every PCR assembly reaction in several copies to end up with 3 μg of each VH-VL assembly (see also Fig. 4 and Table I).

8. Pool the VH-Vκ assembly products and, in another tube, the VH-Vλ assembly products.

9. Purify using the QIAquick PCR purification kit (Qiagen).

10. Double digest the DNA fragments of step 9 (keeping the two pools separate) with enzymes *NcoI/NotI* (*NcoI*: 2 U/μg DNA; *NotI*: 4 U/μg DNA; DNA concentration in the digestion solution: lower than 0.1 μg/μl). Incubate for 4 hr at 37°.

11. Purify the digested fragments (QIAquick PCR purification kit, Qiagen).

12. If not used immediately, freeze the purified *NcoI/NotI* inserts in liquid nitrogen and store at −80°. Keep the VH-Vλ double-digested insert and the VH-Vκ double-digested insert separate.

Ligation

13. Ligate 30 μg of *NcoI/NotI*-digested vector with 20 μg of *NcoI/NotI*-digested VH-Vλ insert and, in a separate tube, 30 μg of *NcoI/NotI*-digested vector with 20 μg of *NcoI/NotI*-digested VH-Vκ insert. Use 100 U/μg DNA of T4 DNA ligase (Eurogentec, Seraing, Belgium) and incubate at 15° for 14–18 hr.

Note: It is advisable, from this point to the end, to split the building of the library in two parts and perform it on different days for a "VH-Vλ sublibrary" and for a "VH-Vκ sublibrary."

14. Purify the DNA by phenol extraction followed by ethanol/sodium acetate precipitation and two 70% ethanol washes, according to standard protocols. Finally, dissolve the precipitated DNA of each of the two ligations in 500 μl of sterile water.

Electrocompetent E. coli TG1 Cells

15. Prepare electrocompetent *E. coli* TG1 cells as follows.
   a. Inoculate 2 liters (for each of the two sublibraries) of 2× TY with 1/100 volume of an overnight culture of TG1 bacteria.
   b. Grow to OD 0.5–0.7.
   c. Chill the flask for 15–30 min on ice.
   d. Spin at 4000*g* maximum for 15 min in a prechilled rotor.
   e. Resuspend the cells in the original volume of ice-cold 1 m*M* HEPES, pH 7.0, in water, 5% glycerol.
   f. Spin as before.
   g. Resuspend the cells in half of the original volume of ice-cold 1 m*M* HEPES, pH 7.0, in water, 5% glycerol.
   h. Wash in 40 ml 10% glycerol plus 1 m*M* HEPES, pH 7.0.
   i. Resuspend in 4 ml 10% glycerol.

Electroporation of E. coli TG1 Cells with the DNA Construct

16. Mix 500 μl of DNA of each of the two ligations from step 14 with 4 ml of fresh electrocompetent *E. coli* TG1 cells. The efficiency of transformation should be $>5 \times 10^8$ transformants/μg of plasmid DNA.

17. Aliquot in prechilled 2-mm electroporation cuvettes (Eurogentec) (50 μl/cuvette).

18. Perform electroporation (Gene Pulser II Electroporation System, Bio-Rad, CA), Set parameters: capacitance 25 μF; voltage 2.5 kV; and resistance 200 Ω.

19. Immediately after every electroporation, add 70 μl of sterile 2× TY-Amp-Glu and plate in a 20-cm-diameter plate (2× TY-Amp-Glu) by gently spreading with a sterile glass loop in half of the plate's surface (every plate is used for the plating of two electroporations).

20. To calculate the size of the library, make appropriate dilution series of transformed bacteria after several electroporations and plate 2× TY-Amp-Glu plates of standard size.

Size of the library = (No. of colonies in a plate) × (dilution factor)
× (No. of electroporations)

The number of colonies is the average number of colonies obtained from the plating of bacteria after several electroporations.

21. Incubate the plates overnight at 30°.

Rescue and Aliquoting of the Library

22. Add 5 ml of (2× TY-Amp-Glu-25% glycerol)/plate and gently loosen bacteria with a sterile glass spreader until a homogeneous suspension

is obtained. Mix the suspensions obtained from all the plates of the same sublibrary, keeping the VH-Vλ sublibrary separated from the VH-Vκ sublibrary.

23. Aliquot 500 μl/sterile tube; freeze in liquid nitrogen; store at −80°.

Library Quality Control. Not only the size of the library, but also the quality is important for its performance. To this aim, three more parameters have to be checked:

1. Number of clones whose phagemid contains the scFv insert (should be >95%)
2. Number of clones expressing phages carrying scFv (should be >90%)
3. Number of clones expressing soluble scFv (should be >70%)

These parameters can be checked easily following standard techniques, PCR screening of individual colonies (to detect the presence of the scFv coding sequence), and dot blot analysis (to detect both phages and soluble scFvs obtained by the screened colonies). See Pini *et al.*[30,38] for appropriate protocols.

Library Storage and Phage Preparation

The ETH-2 library is stored at −80° as bacteria harboring phagemid DNA in 2× TY-Amp-Glu-25% glycerol. From this primary library, phages displaying the recombinant antibodies have to be produced in order to proceed further in the selection of those phages carrying genotypes coding for the phenotype of interest. To this aim, a primary library is grown, infected with helper phage, and the phages harvested from bacterial supernatant. It is recommended that a large number of phage aliquots be prepared from the primary aliquot.

1. Store the ETH-2 library at −80°.
2. Inoculate one glycerol stock of each of the two sublibraries (VH-Vλ sublibrary and VH-Vκ sublibrary) into 100 ml 2× TY-Amp-Glu to obtain an initial OD of 0.1.
3. Grow bacteria until OD 0.4–0.5.
4. Infect each of the 100-ml cultures with helper phage (1 ml of >10^12 tu/ml VCS-M13).
5. Spin down the infected bacteria at 3300g for 10 min. Gently resuspend the pellet in 2 × 500 ml of 2× TY-Amp-Kan in 2-liter flasks (total: four flasks, two for each sublibrary).
6. Grow bacteria at 30° overnight.
7. Spin down the culture at 10,800g for 20 min.
8. Transfer the supernatant to a new tube.

9. Precipitate the phages by the addition of 1/5 volume of a solution of 20% (w/v) polyethylene glycol 6000/2.5 M NaCl (PEG/NaCl) followed by mixing and incubation at 4° or on ice for at least 1 hr.
10. Spin at 10,800g for 20 min.
11. Resuspend the pellet in 40 ml of water.
12. Add 1/5 volume PEG/NaCl, mix, and leave for a minimum of 20 min on ice or at 4°.
13. Spin 10,800g for 10 min and remove the supernatant.
14. Respin briefly and aspirate off any residue of PEG/NaCl.
15. Resuspend the pellet from each sublibrary in 20 ml of 10% glycerol in PBS (50 mM phosphate, pH 7.4, 100 mM NaCl) and aliquot it 500 μl/tube. In this way, 40 phages' aliquots of the VH-Vλ sublibrary and 40 aliquots of the VH-Vλ sublibrary are obtained.
16. Titrate phages. Each aliquot should contain $>10^{12}$ tu and will be sufficient for at least one selection.
17. Store aliquots at $-20°$.

Frozen phages obtained from the primary library stock constitute an excellent secondary stock: phage particles are resistant and maintain a good infectivity.

Selection of Phages from the ETH-2 Phage-Display Library

The selection of phage–antibodies binding to the antigen of interest can be performed in several ways. Proteins of MW $> 15,000$ can be adsorbed on plastic supports [e.g., immunotubes Maxisorp Nunc (Nalge Nunc Int. Corp., Rochester, NY)]. For smaller antigens, this method of immobilization is usually inefficient and either covalent immobilization on reactive plastic (e.g., Xenobind Covalent Binding Microwell Strips, Xenopore, NJ) or selection with biotinylated antigen should be used. All these procedures are well established and easy to perform. The use of plastic supports is less time-consuming and less expensive; however, the amount of antigen needed for the selection on plastic supports is usually higher. Moreover, antigens immobilized on plastic supports may denature. Biotinylated antigens typically retain their native conformation, and the selection is carried on in solution, allowing the selection of scFvs, which recognizes epitopes in correctly folded antigens. Overbiotinylation of the antigen, however, must be avoided as it can lead to unaccessibility of relevant epitopes.

Selection on Plastic Supports (Immunotubes or Reactive Plastic Plates for Covalent Immobilization)

1. Coat the plastic support with the antigen. The efficiency of coating depends on antigen concentration, buffer, and temperature. Usually 10–100 μg/ml antigen in PBS, pH 7.4, is incubated overnight at room temperature to

coat immunotubes. For the coating of Xenobind plastic plates, the condition usually used is 5–20 μg/ml antigen in phosphate or carbonate buffer at a pH above its isoelectric point at 37° for 2 hr.

2. Rinse the plastic support three times with PBS.

3. Block with 4 ml 2% (w/v) skimmed milk in PBS (MPBS) at room temperature for 2 hr.

4. Rinse the plastic support three times with PBS.

5. Add >10^{12} tu phage library to the plastic support and an equal volume of 4% MPBS.

6a. If using immunotubes, seal the tube with parafilm and mix by repeated inversion at room temperature for 30 min. Let the immunotube standing upright at room temperature for 1.5 hr and then throw away the unbound phages in the supernatant.

6b. If using reactive plastic plates for covalent immobilization, leave at room temperature for 30 min.

7. Rinse the plastic support 10 times with PBS–0.1% Tween 20 and then 10 times with PBS. Each washing step is performed by pouring buffer in and then immediately out.

8. Elute the bound phages by adding 1 ml 100 mM triethylamine (0.7 ml triethylamine in 50 ml H_2O).

9. Mix by inverting the tube or leave the plate at room temperature for 5 min.

10. Pour the eluted phages directly into a tube containing 0.5 ml 1 M Tris–HCl, pH 7.4. Phages can be kept on ice if the further step cannot be performed immediately.

11. Proceed as described in "Growth of Libraries after Selection."

For successive rounds of panning, repeat the procedure just described using 1 ml of phages amplified from the previous round of panning.

Selection Using Biotinylated Antigens. Here the phage antibodies react with biotinylated antigen in solution, and the complex is then captured on streptavidin-coated paramagnetic beads (Dynabeads M-280 Streptavidin, Dynal, Oslo, Norway).

The antigen first needs to be biotinylated with NHS-SS-Biotin (Pierce) according to the manufacturer's instructions.[17] Selections are carried out in Eppendorf tubes; for each step in which phages are present, glass Pasteur pipettes are used and discarded every time after their use; separation of beads from solutions is performed with a Magnetic Particle Concentrator (MPC-E, Dynal) according to the manufacturer's instructions.

1a. Mix together in MPBS (in order to have a final 2.5% MPBS solution): 10^{12}–10^{13} tu/phages in PBS and biotinylated antigen to a final concentration of 10^{-7} M.

2a. Gently rotate for 1 hr at room temperature.

1b. Block 100 μl/selection of streptavidin M-280 dynabeads (corresponding to 6.7 × 10^7 dynabeads) in 600 μl MPBS 5% for >15 min at room temperature.

2b. Separate the beads from the MPBS, remove the supernatant, and resuspend the beads in 200 μl PBS.

3. Mix the preblocked beads with the incubated phages–biotinylated antigen.

4. Gently rotate for 15 min at room temperature.

5. Separate the beads and remove the supernatant.

6. Wash five times with 1 ml PBS–0.1% Tween 20.

7. Wash five times with 1 ml PBS.

8. Finally, elute the phage from the beads by adding 1 ml 20 mM DTT in PBS and rotate for 5 min at room temperature.

9. Proceed as described in "Growth of Libraries after Selection."

For successive rounds of panning, repeat the procedure just described, mixing 750 μl of phages amplified from the previous round of panning with 10^{-7} M biotinylated antigen.

Note: The "stringency" of the selection depends on the coating antigen concentration, phage incubation times, number of washing steps, concentration of the detergent in the washing buffer, and use of a molar excess of unbiotinylated antigen in selection performed with biotinylated antigens. By varying these parameters, it is possible to promote the selection from the library of high-affinity binders.[30,48–50]

Growth of Libraries after Selection

After elution from the antigen, phages need to be amplified by bacterial infection and concentrated before using them for further rounds of selection. The titer of the eluted phages, which provides an information on the performance of the selection, also needs to be determined.

1. Infect 10 ml of exponentially growing TG1 cells with the collected phages (keep a small amount of phages for trace infections of HB2151, which you may want to perform, as discussed later).

2. Titrate eluted phages by plating dilution series of phage-infected bacteria on 2× TY-Amp-Glu plates and counting the number of colonies after an overnight incubation.

3. Spin down the remaining infected TG1 cells at 3300g for 10 min.

[48] R. Schier, R. F. Balint, A. McCall, G. Apell, J. W. Larrick, and J. D. Marks, *Gene* **169**, 147 (1996).

[49] N. M. Low, P. H. Holliger, and G. Winter, *J. Mol. Biol.* **260**, 359 (1996).

[50] W. P. Yang, K. Green, S. Pinz-Sweeney, A. T. Briones, D. R. Burton, and C. F. Barbas III, *J. Mol. Biol.* **254**, 392 (1995).

4. Resuspend the pelleted bacteria in 0.5–1 ml 2× TY, spread on a large 2× TY-Amp-Glu plate, and incubate at 30° overnight or until colonies are visible.

5. Add 5–10 ml 2× TY-Amp-Glu-15% glycerol to the large agar plate and *gently* loosen the bacteria with a glass spreader until a homogenous suspension is obtained.

6. Inoculate 50 ml of 2× TY-Amp-Glu with the rescued bacteria to obtain an OD of 0.05–0.1. Store the remaining bacteria at −80°.

7. Grow until OD 0.4–0.5.

8. Infect 10 ml of the culture with helper phage at a ratio of around 20:1 phage:bacteria (the bacterial concentration is 8×10^8 bacteria/ml when OD 1.0). We typically use 100 μl VCS-M13 ($>10^{12}$ tu/ml) per 10-ml culture.

9. Spin down the infected bacteria at 3300g for 10 min. Gently resuspend the pellet in 100 ml of 2× TY-Amp-Kan.

10. Grow at 30° overnight

11. Proceed by PEG precipitating the phages as already discussed in "Library Storage and Phage Preparation."

12. After the second precipitation of phages, resuspend the pellet in 2 ml of PBS 10% glycerol. Use a portion of these phages for other rounds of panning and store the rest at −20°.

Screening by ELISA

After the third round of panning, it is usually good practice to evaluate the progress of the selection by ELISA. Because soluble scFv fragments are used for most applications, this section describes the protocol used to perform the soluble ELISA assay, which directly allows the identification of those clones able not only to produce phages displaying selected scFvs, but also to produce the soluble form of the antibody. Another possibility is to monitor the progress of selection by phage ELISA (see Harrison et al.[44]).

When using the ETH-2 library for the first time, it is good practice to perform selections using bovine serum albumin (BSA) on immunotubes as a positive control. ELISA screening after three rounds of panning should give >25% positive clones.

The ETH-2 library is cloned in pDNEK (Fig. 5), which appends at the C terminus of the cloned scFv fragments: a phosphorylation site, a FLAG peptidic tag, an $(His)_6$ sequence, and the recognition site for the endopeptidase enterokinase. In ELISA, soluble scFvs are best detected using the anti-FLAG mAb M2 (Sigma, Buchs, Switzerland).

Several kinds of ELISA plates can be used to immobilize the antigen (see also "Selection of Phages from ETH-2 Phage-Display Library").

MicroTest III flexible assay plates (Falcon, Becton-Dickinson Lab-
ware, Oxnard, CA) for coating of antigens used in immunotube-
performed selections. The coating is usually efficient if performed
with 10–100 μg/ml of antigen in PBS, overnight, at room temperature.

Reactive plastic plates for covalent immobilization (Xenobind,
Xenopore)

Streptavidin-coated microtiter plates (Roche Diagnostics, Rotkreuz,
Switzerland) for coating of biotinylated antigens (following manufacturer's instructions)

Unless otherwise specified, volumes used are 100 μl/well and incubations are performed at room temperature.

1. Infect 1 ml culture of exponentially growing HB2151 with 10 μl of
   phage eluted during a certain round of panning (typically the third
   or fourth). Plate in 2× TY-Amp-Glu plates and grow colonies.
2. Inoculate individual colonies in 180 μl 2× TY-Amp-0.1% Glu in
   96-well rigid plates.
3. Grow for 3 hr at 37°, shaking at 220 rpm.
4. With a multichannel pipette, transfer 50 μl from each well onto a
   replica microtiter plate containing 50 μl of 40% glycerol. The resulting glycerol stock can be frozen and used at later stage to rescue
   clones of interest.
5. Add 50 μl 2× TY-Amp containing 10 mM (isopropyl)-β-D-thioga-
   lactopyranoside (IPTG). Continue shaking at 30° for another 16 to
   24 hr.
6. Spin at 1800g for 10 min and use the supernatant in ELISA. *Note:* As
   some antibodies are found preferentially in bacterial supernatants,
   some people freeze and thaw the microtiter plate before centrifugation to release antibody from the periplasmic space.
7. Coat the ELISA plate with the antigen of interest.
8. Rinse wells three times with PBS.
9. Block for 2 hr with 200 μl/well of 2% MPBS.
10. Rinse wells three times with PBS.
11. Add 30 μl 10% MPBS and 80 μl supernatant to each well.
12. Immediately afterward, using a multichannel pipette, add 11 μl
    freshly prepared developing mix: 10 μg/ml anti-FLAG M2 anti-
    body + antimouse HRP (Sigma, A-2554; 1/100 dilution) in 2%
    MPBS.
13. Incubate for 40–60 min.
14. Wash three times with PBS + 0.1% Tween 20.
15. Wash three times with PBS.

16. Develop with HRP substrate solution, e.g., with ready-to-use BM blue POD-soluble substrate (Roche Diagnostics). Positive clones will develop a blue color.

Preparation of Antibody Fragments

For many applications, it is convenient to purify several milligrams of antibody fragments from clones that are positive in ELISA screening.

Cultures and Harvesting of scFvs from Bacterial Supernatant

1. From the glycerol replica plate made in step 4 of "Screening by ELISA," choose those clones that give a positive result in ELISA screening, grow them, and make several glycerol stocks of exponentially growing bacteria (final glycerol concentration: 20%).
2. Inoculate 1 liter 2× TY-Amp-0.1% Glu with a 1-ml glycerol stock aliquot.
3. Grow to OD 0.8–0.9 and induce with 1 mM IPTG.
4. Grow overnight at 30°.
5. Spin down at 10,800g for 20 min and collect the antibody-containing supernatant.

Purification of Antibody Fragments. Antibody fragments from the *E. coli* supernatant can be purified by column chromatography.

For recombinant antibodies selected from the ETH-2 library, it is possible to choose among

Immobilized metal affinity chromatography on Ni-NTA Resin (Qiagen)

Chromatography on protein A-Sepharose resin (protein A Sepharose 4 Fast Flow, Pharmacia Biotech, Uppsala, Sweden)

Chromatography on anti-FLAG M2-Sepharose column (Kodak or other suppliers)

Affinity chromatography on antigen resin

This section describes purification using a protein A column, which is easy and reliable. Throughout the purification, try to keep the protein at 4° or on ice all the time and, if possible, to perform the purification in a cold room.

1. Preswell the protein A-Sepharose in PBS.
2. Fill a chromatography column with the appropriate amount of protein A-Sepharose (depending on culture size and resin capacity).
3. Equilibrate the column with PBS.
4. Load antibody preparation onto the column.
5. Wash with 20 column volumes of PBS.
6. Wash with 20 column volumes of PBS–0.5 M NaCl.
7. Elute the protein with about 3 column volumes of either 0.2 M

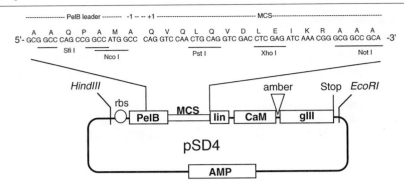

FIG. 6. Schematic representation of the phagemid vector pSD4 for display of calmodulin-tagged enzymes on filamentous phage. Rbs, ribosome-binding site. PelB, leader peptide; MCS, multiple cloning site; CaM, calmodulin; gIII, gene III; AMP, ampicillin resistance gene. Sequence of pSD4 multiple cloning site.

glycine, pH 3.0, or 100 mM triethylamine, collecting 1-ml fractions into 0.2 ml 1 M Tris–HCl, pH 7.4, and mixing immediately.

8. Measure the OD at 280 nm (for a scFv OD = 1 corresponds roughly to 0.8 mg/ml of protein) and dialyze the sample overnight against PBS at 4°.

Methods for Phage–Enzyme Selections

This section contains protocols for the display and use of enzymes on the surface of calmodulin-tagged phages (see Introduction). In the first part, methods for the cloning of enzyme–calmodulin fusions in suitable phagemid vectors are presented, with references to methods for the evaluation of the efficiency of enzyme display on phage. In the second part, selection methodologies for two model enzymes displayed on calmodulin-tagged phage (a transferase and an endopeptidase) are reported.

Display of Enzyme–Calmodulin Fusions

As mentioned previously (see Introduction), two different classes of vectors can be used to express fusion proteins on the surface of a bacteriophage: phage (fd) vectors and phagemid vectors. Although phages produced from phagemids have the disadvantage of mainly displaying helper phage pIII rather than the phagemid fusion pIII,[5,9] phagemids are easier to use for cloning and therefore typically yield larger libraries.

For this reason, this section describes the display of enzyme on phages on cloning on a phagemid vector, pSD4 (Fig. 6; Demartis *et al.*[12]), in which a 16 amino acid linker and the *Xenopus laevis* calmodulin gene have been inserted between the MCS and the gIII.

Cloning Strategy

1. PCR amplify the gene of the enzyme to be displayed on phage with suitable DNA primers (for oligonucleotide design, see Demartis *et al.*[12]) in order to insert the *Nco*I and *Not*I restriction sequences at the extremities of the enzyme's gene (Fig. 6).

2. Clone the *Nco*I/*Not*I digested fragment in frame into *Nco*I/*Not*I-digested pSD4, according to methods analogous to the ones of "Library Construction." The resulting bacterial colonies will be used for phage preparation (see later).

ELISA Characterization of Enzymes Displayed on Calmodulin-Tagged Phage

 a. Detecting the Display of Calmodulin on Phage

3. Prepare phages displaying the enzyme–calmodulin fusion according to procedures mentioned in "Library Storage and Phage Preparation."

4. Incubate for 10 min at room temperature streptavidin-coated microtiter plates (Roche Diagnostics) with 100 μl of a 10^{-7} M solution of biotin-labeled calmodulin-binding peptide biotin-CAAARWKKAFIAVSAAN RFKKIS (Montigiani *et al.*[43]) in TBSC (50 mM Tris, pH 7.4, 100 mM NaCl, 1 mM CaCl$_2$).

5. Wash three times with TBSC.

6. Block with 2% (w/v) skimmed milk in TBSC (2% MTBSC).

7. Add to each well 30 μl 10% MTBSC and 80 μl of phages (10^{12} tu/ ml). Incubate for 30 min at room temperature.

8. Wash five times with TBSC + 0.1% Tween 20.

9. Wash five times with TBSC.

10. Add horseradish peroxidase-conjugated anti-M13 antibody (diluted 1 : 2000 in 2% MTBSC; Pharmacia Biotech, Piscataway, NJ). Incubate for 20 min at room temperature.

11. Wash five times with TBSC + 0.1% Tween 20.

12. Wash five times with TBSC.

13. Detect plate-bound peroxidase with ready-to-use BM-Blue soluble substrate (Roche Diagnostics).

Note: As a negative control assay, test the absence of phage binding in the presence of the Ca^{2+} chelator EDTA in a similar fashion but replacing TBSC with TBSE (50 mM Tris, pH 7.4, 100 mM NaCl, 20 mM EDTA). Moreover, the absence of binding if the biotinylated peptide is omitted should be tested.

 b. Detecting the Display of the Enzyme on Phages

14. Coat microtiter plates (MicroTest III flexible assay plate, Falcon)

with 5 μg/ml of antibody (or 40 μg/ml of scFv) specific for the enzyme in PBS. Incubate overnight at room temperature.

15. Wash three times with PBS.
16. Block for 2 hr at 37° with 2% MPBS.
17. Proceed as described from step 5 to step 11.

Note: As mentioned earlier in this article, phagemid vectors typically yield phage particles displaying 0 or 1 pIII minor coat proteins fused to the protein of interest. Display efficiency is limited by competition with pIII produced by the helper phage and by proteolysis.

Several methods have been used to quantitate protein display on phage as pIII fusion:

1. Western blot analysis with anti-pIII monoclonal antibody.[14]
2. Use of cleavable helper phage and proteolytic digestions.[12,13]
3. Determination of phage titer before and after a chromatographic phage purification using affinity reagents specific for the enzyme or the calmodulin moiety.[12]
4. Quantitative determination of catalytic activity of phage–enzymes.[51]
5. Dilution series in phage–ELISA assays, performed with different phage–enzyme preparations containing the same number of transforming units.[12]

Selection Methodologies

As an illustrative example, selection procedures with a transferase [glutathione *S*-transferase (GST)] and an endopeptidase (genenase[52]) are reported here (Fig. 7). From these two essentially similar methodologies, it should be possible to devise selection schemes for other enzymes.

The methods presented describe the recovery of monoclonal phage enzymes performing the reaction of interest. Selection with enzyme libraries is identical, but feature the use of populations of phages displaying different enzymatic activities.

Selection with GST–Calmodulin–Phage. For methods to display GST calmodulin tagged on phage, see previous section and Demartis *et al.*[12]

1. In a final volume of 20 μl, preblock GST–CaM–phage (10^{10} tu) in TBSC containing BSA (10 mg/ml) for 10 min.
2. Add phages to a 10^{-7} *M* GSH-pep solution in 180 μl TBSC and mix. Incubate for 30 sec. At this stage the BSA concentration should be 1 mg/ml.

[51] C. I. Wang, Q. Yang, and C. S. Craik, *Methods Enzymol.* **267**, 52 (1996).
[52] P. Carter, B. Nilsson, J. P. Burnier, D. Burdick, and J. A. Wells, *Proteins* **6**, 240 (1989).

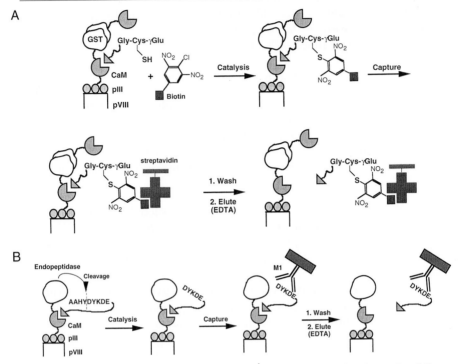

FIG. 7. Selection strategies for the rescue of catalytic activities displayed on calmodulin-tagged phage. (A) Selection of glutathione S-transferase (GST) activity and (B) selection of an endopeptidase activity. In strategy A, the biotinylated reaction product is captured using a streptavidin-coated support. In strategy B, the product of reaction (a cleaved peptide) is isolated using the M1 monoclonal antibody, which recognizes the DYKDE sequence when it is N-terminally located. For further details, see Demartis et al.[12]

3. Add 20 μl of 5 mM biotinylated 4-chloro-3,5-dinitrobenzoic acid derivative[12] and mix. Incubate at room temperature for 15 sec.

4. To the reaction mixture add 1/4 volume PEG solution, mix, and incubate on ice for 15 min.

5. Spin down phages at 10,000g for 2 min. Aspirate off the supernatant, resuspend phage pellet in 200 μl TBSC, and repeat step 4.

6. Spin down phages at 10,000g for 2 min. Aspirate off the supernatant and resuspend phage pellet in 200 μl TBSC containing 3% milk powder and 1 mg/ml BSA.

7. Add reaction mixture to 50-μl streptavidin-coated M280 dynabeads (Dynal), mix, and incubate for 10 min at room temperature with slight stirring.

8. Capture magnetic beads and aspirate off the supernatant.
9. Wash five times with TBSC + 0.1% Tween 20.
10. Wash three times with TBSC.
11. In order to elute phages, add 500 μl TBSE, mix, and incubate for 5 min at room temperature.
12. Aspirate off the supernatant and saturate EDTA with 100 μl 1 M CaCl$_2$ solution.
13. Infect 1 ml exponentially growing TG1 $E.$ $coli$ cells and then plate on 2$\times$ TY-Amp-Glu plates.
14. Incubate overnight at 30–32°.

Incubation with 10 mg/ml BSA in step 1 reduces the background due to the stickiness of phages by a factor of 10–20. Reduction of background is also observed employing milk powder as a blocking agent in step 6.

Selection with Genenase–Calmodulin–Phage. For methods to display genenase calmodulin tagged on phage, see previous section and Demartis et al.[12]

1. Incubate the genenase–CaM–phage (10^{10} tu) in 200 μl of buffer Z (0.5 mM dithiothreitol, 200 mM NaCl, 50 mM Tris–HCl, pH 8.0, 1 mM CaCl$_2$, 50 mg/liter phenylmethylsulfonyl fluoride) with 1 mg/liter peptide GAAHY**DYKDE**GGGAAARWKKAFIAVSAAN RFKKIS at 37° for 15 min.
2. To the reaction mixture add 1/4 volume PEG solution, mix, and incubate on ice for 15 min.
3. Spin down phages at 10,000g for 2 min. Aspirate off the supernatant and resuspend phage pellet in 100 μl TBSC containing 3% BSA.
4. Add reaction mixture to M1–streptavidin–Dynabeads (Dynal), mix, and incubate for 30 min with slight stirring. These beads are prepared by mixing 50 μl of streptavidin-coated M280 dynabeads with 500 μl 3% BSA/TBSC containing 0.4 μg biotinylated M1 anti-FLAG antibody, followed by washing.[12]
5. Capture magnetic beads on and aspirate off the supernantant.
6. Wash five times with TBSC + 0.1% Tween 20.
7. Wash three times with TBSC.
8. In order to elute phages, add 500 μl TBSE, mix, and incubate for 5 min at room temperature.
9. Aspirate off the supernatant and saturate EDTA with 20 μl CaCl$_2$ 1 M solution.
10. Infect 1 ml exponentially growing TG1 $E.$ $coli$ cells.
11. Plate on 2$\times$ TY-Amp-Glu plates.
12. Incubate overnight at 30–32°.

[30] Use of an Lpp-OmpA Fusion Vehicle for Bacterial Surface Display

By Charles F. Earhart

Introduction

Anchoring of homologous or heterologous soluble polypeptides on the bacterial surface is of interest for both practical and basic science reasons. The many possible biotechnology applications can be divided into several categories, including (i) live vaccine development, in which nonvirulent living bacteria exhibiting the antigen of interest are likely to result in an unusually effective vaccine; (ii) display of antibody (Fab or scFv) and peptide libraries that can be screened by affinity chromatography or fluorescence-activated cell sorting (FACS) isolation of bacteria expressing the desired polypeptide; (iii) production of whole cell adsorbents for purification purposes and as solid-phase reagents for diagnostic and analytical purposes; and (iv) expression of catalytic activity on the bacterial surface. Surface expression systems also provide an approach to study basic questions concerning protein kinesis and membrane biogenesis, including the roles of export machinery and stress responses in these processes. Because of their many possible uses, a variety of surface expression systems, all of which employ chimeric proteins, have been developed.[1-4] The vehicle featured here, Lpp-OmpA, is unusual in that it was developed to expose an entire soluble polypeptide, the passenger protein, in active form to the external milieu.

The Lpp-OmpA vector is designed for use in the gram-negative bacterium *Escherichia coli;* the rationale for its dibrid form stems from the architecture of the *E. coli* envelope and the properties of the major lipoprotein (Lpp) and OmpA proteins. The envelope of gram-negative cells consists of (i) a cytoplasmic membrane, (ii) the periplasm, in which the peptidoglycan cell wall is located, and (iii) the outer membrane (OM), a unique structure composed of proteins embedded in an asymmetric lipid bilayer containing phospholipid in the inner leaflet and lipopolysaccharide (LPS) in the outer leaflet. In the Lpp-OmpA vector, the Lpp domain serves to

[1] M. Little, P. Fuchs, F. Breitling, and S. Dubel, *Trends Biotechnol.* **11**, 3 (1993).
[2] G. Georgiou, C. Stathopoulos, P. S. Daugherty, A. R. Nayak, B. L. Iverson, and R. Curtiss III, *Nature Biotechnol.* **15**, 29 (1997).
[3] S. Stahl and M. Uhlen, *Trends Biotechnol.* **15**, 185 (1997).
[4] W. Chen and A. Mulchandani, *Trends Biotechnol.* **16**, 71 (1998).

target and anchor the tribrid to the OM and the OmpA domain is required for surface expression of the passenger. The relevant properties of these two proteins follow immediately. The Lpp protein of *E. coli* is an OM protein and is therefore synthesized with a signal sequence and exported, numerically the most abundant protein of *E. coli,* present as a trimer, and is positioned in the OM such that most of the protein is located in the periplasm. (One-third of Lpp molecules are attached covalently to peptidoglycan, and no portion of the protein can be detected on the external outer membrane face.) A fusion protein consisting of the Lpp signal sequence and the first nine amino acids of mature Lpp linked to the normally soluble, periplasmic protein TEM β-lactamase (Bla) was localized to the OM.[5] This ability of the Lpp amino terminus to target passenger proteins to the OM is utilized in the Lpp-OmpA construct. In keeping with the OM topology of Lpp, the Bla of the Lpp-Bla fusion was in the periplasm and any passengers anchored to OM by the Lpp domain are expected to be positioned similarly. This fact accounts for the OmpA domain of the vector, which is necessary to translocate the passenger protein through the OM for surface display. OmpA is a prevalent, monomeric OM protein with a transmembrane domain consisting of an eight-stranded β barrel[6,7]; the region employed in the vector (mature amino acids 46 to 159) includes five transmembrane segments. (Fusions to OmpA alone are not stably surface expressed; the last OmpA transmembrane segment is required for stable OM association[8,9] and fusions to its carboxy terminus would be in the periplasm.)

The range of proteins capable of surface display with Lpp-OmpA is still being determined. In initial experiments, passenger proteins displayed using the Lpp-OmpA chimera were all known to be capable of passage through cytoplasmic membrane: they include Bla, single chain Fv fragments (scFv), an exported organophosphorous hydrolase (OPH), and *Cellulomonas fimi* exoglucanase Cex and its cellulose-binding domain (CBD$_{cex}$). The monomeric periplasmic-binding protein for ferrienterobactin (FepB) was probably anchored on the surface, but the homodimeric periplasmic protein alkaline phosphatase (PhoA) was not surface expressed. (No protein with quaternary structure has been exposed to the surface with any display technique.) That the Lpp-OmpA vehicle can anchor and translocate cytoplasmic proteins successfully was demonstrated with the passenger thiore-

[5] J. Ghrayeb and M. Inouye, *J. Biol. Chem.* **259,** 463 (1984).
[6] H. Vogel and F. Jahnig, *J. Mol. Biol.* **190,** 191 (1986).
[7] A. Pautsch and G. E. Schulz, *Nature Struct. Biol.* **5,** 1013 (1998).
[8] M. Struyve, M. Moons, and J. Tommassen, *J. Mol. Biol.* **218,** 141 (1991).
[9] M. Klose, F. Jahnig, I. Hindennach, and U. Henning, *J. Biol. Chem.* **264,** 21842 (1989).

doxin (TrxA).[10] Finally, and remarkably, passenger proteins displayed by the Lpp-OmpA vehicle retained their normal activity in all cases.

Plasmids and Vehicles

Construction

The plasmid encoding the original Lpp-OmpA vector (pTX101) is a derivative of pJG311.[11] (pJG311, a pBR322 derivative, is identical to pJG310,[5] except that the EcoRI site downstream of bla has been removed.) A 342-bp SphI–HpaI fragment from OmpA-encoding pRD87[12] was blunt ended and ligated into EcoRI-digested and blunted pJG311 to construct pTX101. The resulting fusion protein contains the signal peptide and first nine amino acids of Lpp, the linker amino acids G and I, OmpA amino acids 46–159, linker amino acids G, I, P, and G, and mature Bla. A large number of plasmids encoding the Lpp-OmpA vector are available; they vary in copy number, fusion protein regulatory region, OmpA domain, presence of lacI[q], antibiotic resistance, and compatibility group. It is not practical to describe all of these plasmids, but for initial constructions the plasmids pSD192 and pSD195[13] are recommended. pSD195 differs from pSD192 in that it lacks a PstI fragment so that it has just one EcoRI site and is sensitive to chloramphenicol. Both plasmids encode an scFv protein, whose hybrid gene contains many useful unique six cutter restriction sites.

OmpA (Middle Domain) Considerations

The OmpA domain of the vector was chosen because it was believed, and has since been proved,[7] that it contained five transmembrane segments of an abundant monomeric OM protein. If the first membrane-spanning sequence is attached to Lpp at the OM periplasmic face and if the five transmembrane segments have the same topology in tribrids as in OmpA, passenger domains at the carboxy terminus should be localized on the surface. This OmpA domain has fulfilled expectations in displaying passengers, but evidence shows that it was a fortuitous choice initially and that

[10] D. S. Stephens, M. Oak, A. Atunez, S. Vokes, and C. F. Earhart, unpublished observations (1999).
[11] J. A. Francisco, C. F. Earhart, and G. Georgiou, Proc. Natl. Acad. Sci. U.S.A. **89**, 2713 (1992).
[12] R. Freudl, H. Schwarz, M. Klose, N. R. Movva, and U. Henning, EMBO J. **4**, 3593 (1985).
[13] P. S. Daugherty, G. Chen, M. J. Olsen, B. L. Iverson, and G. Georgiou, Prot. Eng. **11**, 825 (1998).

it can be improved. Six additional sandwich fusions consisting of the Lpp anchor, varying lengths of the OmpA domain starting at amino acid 46, and Bla as the reporter enzyme were constructed by unidirectional exonuclease III digestion of the *ompA* sequence, restriction digestion, and polymerase chain reaction mutagenesis. Only one of the six oriented Bla to the surface and its 21 amino acid OmpA segment constituted a single β strand.[14] Results showed that not all OmpA sequences work and that the OmpA sequences do not assume their normal conformation when sandwiched between Lpp and passenger proteins.

In obtaining surface expression of OPH, improved results were observed when the Lpp-OmpA OPH fusion contained the OPH signal sequence, i.e., the original proprotein contained one signal sequence at its amino terminus and a second between the OmpA domain and mature OPH. The generality of this observation remains to be determined.

Regulation of Fusion Protein Synthesis

Genes encoding the original Lpp-OmpA passenger fusions were controlled by the hybrid promoter/operator *lpplac* and were therefore inducible by isopropylthiogalactoside (IPTG). The hybrid *lpplac* promoter is very strong; even uninduced and in a *lacI^q*-containing host the synthesis of 30000 proteins/cell/division cycle has been reported.[15,16] Not only was this regulatory system not tightly repressed, but it was also difficult to modulate precisely; near maximum induction was observed over a wide range of IPTG concentrations. An additional concern was that the host strains were not *lacY;* therefore, when low expression levels were observed, it was not clear if this resulted from full induction of a small percentage of the population or if the entire population was partially induced.[17] In most recent constructs, therefore, other promoter/operator combinations have been used. Richins *et al.*[18] reported tight regulation of a Lpp-OmpA-OPH fusion using a *tac* promoter (hybrid *trp* and *lac* promoter). For enhanced repressor synthesis, the tribrid-encoding plasmid also carried a *lacI^q* gene, as did the host *E. coli* strain (JM105). The *tac* promoter has also been used successfully for precise control of thioredoxin (TrxA) and Bla fusions.[19] In accord with the findings of Richins *et al.,*[18] both chromosomal and plasmid copies of

[14] G. Georgiou, D. L. Stephens, C. Stathopoulos, H. L. Poetschke, J. Mendenhall, and C. F. Earhart, *Prot. Eng.* **9,** 239 (1996).
[15] G. Chen, J. Cloud, G. Georgiou, and B. L. Iverson, *Biotechnol Prog.* **12,** 572 (1996).
[16] P. S. Daugherty, M. J. Olsen, B. L. Iverson, and G. Georgiou, *Prot. Eng.* **12,** 613 (1999).
[17] D. A. Siegele and J. C. Hu, *Proc. Natl. Acad. Sci. U.S.A.* **94,** 8168 (1997).
[18] R. Richins, I. Kaneva, A. Mulchandani, and W. Chen, *Nature Biotechnol.* **15,** 984 (1997).
[19] M. Oak, M. A. thesis, The University of Texas at Austin (1999).

the *lacI*^{q} were required; for TrxA and Bla fusions, the plasmid-borne *lacI*^{q} was on a separate plasmid (pREP4) (QIA expression kit, Qiagen, Inc., CA). The state of the *lacY* gene did not affect regulation.

Other workers[16] have used the *tet* and *araBAD* promoters to obtain tight repression of Lpp-OmpA passenger protein synthesis. In addition, they monitored the heterogeneity of protein expression in cell populations by FACS. Lpp-OmpA-mediated surface expression of an scFv antibody on individual cells was determined using an appropriate antigen or hapten labeled with a fluorescent dye. At a submaximal inducer concentration, tribrid expression in the cell population was heterogeneous, whether under *tet* or *araBAD* control. Different levels of homogeneous expression were obtained using saturating inducer concentrations in combination with variations in plasmid copy number and time of induction.

Outer Membrane Integrity Assays

An important consideration in determining the extent of surface expression is OM integrity as many surface expression assays are unreliable if the OM has been structurally compromised. A quick indication of possible membrane destabilization is accomplished by comparing the growth of induced versus uninduced cells. The greater the deviation from the growth rate of uninduced cells, the more likely the OM is being damaged by tribrid induction. Additional membrane integrity assays to assess possible tribrid-induced damage are listed later. Some (RNaseI leakage, disc agar diffusion tests) are suitable only when induction is neither lethal nor required for surface display.

Periplasmic Leakage Assay

The small periplasmic enzyme RNaseI is released spontaneously from cells with weakened OMs.[20] Leakage of RNaseI from individual colonies is monitored by a plate assay for RNA digestion; colonies with a destabilized OM display a clear halo after acid precipitation of the RNA present in the top agar. Conditions have not been found that permit use of this assay when tribrid induction is lethal; it is therefore suitable for situations in which some tribrid synthesis occurs in the absence of induction. Standard *E. coli* K-12 strains differ in leakiness even in the absence of tribrid-encoding plasmids and this leakiness varies significantly with growth temperature. Generally, RNaseI release is increased at lower temperatures.[21]

[20] J. Lopes, S. Gottfried, and L. Rothfield, *J. Bacteriol.* **109**, 520 (1972).
[21] C. Stathopoulos, G. Georgiou, and C. F. Earhart, *Appl. Microbiol. Biotechnol.* **45**, 112 (1996).

Disc Agar Diffusion Method for Sensitivity to Detergents, Lysozyme, and Antibiotics

Discs containing specific antibiotics[21] (Difco Laboratories) are placed on LB/agar plates overlaid with soft agar seeded with 0.1 ml of overnight culture. Discs containing ionic detergents (2%) such as SDS, sarkosyl, and deoxycholate and the enzyme lysozyme (1 mg/ml) are also employed. Sensitivity is indicated by clear zones of growth inhibition around the disc and the zone diameter can be measured for comparative purposes. Antibiotic sensitivity data alone are no assurance that the OM is unaltered, however, as some leaky mutants (e.g., strains bearing *lpp* mutations) have unaltered sensitivities to most antibiotics but are detergent sensitive. This diffusion assay, like the periplasmic leakage assay, is generally not suitable for induced cells.

Spectrophotometric Assay of SDS and EDTA Sensitivity

Cells are suspended in buffer and incubated for several hours in the presence of SDS (0.1%) or EDTA (2 mM).[22] Absorbance at 595 nm is measured over time; extensive lysis of sensitive cells occurs in the first 30 min. The assay is quick and has the advantage that induced as well as uninduced cells can be tested.

Assays for Surface Expression

There are three general types of assays for surface display in common use: (i) immunochemical assays, (ii) protease sensitivity assays, and (iii) comparison of the passenger protein activity of whole cells with that in lysates. Additional assays appropriate for a specific passenger protein are also feasible. If the tripartite fusion protein is stable, there is no difficulty in detecting its presence in the OM. The major complication in most assays is distinguishing whether the passenger domain is in the periplasm or the external milieu. Because Lpp-OmpA passenger chimeras alter OM permeability, molecules such as proteases and enzyme substrates to which the normal OM presents a severe permeability barrier may reach the periplasm of induced cells readily. Extensive controls are necessary if a definitive statement regarding expression on the cell surface as contrasted to available to molecules in the environment is to be made.

Immunochemical Assays

It is advantageous to have antibodies against the passenger polypeptide available as the majority of surface expression assays require such antibod-

[22] G. Georgiou, M. L. Shuler, and D. B. Wilson, *Biotechnol. Bioeng.* **32,** 741 (1988).

ies and the two alternative classes of surface assays are more susceptible to artifacts. Antibodies against peptidoglycan (Chemicon Inc.) provide an important control for immunochemical assays; labeling of cells with these antibodies should occur only if the OM is weakened severely and indicates that all antibodies would have access to the periplasm.

Immunofluorescence Microscopy. Standard procedures have provided evidence that passenger proteins Bla,[11] (CBD$_{cex}$),[23] and a single chain Fv antibody fragment against digoxin (scFv[dig])[24] are surface localized by the Lpp-OmpA vector. Rhodamine-conjugated secondary antibodies were used for Bla and Cex, whereas only fluorescein-conjugated digoxin was necessary for scFv(dig) assays. Controls include treatment of washed cells with proteases prior to addition of the primary antibody and use of cells with a plasmid encoding only an Lpp passenger dibrid, which should be localized in the periplasm. Concern regarding the validity of immunofluorescence microscopy results has been raised[21]; e.g., passenger PhoA, which is not surface expressed, gave a weak positive signal when examined by this technique.

Whole Cell ELISAS. This standard technique showed surface expression of Cex and CBD$_{cex}$.[23] Controls consist of cells synthesizing soluble, not membrane-associated, proteins.

Immunoelectron Microscopy. Although more cumbersome than other procedures, this assay may be the most reliable.[14,21] Washed cells are exposed to antipassenger antibody and incubated with protein A conjugated to 10-nm colloidal gold particles, and then intact cells are visualized by electron microscopy. Antipeptidoglycan antibodies provide both a negative control and, after permeabilization of cells by treatment with Tris–EDTA (250 mM Tris–HCl, pH 7.4, 20 mM EDTA, 30 min at 4°), a positive control.

Flow Cytometry. That scFv(dig) is surface expressed by the Lpp-OmpA vector was shown by reacting cells with digoxin conjugated to fluorescein (digoxin-FITC) and then analysis by fluorescence-activated cell sorting (FACS).[23] Cells pretreated with an excess of free digoxin and cells expressing Lpp-OmpABla provided negative controls.

Protease Sensitivity Assays

Passenger proteins on the cell surface should be accessible to, and therefore degraded by, externally added proteases. Trypsin, papain, proteinase K, and pronase E have all been employed in such experiments; degradation can be measured by loss of passenger enzyme activity or disap-

[23] J. A. Francisco, C. Stathopoulos, R. A. J. Warren, D. G. Kilburn, and G. Georgiou, *Bio/Technology* **11**, 491 (1993).

[24] J. A. Francisco, R. Campbell, B. L. Iverson, and G. Georgiou, *Proc. Natl. Acad. Sci. U.S.A.* **90**, 10444 (1993).

pearance of the tribrid when examined by SDS–PAGE. Lpp-OmpA-fused Bla,[11] OPH,[18] and scFv[23] are all sensitive to protease activity. The best control for such experiments is to note whether OmpA, which is thought to have a large, protease-sensitive, periplasmically located carboxy terminus, is degraded.[25] Degradation of the passenger polypeptide while leaving OmpA intact demonstrates that the protease did not enter the periplasm. Failure to find conditions in which the passenger was preferentially degraded argued against surface exposure of the Bla of several truncated Lpp-OmpA vehicles[14] and prevented a definite conclusion regarding the display of FepB.[26] In contrast, this control provided strong evidence for surface expression of TrxA.[19]

Passenger Protein Activity Assays

Passenger proteins with enzyme activity can be tested for surface expression if their substrate either does not penetrate or diffuse readily through the OM.[11,18,23] These assays are difficult to interpret when the integrity of the OM is compromised, however, and additional, distinct procedures should also be employed. The whole cell activity of surface-expressed Bla, OPH, and Cex was close to that of lysates, approximately nine times higher than that of whole cells without surface expression vehicles, and could be reduced by prior protease treatment. Surface assays can also be devised for passenger proteins with binding activity. For instance, cells expressing Cex or CBD$_{cex}$ were shown to bind cellulose by incubating the cells with cotton fibers and then observing the fibers by light microscopy and scanning electron microscopy.[23] The enterobactin-binding protein FepB bound ferrienterobactin, but because of the small size of enterobactin, this did not prove surface exposure.[26]

Optimization of Surface Display

The usual aim in surface display work is to not only position the passenger protein on the surface but to do so in a way that minimizes both cell damage, as indicated by increased OM permeability and decreased growth rate and viability, and proteolysis of the fusion protein. Many factors must be considered in attempting to achieve this aim. The choice of *E. coli* strain is important and three strains have been used repeatedly: JM109 (*endA recA1 gyrA thi-1 hsdR17 relA1 supE44 Δ (lac-proAB)* F′ *traD36 proAB*

---

[25] J. Tommassen and B. Lugtenberg, *J. Bacteriol.* **157,** 327 (1984).
[26] D. L. Stephens, M. D. Choe, and C. F. Earhart, *Microbiology* **141,** 1647 (1995).

lacI^q lacZΔM15),[27] *SF110 ΔlacX74 galE galK thi rpsL ΔphoA ΔompT degP4)*,[28] and RB791 (*thyA deoC22* IN (*rrnD-rrnE*) *lacI^q* L8).[29] JM109 has the advantages of a plasmid-borne *lacI^q* gene for *lpplac* and *tac* promoter regulation, a defective restriction system, and is *recA,* so recombination among replicons is prohibited. It is, however, slightly leaky at 37° and 30° and very leaky at 22°. SF110 has good OM integrity, exhibiting leakiness only at 22°. It also lacks the OM protease OmpT and the periplasmic protease DegP, which allows enhanced tribrid stability in some cases. RB791 is like SF110 in leakiness and has a chromosomal *lacI^q* gene.

Lower growth temperatures have repeatedly been found to increase both surface display and leakiness. The reasons for this are unclear; the temperature effect may be related to the ability of DegP to act preferentially as a chaperone rather than a protease at low temperatures,[30] the fact that the *E. coli* export apparatus is cold sensitive,[31] or that protein folding is slowed at reduced temperatures. Because of the conflicting effects of temperature, an intermediate temperature (30°) is used frequently.

The effect of media on tribrid expression has not been examined systematically. LB medium, which consists, per liter, of 10 g tryptone, 5 g yeast extract, and 10 g NaCl made to pH 7 with NaOH, is used routinely. For regulatory reasons it must be supplemented with additional carbon and energy sources when the *araBAD* promoter is employed.[32]

Passenger proteins differ greatly in their stability and in the extent their synthesis is detrimental to the cell. In general, the larger the passenger polypeptide, the more it is subject to degradation and the more damage it causes. Therefore, particularly for large passengers, precise control is important and the choice of regulatory system and plasmid copy number is critical. Satisfactory results often require testing several plasmids under a variety of conditions.

Immobilization and Stabilization of Surface-Anchored Tribrids

Enzymes and antibodies have been expressed on the *E. coli* surface as tribrids and their practical use as whole cell biocatalysts or adsorbents can be anticipated. Prior to this, however, procedures to immobilize the cells and stabilize these chimeras are needed so as to permit their repeated use

[27] T. Maniatis, E. F. Fritsch, and J. Sambrook, "Molecular Cloning, A Laboratory Manual," 2nd Ed. Cold Spring Harbor Laboratory Press, Cold Spring Harbor, NY.
[28] F. Baneyx and G. Georgiou, *J. Bacteriol.* **172,** 491 (1990).
[29] R. Brent and M. Ptashne, *Proc. Natl. Acad. Sci. U.S.A.* **78,** 4204 (1981).
[30] C. Spiess, A. Beil, and M. Ehrmann, *Cell* **97,** 339 (1999).
[31] K. J. Pogliano and J. Beckwith, *Genetics* **133,** 763 (1993).
[32] L. M. Guzman, D. Belin, M. J. Carson, and J. Beckwith, *J. Bacteriol.* **177,** 4121 (1995).

with minimal loss of tribrid activity and cell destruction. Two procedures for immobilizing and stabilizing cells expressing tribrids have been described: one using Bla and the other using organophosphorous hydrolase (OPH) as passenger proteins. Freeman et al.[33] described a two-stage chemical cross-linking procedure that used glutaraldehyde (GA) fixation followed by secondary cross-linking with polyacrylamide hydrazide to treat cells expressing Lpp-OmpA-Bla. Cells were killed, had normal morphology, increased cell integrity, and their OM was stabilized. Residual Bla activity was more resistant to thermal denaturation, but the GA treatment reduced its specific activity approximately 85%. Subsequently,[34] GA fixation was conducted in the presence of reversible Bla inhibitors (phenyl boronic acid or sodium borate). Cell stabilization was unaffected and Bla activity was protected significantly, with approximately 40% retained. In addition, a simple immobilization method for GA-treated cells was described: fixed cells were mixed with chitosan-coated cellulose powder. Compared to freely suspended cells, the Bla activity of immobilized cells had a higher thermal stability, the same K_m, but a 40% reduced V_{max}.

The second cell immobilization procedure used cells expressing Lpp-OmpA-OPH.[35] In this procedure, starved/resting cells are adsorbed for 48 hr to nonwoven polypropylene fabric. Adsorption is tight and induced and induced cells adsorb equally well. The cells are alive and capable of being induced for several days after adsorption to the support. OPH activity was completely stable for six bioreactor cycles.

Biotechnology Applications

Several systems employing the Lpp-OmpA vector show promise for immediate biotechnology application. Detoxification of organophosphate compounds such as paraoxon, diazinon, coumaphos, and methyparathion by immobilized E. coli cells expressing Lpp-OmpA-OPH has been accomplished.[35] Bioreactors efficiently hydrolyzed 100% of the model nerve agent paraoxon, with a specific rate of 0.160 mM min^{-1} (g cell dry wt)$^{-1}$ and the stabilized cells could be used repeatedly for weeks. A more general application has been demonstrated using the Lpp-OmpA vector for surface display of large polypeptide libraries, specifically scFv antibody collections. Subsequent screening and selection by FACS, coupled with growth and resorting, have been used to isolate desired clones.[13] Regulation of this system has been optimized[16] so that the initial library diversity can be

[33] A. Freeman, S. Abramov, and G. Georgiou, *Biotechnol. Bioeng.* **52,** 625 (1996).
[34] A. Freeman, S. Abramov, and G. Georgiou, *Biotechnol. Bioeng.* **62,** 155 (1999).
[35] A. Mulchandani, I. Kaneva, and W. Chen, *Biotechnol. Bioeng.* **63,** 216 (1999).

preserved and not lost through decreased viability correlated with high-level expression and overgrowth by nonexpressing or nonfunction cells. Finally, HIV reverse transcriptase has been coupled to the Lpp-OmpA dibrid and the tripartite fusion localized to the OM of an attenuated (Aro⁻) *Salmonella* strain. Following oral administration in mice, mucosal IgA and T-cell responses were elicited.[36] The available evidence, therefore, indicates that the Lpp-OmpA surface display vehicle may function well in several categories of biotechnological application.

Final Comments

Many molecular details of the action of the Lpp-OmpA chimeric vehicle remain to be elucidated, including whether the Lpp export pathway is used (it differs from that of nonlipoproteins in having a different signal peptidase and in requiring a specific chaperone to reach the OM), whether fusion proteins exist as monomers or multimers in the OM, if concurrent LPS synthesis is necessary for OM insertion of the tribrid, which specific proteins involved in folding nascent polypeptides are required and if the same ensemble of these proteins is used with every passenger protein, and to what extent the various tribrids trigger the several bacterial stress response systems. Answers to these questions may provide additional ideas for extending and optimizing the Lpp-OmpA-based surface display procedure. Whether host/Lpp-OmpA-encoding plasmid systems capable of displaying any soluble protein can be developed, regardless of passenger size, topology, normal location, or possibility of quaternary structure, remains to be seen.

[36] M. S. Burnett, M. Hofmann, and G. B. Kitto, submitted for publication.

---

[31] Identification of Bacterial Class I Accessible Proteins by Disseminated Insertion of Class I Epitopes

By Dolph Ellefson, Adrianus W. M. van der Velden, David Parker, and Fred Heffron

The immune system is alerted to danger by the presentation of complexes on the surface of the infected cell. In the case of the mouse, these complexes are composed of antigens derived from the pathogen and specialized proteins of the major histocompatability complex (MHC). Two separate pathways, MHC I and MHC II, each drive cellular and humoral immune responses, respectively. In general, antigens presented in the context of MHC I are either derived from cytoplasmic proteins or, in the case of some professional antigen-presenting cells, via an alternate pathway through a lysosomal compartment.[1,2] Antigens presented in the context of MHC II are generally derived via pinocytotic or phagocytic mechanisms.[1]

Of the many pathogenic bacteria capable of mediating disease in humans and animals, one subset, intracellular pathogens, present unique challenges to researchers attempting to understand bacteria/host cell interactions. Intracellular pathogens are divided into two groups: those that reside within a phagolysosomal compartment (*Salmonella* sp., *Mycobacterium tuberculosis*, etc.) and those that reside within the cytoplasm (*Listeria monocytogenes, Shigella* sp., etc.). Intracellular pathogens adapt to their host cell environment by the selective secretion of proteins designed to alter the normal structural and metabolic machinery of the host cell, thus promoting bacterial survival and avoidance of host immune surveillance. Both phagolysosomal and cytoplasmic intracellular pathogens secrete proteins known to mediate their effects specifically within the host cell cytoplasm.[3–5] Because cytoplasmic localization of the bacterial protein also infers access to the degradative machinery of the host cell proteosome, we have labeled these proteins class I (MHC, HLA) accessible proteins (CAPs). The identification of CAPs has proven laborious and imprecise, as whole genome analysis is of limited value in assigning a function to proteins encoded by bacterial genes with no known homology. Because a substantial proportion of open reading frames derived from whole genome analysis have no known function, a system that allows the identification of CAPs secreted in response to host cell interactions would be an invaluable tool for understanding the many levels of pathogen/host cell interactions.[6] The identification of CAPs may enable the design of better bacterial carrier vaccines and identify whole new classes of potentially useful vaccine target proteins from different pathogens.

To identify *Salmonella typhimurium* CAPs potentially targeted to the host cell cytoplasm, we developed a resolvable Tn5-based transposon, which randomly distributes the MHC I (H-2K^b-restricted) ovalbumin epitope, SIINFEKL, throughout the bacterial chromosome. Epitope-tagged CAPs released from the infecting bacteria are processed by the proteolytic ma-

[1] L. A. Morrison *et al., J. Exp. Med.* **163,** 903 (1986).
[2] J. D. Pfeifer *et al., Nature* **361,** 359 (1993).
[3] G. R. Cornelis and H. Wolf-Watz, *Mol. Microbiol.* **23,** 861 (1997).
[4] C. M. Collazo and J. E. Galan, *Mol. Microbiol.* **24,** 747 (1997).
[5] Y. Fu and J. E. Galan, *Mol. Microbiol.* **27,** 359 (1998).
[6] E. J. Strauss and S. Falkow, *Science* **276,** 707 (1997).

chinery of the host cell and the carried ovalbumin epitope SIINFEKL is presented in the context of H-2K^b on the surface of the host cell. The approach, termed "disseminated insertion of class I epitopes (DICE), contains several inherent strengths in the identification of this important subset of bacterial proteins. First, DICE selection is conditional, and host class I-accessible proteins are isolated as a consequence of being processed and presented in the context of H-2K^b. Second, only in-frame insertions, which do not alter secretory signals, can be recovered. Third, the selection is simple and powerful; interesting strains are recovered quickly from a large population of infected cells by flow cytometry. Fourth, the selection is specific; bacteria cannot be recovered from macrophages that have presented SIINFEKL from nonsecreted intracellular proteins derived by bacterial attrition within in the phagolysosome because these bacteria would not be viable. Fifth, because the DICE transposon encodes a 6× histidine tag, the subcellular location of the protein can be visualized by microscopy, thereby enabling functional and phenotypic inferences to be drawn about proteins with no known homology.[7] Sixth, the protein can be assessed readily as an epitope carrier by attenuating the bacterial strain and immunizing the appropriate animal model. Finally, genes encoding pathogen proteins identified by DICE can be cloned and their protein products assessed for their ability to engender protective responses in an immunized host.

Tn5-DICE Design

Transposition and Resolution of Tn5-DICE

Two primary events are involved in the successful insertion of class I epitopes in the chromosome of *Salmonella:* (1) transposition of the complete mobile genetic element and (2) resolution of an excessive transposon sequence to create a phenotypically distinguishable fusion protein. This system differs from other mutagenesis procedures in that it confers a phenotype in the host cell rather than in the mutated bacterium. Although the transposon is capable of insertion in all six frames, out-of-frame resolved insertions cannot be distinguished because the selection conditions dictate that any insertion must allow read-through of the class I epitope. In addition, sequences critical to preserving the bacterial proteins' trafficking pattern within the host cell must therefore be preserved. Because in-frame insertional events within a gene occur approximately 16% of the time, this system requires a transposon capable of frequent and indiscriminate distribution.

---

[7] J. Zheng, W. Luo, and M. L. Tanzer, *J. Biol. Chem.* **273,** 12999 (1998).

Tn5 satisfies this requirement because of its promiscuous insertion site preference and its high insertional frequency.[8]

Tn5 Transposon

Wild-type Tn5 is a composite transposon that utilizes a conservative strategy as its principal mechanism of transposition.[9] Two nearly homologous insertion sequences, IS50L and IS50R, flank antibiotic resistance genes in wild-type Tn5. IS50R contains transposase genes, and IS50L contains an ochre codon, rendering its sequences inactive.[10] IS50L and IS50R are flanked by two 19-bp ends termed the inside end (I) and the outside end (O). Despite the presence of two insertion sequences in Tn5, functional mini-Tn5 variants have been constructed using only IS50R encoding the transposase, a resistance marker, and the I and O ends.[11]

We surmised that we could take advantage of CAP access to the class I antigen processing and presentation machinery of the host cell to identify the genes encoding these proteins. If the genes can be engineered to carry an appropriate MHC I-restricted epitope, the Salmonella mutant strain could be isolated from a large pool and the gene encoding the CAP identified. Although the total number of CAPS encoded by Salmonella is unknown, we hypothesized that they must represent a small portion of the total Salmonella proteins. Because Tn5 integration is random and therefore capable of insertion in any of six frames, only in-frame-resolved insertions (1/6 of total) would be discernible. Based on these assumptions, we constructed a library of independent epitope insertions large enough to encompass every open reading frame. The high rate of Tn5 transposition, coupled with its lack of insertion site preference, made it the ideal mutagenesis system for this task.

The Tn5-DICE minitransposon consists of a Tn5-transposase and a kanamycin cassette flanked at its 5' and 3' ends by direct repeats of a minimal loxP recombination site (Fig 1).[12] The 5' end of the transposon consists of the Tn5 I end, the H-2K^b-restricted epitope SIINFEKL, a 6× histidine tag, and one loxP site. Tn5-DICE is constructed such that on induction of Cre recombinase, the insertion is resolved at loxP sites. The kanamycin and Tn5-transposase cassettes are segregated to nonreplicating loops and lost. When the insertion is in frame to a gene, the 49 amino acid-

---

[8] D. Biek and J. R. Roth, Proc. Natl. Acad. Sci. U.S.A. **77,** 6047 (1980).

[9] D. E. Berg and M. M. Howe, Mobile DNA, American Society for Microbiology, 1989.

[10] S. J. Rothstein and W. S. Reznikoff, Cell **23,** 191 (1981).

[11] V. de Lorenzo, M. Herrero, U. Jakubzik, and K. N. Timmis, J. Bacteriol. **172,** 6568 (1990).

[12] M. F. Hoekstra et al., Science **253,** 1031 (1991).

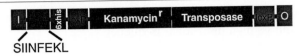

SIINFEKL

FIG. 1. Tn5-DICE transposon. The minitransposon consists of kanamycin and transposase cassettes flanked by direct repeats of the minimal *loxP* recombination site.[12] The entire Tn5-DICE minitransposon is flanked by the IS*50*R *I* and *O* ends. The *I* end, the H-2K^b-restricted ovalbumin epitope SIINFEKL, a 6× histidine site, and *loxP*$_1$ are translationally in frame.

resolved product creates a fusion protein carrying the SIINFEKL epitope (Fig. 2). *Salmonella* proteins that end up in the cytoplasm of the host cell are processed and the SIINFEKL epitope is presented in the context of H-2K^b. The *Salmonella* strain containing the resolved insertion is isolated

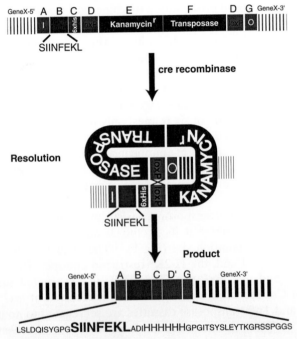

FIG. 2. In-frame resolution of Tn5-DICE. A *Salmonella typhimurium* transconjugant carrying the Tn5-DICE transposon on the F plasmid was used as a donor to generate an insertional library (approximately 120,000 random insertions) in *S. typhimurium* wild-type strain ATCC 14028 via P22 (HT*int*). Arabinose-induced expression of *cre* recombinase resulted in the resolution of Tn5-DICE at *loxP* sites, resulting in a resolved 49 amino acid product. Resolved in-frame insertions of the transposon result in expression of fusion proteins containing SIINFEKL and a 6× histidine tag.

by fluorescence-activated cell sorter (FACS) analysis using the H-2K^b/ SIINFEKL-specific monoclonal antibody 25-D1.2 as a marker.[13]

Identification of CAP Genes

Materials

Strains: F$^+$ *Eschericia coli* (Kan^s, Cam^s, Nal^s),
Salmonella typhimurium (14028, American Type Culture Collection),
E. coli S17λ*pir*,[14] and
P22 phage
Reagents: Phycoerythrin (PE)-conjugated streptavidin
Antibodies: Biotinylated 25-D1.2 monoclonal antibody and FITC-conjugated anti-H-2D^b
Media: Luria broth (LB) agar with appropriate antibiotic(s) (P22 transductants are selected on LB agar plates containing 30 μg/ml kanamycin), RPMI 1640, and fetal bovine serum (FBS)
Mice: C57Bl/6 (H-2^b)
Cells: L929 (Murine fibrosarcoma, American Type Culture Collection)
Equipment: CO$_2$ incubator and fluorescence-activated cell sorter

Methods

Transfer of Tn5-DICE to F'

The Tn5-DICE transposon is transposed onto an F' plasmid using a cotransfer mating selection.

1. An *E. coli* donor strain, which contains both an F' plasmid and the Tn5-DICE-bearing plasmid, pDE510 (*tra*$^-$/*mob*$^-$), is mated with a nalidixic acid-resistant *S. typhimurium* strain.

2. Nalidixic acid and kanamycin-resistant *Salmonella* transconjugants, which now contain F'::Tn5-DICE, are confirmed by a P22 sensitivity test and the ability to transfer the transposon kanamycin marker back into *E. coli* or *Salmonella* recipients at a frequency equal to F' plasmid transfer frequencies.

3. The *Salmonella*-specific bacteriophage, P22, is then used to make a lysate of the *Salmonella* strain containing F'::Tn5-DICE. Because there is no sequence homology to F' in *Salmonella,* the P22 phage lysate can be used to mutagenize a second *Salmonella* recipient (*Salmonella* strain containing

---

[13] A. Porgador *et al., Immunity* **6,** 715 (1997).
[14] S. A. Kinder *et al., Gene* **136,** 271 (1993).

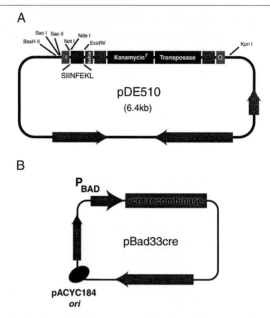

Fig. 3. Plasmids used for DICE analysis. (A) Plasmid carrying Tn5-DICE-resolvable minitransposon. (B) Arabinose-inducible Cre recombinase plasmid pBAD33*cre*.

pBAD33*cre*). The lack of F' homology in the recipient ensures that kanamy-cin-resistant transductants are derived as a result of transposition rather than homologous recombination. Transductants are selected by kanamycin resistance (30 μg/ml) on Luria agar. The Cre recombinase in pBAD33*cre* is under tight regulatory control of the pBAD promoter and mediates resolution and loss of the kanamycin resistance gene and the Tn5 trans-posase gene only when the strain is grown in the presence of arabi-nose (1 mM). The pBAD33*cre* plasmid is unstable and is lost in 3–10 generations when *Salmonella* strains bearing this plasmid are grown without selection (Fig. 3).[15]

4. The pool of *S. typhimurium* mutants is enriched for in-frame inser-tions of the resolved Tn5-DICE transposon within genes encoding secreted effector proteins by FACS as discussed later.

Identification of Strains Containing DICE Insertions by FACS Analysis

Assuming random integration of Tn5-DICE, approximately 1/6 (20,000) mutants of the 120,000 independent Tn5-DICE insertions we generated

[15] L. M. Guzman, D. Belin, M. J. Carson, and J. Beckwith, *J. Bacteriol.* **177**, 4121 (1995).

should contain resolved in-frame insertions. Of this number, many insertions will be in metabolic genes that may be essential. In addition, many insertions will be contained in promoter or noncoding intergenic regions. Of the remaining mutants, far fewer will be contained within CAPs. The precise number of CAPs in *S. typhimurium* is unknown. Because DICE insertions within CAPs may be rare events, a sensitive selection procedure was required. With the appropriate cell marker, FACS enables the isolation of extremely rare mutants.

FACS Analysis of DICE Mutants in Murine Bone Marrow-Derived Macrophages

1. Femurs are harvested from 4- to 6-week-old C57Bl/6 mice (H-2K^b). The femurs are scraped of muscle and connective tissue and are washed with RPMI, and bone marrow is exposed by removing proximal and distal ends with a scalpel.

2. Bone marrow cells are extracted by lavaging each end of the femur with a 3-cc syringe containing a 30-gauge needle and 2 ml of RPMI.

3. Bone marrow cells are then washed three times with RPMI at 37° and resuspended at a density of 1×10^6 cells/ml in RPMI 1640/10% FBS containing 20% L929 medium as a source of granulocytemacrophage colony-stimulating factor (GM/CSF). L929 medium is derived by harvesting L929 media 7 days after growing cells to confluence.[16]

4. The cultures are allowed to differentiate to bone marrow-derived macrophages (BMDM) by culturing the bone marrow cells for 6 days at 37°/5% CO_2.

5. BMDM are prepared for *Salmonella* infection and FACS analysis by scraping the cultures and resuspending them in RPMI 1640/10% FBS in a six-well plate at a density of 1×10^7 cells per well.

6. The pooled library is grown overnight in Luria broth at 37° with shaking.

7. The pooled DICE library is washed three times in RPMI 1640 and suspended in RPMI at a density of 5×10^8 cells ml.

8. Twenty microliters of the resuspended library is dispensed into individual wells of adhered BMDM (MOI = 1). An MOI of 1 or less limits multiple infections within the same BMDM. Typically, a 1% infection rate is expected for *S. typhimurium in vitro*.

9. The cultures are centrifuged for 2 min at 200 rpm to initiate contact and are incubated at 37° for 1 hr.

10. After 1 hr, the cultures are removed and washed three times with phosphate-buffered saline (PBS, pH 7.4) preheated to 37°.

[16] Y. Yamamoto-Yamaguchi, M. Tomida, and M. Hozumi, *Blood* **62,** 597 (1983).

11. The cultures are then overlayed with 3 ml of RPMI 1640/10% FBS containing 50 μg/ml of gentamycin to kill extracellular bacteria and are incubated at 37° for an additional 2 hr.

12. The cultures are then washed three times with PBS preheated to 37°.

13. The cells are scraped from the plate, resuspended in 10 ml of RPMI 1640/1% FBS, and incubated on ice.

14. The cells are labeled for FACS analysis by incubating with FITC-conjugated anti-H-2D^b and biotinylated anti-H-2K^b/SIINFEKL (5 μg, 25-D1.2.[13]

15. The cells are washed three times in PBS (4°) and incubated with 1 μg PE-conjugated streptavidin.

16. BMDM infected with the *Salmonella*-DICE library are sorted by first gating on the forward and side scatter population characteristic for macrophages. Bright red (PE-anti-H-2K^b/SIINFEKL) and bright green (FITC-conjugated anti-H-2D^b) populations, visualized in the double-positive quadrant, are sorted into a 5-ml polypropylene tube containing 2 ml of RPMI 1640/1% FBS.

17. The sorted cells are isolated by centrifugation and lysed in LB/1% Triton X-100.

18. The lysed cells are plated on LB agar and incubated at 37° overnight to recover *Salmonella*-DICE strains.

19. Infected BMDMs lacking CAP insertions may be recovered as a consequence of aggregate formation in the flow-sorted population. To ensure that recovery was to phenotypic expression of H-2K^b/SIINFEKL, the recovered bacterial colonies are counted, pooled, and subjected to two additional rounds of FACS sorting to enrich for *Salmonella* mutants containing CAP insertions. Individual isolates are then subjected to an additional round of FACS analysis to confirm their phenotype.

Sequencing of CAP Genes

In an effort to determine the identity of CAPs containing in-frame SIINFEKL insertions, we attempted to clone flanking DNA sequences by polymerase chain reaction (PCR). However, due to the small target sequence of the resolved Tn5-DICE transposon (147 bp), PCR strategies were prone to nonspecific primer annealing. We therefore opted to construct a unique system that allows specific and efficient identification of CAP genes (Fig. 4). In addition, this system was used to efficiently retransduce Tn5-DICE mutants and reconfirm their phenotypes. A *Kpn*I–*Sac*I fragment of a plasmid carrying the resolved Tn5-DICE transposon (pAV353a) was cloned into an ampicillin-resistant suicide vector (pGP704)[17] to yield plas-

---

[17] V. L. Miller and J. J. Mekalanos, *J. Bacteriol.* **170**, 2575 (1988).

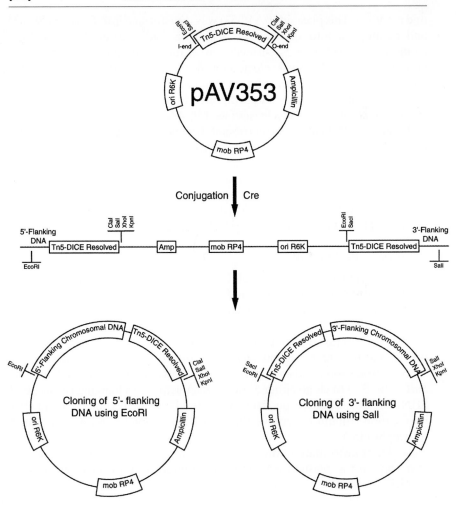

FIG. 4. Sequencing of Tn5-DICE-resolved CAPs. (A) Suicide plasmid pAV353, containing a resolved copy of Tn5-DICE, is conjugated into a naladixic acid-resistant, Cre-expressing Tn5-DICE mutant. (B) An ampicillin- and naladixic acid-resistant transconjugant is obtained via Cre-loxP recombination. (C) Isolated chromosomal DNA is restricted with EcoRI or SalI to clone 5' or 3' sequences flanking the original SIINFEKL insertion, respectively. Religated chromosomal fragments carrying the integrated plasmid pAV353 form functional replicons when transformed into E. coli S17λpir. Plasmid DNA from ampicillin-resistant transformants is isolated and analyzed using the Tn5-DICE-specific primer 5'-GCGGATATCCACCAC-CACCACC-3' (SalI) or 5'-TATGCCCGGGCCGTGGTGGTGG-3' (EcoRI).

mid pAV353. This plasmid (pAV353) was transformed into *E. coli* S17λ*pir*[14] and conjugated into spontaneous naladixic acid-resistant CAP mutants containing pBAD33*cre*. Site-specific integration of plasmid pAV353 at the chromosomal *loxP* site was selected following induction of the Cre recombinase by selecting naladixic acid- and ampicillin-resistant transconjugants. Chromosomal DNA was isolated, digested for 2 hr at 37° with one of several possible restriction endonucleases (see Fig. 4A, pAV353), and ligated overnight at 15°. Religated chromosomal fragments carrying the integrated plasmid pAV353 form functional replicons in *E. coli* S17λ*pir* and carry either 3′ (i.e., *Sal*I) or 5′ (i.e., *Eco*RI) sequences flanking the original SIINFEKL insertion (Fig. 4). Ampicillin-resistant transformants were analyzed further using the Tn*5*-DICE-specific primer 5′-GCGGATATCCAC-CACCACCACC-3′ (*Cla*I, *Sal*I, *Xho*I, or *Kpn*I digests) or 5′-TATGCCCG-GGCCGTGGTGGTGG-3′ (*Eco*RI, *Sac*I digests).

Sequencing Protocol

1. An ampicillin-resistant, nalidixic acid-sensitive doner strain (*E. coli* S17λ*pir*) containing suicide plasmid pAV353 (amp^r *tra*$^+$ *mob*$^+$) is mated to a spontaneous nalidixic acid-resistant, Cre-expressing *S. typhimurium* Tn*5*-DICE mutant.

2. Transconjugants (amp^r nal^r) that carry an integrated copy of plasmid pAV353 at the chromosomal Tn*5*-DICE insertion site are selected.

3. Chromosomal DNA is prepared and restricted with either *Eco*RI or *Sal*I (2 hr, 37°) to clone either 5′ or 3′ DNA sequences flanking the original SIINFEKL insertion, respectively. Digested DNA is absorbed over a DNA purification column to remove the restriction endonuclease and is ligated overnight (15°).

4. On transformation into *E. coli* S17λ*pir*, religated circular fragments that contain the plasmid pAV353 form functional replicons, resulting in ampicilin-resistant transformants.

5. Plasmid preparations from ampicillin-resistant colonies are sequenced using the Tn*5*-DICE-specific primer 5′-GCGGATATCCACCAC-CACCACC-3′ (*Sal*I digest) or 5′-TATGCCCGGGCCGTGGTGGTGG-3′ (*Eco*RI digest).

Tn*5*-DICE: Other Uses

Although we have described a system for the use of the resolvable transposon in which CAPs are identified by incorporation of a classical class I epitope, the transposon has a range of other possible uses. This transposon has been engineered to accept a variety of different elements. For instance, we are using a modified version of Tn*5*-DICE to identify

Salmonella proteins that cycle into the class II (MHC, HLA) pathway. In addition, we are investigating the ability of the transposon to insert green fluorescence protein into genes, thus enabling us to identify *in vivo*-expressed genes by flow-sorting tissue homogenates. The transposon could also be used to modify vaccine carrier strains of *Salmonella* to augment or skew the immune response to the carried antigen by delivering eukaryotic effector proteins such as Jak2 or Tyk2 as CAP fusions. Finally, mutants generated by the transposon could be used to identify tissue-specific *Salmonella* CAPs, potentially useful proteins for regulating the timing of the immune response to carried antigens and thus generate immune responses more amenable to the life cycle of different pathogens.

Acknowledgments

We thank Ron Germain (National Institutes of Health) for his generous contibution of the 25-D1.2 monoclonal antibody and H. G. Bouwer (Veterans Administration Medical Center, Portland, OR) for his valuable assistance in the generation of bone marrow-derived macrophages. In addition, we thank Colin Manoil (Department of Genetics, University of Washington) for his generous gift of transposon reagents.

[32] *Bordetella pertussis* Adenylate Cyclase Toxin: A Vehicle to Deliver CD8-Positive T-Cell Epitopes into Antigen-Presenting Cells

By Pierre Guermonprez, Catherine Fayolle, Gouzel Karimova, Agnes Ullmann, Claude Leclerc, and Daniel Ladant

Introduction

Cytotoxic T lymphocytes (CTL) play a key role in the elimination of cells infected by virus or bacteria, as well as tumor cells that express particular epitopes (tumor associated antigens). This subset of T cells, which express the cell surface glycoprotein CD8, recognizes short peptides (the so-called CD8[+] T-cell epitopes) associated with major histocompatibility complex (MHC) class I molecules. Although CD8[+] T cells secrete cytokines (such as interferon-γ), their main role is to lyse cells that present the relevant peptide–MHC I complexes. Because MHC class I associated peptides are mostly derived from endogenously synthesized proteins, exogenous antigens are generally unable to prime CTL responses.[1,2] These soluble antigens usually enter antigen-presenting cells (APC) by endocytic mechanisms to be eventually targeted to the MHC class II presentation pathway that serves to activate the CD4+ helper T cells.

To circumvent the biological constraints of CTL priming, one approach is to use recombinant bacterial toxins as molecular "Trojan horses" to deliver the antigens into the cytosol of APC. Several bacterial toxins have been shown to be able to deliver CD8+ T-cell epitopes into eukaryotic cell cytosol and therefore to sensitize these cells to be recognized by specific CTL.[3] However, to date, only the anthrax toxin (from *Bacillus anthracis*) and the adenylate cyclase (CyaA) toxin from *Bordetella pertussis* have been effective in priming CTL *in vivo*.[4–6] We have shown that recombinant CyaA toxins carrying CD8+ T-cell epitopes from the nucleoprotein of the lymphocytic choriomeningitis virus (LCMV) or from the V3 region of the HIV glycoprotein are able to prime specific MHC class I-restricted cytotoxic T-cell (CTL) responses.[5,6] Furthermore, mice immunized with the recombinant toxin carrying the LCMV epitope were protected against infection with lethal doses of the LCMV virus.[6]

This article describes the construction and production of recombinant *B. pertussis* CyaA toxins that harbor a single CD8+ T-cell epitope, OVA, derived from chicken ovalbumin.[7] This epitope (amino acid sequence, SIINFEKL, corresponding to residues 257 to 264 of ovalbumin) is the main H-2K^b-restricted CD8+ T-cell epitope derived from ovalbumin when this protein is introduced artificially into or expressed by mouse cells.[7] Analysis of the immunological properties of these recombinant toxins is then described.

Adenylate Cyclase Toxin from *Bordetella pertussis*

Adenylate cyclase is one of the major toxins produced by *B. pertussis,* the causative agent of whooping cough, and plays an important role in respiratory tract colonization.[8] This toxin is able to enter into eukaryotic

---

[1] I. A. York and K. L. Rock, *Annu. Rev. Immunol.* **14**, 369 (1996).

[2] E. Pamer and P. Cresswell, *Annu. Rev. Immunol.* **16**, 323 (1998).

[3] T. J. Goletz, K. R. Klimpel, S. H. Leppla, J. M. Keith, and J. A. Berzofsky, *Hum. Immunol.* **54**, 129 (1997).

[4] J. D. Ballard, R. J. Collier, and M. N. Starnbach, *Proc. Natl. Acad. Sci. U.S.A.* **93**, 12531 (1996).

[5] C. Fayolle, P. Sebo, D. Ladant, A. Ullmann, and C. Leclerc, *J. Immunol.* **156**, 4697 (1996).

[6] M. F. Saron, C. Fayolle, P. Sebo, D. Ladant, A. Ullmann, and C. Leclerc, *Proc. Natl. Acad. Sci. U.S.A.* **94**, 3314 (1997).

[7] M. W. Moore, F. R. Carbone, and M. J. Bevan, *Cell* **54**, 777 (1988).

[8] A. A. Weiss and E. L. Hewlett, *Annu. Rev. Microbiol.* **40**, 661 (1986).

target cells where it is activated by endogenous calmodulin (CaM) to pro-duce supraphysiological levels of intracellular cAMP, thus causing impair-ment of cellular functions.[9] This 1706 residues long protein, encoded by the *cyaA* gene, is synthesized as a protoxin that is converted to the active toxin by palmitoylation of Lys 983, a process dependent on the product of an accessory gene, *cyaC*, adjacent to *cyaA* (for a review, see Ladant and Ullmann[10]). The structural organization of CyaA is shown in Fig. 1. The CaM-dependent adenylate cylase catalytic domain is located within the first 400 amino acids. The carboxy-terminal 1306 residues are responsible for the binding of the toxin to target cell membrane and the subsequent delivery of the catalytic moiety into the cell cytosol. This part also exhibits a weak hemolytic activity due to its ability to form cation-selective channels in biological membranes.[10]

CyaA can invade a large variety of mammalian cell types through a unique mechanism that involves direct delivery of the catalytic domain across the plasma membrane.[10] The translocation of the catalytic domain into the cell cytosol is calcium and temperature dependent. The onset of intoxication is rapid and maximal in 30 min, whereas the hemolytic activity of CyaA exhibits a lag of more than 60 min, suggesting that these two processes occur by different mechanisms: the delivery of the AC domain into cells can be accomplished by toxin monomers, whereas membrane channel formation, leading to hemolysis, requires oligomerization of CyaA.

Exogenous peptides can be inserted into various permissive sites within the catalytic domain of CyaA without hampering its ability to enter into eukaryotic cells.[11,12] One particular permissive site located in the middle of the catalytic domain between amino acid 224 and 225 of CyaA has been used for the construction of recombinant toxins harboring CD8[+] T-cell epitopes. The construction, production, and purification of such toxins are described in detail in the following sections.

Construction of Recombinant CyaA Toxins

All recombinant toxins are produced in *Escherichia coli*. The two basic expression vectors used to construct and express recombinant CyaAs are described in Fig. 1B. They are both derived from a pUC replicon and harbor the two coding regions for CyaC and CyaA polypeptides.[13] The

[9] D. L. Confer and J. W. Eaton, *Science* **217,** 948 (1982).
[10] D. Ladant and A. Ullmann, *Trends Microbiol.* **7,** 172 (1999).
[11] D. Ladant, P. Glaser, and A. Ullmann, *J. Biol. Chem.* **267,** 2244 (1992).
[12] P. Sebo, C. Fayolle, O. d'Andria, D. Ladant, C. Leclerc, and A. Ullmann, *Infect. Immun.* **63,** 3851 (1995).
[13] P. Sebo, P. Glaser, H. Sakamoto, and A. Ullmann, *Gene* **104,** 19 (1991).

A

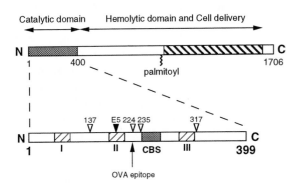

Cyaa protein

Catalytic domain   Hemolytic domain and Cell delivery

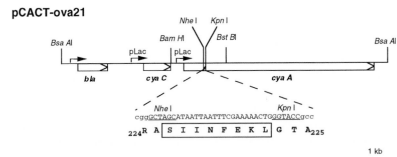

B

pCACT-ova21

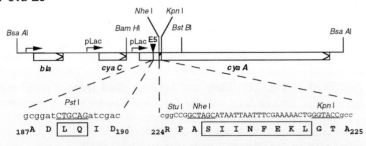

FIG. 1. Schematic representation of the recombinant CyaA toxin and of the expression plasmids. (A) Schematic representation of the CyaA toxin. The upper part represents the full-length CyaA protein with the catalytic domain depicted by the shadowed box and the Gly-Asp-rich repeat region by the hatched box. The site of palmitoylation is also indicated.

cyaC and *cyaA* genes are under the transcriptional control of *lac* promoters and therefore the expression of the corresponding polypeptides can be induced by isopropyl-β-D-thiogalactopyranoside (IPTG). To facilitate construction of recombinant toxins with exogenous epitopes inserted within its catalytic domain, two unique restriction sites, *Nhe*I and *Kpn*I, were inserted within the CyaA-coding region of pCACT plasmids, between codons 224 and 225.[14] Therefore, it is easy to insert any double-strand oligonucleotide encoding the amino acid sequence of a particular CD8$^+$ epitope.

The plasmid pCACT-ova21 encodes the recombinant protein CyaA-OVA, which exhibits full adenylate cylase activity and therefore is cytotoxic. Its construction has been described in detail in Karimova *et al.*[14] The plasmid pCACT-ova-E5 was constructed to express a nontoxic recombinant CyaA protein. For this, a hexanucleotide CTGCAG was inserted into an *Eco*RV site located in the 5' part of the *cyaA* DNA sequence.[15] This results in an in-frame insertion of the dipeptide Leu-Gln between Asp188 and Ile189, which totally abolishes adenylate cyclase activity without affecting the stability of the CyaA protein.[11] During the construction of plasmid pCACT-ova-E5, a new unique restriction site was incorporated in the sequence adjacent to the *Nhe*I site. This results in an additional proline residue inserted between Arg224 of CyaA and the OVA epitope (see Fig. 1B).

All *in vitro* DNA manipulations were performed according to standard protocols.[16]

[14] G. Karimova, C. Fayolle, S. Gmira, A. Ullmann, C. Leclerc, and D. Ladant, *Proc. Natl. Acad. Sci. U.S.A.* **95**, 12532 (1998).

[15] P. Guermonprez, D. Ladant, G. Karimova, A. Ullmann, and C. Leclerc, *J. Immunol.* **162**, 1910 (1999).

[16] J. Sambrook, E. F. Fritsch, and T. Maniatis, "Molecular Cloning: A Laboratory Manual," 2nd Ed. Cold Spring Harbor Laboratory, Cold Spring Harbor, NY, 1989.

The lower part represents an enlargement of the catalytic domain. Hatched boxes labeled I, II, and III represent regions that are essential for catalysis. The dotted box, CBS, corresponds to the main CaM-binding site. White arrowheads indicate the location of permissive sites, and the black arrowhead indicates the position of the E5 mutation that abolishes adenylate cyclase activity. The site of OVA epitope insertion is also shown. (B) Schematic representation of the plasmids used in this work. Open boxes represent the open reading frames of β-lactamase (*bla*), *cyaC,* and *cyaA* genes, and arrows indicate the position of their promoters and direction of transcription. The location of some unique restriction sites is indicated. DNA sequences of the OVA inserts are shown (in capital letters) together with the encoded amino acid sequences (the precise OVA peptide sequence is boxed). In pCACT-ova-E5, the sequence around the site of E5 mutation is displayed: the hexanucleotide insert (capital letters) encoding a new *Pst*I site was inserted in a former *Eco*RV site. The amino acid numbers correspond to that of wild-type CyaA.

Production and Purification of Recombinant CyaA Toxins

Recombinant CyaA proteins are expressed in the *E. coli* strain BLR (Novagen), where they generally accumulate as inclusion bodies,[13] and are purified by a simple two-step procedure. Plasmid pCACT-ova21 (or pCACT-ova21-E5) is cotransformed by electroporation into competent BLR cells, together with plasmid pLacI^Q, a pACYC derivative (compatible with pCACT-ova21) that encodes the Lac repressor (LacI, to repress the expression of *cyaC* and *cyaA* genes). The transformants are selected on Luria-Bertani (LB) plates containing 100 μg/ml ampicillin and 30 μg/ml chloramphenicol. To produce the recombinant proteins, transformed cells are grown at 37° in a 15-liter fermentor in LB medium containing 150 μg/ml ampicillin and 30 μg/ml chloramphenicol until the optical density (OD) at 600 nm reaches 0.8–1. Then 1 mM IPTG is added and after 3 hr of further growth, the cells are harvested by centrifugation.

The harvested cells (about 30 to 50 g wet weight) are resuspended in 20 mM HEPES–Na, pH 7.5, and are disrupted with a French press. After ultracentrifugation, the supernatant (= bacterial extract, Fig. 2, lane 2) is removed. The pellet generally consists of a white layer of inclusion bodies at the bottom overlaid by a brown cushion of lipids and membrane proteins. These latter can be solubilized easily in 8 M urea, 20 mM HEPES–Na, pH 7.5, in a few minutes of vortexing. The inclusion bodies, which contain the

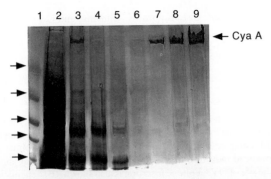

FIG. 2. SDS–PAGE analysis of recombinant CyaA toxins. Fractions collected at various stages of purification of recombinant CyaA-OVA were run on a 5–15% SDS gel and stained with Coomassie blue. Lane 1: molecular mass markers corresponding, from top to bottom, to 94, 67, 43, 30, and 21 kDa; lane 2: bacterial extract; lane 3: urea extract; lane 4: fraction unbound to DEAE-Sepharose; lane 5: first wash of DEAE-Sepharose; lane 6: second wash of DEAE-Sepharose; lane 7: eluted fraction from DEAE-Sepharose; and lanes 8 and 9: proteins eluted from phenyl-sepharose (6 and 18 μg of proteins were run, respectively). A similar profile was obtained for CyaA-OVA-E5.

recombinant CyaA proteins, are solubilized in about 100 ml of 8 M urea, 20 mM HEPES–Na, pH 7.5, by overnight agitation at 4°. After centrifugation (20,000g, 20 min, 14°) the supernatant (= urea extract, Fig. 2, lane 3) is supplemented with 0.14 M NaCl and loaded (at room temperature) on a 25-ml DEAE-Sepharose column equilibrated with 8 M urea, 0.14 M NaCl in 20 mM HEPES–Na, pH 7.5. In these conditions, the CyaA protein, which has a very low isoelectric point (pI 3.8), is entirely retained on the resin while all other contaminants flow through it (Fig. 2, lane 4). After an extensive wash with the same buffer, the CyaA protein is eluted in 8 M urea, 0.5 M NaCl, 20 mM HEPES–Na, pH 7.5 (Fig. 2, lane 7). At this step, the protein is essentially pure as deduced from SDS–polyacrylamide gel analysis, but still contains biological materials that absorb strongly at λ-260 nM (probably nucleotides). To remove these contaminants, a second chromatography on phenyl-sepharose is performed. Proteins eluted from the DEAE-sepharose resin are diluted five times with 20 mM HEPES–Na, 1 M NaCl, pH 7.5, and applied (at room temperature) onto a phenyl-sepharose column (about 20–25 ml packed resin) equilibrated with the same buffer. After washing with 20 mM HEPES–Na, pH 7.5, the toxin is eluted with 8 M urea in 20 mM HEPES–Na. All toxins purified by this method are more than 90% pure as judged by SDS gel analysis (Fig. 2, lanes 8 and 9) and are apparently free of nonproteinaceous contaminants as estimated from UV spectra. CyaA toxin concentrations are determined spectrophotometrically from the absorption at 280 nm using a molecular extinction coefficient of 142,000 M^{-1} cm^{-1} (calculated from the content in amino acids Trp, Tyr, and Phe). Alternatively, protein concentrations are measured with the Bio-Rad Bradford assay reagent using purified wild-type CyaA as a standard.

This procedure has been repeated for more than 30 different recombinant CyaA toxins with high reproducibility. Altogether, the overall recovery from a 15-liter fermentor varies from 50 to more than 200 mg of pure CyaA proteins.

Analytical Methods

Adenylate cyclase activity is measured as described previously[11] in a medium containing 50 mM Tris–HCl, pH 8.0, 6 mM MgCl$_2$, 0.1 mM CaCl$_2$, 0.1 mM cAMP, 5000–6000 cpm of [^{3}H]cAMP, 1.25 μM CaM, and 2 mM [α-^{32}P]ATP (1–2 × 10^5 cpm/assay). One unit of adenylate cyclase activity corresponds to 1 μmol of cAMP formed in 1 min at 30° and pH 8.0. Toxin binding and translocation into sheep erythrocytes are assayed by a trypsin protection assay, essentially as described previously.[14] Alternatively, intoxication of erythrocytes or other cell types is determined by measuring cAMP levels within cells exposed to CyaA toxins.[17]

In Vitro Presentation of OVA CD8⁺ T-Cell Epitope

The ability of recombinant CyaA toxin to deliver the CD8⁺ T-cell OVA epitope to the MHC class I presentation pathway was analyzed in an *in vitro* assay that offers the possibility to study the mechanisms involved in the recognition, by CD8⁺ T cells, of the epitope grafted in CyaA. This assay uses an OVA-specific CD8⁺ T-cell hybridoma, B3Z, produced by Karttunen and co-workers,[18] to monitor the appearance of the OVA epitope/MHC class I K^b complex at the surface of APC. The B3Z hybridoma is cocultured with antigen-presenting cells in the presence of the detoxified toxin, CyaA-OVA-E5, or of the OVA peptide itself as a control. Engagement of the T-cell receptor of B3Z by the OVA/K^b complex triggers the production of IL-2 by the B3Z cells that can be taken as a readout of CD8⁺ T-cell hybridoma activation.

Preparation of Antigen-Presenting Cells

Many different cell types can present CyaA-OVA-E5 to the B3Z T-cell hybridoma, albeit with various efficacies.[15] In this work, APC are inflammatory peritoneal macrophages obtained from C57BL/6 mice (6- to 8-week-old female mice, Iffa Credo, L'Arbresle, France) injected intraperitoneally with 2 ml of thioglycollate (Sanofi Diagnostic Pasteur, Paris). Four to 7 days later, mice are sacrificed, and cells infiltrating the peritoneal cavity (Thio-PEC: thioglycollate-induced peritoneal exudate cells) are harvested by washing with 10–15 ml of cold RPMI 1640. The cells are pelleted and resuspended in complete medium (RPMI 1640 supplemented with 10% fetal calf serum, 2 mM L-glutamine, 10^{-5} M 2-mercaptoethanol, 100 U/ml penicillin, 100 μg/ml streptomycin). In some experiments, we also used macrophages or dendritic cells differentiated from bone marrow precursors in conditioned media containing M-CSF or GM-CSF.[19] Also, fresh APC can be obtained from lymphoid organs such as the spleen.

Antigen Presentation Assays

CD8⁺ T cells (the B3Z CD8⁺ T-cell hybridoma, 10^5 cells/well) are cocultured in flat-bottomed 96-well culture plates for 18 hr in the presence

---

[17] N. Heveker and D. Ladant, *Eur. J. Biochem.* **243**, 643 (1997).

[18] J. Karttunen, S. Sanderson, and N. Shastri, *Proc. Natl. Acad. Sci. U.S.A.* **89**, 6020 (1992).

[19] C. V. Harding, *in* "Current Protocol in Immunology" (J. E. Coligan, A. M. Kruisbeek, D. H. Margulis, E. M. Shevach, and W. Strober, eds.), p. 16.0.1. Wiley, New York, 1997.

of antigen and APC (generally 10^5/well) in 200 μl of complete medium. After the coculture, IL-2 present in the supernatant is measured by the CTLL proliferation method.[20] The supernatant (freezed and thawed to kill any living cells) is added (100 μl) to 100 μl of CTLL-2 cells (10^4/well), which proliferate specifically in response to IL-2. After 48 hr of culture, 1 μCi of [^{3}H]thymidine (NEN Life Science, Boston, MA) is added to each well and the incubation is continued for 18 hr. Cells are then harvested, and incorporated [^{3}H]thymidine is counted in a liquid scintillation counter. Results are expressed in cpm or Δcpm (cpm in the presence of antigen − cpm in the absence of antigen). Recombinant IL-2 (or supernatant containing IL-2) and medium alone are included in the experiments as positive and negative controls, respectively.

As shown in Fig. 3A, Thio-PEC APC from C57BL/6 mice stimulated B3Z T cells efficiently in the presence of CyaA-OVA-E5 but not in the presence of the detoxified toxin CyaA-LCMV-E5, which harbors an irrelevant epitope.[6] The B3Z hybridoma was also stimulated strongly by Thio-PEC in the presence of the synthetic OVA peptide [SIINFEKL, synthesized by Neosystem, Strasbourg, France, and purified by reversed-phase high-performance liquid chromatography (HPLC); stock solutions of peptide were prepared in phosphate-buffered saline (PBS) and stored at −20°], but not in the presence of ovalbumin, as expected.

To study the cellular mechanisms involved in the presentation of CyaA-OVA-E5, we tested the effects of various drugs shown previously to affect antigen presentation.[1,2,21] For these inhibition studies, antigen-presenting cells were first exposed to the drugs and then to the antigens (CyaA-OVA-E5 or OVA peptide). After washing, APC were fixed to stop all processing steps and were then coincubated with B3Z CD8$^+$ T-cell hybridoma to detect the MHC class I–peptide complexes at the surface of the fixed APC.

Thio-PEC APC (10^5/well) are incubated in 0.1 ml of complete medium with the indicated final concentrations of drugs for 1 hr. Then the antigens, CyaA-OVA-E5 or OVA peptide, diluted in 0.1 ml, are added to the wells (0.2 ml final volume) at the final concentration indicated in the continuous presence of the inhibitor. Lactacystin (Tebu, Le Perray-en-Yvelines, France) is dissolved in PBS at 1 mg/ml, brefeldine A (Sigma) is dissolved in methanol at 0.5 mg/ml, and LLnL and LLmL (Sigma) are dissolved in dimethyl sulfoxide at 10 mg/ml. Cycloheximide (Sigma) is dissolved in

[20] K. Bottomly, L. S. Davis, and P. E. Lipsky, in "Current Protocol in Immunology" (J. E. Coligan, A. M. Kruisbeek, D. H. Margulis, E. M. Shevach, and W. Strober, eds.), p. 6.3.1. Wiley, New York, 1991.
[21] A. Craiu, M. Gaczynska, T. Akopian, C. F. Gramm, G. Fenteany, A. L. Goldberg, and K. L. Rock, J. Biol. Chem. **272**, 13437 (1997).

A

IL-2 secretion, cpm

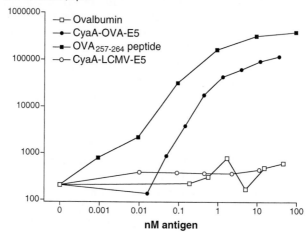

B

% of control response
without drug

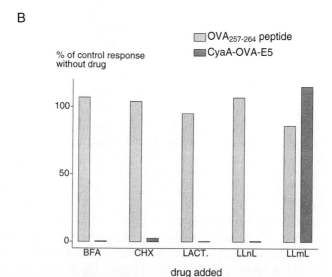

Fig. 3. *In vitro* presentation of CyaA-OVA-E5 to B3Z. (A) B3Z T cells and Thio-PEC APC were cocultured with the indicated concentrations of OVA peptide, chicken ovalbumin, or detoxified recombinant CyaA toxins for 18 hr, and IL-2 production was measured by the CTLL proliferation method, as described in the text. (B) Thio-PEC APC were preincubated for 1 hr with the indicated drugs: brefeldine A, BFA (5 μg/ml); cycloheximide, CHX (10 μg/ml); lactacystin, LACT. (2 μM); and LLnL or LLmL (30 μM). Then, 5 nM CyaA-OVA-E5 or 100 nM OVA peptide was added for a 5-hr pulse in the continuous presence of the drugs. APC were then fixed, and B3Z T cells were added for an 18-hr culture. IL-2 production was measured as described and is displayed as a percentage of the response obtained in the absence of drugs.

water at 2 mg/ml. After 5 hr, APC are fixed with 0.05% glutaraldehyde (for 2 min at 37°) and washed twice, and B3Z T cells are added to the wells and cultured for 18 hr in a 0.2-ml final volume as indicated earlier. IL-2 production is measured as described previously.

As shown in Fig. 3B, the proteasome inhibitors lactacystin and LLnL inhibited CyaA-OVA-E5 presentation. A control tripeptide, LLmL, which does not inhibit the proteasome activity, did not interfere with CyaA-OVA-E5 presentation. These results demonstrate that the processing of the CD8$^+$ OVA epitope inserted in CyaA requires proteasome activity. The stimulation of B3Z by CyaA-OVA-E5 was also sensitive to cycloheximide and brefeldine A (Fig. 3B). This indicates that the neosynthesis and the export of MHC class I molecules are required for presentation of CyaA-OVA-E5. In contrast, the presentation of OVA peptide, which binds directly to the MHC class I molecules at the cell surface, was not affected by these inhibitors, as expected (Fig. 3B). We have also shown that peritoneal macrophages derived from TAP1 (transporter associated with antigen processing) knockout mice[22] back-crossed on a C57BL/6 background did not present CyaA-OVA-E5.[15] Altogether, these results demonstrate that the presentation of CyaA-OVA-E5 to the B3Z CD8$^+$ T-cell hybridoma displays all the characteristics of the classical cytosolic pathway for the presentation of endogenous antigens.

In Vivo Induction of CTL Responses and Cytotoxicity Assays

To examine the capacity of recombinant CyaA toxins harboring the CD8$^+$ T-cell OVA epitope to induce specific cytolytic T-cell responses *in vivo,* C57BL/6 mice are immunized intraperitoneally (ip) on days 0, 21, and 42 with 50 μg of CyaA-OVA or CyaA-OVA-E5, mixed with 1 mg of aluminum hydroxide in 500 μl of PBS.[14,23] Control groups are injected ip with either PBS or wild-type CyaA mixed with 1 mg of aluminum hydroxide in PBS.

Spleen cells are removed surgically from immunized mice 7 days after the last injection. Spleens from three or four mice in each group are pooled. Immune spleen cells are resuspended in complete RPMI medium (see earlier discussion). Then, 25 × 10^6 responder spleen cells from *in vivo*-primed mice are cocultured with 25 × 10^6 irradiated (3300 rad) syngeneic spleen cells in the presence of 1 μg of OVA peptide with 10 ml complete

[22] L. Van Kaer, P. G. Ashton-Rickardt, H. L. Ploegh, and S. Tonegawa, *Cell* **71,** 1205 (1992).
[23] C. Fayolle, D. Ladant, G. Karimova, A. Ullmann, and C. Leclerc, *J. Immunol.* **162,** 4157 (1999).

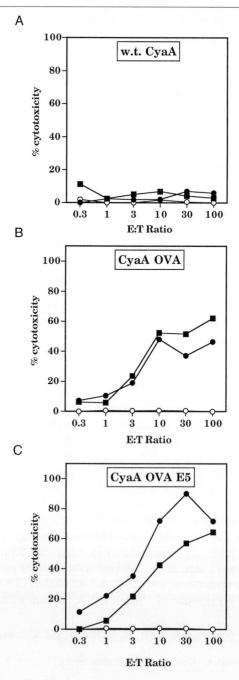

medium at 37° in humidified air with 7% CO_2 for 5 or 6 days. After washing, these effector cells are tested in classical cytolytic assays using, as target cells, either EL4, an H-2^b thymoma cell line, or E.G7, an ovalbumin-transfected subclone of EL4 (both obtained from the American Type Culture Collection).

For peptide sensitization, EL4 target cells are incubated simultaneously with 50 μM of OVA peptide and 100–200 μCi of $Na_2{}^{51}CrO_4$ for 1 hr at 37°. After two washes, 10^4 target cells are incubated with spleen effector cells (from 3×10^3 to 10^6) in round-bottom plates in a final volume of 200 μl. After 4 hr of incubation at 37°, 7% CO_2, 50 μl of supernatant is collected, and the percentage of specific ^{51}Cr release is calculated as follows:

$$\% \text{ specific release} = 100 \times (\text{experimental release} - \text{spontaneous release})/(\text{maximum release} - \text{spontaneous release})$$

Maximum release was determined from supernatants of cells that were lysed by the addition of 10% Triton X-100. Spontaneous release was determined from supernatants of cells incubated with medium alone and was always less than 15%. Results are the mean of duplicate cultures. The standard deviation of duplicate samples was always less than 15–20% of the specific ^{51}Cr release. As EL4 cells express only class I MHC antigens, antigen-specific CTL activity was considered to be restricted to MHC class I molecules.

As illustrated in Fig. 4, a strong peptide-specific cytotoxic response was induced after *in vivo* priming with the recombinant toxins CyaA-OVA and CyaA-OVA-E5. *In vitro*-stimulated spleen cells from CyaA-OVA- and CyaA-OVA-E5-immunized mice did not kill EL4 target cells incubated in the absence of OVA peptide, demonstrating the specificity of the cytotoxic response. Furthermore, splenocytes from mice injected with PBS (not shown) or with control wild-type CyaA did not kill E.G7 or EL4 cells coated with the OVA peptide (Fig. 4). Other experiments have shown that

---

Fɪɢ. 4. Immunization with recombinant adenylate cyclase toxins carrying the OVA epitope induces CTLs that lyse OVA-transfected thymoma cells. C57BL/6 mice were immunized intraperitoneally on days 0, 21, and 42 with 50 μg of wild-type CyaA (a), CyaA-OVA (b), or CyaA-OVA-E5 (c) mixed with 1 mg of aluminum hydroxide in PBS. Splenocytes were restimulated with syngeneic-irradiated splenocytes and assayed for cytotoxic function using ^{51}Cr-labeled EL4 (○), EL4 loaded with 50 μM of the OVA peptide (■), or E.G7 (●) as target cells. Results are reported as percentage of ^{51}Cr release at varying effector : target ratios. Data shown are the mean of duplicate cultures. The SD of duplicate wells was less than 15–20% of the mean.

the cytotoxic activity was mediated by the CD8[+] T-cell subset and was restricted by H-2K[b] MHC class I molecules.[14] In conclusion, these results indicate that the recombinant CyaA-OVA and CyaA-OVA-E5 toxins stimulated high levels of CD8[+] CTL precursors specific for the OVA/H-2[b] MHC I complex.

In Vivo Tumor Protection Studies

In order to determine whether recombinant toxins could induce protective antitumor CTL responses, groups of mice were immunized as described earlier with the recombinant CyaA-OVA and CyaA-OVA-E5 proteins or, as controls, with wild-type CyaA or PBS. Seven days later, the mice were injected sc on the right flank with either a C57BL/6-derived murine melanoma, B16, or the same murine melanoma transfected by chicken ovalbumin, MO5 (kindly provided by L. Rosthein and L. Sigal, Worcester, MA). Each mouse received 2×10^4 B16 or MO5 cells, grown in tissue culture flasks until midlog phase, and washed three times in PBS in a 0.2-ml volume. All experiments included six or eight mice per group and were repeated at least three times. The survival of mice was recorded as the percentage of mice surviving after the tumor graft. Statistical difference was determined using the method of Kaplan and Meier.[24] Statistical significance was determined at the <0.05 level.

In all mice of the control groups (injected with PBS or with wild-type CyaA), MO5 tumors grow progressively and were uniformly lethal by day 50 (Fig. 5a). Immunization with CyaA-OVA or CyaA-OVA-E5 markedly increased the survival of mice to the MO5 tumor graft (Fig. 5a). Most notably, the protective immunity induced by CyaA-OVA or CyaA-OVA-E5 was specific of ovalbumin-expressing MO5 tumor cells; as shown in Fig. 5b, mice immunized with these recombinant toxins were not protected from challenge with the parental melanoma B16 (which does not express ovalbumin). This indicates that the protective immunity induced by recombinant CyaA toxins is antigen specific.

Further experiments indicated that recombinant toxins could be effective as a therapeutic treatment of established MO5 tumors. In these experiments, MO5 cells were first grafted to mice that later received ip injections of CyaA-OVA or CyaA-OVA-E5 (50 μg mixed with 1 mg of aluminum hydroxide in PBS) on days 1, 7, 14, 21, and 32 after the graft. The CyaA-OVA-E5 toxin provided a good therapeutic effect with 6/15 mice surviving

---

[24] E. L. Kaplan and P. Meier, J. Am. Statis. Assoc. **53**, 457 (1958).

A

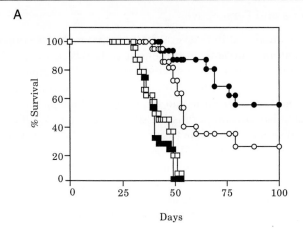

B

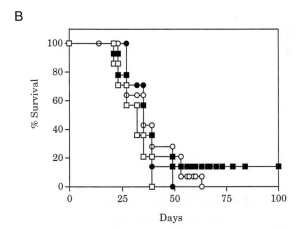

Fig. 5. Immunization with CyaA-OVA and CyaA-OVA-E5 induces protective immunity to the OVA-transfected melanoma MO5. C57BL/6 mice (eight mice per group) were immunized with PBS (□) or 50 µg of wild-type CyaA (■), CyaA-OVA (○), or CyaA-OVA-E5 (●) mixed with 1 mg of aluminum hydroxide in PBS. Seven days after the last immunization (day 0), mice were grafted with 2×10^4 MO5 (a) or B16 (b) tumor cells. Mice were then followed for tumor growth and survival. The percentage of survival was recorded as the percentage of mice surviving after the tumor graft. Representative data are shown. The experiment was reproduced three times with similar results.

at day 100 after tumor graft (all mice treated with PBS or wild-type CyaA died by day 60).[23] This suggests that recombinant CyaA toxins harboring tumor associated antigens might find clinical applications in an antitumor vaccination strategy.

Conclusions and Perspectives

The Cyaa toxin exhibits several features that makes it a potential vector to deliver epitopes into antigen-presenting cells. It is easy to produce in large quantities and a genetically detoxified variant still exhibits CTL priming capabilities. Work has shown that large polypeptide fragments (up to 200 residues) could be inserted within the catalytic domain and delivered into cell cytosol. Also, several permissive sites have been identified previously. Therefore, it should be feasible to construct recombinant toxins with multiple epitopes that could be used as multivalent vaccines. Further work will focus on the molecular mechanisms that underlie the CTL priming activity of this engineered toxin. In conclusion, the genetically detoxified recombinant Cyaa toxin might constitute an attractive nonreplicative vector to induce cellular immunity toward pathogens or tumor cells.

Acknowledgments

We thank N. Shastri for the B3Z cell line and L. Rosthein and L. Sigal for MO5 murine melanoma. Financial support came from the Institut Pasteur, the Centre National de la Recherche Scientifique (URA 2185), the Association pour la Recherche contre le Cancer (ARC), the Association Nationale pour la Recherche sur le SIDA (ANRS), and a Pasteur-Weizmann grant to C.L.

[33] Use of Anthrax Toxin Fusions to Stimulate Immune Responses

By Christopher C. Zarozinski, R. John Collier, and Michael N. Starnbach

Introduction

All viruses and many bacterial pathogens spend a large portion of their life cycle within the cytoplasm of mammalian cells. While they remain in this intracellular niche, they can replicate safely hidden from humoral immune responses. To respond to these organisms, the immune system has evolved potent effector cells termed cytotoxic T lymphocytes (CTL). CTL become activated upon recognition of pathogen-encoded peptides (often 8–10 amino acids in length) presented on the surface of the host cell in the context of class I major histocompatibility molecules (MHC-I). Activated

CTL respond by lysing the host cell and secreting cytokines, disrupting intracellular replication of the pathogen.[1] After clearance of the organism, a subset of these activated CTL differentiate into long-lived memory CTL that can respond more effectively on reexposure to the same organism.[2]

In order to protect individuals against these intracellular pathogens, a vaccine should ideally induce memory CTL. However, in order to prime memory CTL, it is necessary to devise a method of translocating the antigen of interest into the cytosol of living cells *in vivo*. This goal has been accomplished using live viral[3] and bacterial[4] vectors, as well as through the use of DNA vaccines.[5] However, the use of naturally occurring or recombinant infectious agents, even if attenuated, raises concerns about safety. Concerns about possible recombination between DNA vaccines and the host genome may also limit the widespread use of DNA vaccine technology.

This article describes the use of a vaccine system based on a bacterial protein toxin. Many bacterial toxins have evolved the ability to bind mammalian cells and translocate into the cytosol of these cells. Native toxins contain catalytic activities that disrupt host cell functions once in the cytosol of cells. We have modified one such toxin, removing the catalytic domain and replacing it with antigens from selected intracellular bacteria and viruses. The resulting fusion protein is able to deliver CTL epitopes to the cytoplasm of treated cells. This article describes the use of this modified toxin, the lethal toxin of *Bacillus anthracis,* to stimulate antigen-specific CTL *in vivo*.

Bacillus anthracis secretes three proteins that together are referred to as anthrax toxin. Individually protective antigen (PA), lethal factor (LF), and edema factor (EF) are nontoxic. However, injection of a mixture of PA and LF (known as lethal toxin) has been shown to be lethal to rats, while the combination of PA and EF (known as edema toxin) produces skin edema.[6] At the molecular level it is known that PA binds to a receptor on the surface of cells, where it is cleaved by a cell surface protease. After cleavage, PA molecules on the cell surface assemble into a heptamer able to bind either LF or EF.[7] The complex is then endocytosed, and during

[1] F. Gotch, A. Gallimore, and A. McMichael, *Immunol. Lett.* **51**, 125 (1996).
[2] R. Ahmed and D. Gray, *Science* **272**, 54 (1996).
[3] S. Pincus, J. Tartaglia, and E. Paoletti, *Biologicals* **23**, 159 (1995).
[4] M. K. Slifka, H. Shen, M. Matloubian, E. R. Jensen, J. F. Miller, and R. Ahmed, *J. Virol.* **70**, 2902 (1996).
[5] C. C. Zarozinski, E. F. Fynan, L. K. Selin, H. L. Robinson, and R. M. Welsh, *J. Immunol.* **154**, 4010 (1995).
[6] J. L. Stanley and H. Smith, *J. Gen. Microbiol.* **26**, 49 (1961).
[7] A. M. Friedlander, *J. Biol. Chem.* **261**, 7123 (1986).

endosome acidification the LF or EF is translocated into the cytoso.[8] In the cytoplasm of the cell LF has a currently unknown enzymatic activity that leads to cytokine dysregulation and macrophage death,[9] whereas EF has an adenylate cyclase activity that raises cAMP levels.[10] It has been shown that the amino-terminal 255 residues of LF (LFn) are capable of binding to PA and being translocated into the cytoplasm.[11] LFn lacks the C-terminal enzymatic domain and therefore is harmless when translocated into cells. It has been shown that heterologous proteins can be fused genetically to LFn and, when mixed with PA, can enter the cytoplasm of cells.[12] This article describes the methods used to fuse CTL epitopes to the LFn molecule and the use of these fusion proteins to induce antigen-specific CTL *in vivo*.

Expression Plasmid Construction

The first step in generating an anthrax toxin fusion protein is to construct a protein expression plasmid encoding the desired LFn-CTL epitope fusion. This section details the construction of a plasmid encoding LFn fused to an 8 amino acid H-2D^b-restricted CTL epitope from chicken ovalbumin (OVA$_{257-264}$).[13] However, the techniques described may be used to fuse any other CTL epitope to LFn or may be modified to fuse larger proteins to LFn (see later).

A DNA insert is first generated using polymerase chain reaction (PCR). The LF gene is encoded on the *B. anthracis* plasmid pX01, and this plasmid is used as the template. The upstream primer includes, 5' to 3', an *Nde*I restriction site and sequence identical to the 5' end of the LF gene (Fig. 1). The sequence of the downstream primer is complementary to the last six codons of the LFn gene, followed by the sequence encoding the CTL epitope of choice (in this case OVA$_{257-264}$), and ending with two stop codons and a *Bam*HI restriction site. It is prudent to change the codons that encode the amino acids in the CTL epitope to those most commonly used by *Escherichia coli* so that the fusion protein may be expressed optimally.

[8] J. C. Milne, D. Furlong, P. C. Hanna, J. S. Wall, and R. J. Collier. *J. Biol. Chem.* **269**, 20607 (1994).

[9] P. C. Hanna, B. Kruskai, R. A. Ezekowitz, B. Bloom, and R. J. Collier, *Mol. Med.* **1**, 7 (1994).

[10] S. H. Leppla, *Proc. Natl. Acad. Sci. U.S.A.* **79**, 3162 (1982).

[11] N. Arora and S. H. Leppla, *J. Biol. Chem.* **268**, 3334 (1993).

[12] J. C. Milne, S. R. Blanke, P. C Hanna, and R. J. Collier, *Mol. Microbiol.* **15**, 661 (1995).

[13] J. D. Ballard, A. M. Dolling, K. Beauregard, and R. J. Collier, *Infect. Immun.* **66**, 615 (1998).

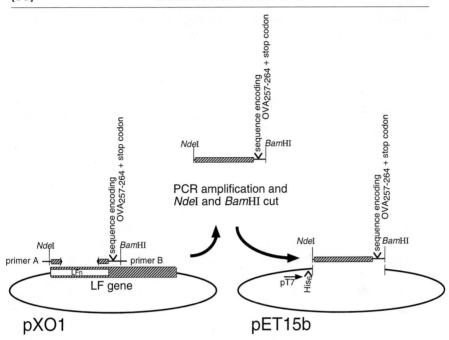

primer A: 5'-GGAAGGAACATATGGCGGGCGGTCATGGTGATGTA-3'

primer B: 5'-CCGGCGGATCCTTATCACAGTTTTTCGAAGTTGATGATGGAGGATAGATTTATTTCTTGTTC-3'

FIG. 1. Strategy for construction of LFn–epitope fusions.

The primers also have additional nucleotides on the 5' end to facilitate recognition by the *Nde*I and *Bam*HI restriction enzymes. The 50-μl PCR reaction volume consists of 47 μl PCR supermix (GIBCO-BRL), 0.02 μg of pX01 template, and 2 μg of each of the primers described earlier. The samples are then amplified in a thermal cycler using reaction conditions of template denaturation at 94° for 1 min, primer template annealing at 55° for 1 min, and DNA synthesis at 72° for 3 min. This is repeated for 29 cycles followed by an additional 10 min at 72° to fill in any gaps in the sequence. The product of this reaction should contain, 5' to 3', an *Nde*I site, the codons encoding the 255 amino acids of LFn, the sequence of the CTL epitope of interest, two stop codons, and a *Bam*HI site. After amplification, 5 μl of the reaction is visualized by electrophoresis through a 1% agarose gel and staining with ethidium bromide. If the appropriate 0.7-kb band is present, the remaining reaction is then purified using a QIAquick PCR purification kit (Qiagen).

This PCR product is then prepared for ligation into the *Nde*I and *Bam*HI sites of the pET15b expression vector (Novagen). The pET15b vector is used because fusion proteins expressed in this vector contain six histidine residues (His-tag) at their amino terminus, allowing for one-step protein purification over an Ni^{2+}-charged column as described later. In addition, the control of high-level recombinant protein expression using pET15b is inducible in *E. coli* by the addition of isopropyl-β-D-thiogalactopyrano-side (IPTG).

Both the PCR product and the pET15b vector must be cut with *Nde*I and *Bam*HI to prepare them for ligation. This is accomplished by separately digesting 2 μg of the pET15b vector, and 48 μl of the PCR product with *Nde*I, for approximately 2 hr in a volume of 100 μl. After 2 hr, 20 μl of 10× *Bam*HI reaction buffer, 78 μl of doubly distilled H$_2$O, and 2 μl *Bam*HI restriction enzyme are added to each reaction tube and incubated overnight at 37°. The overnight incubation is recommended in order to ensure that as much of the vector and PCR products are digested as possible. The efficiency of the pET15b vector digestion can be monitored by running uncut vector, vector cut with *Nde*I alone, and vector cut with both *Nde*I and *Bam*HI on a 1% agarose electrophoresis gel. After restriction enzyme digestion, the samples are subjected to phenol/chloroform extraction to remove the restriction enzymes. The DNA is then ethanol precipitated and resuspended in 10 μl of doubly distilled H$_2$O.

In order to ligate the digested PCR product into the pET15b vector, the following reaction is performed. One microliter of a 1:10 dilution of the digested pET15b vector is added to 15 μl of doubly distilled H$_2$O, followed by 1 μl of T4 ligase (Promega) and 2 μl of 10× ligase buffer. Finally either 1 μl of PCR product is added to the reaction mix or, for a mock ligation reaction, 1 μl of doubly distilled H$_2$O is added. This reaction is then incubated overnight a 16°, and the DNA is ethanol precipitated and used to transform *E. coli* XL-1 Blue (Stratagene).

Next it is necessary to screen colonies of transformants to determine which ones have been transformed with the pET15b plasmid containing the LFn–epitope fusion insert. Several colonies are selected and individually grown overnight in 2-ml cultures at 37° in Luria broth (LB). Plasmids are isolated from these cultures and digested with *Nde*I and *Bam*HI. If the plasmid contains the proper PCR product insert, two fragments should be visible on a 1% agarose electrophoresis gel: a 5.7-kb vector fragment and a smaller 0.7-kb insert. Once a plasmid has been found that contains the LFn insert, the insert should be sequenced in order to confirm that the fusion between LFn and the CTL epitope is correct and in frame.

Construction of Other Modified Forms of LFn

As mentioned earlier, any CTL epitope should be able to be fused to LFn using the protocol described earlier. The only practical limitation to this method is the addition of a CTL epitope that might render the anthrax toxin fusion protein insoluble or prevent its PA-dependent translocation. In some circumstances, such as when the precise CTL epitope is not known, one may want to fuse a whole protein to the LFn molecule. To construct these fusion proteins, PCR primers are designed so that the gene one wishes to fuse to LFn is amplified with the addition of a 5' *Bam*HI site, a 3' stop codon, and a terminal 3' *Bam*HI site. By flanking the gene of interest with *Bam*HI sites, the PCR products can be digested with *Bam*HI and then cloned into the *Bam*HI site in the vector pLFn1. pLFn1 is a plasmid in which the gene encoding LFn has been cloned into the *Nde*I (5') and *Bam*HI (3') sites of pET15b. As a result, any sequence cloned into the *Bam*HI site of pLFn1 will be fused to the LFn gene. To keep the fused protein in frame with LFn, the 5' primer must be designed so that the six nucleotides of the *Bam*HI recognition site represent two codons upstream of the fused protein. After ligation and transformation into *E. coli* XL-1 Blue, diagnostic digests must be performed in order to determine if the PCR product has been inserted in the correct 5' to 3' orientation. Plasmids containing the insert in the proper orientation are then sequenced for conformation and used to transform *E. coli* BL21[14] for expression of the fusion protein (see later). Using this approach, any protein should be able to be fused to the modified anthrax toxin. Some larger fusion proteins are insoluble following expression in *E. coli,* and others are not translocatable by LFn.[15] It has been difficult to predict which fusions will fail for these reasons. While it is not clear at this time exactly how large a protein can be translocated by LFn, full-length LF is over 60 kDa larger than LFn, suggesting that proteins as large as 60 kDa are translocatable.

Using the anthrax toxin protein fusion system it is also possible to disulfide link commercially synthesized peptide epitopes to LFn. Variations of this disulfide linkage technique might allow LFn to be linked to molecules such as RNA, lipids, or modified amino acids. These fusions could not be constructed using the genetic methods described earlier. CTL epitopes are fused to LFn using a modified form of LFn protein in which the serine at position 255 has been changed to a cysteine (LFn-cys). LFn-cys, just like LFn, functions to translocate in a PA-dependent manner.[16] To prepare

[14] F. W. Studier and B. A. Moffatt. *J. Mol. Biol.* **189,** 113 (1986).
[15] J. Wesche, J. L. Elliot, O. O. Falnes, S. Olsnes, and R. J. Collier, *Biochemisty* **45,** 15737 (1998).
[16] J. D. Ballard, R. J. Collier, and M. N. Starnbach, *Infect. Immun.* **66,** 4696 (1998).

Lfn-cys for linkage, it is exchanged into 20 mM Tris–HCl, pH 7.5, using a PD-10 gel filtration column. This removes reducing agents used in the storage of LFn-cys. A commercially synthesized peptide containing a CTL epitope to be linked to LFn-cys is designed with a free terminal cysteine. The peptide is combined with the LFn-cys in molar ratios of 0, 1, 10, 50, and 100 to 1 (linkage partner to LFn-cys), and the mixture is incubated overnight at 4°. The samples are then analyzed by SDS–PAGE gel electrophoresis to determine the ratio producing the highest yield of heterodimer product. This reaction is then passed over a PD-10 column to remove unincorporated linkage partners and then aliquoted and stored at −20°. Use of the linked product is the same as for fusion proteins.

Fusion Protein Induction and Purification

After a plasmid encoding the desired fusion sequence is found, it is necessary to test for expression of the protein in *E. coli* harboring the plasmid. For expression, the plasmid is purified from the *E. coli* XL-1 Blue strain and transformed into the *E. coli* BL21 strain. BL21 expresses T7 RNA polymerase under the control of an IPTG-inducible promoter. After transfection of BL21 with the plasmid, an individual colony can be selected for a small-scale induction. *Escherichia coli* BL21 containing a pET15b–LFn fusion are grown in 10 ml of LB, containing 50 μg/ml ampicillin, until the optical density of the culture at 600 nm is between 0.6 and 0.8. At this point, 500 μl of the culture is removed and placed on ice. The expression of the anthrax toxin fusion protein is induced in the remainder of the culture by the addition of IPTG to a final concentration of 1 mM. After 3 hr, 500 μl of the culture is removed and stored on ice. Both pre- and postinduction samples are centrifuged to pellet the bacteria, and the cells are then resuspended in SDS–PAGE sample buffer. The samples are separated on a 12% SDS–PAGE gel and stained with Coomassie blue. The gel is examined to determine if a protein is seen at 32 kDa (when fusing single 8–10 amino acid epitopes) that is more abundant in the induced versus the noninduced sample. If an induced protein is seen, then a large-scale induction is prepared. In a large-scale induction, 10 ml of an overnight culture of the BL21 colony containing the LFn-encoding plasmid is added to 500 ml of LB and induced with IPTG as detailed previously. The entire culture is then pelleted and resuspended in 5 ml of 1× binding buffer containing 5 mM imidazole, 0.5 M NaCl, and 20 mM Tris–HCl, pH 7.9. Bacteria are then disrupted by sonication. At this point, the material can be stored at −20° for later purification. The sonicate is then centrifuged, the supernatant is saved, and the pellet is resuspended in 5 ml of 1× binding buffer, sonicated, and centrifuged again. The supernatants are then pooled and passed over an equilibrated Ni^{2+}-charged column (Novagen) that has been prepared as specified by the manufacturer. The recombinant LFn is then eluted from the column by the addition of 2 ml of elution buffer containing 0.5 M imidazole in 20 mM Tris–HCl, pH 7.5. Excess imidazole is removed by passing the purified protein over a PD-10 gel filtration column. The protein is then eluted in 20 mM Tris–HCl, pH 8.0, and the concentration is determined using a Coomassie blue protein assay (Pierce).

Immunization of Mice with Anthrax Toxin Fusion Protein

To determine if the anthrax toxin fusion protein primes an antigen-specific CTL response *in vivo*, the fusion protein is used to immunize mice. Groups of mice that possess the MHC class I molecule appropriate for the presentation of the CTL epitope are immunized intraperitoneally (ip) with a mixture of the LFn fusion protein and PA. As described earlier, anthrax toxin is a bipartite toxin in which PA is required for LFn fusion toxin delivery. Each experimental mouse is injected with 30 pmol of the LFn fusion mixed with 6 pmol of PA. One control group of mice is immunized with PBS and another control group is immunized with the LFn fusion protein in the absence of PA. Fourteen days after immunization, CTL assays can be performed to determine if the LFn fusion primes antigen-specific CTL *in vivo*.

Assays for CTL Activity

To prepare stimulator cells, spleen cells from naive mice are disassociated into a single cell suspension and washed with RPMI 1640 (GIBCO-BRL) supplemented with 10% fetal calf serum, penicillin (50 units/ml), streptomycin (50 μg/ml), gentamicin (50 μg/ml), HEPES (5 mM final concentration), and 2-mercaptoethanol (50 nM final concentration) (RP-10). Enough mice are used so that approximately 4×10^7 spleen cells will be available for each of the experimental and control mice. These cells are then resuspended in 2 ml of RP-10 and placed in a 50-ml conical tube where they are treated with sterile antigenic peptide (10 μM final concentration) for 1 hr at 37° in a humidified 7% CO_2 incubator. The cap of the conical tube should remain loose during the incubation. These cells will serve as antigen-presenting cells (APC) for splenocytes from vaccinated and control animals. After peptide treatment the APCs are irradiated with 2000 rads and washed twice with 10 ml of RP-10. These cells are then aliquoted into T-25 tissue culture flasks so that each flask contains approximately 4×10^7 stimulator cells. Spleens from the vaccinated and control mice are then individually harvested and disassociated into single cell suspensions. These cells are washed twice with RP-10, and 4×10^7 cells from each mouse are added to one of the flasks containing the stimulators. The flasks are then filled to a total volume of 20 ml with RP-10 and incubated upright for 5 days in a 37° humidified 7% CO_2 incubator. After the incubation, half the culture is removed, washed once with RP-10, and used as effector cells in a ^{51}Cr release CTL assay.

To prepare the ^{51}Cr release CTL assay, 1×10^6 MHC matched target cells are incubated for 1 hr in 100 μl of RP-10 medium, in either the presence or the absence of 10 μM antigenic peptide. One hundred microcuries of sodium [^{51}Cr]chromate is also added to the tubes prior to incubation. Common target cells used are P815 cells for H-2^d and EL-4 cells for H-2^b. After incubation, targets are washed three times with RP-10 to remove unbound peptide and excess extracellular radiolabel, and then resuspended in 10 ml of RP-10. Dilutions of cells from T-cell cultures (effectors) are prepared in 100 μl RP-10 and mixed with 10^4 of the labeled target cells (in 100 μl) in U-bottom 96-well plates. Spontaneous and maximum lysis of target cells is determined by incubation of 10^4 labeled target cells with 100 μl of RP-10 and 1% Triton X-100, respectively. After 4 hr the 96-well plates are centrifuged at 1000g, and 100 μl of the supernatant is harvested and assayed for ^{51}Cr release using a gamma counter. Percentage specific lysis can be determined using the formula 100 × (experimental release − spontaneous release) ÷ (maximum release − spontaneous release).

Concluding Remarks

One advantage of the anthrax toxin fusion protein vaccine system described here is that it induces antigen-specific CTL *in vivo* in the absence of adjuvant.[17] Other bacterial toxin delivery systems, such as those that use shiga-toxin,[18] *Pseudomonas exotoxin*,[19] or *Bordetella pertussis* adenylate cyclase,[20] either require adjuvants or have not been demonstrated to function *in vivo*, thus limiting their potential use in humans. This anthrax toxin fusion protein system may also have potential *in vitro* applications. One of the goals of adoptive immunotherapy is to expand patient CTL *in vitro* so

[17] J. D. Ballard, R. J. Collier, and M. N. Starnbach, *Proc. Natl. Acad. Sci. U.S.A.* **93**, 12531 (1996).

[18] R. Lee, E. Tartour, P. van der Bruggen, V. Vantomme, I. Joyeux, B. Goud, W. H. Fridman, and L. Johannes, *Eur. J. Immunol.* **28**, 2726 (1998).

[19] J. J. Donnelly, J. B. Ulmer, L. A. Hawe, A. Friedman, X. Shi, K. R. Leander, J. W. Shiver, A. I. Oliff, D. Martinez, D. Mongomery, and M. A. Liu, *Proc. Natl. Acad. Sci. U.S.A.* **90**, 3530 (1993).

[20] M. F. Saron, C. Fayolle, P. Sebo, D. Ladant, A. Ullmann, and A. LeClerc, *Proc. Natl. Acad. Sci. U.S.A.* **94**, 3314 (1997).

that large numbers of antigen-specific lymphocytes can be returned to the individual, mediating a protective effect. The use of attenuated viruses for lymphocyte stimulation *in vitro* raises concerns when used in immunocompromised patients. Noninfectious anthrax toxin fusion proteins may be an ideal substitute to prepare stimulator cells.

The anthrax toxin fusion protein system also has potential in applications unrelated to vaccine delivery. For example, it may be possible to fuse LFn to molecules or enzymes that function intracellularly, such as those involved in signal transduction. This may allow for the precise intracellular manipulation of these complex biochemical pathways.

[34] Use of Fusions to Viral Coat Proteins as Antigenic Carriers for Vaccine Development

By PETER G. STOCKLEY and ROBERT A. MASTICO

Introduction

Vaccination remains the most successful form of medical intervention against disease to date, hopefully, having led to the purposeful eradication in the human population of smallpox and, shortly, poliomyelitis. Despite major successes against endemic diseases such as poliomyelitis and diphtheria since the 1950s, there are still major targets for which effective vaccination strategies are not yet at hand, such as rotaviruses, HIV, and malaria. Fortunately, our understanding of the structural molecular biology underlying the immunological responses of organisms to foreign pathogens has been transformed since the determination of the X-ray crystal structure for the first spherical viruses in the 1980s. We now have high-resolution structures for antigen–antibody complexes,[1] virus–antibody complexes,[2] antibody–peptide complexes,[3] HLA–peptide complexes,[4] and even the ternary complexes with the T-cell receptor.[5,6] There is therefore great expectation that identification of the correct epitope and its presentation to the

[1] A. G. Amit, R. A. Mariuzza, S. E. Phillips, and R. J. Poljak, *Science* **233,** 747 (1986).
[2] E. A. Hewat, T. C. Marlovits, and D. Blaas, *J. Virol.* **72,** 4396 (1998).
[3] R. L. Stanfield, T. M. Fieser, R. A. Lerner, and I. A. Wilson, *Science* **248,** 712 (1990).
[4] P. J. Bjorkman and P. Parham, *Annu. Rev. Biochem.* **59,** 253 (1990).
[5] K. C. Garcia, M. Degano, R. L. Stanfield, A. Brunmark, M. R. Jackson, P. A. Peterson, L. Teyton, and I. A. Wilson, *Science* **274,** 209 (1996).
[6] D. N. Garboczi, P. Ghosh, U. Utz, Q. R. Fan, W. E. Biddison, and D. C. Wiley, *Nature* **384,** 134 (1996).

immune system in an appropriate way will dramatically extend the utility of vaccination.

Epitope Identification

Comparison of the X-ray crystal structures of simple spherical plant viruses and their animal picornavirus counterparts revealed an underlying conservation of the tertiary fold of the protein subunits enclosing the viral RNAs: the Swiss role topology first seen in the structure of tomato bushy stunt virus. However, the animal viruses also contained large loops of polypeptide inserted into this conserved framework and these loops appeared to contain the immunodominant epitopes of the virus.[7] The implication seemed clear that viruses present immunodominant epitopes in the form of constrained surface loops, which do not themselves play important structural roles, hence they can alter readily to escape immune surveillance without loss of function. A similar story emerged from the analysis of the structure of the major protein antigen, hemagglutinin, from influenza virus.

These looped regions of polypeptide act as B-cell epitopes readily accessible on the outer surface of globular proteins to IgG molecules and B-cell receptors, which leads to the production of antibody molecules of increasing refinement and affinity. Such B-cell epitopes can be conformation dependent, i.e., they require the native structure of the antigen to be effective. This allows some epitopes to consist of discontinuous regions of the polypeptide chain juxtaposed by tertiary folding. Alternatively, they can also be simply sequence dependent and consist essentially of continuous stretches of polypeptide. Successful vaccination requires stimulation of both humoral, B-cell-mediated and cellular, T-cell-mediated immune responses. The latter involves processing of polypeptide antigens into short peptides, which are then presented to T-cell receptors as a complex with the molecules of the major histocompatibility locus. Such T-cell epitopes are between 8 and 14 residues in length, depending on which type of T-cell response is being stimulated. Automated peptide-scanning techniques now exist to identify B- and T-cell epitopes for antigens of known primary sequence.[8]

Epitope Presentation

Short peptides are short lived in the bloodstream and are in general poorly antigenic. Similarly, recombinant subunits are usually only weakly

---

[7] M. G. Rossmann and J. E. Johnson, *Annu. Rev. Biochem.* **58**, 533 (1989).
[8] R. H. Meloen *et al.*, "Pepscan to Determine T and B Cell Epitopes." CRC Press, Boca Raton, FL, 1995.

antigenic unless they have the ability to assemble into virus-like particles (VLPs), such as those used to immunize against hepatitis B and in development for human cervical carcinomas caused by human papilloma virus. Alternatively, epitopes can be synthesized chemically and coupled to larger carrier molecules, such as keyhole limpet hemacyanin, expressed at the surface of prokaryotic or eukaryotic cells, and subcellular organelles, or inserted into chimeric live viruses. The live virus class divides into those viruses capable of replication in the vaccinated host, such as vaccinia and poliomyelitis, and those that cannot, such as filamentous bacteriophage and the plant viruses, such as TMV and cowpea mosaic virus. The use of RNA bacteriophage,[9,10] particularly MS2, to present epitopes at the surface of a largely nucleic acid-free viruslike particle, is described in this article. These phage are in widespread use as research tools, and many laboratories have access to the reagents needed to construct and express such phage coat protein chimeras.

Preparation of Chimeric Expression Vector

The production of chimeric VLPs depends on finding sites of insertion for epitopes of interest that do not disrupt the processing and assembly of the particle and present the new sequence in a defined structural context. For some systems, N- or C-terminal extensions are the preferred location; however, such locations suggest that the epitope might not be constrained at one end. In RNA bacteriophages, the preferred site of insertion is at the top of the N-terminal β-hairpin (between residues 14 and 15) (Fig. 1). This is the most radially distal feature in the assembled $T = 3$ viral shell.[11] Appropriate modification of the recombinant coat protein gene allows insertion to occur between residues G14 and T15, hopefully allowing presentation in the context of a surface-accessible constrained loop on every subunit of the VLP. Note that insertion one or two amino acids away from this site leads to chimeric proteins that appear to have lost the ability to self-assemble,[12] although, surprisingly, covalent fusion of the two subunits in a coat protein dimer can restore the tolerance of inserts.[13]

[9] R. A. Mastico, S. J. Talbot, and P. G. Stockley, J. Gen. Virol. **74**, 541 (1993).
[10] K. G. Heal et al., Vaccine **18**, 251 (1999).
[11] R. Golmohammadi, K. Valegard, K. Fridborg, and L. Liljas, J. Mol. Biol. **234**, 620 1993).
[12] T. Kozlovskaya et al., Mol. Bio. **22**, 584 (1988).
[13] D. S. Peabody, Arch. Biochem. Biophys. **347**, 85 (1997).

Sites of epitope insertion

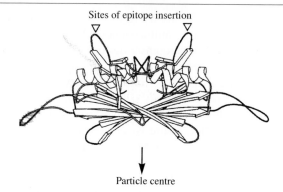

Particle centre

FIG. 1. Structure of the chimeric coat protein subunit. The MS2 coat protein dimer is shown as a ribbon with an arrow indicating the center of the assembled particle. Inverted arrowheads indicate the site of insertion of foreign epitopes in the N-terminal β hairpin.

Expression Studies

Cloning of MS2 Coat Protein Gene

The MS2 coat protein gene is cloned into an *Escherichia coli* plasmid prior to site-directed mutagenesis to create chimeric molecules. Our initial clone is obtained by production of cDNA from a sample of wild-type MS2 RNA provided by P. Kaesberg and P. N. Shaklee (Wisconsin), but the phage can also be obtained commercially from the American Type Culture Collection. The first strand of the cDNA is primed using a 21 nucleotide synthetic oligonucleotide corresponding to the region immediately 3' to the coat protein stop codon. This oligonucleotide also contains a noncomplementary nucleotide, thus introducing a unique *Xho*I restriction site into the resultant cDNA. After annealing at high temperature, the primer–RNA template is extended by the addition of AMV reverse transcriptase and NTPs and incubation at 42° for 1 hr before treatment with alkali to destroy the template strand. The second strand of cDNA is primed from an oligonucleotide complementary to the region immediately 5' of the coat protein gene and encompassing the *Xba*I restriction site at base 1304. Second strand extension is carried out using the Klenow fragment of DNA polymerase I. cDNA synthesis is monitored at each step by radiolabeling of the appropriate oligonucleotides and gel electrophoresis of the reaction products.[14]

The double-stranded cDNA is then extracted with phenol and ethanol

---

[14] J. Sambrook, E. F. Fritsch, and T. Maniatis, "Molecular Cloning: A Laboratory Manual," 2nd Ed. Cold Spring Harbor Laboratory Press, Cold Spring Harbor, NY, 1989.

precipitated before being restricted with both *Xho*I and *Xba*I, and the resultant 434-bp fragment is purified by gel electrophoresis and then ligated into a suitable *tac* promoter region of an appropriately restricted and dephosphorylated plasmid backbone.[15] We initially used a derivative of pGW11.[9,16] The ligation mixture is then used to transform *E. coli* K12 TG1 Δ(*lac-proAB*), *supE, thi, hsdD5* (F' *traD36, proA+B+, lacIq, lacZM15*) to ampicillin resistance. Positive clones are identified by screening miniprep plasmid DNA with the appropriate restriction enzymes.

Construction of MS2 Chimera Vectors

For mutagenesis of the wild-type coat protein gene, the 434-bp *Xba*I–*Xho*I fragment is subcloned into the polylinker of M13mp18. Oligonucleotide-directed mutagenesis is used to introduce two nucleotide changes in the wild-type sequence (A1380T and T1383C), resulting in a unique *Kpn*I restriction site at position 1378 relative to the MS2 genome sequence (Fig. 2), and is carried out using the method described by Sayers *et al.*[17] Engineering of the new site is such that it does not alter the coding sequence of the MS2 coat protein gene. (More recently we adopted the polymerase chain reaction mutagenesis approach for site-directed changes such as this.[18]) The mutant coat protein gene is then reintroduced into the *tac* expression vector or a variant carrying a copy of the bacteriophage A protein gene between the promoter and the coat protein gene. The latter construct confers the property of high level constitutive expression in the downstream coat protein gene.[19,20]

Cloning Chimeric Coat Protein Genes

Short epitopes are introduced into *Kpn*I-restricted expression vectors by ligation of annealed synthetic oligonucleotides encompassing the sequence of interest and carrying single-stranded extensions to anneal at the *Kpn*I site. Larger inserts (>20 amino acids) are introduced by annealing two synthetic oligonucleotides encompassing the epitope of choice, converting them to a double-stranded molecule by treatment with Klenow polymerase, followed by self-ligation and subsequent *Kpn*I restriction to yield unit length molecules with the appropriate flanking sequences before ligation to the

[15] E. Amann *et al., Gene* **25,** 1677 (1983).
[16] M. C. M. Smith *et al., Mol. Micro.* **3,** 23 (1989).
[17] J. R. Sayers *et al., Nucleic Acid Res.* **16,** 781 (1988).
[18] R. M. Horton and L. R. Pease, in "Directed Mutagenesis" (M. J. McPherson, ed.). Oxford Univ. Press, New York, 1991.
[19] R. A. Kastelein *et al., Gene* **23,** 245 (1983).
[20] M. H. de Smit and J. van Duin, *Prog. Nucleic Acids Mol. Biol.* **38,** 1 (1990).

Fig. 2. Construction of chimeric MS2 coat protein genes. The coding sequence in the region of the β hairpin is shown together with the amino acid sequence encoded.

	5′	8	9	10	11	12	13	14	15	16	17	18	19	20							3′
Wild-type codon No.		8	9	10	11	12	13	14	15	16	17	18	19	20							
Amino acid		V	L	V	D	N	G	G	T	G	D	V	T	V							
Codon		GTT	CTC	GTC	GAC	AAT	GGC	GGA	ACT	GGC	GAC	GTG	ACT	GTC							
Oligos inserted								*Kpnl*													
		GTT	CTC	GTC	GAC	AAT	GGC	GGT	ACC	GGC	GAC	GTG	ACT	GTC							
HA9 Epitope	5′ .C		TAC	CCA	TAC	GAC	GTT	CCA	GAC	TAC	GCT	GGT	AC. 3′								
			Y	P	Y	D	V	P	D	Y	A	G									
IgE Epitope	5′ .C		TTC	GTT	TTC	TTC	GGT	TCT	GGT	AAA	ACT	AAA	GGT	AC. 3′							
			F	V	F	F	G	S	G	K	T	K									
Mal Epitope	5′ .C		AAT	GCA	AAC	AAC	CCG	AAT	GCA	AAC	CCG	AAT	GCA	AAC	CCG	AC. 3′					
			N	A	N	P	N	A	N	P	N	A	N	P							
HPV L1 Epitope	5′ .C		CCT	ATG	GAT	ACC	TTC	ATC	GTT	AGC	GCT	AAC	GCA	CCT	AGC	AGC	ACC	CCT	ATC	TTT	AC. 3′
			P	M	D	T	F	I	V	S	A	N	A	P	S	S	T	P	I	K	
HPV L2 Epitope	5′ .C		AAG	GGT	AGC	CCT	TGT	ACC	AAC	GGT	GCT	GTT	AAC	CCT	TGT	CTC	GAC	CTC	GGT	GAC	AC. 3′
			K	G	S	P	C	T	N	V	A	V	N	P	C	L	D	L	G	D	
HIV gp120	5′ GGT ⋯	ACC	AAC	AAC	ACT	CAG	ATC	CGT	ATC	CGT	ATC	GTT	CGT	ATC	CAG	CGT	GGT	CCA	GCA	CGA	⋯ CCA 3′
Epitope	3′ ⋯		N	N	T	Q	I	R	I	R	I	V	R	I	Q	R	G	P	A	R	⋯ TGG 5′

vector. Transformants are checked for inserts by restriction analysis and then the plasmid DNA is sequenced. This also confirms the orientation of clones of larger inserts. This strategy results in the duplication of two MS2 codons (for G14 and T15), i.e., inserted epitopes are extended by two residues so that the sequence becomes N-. . .GT-epitope-GT. . .-C.

Purification Strategy

Single colonies of *E. coli* TG1 cells transformed to ampicillin resistance using the recombinant chimeric expression plasmid DNAs are transferred into 5 ml of 2 TY (tryptone and yeast extract broth)[14] containing 100 μg/ml of ampicillin, and cultures are grown overnight at 37°. Aliquots (3 ml) of these cultures are inoculated into 500 ml of the same media and these are grown to A_{600} of approximately 0.4–0.6 OD units before protein expression is induced by the addition of solid isopropyl-β-D-thio-galactopyranoside (IPTG) to a final concentration of 1 mM and incubation is continued overnight. The final cell density, estimated by A_{600}, is approximately 2.5 $\times$ 10^9 cells/ml.

Cells are harvested by centrifugation, and pellets are resuspended in 2–3$\times$ pellet volume using 50 mM Tris–HCl, 0.5% (w/v) Sarkosyl, pH 6.5, before the cells are disrupted by sonication, maintaining the temperature at 4° using an ice bath. Total sonicates are loaded onto SDS–polyacrylamide gels (18%, w/v).[21] Electrophoresis is performed at 95 V/120-14 mA for 8–10 hr, and the gels are stained with a 0.5% (w/v) Coomassie Brilliant Blue R250 (CBB) solution in water : methanol : acetic acid (3 : 2 : 0.2; v/v) for 8 hr and then destained in the same solution, omitting the dye. Cells expressing the recombinant coat protein are then stored as a stock suspension at −20° in 30% (v/v) glycerol until further use. *Note:* wild-type MS2 coat protein is extremely soluble in acetic acid, and amounts of the protein on SDS–PAGE can be underestimated significantly unless fresh methanolic staining solutions are used.

Purification Scheme

Purification is based on isolation of assembled chimeric proteins. Transmission electron microscopy of negatively stained samples suggest that chimeric proteins are still able to self-assemble into $T = 3$ shells similar to the wild-type phage.[11,22] This is only practical on a large scale, when the majority of the chimeric protein remain soluble, which it does not for

[21] H. Schägger and G. von Jagow, *Anal. Biochem.* **166,** 368 (1987).
[22] K. Valegård *et al., Nature* **345,** 36 (1990).

epitopes longer than 12 residues (see later). Cells expressing the recombinant coat protein are grown at 28° in order to reduce the formation of inclusion bodies and are induced as described earlier. Generally, cells are grown in 4- to 8-liter batches (8–16 × 500-ml cultures) and harvested by centrifugation (2455g, 30 min). Pellets are resuspended with 2–3 volumes of 50 mM HEPES, 100 mM NaCl, 10 mM dithiothreitol (DTT), 5 mM EDTA, 1 mM phenylmethylsulfonyl fluoride (PMSF) (from a stock solution of 100 mM PMSF prepared in ethanol), pH 7.4 (buffer A), and stored on ice.

The following purification procedure for the recombinant coat protein is carried out at 0–4°, and SDS–PAGE is used to follow the purification procedure, unless stated otherwise.

Step 1: Extraction. Resuspended cells from a 4-liter culture are sonicated 15 × 30 sec with a Model MSE 9-66 sonifier set at maximum power using a macro tip. The sonicate is centrifuged twice (26,890g) for 15 min and decanted, and the soluble supernatant (TS) is retained.

Step 2: Removal of Nucleic Acids. Magnesium acetate is added to the TS to a final concentration of 6 mM following the addition of DNase 1 to 10 μg/ml. Using a magnetic stirrer, the mixture is stirred for 30 min at 20°, followed by centrifugation at 1088g for 30 min. The supernatant (TDS) is retained.

Step 3: Ammonium Sulfate Precipitation. The TDS is equilibrated for 10 min in an ice bath and then, while stirring as described earlier, ammonium sulfate is added to a final concentration of 1.21 M. After 1–2 hr the sample is centrifuged at 1088g for 30 min and the supernatant is decanted. Pellets (AS) are resuspended in 1–2× volume of buffer A.

Step 4: Isoelectric Point Precipitation. The AS fraction, which contains the recombinant coat protein, is dialyzed overnight against two changes of 4 liters of mM sodium acetate, pH 4.7, and then centrifuged at 1088g for 30 min. Experiments using isoelectric focusing of the recombinant coat proteins suggest that their isoelectric points are similar to those of wild-type MS2 (~pH 3.5).[23] Dialysis from pH 7.4–4.7 precipitates most of the contaminating membrane components,[24] leaving the recombinant coat protein as the soluble fraction; the supernatant postdialysis (PD) is saved.

Step 5: Sucrose Density Gradient. A 60-ml linear sucrose density gradient [15–40% (w/v) sucrose with a 60% (w/v) sucrose cushion] is prepared using 10 mM HEPES, 100 mM NaCl, 0.01% (v/v) diethylpyrocarbonate (DEPC), pH 7.4. Aliquots (5–7 ml) of the PD are overlaid onto the gradient, and

---

[23] L. R. Overby, G. H. Barlow, R. H. Doi, M. Jacob, and S. Spiegelman, *J. Bacteriol.* **91,** 442 (1966).

[24] T. E. Creighton, *in* "Protein Structure: A Practical Approach" (T. E. Creighton, ed.). Paston Press, Norfolk, England.

the samples are centrifuged for 4–5 hr at 28,000 rpm using a Beckman 45 Ti rotor. Gradients are fractionated by collecting 1-ml fractions from the bottom of the tube at a flow rate of 1 ml/min and monitored using absorbance at 280 nm. Peak fractions containing the recombinant assembled coat protein (sedimentation coefficient ~43S) (PS) are retained.

Step 6: Size-Exclusion Chromatography. The PS fraction is loaded (2 ml of approximately 0.5 mg/ml recombinant coat protein) onto a Bio-Gel A1.5M column (5.0 × 160 cm), which has been equilibrated previously with 10 mM HEPES, 100 mM NaCl, pH 7.4. Fractions (1 ml) are collected at a 0.30- to 0.35-ml/min flow rate, and peak 280-nm fractions are pooled (PB).

Step 7: Final Concentration. The pooled PB is concentrated by precipitation with 1.85 M ammonium sulfate overnight. The sample is centrifuged at 35,000 rpm for 2 hr using a Beckman 45 Ti rotor, the supernatant is decanted, and the pellet is resuspended overnight in 2 ml of 10 mM HEPES, 100 mM NaCl, 0.02% (w/v) NaN₃, pH 7.4.

Immunoaffinity Purification

Alternatively, chimeric VLPs can be purified by immunoaffinity chromatography as follows.

Step 1: Purification of Wild-Type MS2 Coat Protein from Phage. MS2 bacteriophage is grown and purified by the standard method.[25] The production of disassembled coat protein units follows the classical procedure of Sugiyama et al.[26] in which dissociation of the wild-type MS2 bacteriophage particle into soluble coat protein subunits and an insoluble RNA–A protein complex is achieved by treatment with 2 volumes of ice-cold 18 M acetic acid. After 1 hr at 0–4°, the solution is centrifuged at 1088g for 30 min, removing insoluble material, and then the sample is dialyzed extensively against 1 mM acetic acid at 4°. The reassembly of coat protein subunits into VLPs is achieved by dialysis against 20 mM HEPES, 0.1 M DTT, pH 7.4. Unassembled components and VLPs are separated using size-exclusion chromatography as described earlier for soluble chimeric particles.

Step 2: Production of Rabbit anti-MS2 Antibodies. The production of rabbit anti-MS2 coat protein antibodies follows the methods described by Harlow and Lane[27] but with the following alterations. Two shaved backed New Zealand white female rabbits (2 kg) are injected at three sites intradermally with 2 mg of wild-type MS2 bacteriophage coat protein, purified from phage as described earlier, in a mixture of Freund's complete adjuvant

[25] H. J. Strauss and L. R. Sinsheimer, *J. Mol. Biol.* **7**, 43 (1963).
[26] T. Sugiyama *et al., J. Mol. Biol.* **25**, 455 (1967).
[27] E. Harlow and D. Lane, "Antibodies: A Laboratory Manual." Cold Spring Harbor Press, Cold Spring Harbor, NY, 1988.

(1:1, v/v) and emulsified by sonication for primary immunization. Next, a secondary booster is administered subcutaneously at three sites after 3 weeks using 2 mg of the same purified stock in a mixture of Freund's incomplete adjuvant (1:1, v/v), emulsified as described previously. After 5 weeks, a 1-ml bleed is taken for titer check using enzyme-linked immunosorbent assay (ELISA) analysis (see later), and a third booster of 2 mg purified stock solution in normal saline solution is administered intravenously. In the sixth week, and after confirmation of a titer against self using the ELISA test, a 12-ml bleed from both rabbits is taken and the serum is separated via centrifugation and stored at 4° for further purification.

Step 3: Isolation of Total IgGs from Rabbit Anti-MS2 Sera. Sera, which are separated from whole blood, are treated with caprylic acid using the method of Steinbuch and Audran[28] for the isolation of a total IgG mixture. Next, the IgG sample is passed through a Pharmacia Protein G affinity column to remove trace contaminants[29,30]; IgG-containing samples are aliquoted and stored at −20°.

Step 4: Separation of Rabbit Anti-MS2 IgG from Total Affinity-Purified IgG Pool. The IgG mixture is thawed on ice, and 10 mg of total rabbit anti-MS2 IgG is applied to the MS2 coat protein affinity matrix, which is prepared as follows. MS2 bacteriophage coat protein, isolated as described previously, is immobilized onto Bio-Rad's Affi-Gel 10/15 N-hydroxysuccinimide ester-activated agarose gel bead support. Two milligrams of MS2 bacteriophage coat protein/ml of gel matrix is coupled using the method of Osborne et al.[31] The column is stored at 4° in 10 mM sodium phosphate, 100 mM NaCl, 0.05% (w/v) NaN₃, pH 7.4. Affinity chromatography is performed following the method of Phillips et al.[32] The affinity-purified isolated rabbit anti-MS2 IgG is pooled, ammonium sulfate precipitated, and collected by centrifugation and stored at 4°.

Step 5: Preparation of Rabbit Anti-MS2 Bacteriophage Coat Protein IgG Affinity Column. Affinity-purified rabbit anti-MS2 bacteriophage coat protein IgG, isolated as described earlier, is covalently cross-linked to Bio-Rad's hydrazide-activated agarose beads, Affi-Gel HZ. Eight milligrams of rabbit anti-MS2 bacteriophage coat protein IgG is immobilized at the Fc region of the antibody in 50 mM sodium acetate, pH 5.5, following the methods outlined.[33]

[28] M. Steinbuch and R. Audran, *Arch. Biochem. Biophys.* **134,** 279 (1969).
[29] D. Baxby, B. Getty, N. Blundell, and S. Ratcliffe, *J. Clin. Microbiol.* **19,** 566 (1984).
[30] J. W. Gnann, L. L. Smith, and M. B. Oldstone, *Methods Enzymol.* **178,** 693 (1989).
[31] L. C. Osborne et al., *Infect. Immunol.* **23,** 80 (1979).
[32] D. J. Phillips et al., *J. Chromotogr.* **536,** 95 (1991).
[33] W. L. Hoffman and D. J. O'Shannessy, *J. Immunol. Methods.* **112,** 113 (1988).

Step 6: Immunoaffinity Purification Scheme. Clarified cell lysates are obtained as described earlier, and the supernatant, containing the recombinant coat protein of interest, is dialyzed against 20 mM sodium phosphate, pH 7.4 (DB), overnight at 4° with two changes of buffer. The following chromatographic procedure is carried out in Bio-Rad disposable Poly-Prep columns (0.8 × 4 cm) packed with rabbit anti-MS2 bacteriophage coat protein IgG affinity matrix prepared as described previously. The column is equilibrated with DB, and 2× column volume (CV) of supernatant is added to the affinity matrix. The affinity column is incubated at room temperature for 30 min. The affinity matrix is then washed with 30× CV of DB; washed again with 30× CV of 10 mM sodium phosphate, 100 mM NaCl (pH 7.4); eluted with 5× CV of 20 mM acetic acid, 200 mM NaCl (pH 3.0); and regenerated with 20× CV of DB for a second round of chromatography.

The eluted coat protein, identified by the absorbance profile at 280 nm, is immediately adjusted to pH 7.0 with 1 M Tris–HCl (pH 9.0), and the samples are concentrated via 1.85 M ammonium sulfate precipitation overnight at 4° and collected by centrifugation at 35,000 rpm using a Beckman 45 Ti rotor for 2 hr. The pellet containing the coat protein is resuspended gently in 10 mM HEPES, 100 mM NaCl, 0.02% (w/v) NaN$_3$, pH 7.4, and stored at 4° until further use.

Refolding

Although a fraction of each of the chimeric molecules could be recovered in the soluble fraction after brief overexpression of *tac* plasmids lacking the A protein gene, constitutive expression or high levels of overexpression of chimeras with inserts larger than 12 amino acids in size appeared to result in significantly reduced solubility. Insert size and solubility were roughly inversely related, although expression levels were in general unaffected by inserts in the range listed in Fig. 2. Examination of the "insoluble" chimeras *in situ* by electron microscopy of thin sections of embedded cells suggested that the chimeras had assembled into VLPs, which had then formed semicrystalline lattices similar to those seen in wild-type phage infections. The generation of insoluble inclusion bodies of recombinant proteins and chimeric virus proteins, such as herpes simplex and yeast Ty, is not uncommon and can be exploited as a purification feature.[34,35] The approach taken is selective solubilization with chaotropic agents. We have

[34] M. J. Geisow, *Tib. Tech.* **9**, 368 (1991).
[35] E. M. Gilmore *et al.*, *AIDS* **3**, 717 (1989).

used a wide range of agents, such as detergents and denaturants, and found that the most successful strategy is denaturation in urea.

Urea Solubilization of Chimeric VLPs

Recombinant bacterial cultures expressing insoluble or only slightly soluble chimeras are processed through stage 1 described earlier. At that point, low-speed centrifugation is used to fractionate insoluble and soluble fractions, with the insoluble fraction being resuspended and washed in 50 mM HEPES, 100 mM NaCl, 10 mM DTT, 5 mM EDTA, pH 7.4. Then solid urea is added to 6 M, and the chimeric proteins are solubilized by stirring at 4° overnight. The sample is clarified by centrifugation (27,000g), 4°, 30 min, before the supernatant is decanted and diluted to 1 M urea by the addition of 50 mM HEPES, pH 7.4. The extracted proteins are then fractionated on a Bio-Gel A1.5M column equilibrated in 50 mM HEPES, 1 M urea, pH 7.4. Column fractions are monitored by SDS–PAGE, and the included coat protein fractions are pooled. At this stage, chimeric coat protein subunits can represent up to 95% of the protein in the sample, which is then desalted rapidly into 20 mM HEPES, 100 mM DTT, pH 7.4, over a Pharmacia HR10/10 column. The sample is then allowed to reassemble overnight at 4° before being applied to a size-exclusion column (Pharmacia Superose-12) equilibrated in the same buffer. Fractions containing the solubilized, reassembled VLPs are then pooled.

Characterization of Chimeric VLPs

Chimeric VLP proteins can be analyzed by standard protein chemistry techniques, such as electrospray mass spectroscopy to confirm their identity. Alternatively, Western blotting with antiepitope-specific antibodies can be used to confirm the isolation of a chimeric product of the correct molecular weight with the desired properties. The structure of the VLP particles can be analyzed by electron microscopy using negatively stained samples[26] or by centrifugation in sucrose density gradients. The structure of chimeric particles, their effectiveness as presentation systems for B- and T-cell epitopes, and conformations of the inserted epitopes can be analyzed as follows. Note: the Plasmodium falciparum LSA-1 T1 epitope, a 24 amino acid insert produced by the two overlapping oligonucleotide procedures (see earlier discussion), was chosen to analyze presentation of a T-cell epitope. It is known to contain human and putative murine HLA class I binding motifs.[36,37] The LSA-1 protein comprises a large central repeat region

---

[36] J. Sidney, H. M. Grey, S. Southwood, E. Celis, P. A. Wentworth, M. F. del Guercio, R. T. Kubo, R. W. Chestnut, and A. Sette, Hum. Immunol. **45,** 79 (1996).

[37] K. Falk et al., Nature **351,** 290 (1991).

flanked by relatively invariant nonrepeat regions. The T1 peptide is located within the N-terminal nonrepeat and was identified as a potential T-cell epitope by amphipathicity analysis.[38]

Immunoaffinity Chromatography

Clarified cell extracts containing chimeric VLPs can be applied directly to the anti-MS2 IgG immunoaffinity matrix described previously. By comparing the amount of each chimeric VLP retained by a fixed amount of resin to the amount of wild-type empty capsids, it is possible to infer that increasing the size of the inserted epitope reduces the accessibility of the underlying MS2 epitopes (Table I), as expected (Fig. 1).

B-Cell Epitope Presentation: Production of Mouse Antiprotein Antibodies

The immunization of TUXCS Number One mice (mixed outbred population, 12–16 g) used the following regime to investigate the dose needed for a response against chimeric coat protein constructs. Five mice per dosage are inoculated subcutaneously at one site after a 50-μl bleed from the marginal tail vein on day 0 with dosages (5 ng, 20 ng, 200 ng, 100 μg, 250 μg, 500 μg) of HA9 chimeric construct as the immunizing antigen absorbed to Alhydrogel. On days 7, 14, 21, 26, and 42, 50-μl test bleeds are taken for ELISA screening of antibody activity (see later), and mice are boosted with their respective selected doses on days 14, 21, and 26 as described previously. Mice are sacrificed on day 42 by a cardiac punch bleed while administering Halothane as a general anesthetic.

Specificity and Effect of Adjuvant

The immunization regime for the analysis of specificity of the HA9 chimeric construct and requirement for adjuvant (aluminum hydroxide; Alhydrogel) used the same schedule as outlined earlier but with the following changes. Female BALB/c mice (inbred population 8–12 g, five mice per antigen set) are immunized subcutaneously at one site with the following antigen sets: HA9 chimeric VLP ± absorption to Alhydrogel; wild-type MS2 empty capsids ± absorption to Alhydrogel, KLH-HA9 (keyhole limpet hemocyanin cross-linked covalently to a synthetic peptide, encompassing the hemagglutinin epitope) by the method of McCray and Werner[39] ± absorption to Alhydrogel, KLH ± absorption to Alhydrogel; and HA9 peptide ± absorption to Alhydrogel.

[38] M. Connelly, C. L. King, K. Bucci, S. Walters, B. Genton, M. P. Alpers, M. Hollingdale, and J. W. Kazura, *Infect. Immun.* **65**, 5082 (1997).
[39] J. McCray and G. Werner, *Methods Enzymol.* **178**, 679 (1989).

Enzyme-Linked Immunosorbent Assay

The ELISA protocol is based on standard methods.[40] The antigen of interest [50 μl of a 50-μg/ml stock solution in 10 mM sodium phosphate, 100 mM NaCl, pH 7.4, phosphate-buffered saline (PBS)] is dispensed into every well of a flat-bottomed 96-well polyvinyl chloride or polystyrene microtiter plate using a multichannel pipetter, and the plate is incubated overnight at 4°. Each well is then washed four times with a solution of PBS containing 0.1% (v/v) Tween 20 and aspirated using a multichannel washer/aspirator. The wells are then treated with bovine serum albumin [1% (w/v) in PBS] for 1 hr at 25° to block nonspecific binding sites, washed to remove excess solution, and aspirated as described earlier. The primary antibody is diluted 1:10 with PBS, and the titer of this solution is determined by successive 1:1 dilutions along a row of wells. A 50-μl aliquot is applied to each well, plates are incubated at 25° for 2 hr, washed/aspirated as described earlier, and the secondary antibody (e.g., goat antimouse IgG-conjugated horseradish peroxidase) is added to each well, repeating the incubation procedure described previously. After aspiration, a 50-μl aliquot of developer solution [100 mM sodium citrate, 100 mM sodium acetate, pH 5.0, containing 0.04% (w/v) 1,2-benzenediamine and 0.004% (v/v) hydrogen peroxide] is added to each well and incubated for 10 min at 25°. The enzyme reaction is quenched by the addition of 25 μl of 2 M sulfuric acid. After 5 min the A_{495} values are measured.

Alterations in the procedure for covalent cross-linking of peptide to polystyrene microtiter plates follow the procedure outlined,[41,42] with the only change being that the free radical initiation of the plates utilized UV light (254 nm, 7000 lumens for 30 hr).

Partial Purification of T-Cell Epitope Chimeric Coat Protein Prior to Immunization

The solubility of the LSA-1 T1 chimera is examined by sonication of a 10-ml overnight culture in buffer A (50 mM HEPES/100 mM NaCl/5 mM EDTA/10 mM DTT, pH 8.0), followed by centrifugation to separate the pellet and supernatant fractions. The chimera is only partially soluble, as would be anticipated from previous observations.[9] Because T-cell epitope presentation is not thought to be dependent on soluble forms of the antigen, an alternative purification strategy was adopted for this construct. Pellets

[40] E. Engvall and P. Perlmann, *Immunochemistry* **8,** 871 (1971).
[41] P. H. Larsson, S. G. Johansson, A. Hult, and S. Gothe, *J. Immunol. Methods* **98,** 129 (1987).
[42] R. Jemmerson, *Proc. Natl. Acad. Sci. U.S.A.* **84,** 9180 (1987).

are treated by stirring overnight at 4° with 1% Triton X-100/0.5% (w/v) sodium deoxycholate in buffer A to remove a substantial portion of the contaminating *E. coli* proteins.

Immunization for T-Cell Responses

BALB/c (H-2d) inbred strain mice (Harlan Olac, Bicester, UK) are used when 8–10 weeks old. Female mice are injected subcutaneously in the scruff of the neck with 50 µg total protein of recombinant LSA-1 T1 (MS2-T1) or purified wild-type empty capsids resuspended in sterile PBS and emulsified 1 : 1 in Freund's adjuvant (Sigma, Poole, UK). Primary and secondary injections are given on day 0 in Freund's complete adjuvant and on day 37 in Freund's incomplete adjuvant, respectively.

Spleen Cell Proliferation Assay

Splenocytes from mice immunized with MS2 native or recombinant protein are restimulated *ex vivo* with homologous peptide and assayed for proliferation and production of type 1 and 2 cytokines and their corresponding Ig isotypes. On day 51 postprimary immunization (14 days after boosting), designated mice (six per group) are sacrificed, spleens aseptically removed, and single cell suspensions in RPMI 1640 (Gibco, Paisley, UK) supplemented with 10% (v/v) heat-inactivated fetal calf serum (complete medium) prepared as described previously.[43] Red blood cells are lysed with 0.17 M Tris-buffered ammonium chloride. Viability determined by trypan blue exclusion is routinely >95%. T- and B-cell function is measured *ex vivo* by the ability of spleen cells to respond to the mitogens concanavalin A (Con A) and lipopolysaccharide (LPS) (Sigma), respectively, as well as to MS2-T1 and wild-type empty capsids, as determined by proliferation assay.[44] Responder cells are adjusted to a final concentration of 5×10^6/ml, and 100-µl aliquots are placed in 96-well flat-bottom tissue culture plates (Nunc, Roskilde, Denmark), to which are added 100 µl volumes of complete medium alone or containing one of the following: 30 µg/ml MS2 coat protein, 25 µg/ml LSA-1 T1 peptide, 1 µg/ml Con A, or 25 µg/ml LPS.[43] Cultures are incubated for 3 or 6 days [37°, 5% (v/v) CO_2]. [^{3}H]Thymidine [1 µCi (18.5 KBq] (Amersham Int., Little Chalfont, UK) in 10 µl medium is added to each well for the final 18 hr, and radioactive incorporation is measured by standard liquid scintillation counting. The stimulation index (SI) is calculated as (mean cpm value of cells + antigen/mitogen − mean

[43] A. W. Taylor-Robinson and R. S. Phillips, *Immunology* **77**, 99 (1992).
[44] A. W. Taylor-Robinson and R. S. Phillips, *Eur. J. Immunol.* **24**, 158 (1994).

cpm value of cells in medium alone)/mean cpm value of cells in medium alone. An SI ≥2 is considered a positive response.

Cytokine Measurement

Levels of the cytokines interleukin (IL)-4, IL-10, IL-12, interferon (IFN)-γ, and tumor necrosis factor (TNF)-α in 6-day spleen cell culture supernatants (derived as described earlier) are quantified by two-site sandwich ELISA[44] using DuoSeTTM-matched antibodies (Genzyme, West Malling, UK.) and following manufacturer's instructions. Reactivity is visualized using 3,3′,5,5′-tetramethylbenzidine (TMB) in 0.05 M phosphate–citrate buffer and 0.014% (v/v) hydrogen peroxide (Sigma) as substrate. Optical densities are determined at 450 nm using an Emax plate reader (Molecular Devices, Crawley, UK). Recombinant murine cytokines (Genzyme) are used for calibration. Control samples of spleen cell culture supernatants from naive, nonimmunized mice, derived under identical conditions to experimental samples, showed low background cytokine levels to specific antigen (<5 ng/ml; SI ≪2).

Collection of Serum

Mice sacrificed on day 51 postprimary immunization are exsanguinated by cardiac puncture. Blood is collected in glass capillary tubes and allowed to clot at 37° and 4° successively, and then serum is removed and stored at −20° as pooled samples from six mice per group.

Antibody Isotype Measurement

Levels of Ig isotypes in sera are quantified by single-site ELISA by modification of a previously published method.[45] Partially purified MS2 coat protein (prepared as described earlier) is used as a coating antigen at a concentration of 25 μg/ml. The antigen is diluted in PBS and applied to round-bottom 96-well ELISA plates (Nunc Maxisorp). Excess binding sites are blocked with a solution of 3% (w/v) bovine serum albumin in PBS (pH 7.4). Serum samples are titered out from a starting dilution of 1/50. Antigen-specific antibodies are detected using rat antimouse horseradish peroxidase-conjugated monoclonal antibodies specific for IgG_1 and IgG_{2a} (Serotec, Oxford, UK). Serum obtained from naive BALB/c mice provide a background response value. Reactivity is visualized using TMB and hydrogen peroxide as described earlier.

---

[45] A. W. Taylor-Robinson *et al., Science* **260,** 1931 (1993).

TABLE I
IMMUNOAFFINITY CHROMATOGRAPHY OF CHIMERIC
VLPs LISTED IN FIG. 2

Sample	% Retained
MS2 wild-type	100
HA9	60
IgE	35
Mal	24
HPV L1	15
HPV L2	15
HIV gp120	8

Conclusions

The immunoaffinity chromatography experiments (Table I) suggest that as expected from the three-dimensional structure of the phage particle, insertion of peptide epitopes within the β hairpin blocks access of the anti-MS2 IgG molecules to their epitopes on the wild-type virus. Because it is important to consider secondary immune responses against the epitope carrier in a vaccine, this result is important because it suggests that with enhanced formulations it would be possible to mask B-cell antigenic carrier sites. ELISA analysis of antisera generated against one of the chimeras shown in Fig. 2 (HA9) suggests that even with a short epitope a significant fraction of the immune response is directed against the inserted peptide. Note that it is difficult to interpret the difference between titers for the HA9 construct and wild-type capsids (Table II) as due to specific reactivity, as some of the responses could be directed against the MS2–epitope junction sequence. However, results with KLH-HA9 and HA9 peptides as the antigen do suggest a generation of specific antibodies. As a probe of the

TABLE II
ELISA ASSAYS OF ANTICHIMERIC HA9 VLP SERA
AND Cas-125 MONOCLONAL

	ELISA titer	
Antigen	Mouse anti-HA9 VLP	Cas-125
HA9 VLP	1 : 7500	1 : 379
Wild-type capsid	1 : 3200	
KLH-HA9	1 : 700	1 : 3650
HA9 peptide	1 : 950	

conformation of the epitope being presented, titers of a monoclonal antibody (Cas-125) against the HA9 chimera or the KLH-HA9 conjugate were determined. The Cas-125 antibody has a high affinity for the epitope and the interaction can be used to immunoaffinity purify epitope-tagged proteins.[46] Its three-dimensional structure in complex with free peptide has also been determined.[47] This suggests that it would not be easy to make the complex seen in the crystal for an epitope in a constrained loop. ELISA data support this hypothesis, the titer against HA9 conjugated covalently to KLH at one end, and hence relatively free to adopt the conformation seen in the complex with the antibody, drops 10-fold when the same epitope is presented on the chimera. These data suggest that RNA phage coat proteins can be used to present short foreign epitopes in a constrained loop to the immune system, which may be a suitable system for a future vaccine formulation.[9]

T-cell epitope responses are also encouraging.[10] Splenocytes derived from wild-type empty capsid immunized mice elicited a significant proliferative response on homologous stimulation *ex vivo* (30 mg/ml coat protein). Cellular proliferation increased between days 3 and 6 poststimulation, suggesting that this response was specific for MS2 coat protein and not due to any residual *E. coli*-derived mitogenic contamination following protein purification. Wild-type MS2 elicited a significant but much reduced proliferation response in cells extracted from MS2–T1-immunized mice. The LSA-1 T1 chimeric coat protein is of course only partially soluble, which may explain the decreased proliferation observed. Following stimulation with Con A and LPS, splenocytes from each experimental group showed appreciable proliferation, denoting both T- and B-cell responsiveness.

Marked upregulation of the type 2 cytokines IL-4 and IL-10, as well as the type 1 cytokine IFN-γ (seen as a 5.9-fold increase), were observed for wild-type MS2-derived splenocytes with respect to similarly stimulated cells from naive mice. For two other type 1 cytokines, no significant difference in secretion of TNF-α was observed between these two groups, but the production of IL-12 by cells from wild-type MS2-immunized mice was downregulated compared to controls (seen as a 2.4-fold decrease). As may have been anticipated, cytokine profiles of splenocytes from each experimental group produced largely type 1 and type 2 responses, respectively, following stimulation with Con A and LPS.

Splenocytes derived from MS2–T1-immunized mice elicited significant LSA-1 T1-specific proliferation ($p < 0.05$). The peptide specificity of this

[46] H. L. Niman, R. A. Houghten, L. E. Walker, R. A. Reisfeld, I. A. Wilson, J. M. Hogle, and R. A. Lerner, *Proc. Natl. Acad. Sci. U.S.A.* **80**, 4949 (1983).
[47] J. M. Rini et al., *Science* **255**, 959 (1992).

response was gauged by the negative SI attained with samples of naive and wild-type MS2-immunized mice tested in parallel with the MS2–T1 samples. LSA-1 T1-stimulated splenocytes from MS2–T1-immunized mice upregulated the type 1 cytokines IL-12, IFN-γ, and TNF-α and downregulated the type 2 cytokines IL-4 and IL-10 when compared to similarly stimulated cells from naive mice. In particular, substantial levels of IFN-γ were observed (seen as a 233-fold increase over the control). This cytokine profile strongly suggests that the MS2–T1 chimeric protein elicits a predominantly type-1 cytokine response, indicative of Th1/Tc1 cell activation.

Acknowledgments

We thank Professor Michael G. Hollingdale, Drs. Rachael Hill and Andrew Taylor-Robinson, and Ms Karen Heal for their helpful discussions of epitope presentation in the MS2 system.

Author Index

Numbers in parentheses are footnote reference numbers and indicate that an author's work is referred to although the name is not cited in the text.

A

Abboud, C. N., 135
Abney, C. C., 257, 266(18), 320
Abrahmsen, L., 256
Abramov, S., 515
Abremski, K., 41
Acker, J., 261
Ackerman, S. H., 450, 456
Adams, C. W., 470, 483
Adams, G. P., 468, 477
Adams, S. R., 289
Adenot, P., 172
Ahmed, R., 543
Ahn, D. H., 429
Aikawa, M., 219
Aingorn, H., 429
Airenne, K. J., 259, 311
Ajchenbaum, F., 259
Akamatsu, Y., 484
Akerman, B., 441, 443(9), 445(9), 451
Åkerman, K., 308(18), 309, 310(18)
Akhavan, H., 176
Akopian, T., 535
Alam, J., 175, 188(1)
Alban, C., 443, 445, 447(17), 448(17), 451
Albertini, R. J., 134
Alberts, A. W., 278
Albertson, H., 310
Albrecht, G., 178
Aldredge, T., 399
Alimonti, J., 429
Allan, P. W., 145
Allemanni, G., 484
Allen, D., 470, 483(24), 484
Allet, B., 35
Allewell, N. M., 363
Allison, T. J., 483, 484(17), 496(17)

Altenberg, G. A., 317, 320(25)
Alter, D., 205
Altman, J. D., 430, 432(7)
Amann, 554, 557(14)
Amersdorfer, P., 461, 483
Amit, 551
Ammerer, G., 107
Amrani, N., 116
Anasri, A. A., 221
Andersen, B., 247, 249(14)
Andrew, S. M., 470
Anfinsen, C. A., 363, 364
Anfinsen, C. B., 363
Angelichio, M. J., 93, 94
Anke, T., 201
Antin, P. B., 218, 219(39)
Aoki, Y., 441, 443(8), 445(8), 451
Aoyama, S., 167, 171
Apell, G., 483, 497
Apone, S., 256
Apreda, B., 311
Arai, K., 257, 270(19)
Arends, J. W., 461, 483
Argaraña, C. E., 207(5), 305
Arnheim, N., 15
Arora, N., 544
Arosio, P., 310, 311
Artymiuk, P. J., 441
Arvola, M., 246
Ashburner, M., 147(10), 148
Ashton-Rickardt, P. G., 537
Aslandis, C., 350
Assouline, Z., 429
Athappily, F. K., 446, 447(32)
Atunez, A., 508
Atwell, J. L., 469
Aubin, R., 206
Audran, R., 560

571

Chua, N. H., 246
Chumakov, A., 261
Chun, K., 44
Chun, S. Y., 313
Chute, I. C., 380, 383(9), 399(9), 402(9), 403(9), 405(8), 406(8), 415(8), 418(8)
Cirillo, D. M., 53
Clark, E. D. B., 312
Clark, I., 146
Clark, R., 121
Clay, F. J., 115
Clement, J. M., 313
Cline, J., 360
Cloud, J., 509
Coates, K., 259
Cobb, M. H., 251, 252
Coda, A., 311, 457
Coffin, J. M., 222
Cohen, B., 327
Cohen, H. D., 453, 456(50)
Cohen, I. R., 429
Cohen, J., 204, 206(13)
Cohen, S. N., 5, 26
Coia, G., 309(24), 310, 463
Colas, P., 327
Colbere-Garapin, F., 134(12), 135
Cole, M. S., 469, 484
Cole, P. A., 383
Coleman, R., 218, 219(40)
Coleman, S. T., 118
Colin, T., 255
Collado-Vides, J., 15
Collazo, C. M., 517
Collier, R. J., 74, 86(6), 528, 542, 544, 547, 550
Collins, M. E., 448, 451, 456(39)
Collins-Racie, L. A., 322, 323, 328(10), 330, 355
Comb, D. G., 385
Comerford, S. A., 260
Confer, D. L., 529
Cong, S., 385
Conneley, M. O., 305
Connelly, 562
Connelly, P. R., 363
Conner, C. P., 80
Conner, G. E., 313, 318, 318(16), 320(16)
Conry, L., 121
Consler, T. G., 449, 451(41, 42), 456(41, 42), 458(41, 42)
Cook, J. L., 175, 188(1)

Cook, R., 399
Coon, M. J., 448, 450, 456(37)
Cooney, A. J., 260
Cormack, B. P., 49, 53
Cormier, M. J., 165, 166, 168, 168(6, 11)
Cornelis, G. R., 517
Corti, A., 311
Cosman, C. L., 29
Cossart, P., 74
Cottrell, T. J., 318
Covarrubias, M., 463, 467(24)
Cox, J. P., 461, 462(5), 489
Crabb, D., 204, 205(14)
Craig, F. F., 222
Craik, C. S., 503
Craiu, A., 535
Crameri, A., 53
Crawford, F., 430, 433(8)
Crawford, R. S., 468
Creagh, A. L., 418, 429(2)
Creighton, 558
Cresswell, P., 528, 535(2)
Creton, R., 171
Cribbs, D. L., 268
Crick, F. H., 4
Cronan, J. E., Jr., 430, 440, 441, 443(6), 445, 445(6), 446, 446(11, 25), 447(11, 25, 31), 448(6), 449, 449(11, 24), 450, 451, 451(11, 44), 452(11, 24, 27), 453(11), 454(11, 24), 455(15), 456, 456(24, 25, 44), 457(11), 458(11, 44)
Crosby, W. L., 461, 462(5), 483, 484(17), 496(17)
Crosier, W. J., 121
Crothers, D. M., 464
Crowe, J., 246, 247(12), 248, 250(12)
Crowl, R. M., 218, 219(41)
Cruise, K. M., 255
Cubitt, A. B., 49, 53, 289
Cuello, A. C., 469
Cull, M. G., 430
Cullen, B. R., 159, 160, 161, 161(2), 188, 189
Culp, L. A., 177, 188(16)
Culver, K. W., 135
Cumber, A. J., 467
Cummings, C. E., 470
Curby, R. J., 31
Curnis, F., 311
Curtiss, R., 506

Gentz, R., 245, 246, 247(12), 248, 250(12), 464
George, A. J., 462, 463, 470
Georgiou, G., 277, 506, 508, 509, 510, 510(16), 511, 511(14, 21), 512, 512(11, 21), 513, 513(11, 14, 23), 514, 515, 515(13, 16)
Geppert, T. D., 251, 252
Gerardi, J. M., 217(36), 218
Gerber, L., 160
Gerhart, J. C., 461, 483
Gerngross, U. T., 418
Ghebert, L., 456
Gheradi, E., 461, 462(5)
Ghigo, J. M., 6
Ghrayeb, J., 507, 508(5)
Gibson, K., 44
Gibson, K. M., 441, 445(10)
Gibson, R., 399
Gietz, D., 114
Giladi, M., 97
Gilbert, K., 399
Gilkes, N. R., 418, 429, 429(2)
Gillespe, P. G., 455, 456(54)
Gilley, G. G., 309(24), 310
Gilmore, 561
Giniger, E., 146
Giovannoni, L., 483, 485(12), 501(12), 503(12), 504(12), 505(12)
Gladyshev, V. N., 293
Glaser, P., 529, 532(13)
Glasner, J. D., 15
Glazer, A. N., 306, 308(8)
Gleeson, C., 85
Glockshuber, R., 276
Glotzer, M., 120
Gmira, S., 531, 533(14), 537(14), 540(14)
Gnann, 560
Goddard, A., 470, 483
Goebl, M., 44
Goeden, M. A., 15
Goffeau, A., 118
Gold, L., 322
Goldberg, A. L., 535
Goldenberg, S., 257
Goldstein, M. A., 418, 419, 420(3, 5), 421
Goletz, T. J., 528
Golmohammadi, 554, 557(11)
Goodman, A. E., 53
Goossens, P. L., 74
Gope, M. L., 305
Gorick, B. D., 465, 485

Gorman, C., 202, 203, 206(7, 8), 218, 219, 219(8, 38), 220, 220(8)
Goss, N. H., 445
Gotch, F., 543
Goto, T., 146, 148, 165
Gotschlich, E. C., 296
Gotter, S., 309(23), 310, 467
Gottfried, S., 510
Goud, B., 550
Gould, S. J., 167
Goulder, P. J. R., 430, 432(7)
Gove, C., 178
Goyffon, M., 167
Gradl, G., 261
Graham, D. E., 218, 219(38)
Graham, F. L., 134
Gramm, C. F., 535
Grandea, A. G. D., 313, 314(18)
Grandison, P., 261
Grant, K. L., 313, 323, 324, 328(6, 10), 330, 339, 355
Gravel, R. A., 441, 443(9), 445(9, 10), 446, 447(26), 451
Graves, P. N., 267
Grawunder, U., 302, 303
Gray, D., 543
Green, K., 497
Green, N. M., 305, 306(1, 2), 437, 455, 456(52)
Green, P., 209, 210(26), 215(26)
Greene, R., 482
Greener, A., 260, 352
Greenough, T., 164, 188
Greenwood, J. M., 429
Gregerson, D. S., 294
Gregor, J., 15
Greiner, E., 441
Grierson, A., 261
Griffiths, A. D., 461, 462(5), 465, 480, 482, 483, 483(4), 484(17), 485, 489(20), 496(17)
Grimwade, B., 133
Grishina, I., 327
Grisshammer, R., 246, 249(11), 253, 434
Gritz, L., 136
Grivell, A. R., 121
Groen, G., 365
Groisman, R., 178
Gross, L. A., 289
Gruen, C. L., 469
Grütter, M. G., 291

Subject Index

A

Adenylate cyclase toxin, *see* CyaA-OVA vector

Aequorin fusion proteins
assays
extraction and regeneration, 174
luciferins
sources, 172
stock solutions, 173
luminescence measurement, 174
transiently-transfected cells, 174
calcium indicator, 170
coelenterazine substrates, 170–171
genes and expression systems, 166–167
regeneration from apoaequorin, 171
structure, 170–171
substrate specificity, 169

Alkaline phosphatase fusion proteins, *see also* Secreted placental alkaline phosphatase reporter
applications, 36
assay in permeabilized *Escherichia coli*, 46–47
enzyme-linked immunosorbent assay using StrepTactin-coated microtiter plates, 297
phoA fusion transposons
IS*phoA*/hah insertion in chromosomal genes
cell growth and infection, 43
generating 63 codon insertions, 43
host specificity, 41
insertion site identification with polymerase chain reaction, 44–46
strains and plasmids, 41
IS*phoA*/in insertion in cloned genes
generating 31 codon insertion mutations, 41
infection, 39–40
λTn*phoA*/in growth, 39
materials, 39
plasmid isolation, 40
screening, 40

target gene, 39
transformation, 40
overview, 36–37
signal sequence and activity upon export, 6, 35–36, 97

3-Amino-1,2,4-triazole, imidazoleglycerolphosphate dehydratase inhibition in yeast fusion reporters, 107, 110–111, 115–118

Anthrax toxin fusion proteins
advantages in vaccination, 550–551
applications, 551
components of toxin, 543–544
cytotoxic T lymphocyte
activity assay, 549–550
epitope fusion, 547
memory cells in vaccination, 543
response, 542–543
expression in *Escherichia coli*
cell growth and induction, 548
plasmid construction, 544–546
purification of histidine-tagged proteins, 548–549
immunization of mice, 549
LFn
amino terminal sequence in constructs, 544
disulfide linking of epitopes, 547–548

Antibody fragments
applications and prospects, 479
bispecific diabody
agglutination assay for multivalency, 477–478
engineering, 469–470
sandwich enzyme-linked immunosorbent assay for bispecificity, 478
expression in *Escherichia coli*
cell growth and induction, 471–472
harvesting
periplasm fragments, 473
supernatant fragments, 472
inclusion bodies, 462
materials, 470–471

601